Advances in the Theory of Computational Mathematics
Volume 3

RECENT TRENDS IN NUMERICAL ANALYSIS

Advances in the Theory of Computational Mathematics
Marcin Paprzycki (Editor)

Recent Trends in Numerical Analysis
Donato Trigiante
ISBN 1-56072-885-X

High Performance Algorithms for Structured Matrix Problems
Peter Arbenz, Marcin Paprzycki, Ahmed Sameh and Vivek Sarin
ISBN 1-56072-594-X

We intend to publish books from any area of mathematics or computer science involved with computation broadly understood.
We plan to publish four types of high quality manuscripts:

- monographs
- topic-oriented volumes consisting of contributed papers devoted to the current research
- topic-oriented volumes consisting of invited overview papers, summarizing the state of the art
- proceedings from selected Workshops/Conferences/Minisymposia

To discuss a possible contribution, please contact the Managing Editor:

Advances in the Theory of Computational Mathematics
Volume 3

RECENT TRENDS IN NUMERICAL ANALYSIS

EDITED BY
DONATO TRIGIANTE

Nova Science Publishers, Inc.
Huntington, New York

Senior Editors: Susan Boriotti and Donna Dennis
Office Manager: Annette Hellinger
Graphics: Wanda Serrano and Dorothy Marczak
Information Editor: Tatiana Shohov
Book Production: Cathy DeGregory, Kay Seymour, Jennifer Vogt and Lynette Van Helden
Circulation: Ave Maria Gonzalez, Mike Hedges, Ron Hedges and Andre Tillman

Library of Congress Cataloging-in-Publication Data
available upon request

ISBN 1-56072-885-X

227 Main Street, Suite 100
Huntington, New York 11743
Tele. 631-424-6682 Fax 631-425-5933
e-mail: Novascience@earthlink.net
Web Site: http://www.nexusworld.com/nova

Printed in the United States of America

Dedicated to Professor Ilio Galligani on the occasion of his 65th birthday

CONTENTS

PREFACE

In 1970 I was a young theoretical physicist and some months earlier I had moved to the Institute of Physics at the University of Bari. My activity consisted in providing theoretical support to an active group of experimental physicists operating in the field of the Solid State. My duties also required the use of the computer for the calculation of the energy levels of electrons in some crystalline compounds. Having followed a good course of studies at the University of Turin, I was well learned in the developments of Quantum Mechanics, but I did not have any knowledge in numerical calculations. Consequently, I used to waste the greatest part of my time at the computer center (terminals were still to come at that time) trying to understand why the energy levels supplied by the calculations were so much different than those expected. In one of the many moments of frustration, a friend informed me of the presence at the Institute of an expert in numerical calculation, whom I could ask for an aid. I called him up for an appointment. The same day I met for the first time Ilio Galligani. The meeting did not last less than three hours. We spoke about my problem and the finite elements method, his main interest at that time. Very soon, however, enthusiasm carried him on to speak about Numerical Analysis as a whole, his great passion. The most striking characteristic of Galligani, in fact, is the enthusiasm for his job. In the following years I have known the majority of the Numerical Analysts from all over the world: very few of them have seemed to me to have an enthusiasm for the whole of Numerical Analysis comparable to that of Ilio.

Enthusiastic people always have a fascinating effect on the young and such was the effect of Galligani on me.

In the following days my energy levels fell into their right positions. Galligani followed with interest the vicissitude and this revealed to me another aspect of his personality, confirmed later on by many other episodes: the altruism and the deep respect for the others. The successive year he was called to direct the Institute for the Application of Calculus (IAC) in Rome, but he continued to teach a course in Bari at the Information Sciences, an academic institution whose creation saw him as one of the main contributors. I attended one of his courses. His teaching was indeed bright but, above all, it was passionate, enriched by examples and anecdotes, typical of the teacher of great experience. Considering the large number of students who attend his courses and who write their theses with him, I am convinced that his teaching has more and more improved over the years.

The INAC, later on named IAC, was founded in the forties by M. Picone. Famous Italian and foreign mathematicians such as L. Cesari, W. Gröbner, W. Gautschi, to name a few, worked there and it achieved a worldwide resonance, unfortunately diminished in the last years

The choice to entrust the Institute to Galligani revealed to be excellent. In fact, he showed great organizational skills. In a few years he succeeded in providing the institute with a new and adequate location; he launched important projects of Applied Mathematics and, through a wise program of invitations, he made the IAC one of the cultural centers of Italian applied mathematics. In those years I knew many personalities of our field at the IAC, among which I remember L. Collatz, I.J. Schoenberg, W. Gautschi, J. Butcher, P. Lancaster, D. Greenspan, W. Miranker, G. Meinardus, T. Bui.

I often went to visit him and every time I was overwhelmed observing his rhythms of work. In the evenings we discussed mathematics until late, often until midnight, taking long walks in the neighborhood of Porta Pia, near his home. We used to discuss mainly about the research plans that he was launching at the IAC, among which I would like to remember (mainly because many people he trained on these topics still continue to cultivate them) numerical fluid dynamics, electrodynamic models of the heart, statistics, optimization, numerical linear algebra, differential equations and computer graphics. Those who do not know Galligani personally will have difficulty to believe that he did not limit itself to be the manager, but that he was an expert in all the above mentioned fields. His publications and the lectures (memorable was that one on numerical fluid dynamics at Cortona in 1978) he delivered in those years demonstrate it abundantly.

Later on, having been appointed chair of Theory and Applications of Computing Machines, after a short period at the University of Turin, he moved to Bologna, where he still teaches. To say that in a few years he created one of the main Italian groups of Numerical Analysis would be reductive. In fact his deep intuition guided him to create one of the first laboratories of Scientific Calculus.

The eighties saw him in the front line in promoting, both at his university and among the Italian community of Numerical Analysis, new topics of research such as, for example, ill-conditioned problems, computer-aided didactics, approximation by means of splines, problems of industrial mathematics, image reconstruction. The advent of supercomputers saw him as an enthusiastic organizer and investigator of the new algorithms: it is not by chance that the first one of such computers in Italy was installed in Bologna.

In those years he became the director of the National Group of Mathematical Informatics (GNIM) of the National Research Council. His experience as organizer, gained in the previous decade, proved of great usefulness in this function as well. In fact, in a short time he realized a wide plan which permitted, among the rest, to equip all the Italian groups of Numerical Analysis with a laboratory of Scientific Calculus, giving a decisive impulse to the Italian researches in the field. In the nineties his personal activities of research moved towards parallel computing, optimization, computer graphics and the solution of large size problems.

A few years ago, and again in the last year, he was elected member of the Directive Council of the Institute of High Mathematics (INDAM) where his work is a lot appreciated.

I have pointed out his skills as an organizer, researcher and teacher, but other aspects of his versatile personality are equally remarkable. Since such aspects concern his private sphere, however, I am convinced that he would not appreciate them to be mentioned. I can only say that he listens to all with great patience, dispensing to everyone his advises and, above all, his always-innovative ideas. His office is always full of students and colleagues, sign of the great esteem which he enjoys.

Personally, during so many years of acquaintance, I have learned to appreciate, over all, his exceptional interior strength.

Congratulations Ilio, *ad maiora, semper.*

D. T.

INTRODUCTION

The contributions for this volume, dedicated to honor the 65th birthday of Professor I. Galligani, have been numerous and cover a wide range of topics of the current Numerical Analysis and of its applications. The papers are presented in alphabetical order, according to the first-named author.

Below, we provide a brief presentation of each contribution. For sake of conciseness, and for allowing a convenient reference, we have grouped the twenty three contributions as follows, although many of them are of interdisciplinary nature:

1. Approximation
2. Numerical Linear algebra
3. Numerical Methods for Differential Equations
4. Optimization
5. Parallelism
6. Scientific Tools.

Approximation. In the paper of Casciola and Morigi the problem of the geometric modeling has been considered. The authors use the non-uniform rational B-splines (NURBS) and they show their use in an interactive system for computer graphics.

The problem of the reconstruction of images and the definition of appropriated filters is discussed in the papers written by Barnabei, Guerrini, Montefusco, and Bacchelli, Cotronei, Lazzaro, Puccio, even though the approaches are different. In the former article algebraic tools and banded recursive matrices are specially used, while in the latter the authors use multiwavelets.

The construction of bivariate quasi-interpolant splines is dealt in the article of Dagnino and Lamberti, where it is also presented a corresponding computational procedure.

In the contribution of Conti, Morandi and Sestini the mixed problem of interpolation and approximation of curves by means of mixed splines functions is treated.

The article of Li, Chang and Bui, finally, deals with the problem of the definition of blended curves in 3D by means of the finite elements numerical solution of differential equations.

Finally, in the contribution of Mastroianni and Monegato it is shown that, for a certain class of functions, truncated Gauss-Laguerre formulae perform better than standard ones.

Numerical Linear Algebra. The contribution of Broyden and Vespucci introduces an interesting historical overview of the conjugated gradient method from a particular point of view that allows the generalization of the method to the unsymmetrical case.

In the contribution of Spedicato and Li, methods in the ABS class for solving linear systems of large dimensions are analyzed from the point of view of memory requirements. They introduce some new storage efficient implementations of such methods. The relations among the families of conjugated direction methods and ABS methods are also exploited.

The contribution of Bertaia, Morigi, Loli Piccolomini, Sgallari and Zama concerns the use of projection methods for solving the sparse, large size linear systems deriving from the discretization of integral equations that occur in many problems of image restoration.

In the paper of Van Der Mee, Rodriguez and Seatzu, some numerical methods for the spectral factorization of bi-infinite Toeplitz matrices are reviewed. The use of such methods for the solution of semi-infinite block Toeplitz systems is also discussed.

The contribution of He and Ricci reflects the current tendency to gain classical and new properties of special functions by means of a matrix formulation of the problems. In the paper the Fibonacci and Lucas polynomials are studied and generalized.

Finally, the contribution of Peluso and Sgura concerns the definition of new numerical methods for the symmetric inverse eigenvalue problem, which are based on the use of traces of matrices in connection with the Newton method.

Numerical Methods for Differential Equations. The contribution of Bellen, Maset and Torelli concerns the problem of the numerical solution of equations with delay in the case where the delay is variable and vanishing at the origin. It is an important applicative problem (pantograph) that demands for a particular numerical treatment.

The problem of stability for neutral delay equations is studied in the paper of Guglielmi, where a geometric characterization of the stability regions of θ-methods applied to such equations is also given.

When a large system of ODEs has to be solved, besides the classical requirements of convergence and stability, numerical methods must satisfy additional properties whose importance are especially appreciated when the solutions are to be obtained in a reasonable time. To this problem is dedicated the paper of Brugnano and Trigiante, where a class of numerical methods is studied from this point of view.

The contribution of Brezzi and Marini regards the solution of PDEs in a variational form by means of the so called three-field formulation. The robustness of the approach and the possibility to use powerful preconditioners are discussed.

In the contribution of Funaro, the elliptic problem in three dimensions is treated by means of suitable spectral methods. The efficiency of the approach is also shown by experimental results.

The paper of Comincioli, Naldi, Scapolla and Toscano introduces some multiscale problems and discusses the main numerical techniques used to treat them, based on asymptotic preserving relaxation schemes.

Finally, to the study of multiscale problems is also dedicated the article of Pareschi and Russo. In this paper, the authors propose the solution of the multiscale problem by means of implicit-explicit methods of Runge-Kutta type, whose properties are also discussed.

Optimization. In the contribution of Ruggiero and Zanni, new projection methods for the solution of large convex quadratic programming are introduced. The novelty of the proposed methods consists in the automatic determination of a variable projection parameter, which proves to be very efficient and reliable.

Parallelism. Although this aspect of the numerical methods is dealt in many contributions, in the article of Murli, D'Ambra and D'Amore, the topic is discussed in a very detailed way, by relating it to other aspects of the more general subject of Problem Solving, of which Scientific Calculus is only a facet, though the main one.

Scientific Tools. In the article of Sofroniu and Spaletta it is shown, on a concrete example (the construction of Runge-Kutta methods), how a powerful scientific tool such as Mathematica, can be of great aid also in the theoretical study of numerical methods.

MULTIWAVELETS AND CONSTRUCTION OF BIORTHOGONAL K-BALANCED MULTIFILTERS

S. BACCHELLI*, M. COTRONEI†, D. LAZZARO‡, AND L. PUCCIO§

Abstract. The aim of this work is to present an overview on multiwavelets together with a discussion of the problems connected to their practical applications. It summarizes some results produced by the authors in last years and describes new developments.

Multiwavelets, namely, vector-valued wavelet functions, are a new addition to the classical wavelet theory that have revealed to be successful in practical applications, such as signal and image compression. In fact, multiwavelets possess several advantages in comparison to scalar wavelets, since a multiwavelet system can simultaneously provide perfect reconstruction while preserving orthogonality, simmetry, a high order of approximation (vanishing moments), etc. Nevertheless, multiwavelets differ from scalar wavelet systems in requiring two or more input streams to the multiwavelet filter bank. This initialization step is called prefiltering of the discrete-time data. The use of prefiltering can create some problems in the implementation of discrete multiwavelet transform, as prefilters can increase the computational cost of the whole analysis-synthesis algorithm, and can destroy the good properties of the multifilter bank. It is possible to overcome this problem by introducing the concept of *balanced multiwavelets*, multiwavelet pairs which are more suitable for image processing applications. In this work, after recalling some basic facts on the theory and discuss about the problems which arise with it, we present a construction of balanced biorthogonal multiwavelets by means of the lifting scheme.

Key words. Multi-scaling function, multiwavelet, balancing order, lifting scheme.

AMS subject classifications. 42C15, 65D20, 65Y20

1. Introduction. Wavelet functions are a useful tool for signal processing applications such as image compression and denoising.

Recently, much interest has been generated by the study of multiwavelets, namely, wavelets generated by vector-valued scaling functions [16, 26, 23, 7, 14, 24, 9]. This interest is motivated by the fact that, realizable as matrix-valued filter banks leading to wavelet bases, multiwavelets can be constructed offering simultaneous orthogonality, symmetry, short support, good performance at the boundaries, a high order of approximation and a high number of vanishing moments, etc., while all real-valued scalar wavelets can never possess all the above properties at the same time. For example, a real-valued scalar wavelet-based system which is different from the Haar one can not possess orthogonality and symmetry at the same time.

Thus, multiwavelets offer the possibility of superior performance for image processing applications, compared to scalar wavelets. This is due to the fact that, unlike the scalar case, in a multiwavelet setting some extra degrees of freedom are allowed, since we are working with matrix rather than scalar coefficients.

However, multiwavelets lack some attributes that scalar wavelets, instead, possess. In particular, unlike the scalar case, approximation order k of the multiscaling function does not imply that the corresponding low-pass multifilter preserves discrete-time polynomials up to the degree $k-1$, while the high-pass multifilter annihilates them.

Indeed in the processing of discrete time signals, complications arise for multiwavelet transforms that do not arise for scalar wavelets transforms. In fact, as multiscaling and

*Department of Mathematics, University of Padova, silvia@agamennone.csr.unibo.it
†Department of Mathematics, University of Messina, marianto@dipmat.unime.it
‡Department of Mathematics, University of Bologna, lazzaro@csr.unibo.it
§Department of Mathematics, University of Messina, gina@dipmat.unime.it

multiwavelet bases with approximation order k lack those preservation/cancellation properties, a preprocessing or prefiltering step is fundamentally more important for the discrete multiwavelet transform than for the classical discrete wavelet one.

Therefore, the problem of initialization of multiwavelets, namely, the problem of obtaining the vector input streams from the input signal, is a very important point in multiwavelet-based algorithms, since in applications using multiwavelets the prefiltering step is necessary to obtain an efficient signal or image compression [17, 30, 12, 10, 11]. Recently, to overcome these problems, Lebrun and Vetterli, and Selesnick [20, 25], introduced the concept of balanced multiwavelets. They constructed orthogonal multiwavelet bases whose filters satisfy the discrete version of the approximation and zero-moments properties. Thus, they constructed bases that *do* possess those properties that are normally absent, and in so doing they avoid the use of prefilters in implementing the discrete multiwavelet transform. This is a great advantage, because the preprocessing step can sometimes destroy the good properties a multiwavelet basis is designed to have. Nevertheless, the above authors' request of orthogonality in the multifilter construction implies the resolution of non-linear equations, that are solved by the Gröbner basis method.

Following the previous authors' idea, in order to avoid the complications generated by the above-mentioned non-linearity, in [3] we have given a simple algebraic construction of k-balanced multifilters by means of the well known tool called lifting scheme, which has been introduced by Sweldens in [28]. It provides a simple method for constructing new bases of biorthogonal filters with desirable requested properties, starting from an assigned set of biorthogonal analysis-synthesis filters. This method has been extended to the multifilter case [3]. Starting from an assigned set of perfect reconstruction (PR) analysis-synthesis multifilters, as is shown in [19] in the M-band scalar case, we have naturally obtained a new set of biorthogonal PR multifilters. In so doing, we could use all the degrees of freedom arisen from our construction to impose the k-balancing conditions and other desirable properties on the new filters. Unlike the orthogonal case, where the balancing and orthogonality conditions on the unknown filters give rise to non-linear equations, the balancing conditions applied to the lifted biorthogonal filters generate simple linear equations. All our results have been stated in the algebraic framework of banded block recursive matrices, using this flexible mathematical tool to translate both the k-balancing conditions and other useful properties in terms of simple linear conditions on the new multifilter coefficients.

In [4] we have analyzed the effect of the choice of the shift constant p, which appears in [3] in our definition of k-balancing, on the compression capabilities of the new filters. In fact, shift constant p plays an important role, and it can be used as a further degree of freedom, because numerical experimentations show that the p parameter influences the spectral behaviour of the newly constructed low-pass filter, and in [4] we determined the value of p such that the shape of the Fourier transform associated with the new low-pass matrix coefficients is as close as possible to the optimal shape.

In this work, we present the principal results that we have obtained working with multiwavelets; we discuss the problem of initialization of the input data (prefiltering), and we investigate the design of multifilters, with specified approximation order possessing those preservation/annihilation properties, that are normally not true in a multiwavelet context. For the algebraic preliminaries and notations, the reader is referred to [3].

2. Multi-scaling functions and multiwavelets. As in the scalar case, multiwavelets can be introduced in the context of an extension of the classical multiresolution analysis to the general dimension $r \in \mathbb{N}$ [16].

Let $\boldsymbol{\phi} = [\phi_1\ \phi_2\ \ldots\ \phi_r]^T$ be a vector-valued function belonging to $L^2(\mathbb{R})^r$, $r \in \mathbb{N}$. Define, for $j \in \mathbb{Z}$,

$$V_j = \overline{span\left\{2^{j/2}\phi_i(2^j \cdot -k) : 1 \le i \le r,\ k \in \mathbb{Z}\right\}}. \tag{2.1}$$

DEFINITION 2.1. *$\boldsymbol{\phi}$ is called a multiscaling function if the spaces defined in* (2.1) *satisfy the following properties:*

(i) $\cdots \subset V_{-1} \subset V_0 \subset V_1 \subset \cdots$;
(ii) $\overline{\bigcup_{j\in\mathbb{Z}} V_j} = L^2(\mathbb{R})$;
(iii) $\bigcap_{j\in\mathbb{Z}} V_j = \{0\}$;
(iv) $f(\cdot) \in V_j \Leftrightarrow f(2\cdot) \in V_{j+1}, \quad \forall j \in \mathbb{Z}$;
(v) *the family*

$$\{\phi_i(\cdot - k) : 1 \le i \le r,\ k \in \mathbb{Z}\}$$

is a Riesz (stable) basis for V_0, i.e. two constants A and B exist, $0 < A \le B < \infty$, such that

$$A\|\mathbf{c}\|^2_{\ell^2(\mathbb{Z})^r} \le \left\| \sum_{k\in\mathbb{Z}} \sum_{i=1}^{r} c_i^k \phi_i(\cdot - k) \right\|^2_{L^2(\mathbb{R})} \le B\|\mathbf{c}\|^2_{\ell^2(\mathbb{Z})^r} \tag{2.2}$$

for every sequence $\mathbf{c} = \{\mathbf{c}_k\}_{k\in\mathbb{Z}} \in \ell^2(\mathbb{Z})^r$.
In this case we say that the multiscaling function $\boldsymbol{\phi}$ generates a multiresolution analysis (MRA) of multiplicity r for $L^2(\mathbb{R})$.

Since the functions $\phi_1(\cdot), \ldots, \phi_r(\cdot)$ belong to V_0 which is a subset of V_1, it follows that a sequence of matrices $\{\mathbf{H}_k\}_{k\in\mathbb{Z}} \in \ell^2(\mathbb{Z})^{r\times r}$ exists such that

$$\boldsymbol{\phi}(x) = \sqrt{2} \sum_{k\in\mathbb{Z}} \mathbf{H}_k \boldsymbol{\phi}(2x - k). \tag{2.3}$$

The previous equation is called the *matrix two-scale relation* or the *matrix refinement equation* of the multi-scaling function $\boldsymbol{\phi}$.

Now for every $j \in \mathbb{Z}$, let W_j be the complementary space of V_j, such that $V_{j+1} = V_j \oplus W_j$ and $V_j \cap W_j = \{0\}$. Let $\boldsymbol{\psi}$ a vector-valued function $\boldsymbol{\psi} = [\psi_1, \ldots, \psi_r]^T \in L^2(\mathbb{R})^r$, such that

$$W_j = \overline{span\{2^{j/2}\psi_i(2^j \cdot -k) : 1 \le i \le r,\ k \in \mathbb{Z}\}}.$$

DEFINITION 2.2. *The above-defined vector-valued function $\boldsymbol{\psi}$ is a multiwavelet if, for every $j \in \mathbb{Z}$, the family*

$$\{\psi_i(\cdot - k) : 1 \le i \le r,\ k \in \mathbb{Z}\}$$

is a Riesz basis for W_0.

In this case, for each $j \in \mathbb{Z}$, the set of wavelet functions $\{\psi_{i,j,k} : 1 \le i \le r, k \in \mathbb{Z}\}$, where $\psi_{i,j,k}(\cdot) = 2^{j/2}\psi_i(2^j \cdot -k)$, is a Riesz basis for W_j, and the collection $\{\psi_{i,j,k} : 1 \le i \le r, j, k \in \mathbb{Z}\}$ is a Riesz basis for $L^2(\mathbb{R}) = \oplus_{j\in\mathbb{Z}} W_j$.

As above, since the functions $\psi_1(\cdot), \ldots, \psi_r(\cdot)$ belong to $W_0 \subset V_1$, than another sequence of $r \times r$ matrices $\{\mathbf{W}_k\}_{k \in \mathbb{Z}}$ exists such that:

$$\psi(x) = \sqrt{2} \sum_{k \in \mathbb{Z}} \mathbf{W}_k \phi(2x - k) \tag{2.4}$$

and this is the multiwavelet two-scale relation.

There is another way of writing the two-scale relations, using a more compact notation, and, as background, the theory of recursive matrices.

Let $R(t^2, \mathcal{H}), R(t^2, \mathcal{W})$ be the block Hurwitz matrices whose generating functions are, respectively, the Laurent series:

$$\mathcal{H} = \sum_k \mathbf{H}_k t^k, \quad \mathcal{W} = \sum_k \mathbf{W}_k t^k.$$

Then, matrix versions of the two-scale relations can be given:

$$\mathbf{\Phi}(x) = \sqrt{2}\, R(t^2, \mathcal{H})\, \mathbf{\Phi}(2x), \tag{2.5}$$

$$\mathbf{\Psi}(x) = \sqrt{2}\, R(t^2, \mathcal{W})\, \mathbf{\Phi}(2x), \tag{2.6}$$

where $\mathbf{\Phi}(x)$ and $\mathbf{\Psi}(x)$ are the bi-infinite block vectors:

$$\mathbf{\Phi}(x) = \begin{bmatrix} \vdots \\ \phi(x+1) \\ \phi(x) \\ \phi(x-1) \\ \vdots \end{bmatrix}, \quad \mathbf{\Psi}(x) = \begin{bmatrix} \vdots \\ \psi(x+1) \\ \psi(x) \\ \psi(x-1) \\ \vdots \end{bmatrix}.$$

For multi-scaling functions and multiwavelets with finite support, a finite number of two-scale coefficients is involved. In this case all the bi-infinite matrices involved are banded and associated with Laurent matrix polynomials.

Different types of multiwavelets can be characterized, depending on the nature of the multiwavelet bases and of the associated MRA.

2.1. Semi-orthogonal multiwavelets. In this case the spaces W_j are the *orthogonal* complement of V_j into V_{j+1}. This implies that the spaces W_j are mutually orthogonal.

The two conditions

$$V_j \perp W_j, \; W_j \perp W_l, \quad j, l \in \mathbb{Z}, \; j \neq l \tag{2.7}$$

translate, respectively, into the following conditions on the functions:

$$\langle \phi_{j,k}, \psi_{lm} \rangle = 0, \quad j, k, l, m \in \mathbb{Z} \tag{2.8}$$

$$\langle \psi_{j,k}, \psi_{lm} \rangle = 0,, \quad j, k, l, m \in \mathbb{Z}, j \neq l \tag{2.9}$$

and into the following conditions on the filter coefficients:

$$R(t^2, \mathcal{H}) R(t, \mathcal{F}) R(t^2, \mathcal{W})^T = 0 \tag{2.10}$$

where $R(t, \mathcal{F})$ is a block Toeplitz matrix whose generating function is the block Laurent series of the inner products of the functions in V_0, that is:

$$\mathcal{F} = \sum_k \langle \phi(\cdot), \phi(\cdot - k) \rangle t^k.$$

A study of semi-orthogonal multiwavelets has been made in [12], where a method for their construction has been given. As example, in fig. 2.1 we show the the cubic Hermite multi-scaling function and the corresponding semi-orthogonal multiwavelet.

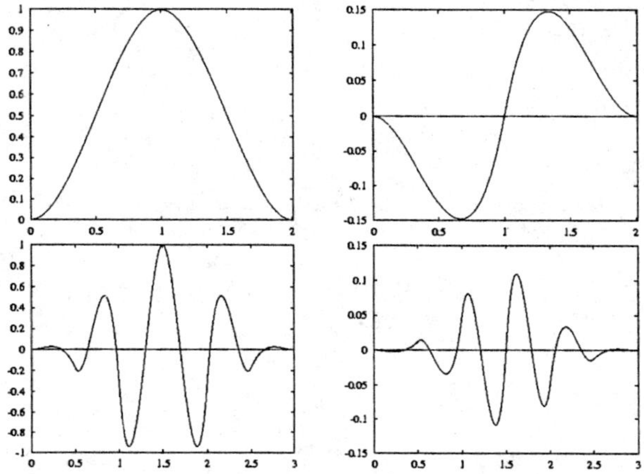

FIG. 2.1. *Cubic cardinal Hermite B-spline multi-scaling function (top) and corresponding semi-orthogonal multiwavelet (bottom)*

2.2. Orthonormal multiwavelets. In this case, in addition to the previous requirements, the families: $\{\phi_{i,j,k} : 1 \leq i \leq r, k \in \mathbb{Z}\}$, $\{\psi_{i,j,k} : 1 \leq i \leq r, k \in \mathbb{Z}\}$ constitute orthonormal bases respectively for the spaces V_j, W_j, $j \in \mathbb{Z}$, that is

$$\langle \phi_{j,k}, \phi_{l,m} \rangle = \delta_{j,l}\delta_{k,m}I, \quad \langle \psi_{j,k}, \psi_{l,m} \rangle = \delta_{j,l}\delta_{k,m}I, \quad j,k,l,m \in \mathbb{Z}. \tag{2.11}$$

In the orthonormal case, we have $\mathcal{F} = I$, then the conditions on the filter coefficients are:

$$\begin{aligned} R(t^2, \mathcal{H})R(t^2, \mathcal{H})^T &= I \\ R(t^2, \mathcal{W})R(t^2, \mathcal{W})^T &= I \\ R(t^2, \mathcal{H})R(t^2, \mathcal{W})^T &= 0. \end{aligned}$$

It is worthwhile to notice that, unlike the semiorthogonal case, the previous conditions are non-linear. Then, the construction of orthonormal multiwavelets is more involved.

An example of orthonormal multiscaling function and multiwavelet is given by the Chui and Lian's ones [7], represented in fig 2.2.

2.3. Biorthogonal multiwavelets. This is the most general case, in which, given a MRA $\{V_j\}$ of multiplicity r generated by ϕ, another MRA $\{\tilde{V}_j\}$, generated by the multi-scaling function $\tilde{\phi}$ is introduced, together with a corresponding multiwavelet $\tilde{\psi}$.

We say that the pairs $\{\phi, \psi\}$, $\{\tilde{\phi}, \tilde{\psi}\}$ are *biorthogonal* or *dual* with respect to each other if

$$\begin{aligned} &\langle \phi_{j,k}, \tilde{\phi}_{l,m} \rangle = \delta_{j,l}\delta_{k,m}I, \quad \langle \psi_{j,k}, \tilde{\psi}_{l,m} \rangle = \delta_{j,l}\delta_{k,m}I, \\ &\langle \phi_{j,k}, \tilde{\psi}_{l,m} \rangle = 0, \qquad\qquad\quad \langle \psi_{j,k}, \tilde{\phi}_{l,m} \rangle = 0, \end{aligned} \qquad j,k,l,m \in \mathbb{Z}. \tag{2.12}$$

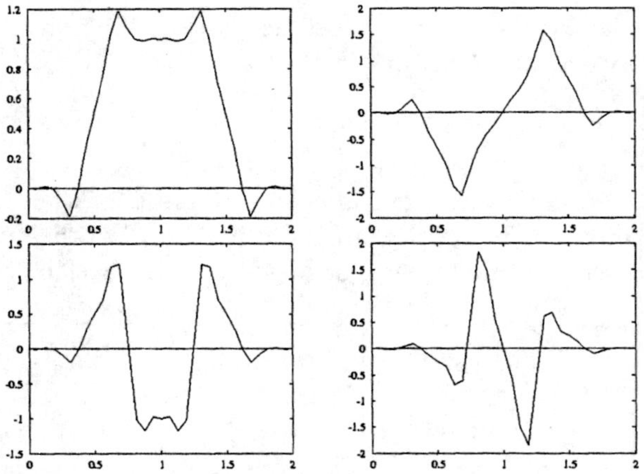

FIG. 2.2. *Chui-Lian orthonormal multi-scaling function (top) and multiwavelet (bottom)*

This implies, in particular, that the two MRA's are connected by the following perpendicularity relations:

$$V_j \perp \tilde{W}_j,\ W_j \perp \tilde{V}_j, V_j \perp \tilde{V}_j,\ W_j \perp \tilde{W}_j. \tag{2.13}$$

In this case, we say that the associate pairs of filters $\{\mathcal{H}, \mathcal{W}\}$, $\{\tilde{\mathcal{H}}, \widetilde{\mathcal{W}}\}$ satisfy the Perfect Reconstruction (PR) property, namely

$$R(t^2, \tilde{\mathcal{H}})^T R(t^2, \mathcal{H}) + R(t^2, \tilde{\mathcal{W}})^T R(t^2, \mathcal{W}) = I. \tag{2.14}$$

This can be given in terms of the following *2-decimated matrix*

$$\Delta_{\mathcal{H},\mathcal{W}} = \begin{bmatrix} \mathcal{H}_0 & \mathcal{H}_1 \\ \mathcal{W}_0 & \mathcal{W}_1 \end{bmatrix},$$

whose elements are the 2-decimated block polynomials related to $\mathcal{H}$, $\mathcal{W}$:

$$\mathcal{H}_k = \sum_i \mathbf{H}_{2i+k} t^i, \quad \mathcal{W}_k = \sum_i \mathbf{W}_{2i+k} t^i.$$

Then we have (see [3] for details) that the couples $\{\mathcal{H}, \mathcal{W}\}$, $\{\tilde{\mathcal{H}}, \widetilde{\mathcal{W}}\}$ have the perfect reconstruction property if and only if one of the two following equivalent conditions is satisfied:

(i) $\Delta^{T*}_{(\tilde{\mathcal{H}},\widetilde{\mathcal{W}})} \times \Delta_{(\mathcal{H},\mathcal{W})} = I$;

(ii) $\Delta_{(\mathcal{H},\mathcal{W})} \times \Delta^{T*}_{(\tilde{\mathcal{H}},\widetilde{\mathcal{W}})} = I$.

Furthermore, a sufficient condition for PR is:

$$\det \Delta_{(\mathcal{H},\mathcal{W})} = \left(\det \tilde{\Delta}^*_{(\tilde{\mathcal{H}},\tilde{\mathcal{W}})}\right)^{-1} = ct^k, \quad c \in \mathbb{R}, k \in \mathbb{Z}.$$

3. Biorthogonal two-channel multifilter bank. Dual pairs of block Laurent polynomials $\{\mathcal{H}, \mathcal{W}\}$, $\{\tilde{\mathcal{H}}, \widetilde{\mathcal{W}}\}$, satisfying the PR condition (2.14), can be associated with the analysis and synthesis phase, respectively, of a two channel FIR (Finite Impulse Response) multifilter bank. In the following we will refer to $\mathcal{H}$, $\mathcal{W}$ as *analysis filters*, and to $\tilde{\mathcal{H}}$, $\widetilde{\mathcal{W}}$ as *synthesis filters.*

With these matrices we can give an algebraic description of the action of the analysis-synthesis system.

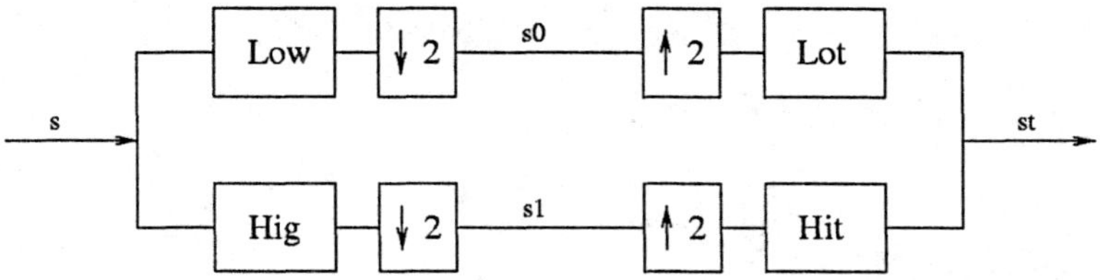

FIG. 3.1. *Analysis-synthesis system*

More precisely, we have:
analysis

$$[\sigma^{(0)}] = R(t^2, \mathcal{H})[\sigma]$$

$$[\sigma^{(1)}] = R(t^2, \mathcal{W})[\sigma]$$

synthesis

$$[\hat{\sigma}] = R(t^2, \tilde{\mathcal{H}})^T[\sigma^{(0)}] + R(t^2, \tilde{\mathcal{W}})^T[\sigma^{(1)}]$$

where $\boldsymbol{\sigma}^{(0)}$ and $\boldsymbol{\sigma}^{(1)}$ represent the output of the analysis phase, while $\hat{\boldsymbol{\sigma}}$ represent the output of the synthesis phase and then of the whole FIR system.

If we denote with $\mathbf{L}$ the operator describing the combined action of the analysis-synthesis system phases, then the PR property implies $\mathbf{L}\boldsymbol{\sigma} = \boldsymbol{\sigma}$.

The construction of a PR multifilter bank can be seen as the starting point of a multiwavelet system. Nevertheless, it is important to recall that, given a set $\{\mathcal{H}, \mathcal{W}, \tilde{\mathcal{H}}, \widetilde{\mathcal{W}}\}$ of PR multifilters then it is possible to define corresponding biorthogonal multiscaling functions and multiwavelets, satisfying two scale relations of the type (2.3) and (2.4), only if, furthermore, $\mathcal{H}$ and $\tilde{\mathcal{H}}$ admit a convergent subdivision scheme [22].

4. Prefiltering. Prefiltering is a fundamental step which precedes the analysis phase of the algorithm described in fig 3.1. Given a set of rN scalar values, it associates r vectors, each of length N. This process can be seen as the application of a (pre)filter to the given data, which vectorizes the input signal to produce a r-dimensional input signal.

A prefiltering step (also called wavelet initialization) is often omitted in the scalar wavelet case, but it is essential in a multiwavelet context. As example we show what happens if, ignoring any prefiltering step, we just apply the analysis phase of the algorithm to the signal (representing an ECG) shown in fig 4.1(a), take the first 100 largest coefficients over 1024, and apply the synthesis phase. Here we take the CL (Chui-Lian) [7] and the GHM (Geronimo, Hardin, Massopust) [14] multifilter banks, and compare their behaviour when the prefiltering step is omitted, and when, instead, the proper prefilter is used. The results are represented in fig 4.1, and show the completely wrong reconstructions that are obtained when no prefilter is used.

On the other hand, the prefilter cannot be arbitrary: as pointed out in [12], the prefiltered data must be as close as possible to the approximation coefficients with respect to

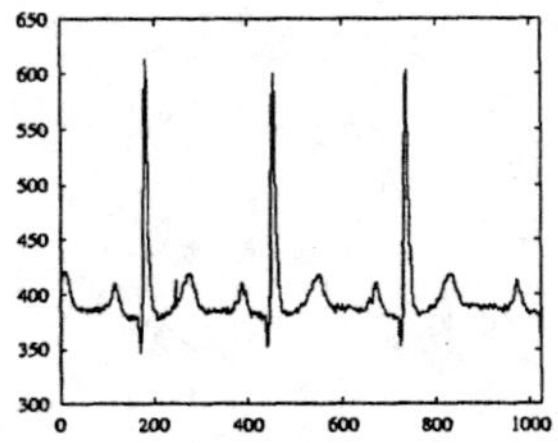

(a) Test signal: an ECG sampled over 1024 points

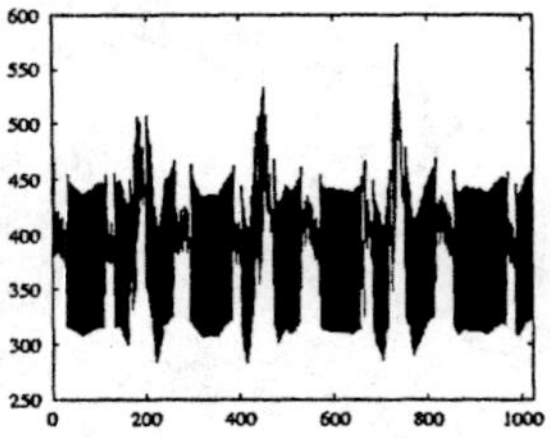

(b) GHM - no prefilter

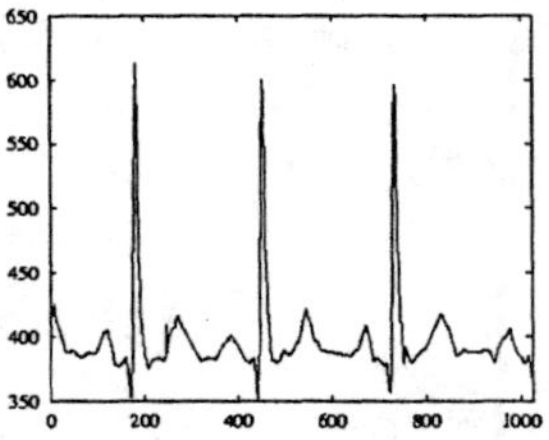

(c) GHM - interpolation prefilter

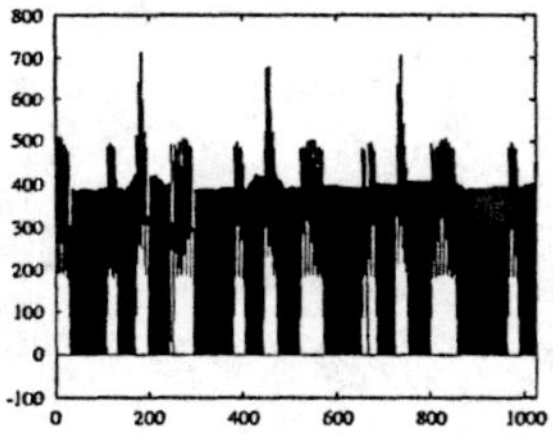

(d) CL - no prefilter

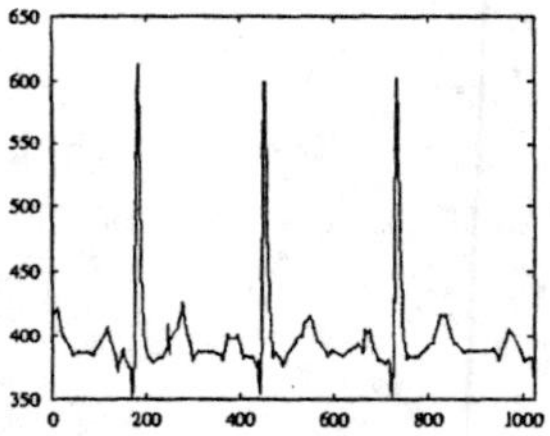

(e) CL - Haar prefilter

FIG. 4.1. *Reconstruction of an ECG signal from 100 largest coefficients analyzed with CL and GHM bases and different prefilters*

the multiwavelet basis one is considering. It turns out that a prefilter is strictly connected with the nature of the multifilter bank.

The use of a prefilter has, nevertheless, the following drawbacks:

1. it increases the computational cost of the entire analysis-synthesis algorithm;
2. it can destroy the good properties of the multifilter bank, for example the symmetry or the orthogonality.

Recently, Lebrun and Vetterli [20] overcame this problems by constructing bases of orthonormal multiwavelets which do not need the prefiltering step. Furthermore, they possess properties that are usually absent in a classical multiwavelet context. They called this bases "balanced". In the next section we give a generalized definition of balanced multiwavelets to the biorthogonal case.

5. Balanced biorthogonal multifilters. As pointed out previously, the balancing procedure is important essentially for two reasons:

1. it allows the construction of multiwavelet bases whose filters satisfy the discrete versions of the approximation and zero-moments properties,
2. the associated decomposition algorithm do not need a prefiltering step.

In [3] we have extended the concept of k-balancing order, introduced in [20], to biorthogonal multifilters.

DEFINITION 5.1. *A pair of filters $\{\mathcal{H}, \mathcal{W}\}$ related to the analysis phase of a FIR system is said to be balanced of order k, (or k-balanced) if there exists at least a real number p such that the following relations hold:*

$$
\begin{aligned}
&R(t^2, \mathcal{H}) \times [\pi_n] = \sqrt{2}\, 2^n [(\pi + p)_n], \\
&R(t^2, \mathcal{W}) \times [\pi_n] = 0,
\end{aligned}
\qquad n = 0, \ldots, k-1, \tag{5.1}
$$

where $[\pi_n]$ and $[(\pi + p)_n]$ are the column vectors

$$
[\pi_n] = \begin{bmatrix} \vdots \\ (-2)^n \\ (-1)^n \\ 0^n \\ 1^n \\ 2^n \\ \vdots \end{bmatrix}, \quad [(\pi + p)_n] = \begin{bmatrix} \vdots \\ (-2+p)^n \\ (-1+p)^n \\ (0+p)^n \\ (1+p)^n \\ (2+p)^n \\ \vdots \end{bmatrix} \tag{5.2}
$$

that can also be considered as r-block vectors and associated to the formal block series

$$
\pi^n = \sum_i \begin{bmatrix} (ri)^n \\ (ri+1)^n \\ \vdots \\ (ri+r-1)^n \end{bmatrix} t^i, \quad (\pi + p)^n = \sum_i \begin{bmatrix} (ri+p)^n \\ (ri+1+p)^n \\ \vdots \\ (ri+r-1+p)^n \end{bmatrix} t^i.
$$

If such is the case, the pair of filters $\{\mathcal{H}, \mathcal{W}\}$ is said to be k-balanced with respect to the constant p. Given a pair of k-balanced filters $\{\mathcal{H}, \mathcal{W}\}$, if furthermore $\mathcal{H}$ admits a convergent subdivision scheme, then the corresponding multiwavelet ψ is said to be balanced of order k.

We have shown that the balancing conditions (5.1), which seem to be given on a whole block Laurent series, can be greatly simplified, since, using the properties of block recursive matrices given in [2], they can just be given on the zero-th terms of the block series involved. Consequently, we have

THEOREM 5.2. *A pair of filters $\{\mathcal{H}, \mathcal{W}\}$ is balanced of order k if and only if*

$$
\sum_{j=l_1}^{l_2} \mathbf{H}_j \begin{bmatrix} (rj)^n \\ (rj+1)^n \\ \vdots \\ (rj+r-1)^n \end{bmatrix} = \sqrt{2}\, 2^n \begin{bmatrix} p^n \\ (p+1)^n \\ \vdots \\ (p+r-1)^n \end{bmatrix}, \tag{5.3}
$$

$$\sum_{j=m_1}^{m_2} \mathbf{W}_j \begin{bmatrix} (rj)^n \\ (rj+1)^n \\ \vdots \\ (rj+r-1)^n \end{bmatrix} = 0, \tag{5.4}$$

for $n = 0, \ldots, k-1$.

The proof of this theorem is given in [3].

6. Construction with the lifting scheme. *Lifting scheme* has been introduced by Sweldens in [28] as a flexible tool for constructing new biorthogonal bases, starting from an assigned set of biorthogonal filters. We have extended this scheme to the multifilter case [3]. Loosely speaking, given a set $\{\mathcal{H}, \widetilde{\mathcal{H}}, \mathcal{W}, \widetilde{\mathcal{W}}\}$ of biorthogonal PR analysis-synthesis multifilters, than the new multifilters

$$\begin{cases} \widetilde{\mathcal{H}}^{new} = \widetilde{\mathcal{H}} + (\mathbf{S} \circ t^2)\widetilde{\mathcal{W}} \\ \mathcal{W}^{new} = \mathcal{W} - (\mathbf{S}^{*T} \circ t^2)\mathcal{H} \end{cases}, \tag{6.1}$$

where $\mathbf{S}$ is any block Laurent polynomial, give rise to a new set $\{\mathcal{H}, \widetilde{\mathcal{H}}^{new}, \mathcal{W}^{new}, \widetilde{\mathcal{W}}\}$ of biorthogonal multifilters. Analogously, by simply changing the roles of the multifilters involved, a new set of biorthogonal filters can be built $\{\mathcal{H}^{new}, \widetilde{\mathcal{H}}, \mathcal{W}, \widetilde{\mathcal{W}}^{new}\}$, starting from the set $\{\mathcal{H}, \widetilde{\mathcal{H}}, \mathcal{W}, \widetilde{\mathcal{W}}\}$, as follows:

$$\begin{cases} \mathcal{H}^{new} = \mathcal{H} + (\widetilde{\mathbf{S}} \circ t^2)\mathcal{W} \\ \widetilde{\mathcal{W}}^{new} = \widetilde{\mathcal{W}} - (\widetilde{\mathbf{S}}^{*T} \circ t^2)\widetilde{\mathcal{H}} \end{cases} \tag{6.2}$$

where $\widetilde{\mathbf{S}}$ is any block Laurent polynomial. The scheme of construction (6.2) is called *dual lifting scheme.*

We also have given some conditions to impose on the unknown filters in order to preserve symmetry/antisymmetry on the newly constructed filters.

We can use the previous flexible scheme in order to construct new biorthogonal multifilters by solving simple linear equations. In fact, unlike the orthogonal case [20, 25], our scheme of construction is based on simple linear conditions.

It is important to notice that, since the balancing conditions affect both the analysis filters, in order to construct new balanced filters, a simple step of lifting (or dual lifting) is not sufficient, because it would modify only one of the two analysis filters. In [3] we have shown that the scheme "dual lifting followed by lifting" gives the only possible way to apply the desired balancing conditions.

The principal steps of our scheme of construction are:

1. Construct the new low-pass filter, with the dual lifting
$$\mathcal{H}^{new} = \mathcal{H} + (\mathbf{S} \circ t^2)\mathcal{W}.$$
2. Choose the shift parameter p and apply the first balancing condition (5.3), up to the desired order k. We obtain a linear system of rk equations. Solve the linear system and find the elements of $\mathbf{S}$.
3. Construct the new high-pass filter, with the lifting
$$\mathcal{W}^{new} = \mathcal{W} - (\widetilde{\mathbf{S}}^{*T} \circ t^2)\mathcal{H}^{new}.$$

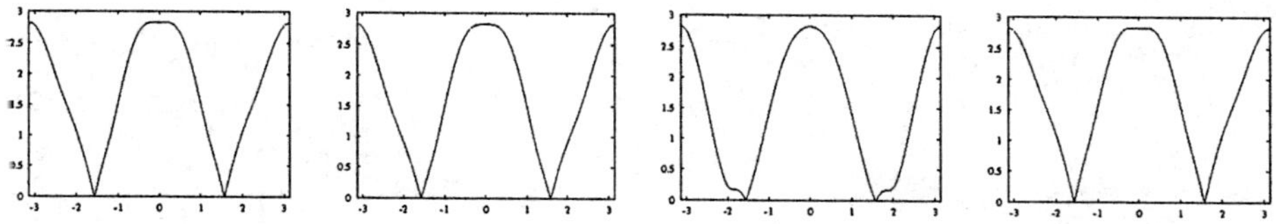

FIG. 7.1. *2-balanced Lazy multifilters. Graphs of* $|\hat{h}^{0,new}| + |\hat{h}^{1,new}|$ *with, from the left to the right,* $p = -11/15, -2/3, -1/4, 1/4$

4. Apply the second balancing condition (5.4) up to k. This gives another linear system of rk equations. Solve the linear system and find the elements of $\widetilde{\mathbf{S}}$.
5. Construct the corresponding dual filters:

$$\widetilde{\mathcal{W}}^{new} = \widetilde{\mathcal{W}} - (\mathbf{S}^{*T} \circ t^2)\widetilde{\mathcal{H}},$$

$$\widetilde{\mathcal{H}}^{new} = \widetilde{\mathcal{H}} + (\widetilde{\mathbf{S}} \circ t^2)\widetilde{\mathcal{W}}^{new}.$$

In [3, 4] we have noted that the shift constant p plays a very important role on the effectiveness of the new filters in image compression applications. In fact, our numerical experimentations have shown that p influences the shape of the Fourier transform, and therefore the spectral behaviour, of the newly obtained multifilters.

7. Examples. All our constructions (which are limited to the case $r = 2$) start with the Lazy multifilters as initial set. They can be defined as follows:

$$2\mathcal{H} = \widetilde{\mathcal{H}} = \sqrt{2}\,I, \quad \mathcal{W} = 2\widetilde{\mathcal{W}} = \sqrt{2}\,I\,t.$$

If we want to construct k-balanced filters $\{\mathcal{H}^{new}, \mathcal{W}^{new}\}$, with k even integer, which maintain the symmetry of the initial multifilters then we must find lifting and dual lifting polynomials of the following type:

$$\mathbf{S}^* = t^{-1}\mathbf{\Pi}\,\mathbf{S}\,\mathbf{\Pi}, \quad \widetilde{\mathbf{S}}^* = t^{-1}\mathbf{\Pi}\,\widetilde{\mathbf{S}}\,\mathbf{\Pi},$$

where $\mathbf{\Pi} = \mathrm{diag}(1, -1)$ and the number of the unknown matrix coefficients is equal to k.

A total of $4k$ equations are given by the k-balancing conditions (5.3),(5.4). and the unknowns are easily found by solving two linear systems.

The obtained families of multifilters depend on the shift parameter p, which strongly influences the behaviour of the Fourier transforms of $h^{0,new} = \{h_k^{0,new}\}_{k\in\mathbb{Z}}$, $h^{1,new} = \{h_k^{1,new}\}_{k\in\mathbb{Z}}$, where:

$$h_{2k+n}^{m,new} = [\mathbf{H}_k^{new}]_{m,n}, \quad k \in \mathbb{Z},\ m, n = 0, 1,$$

making them more or less suitable for application problems.

To explain this last fact, we restrict to the cases $k = 2$ and $k = 4$, and show (figs. 7.1,7.2), for each case, 4 graphs of the sum $|\hat{h}^{0,new}| + |\hat{h}^{1,new}|$, with p varying respectively in $\{-11/15, -2/3, -1/4, 1/4\}$ and in $\{1/5, 1/3, 2/5, 1/2\}$.

It can be seen that the choices $p = 1/4$ and $p = 2/5$ give, visually, a better low-pass behaviour respectively for the 2 and 4-balanced Lazy multifilters.

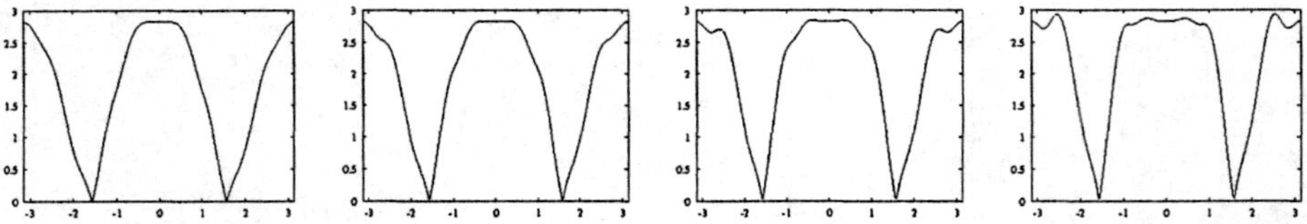

FIG. 7.2. *4-balanced Lazy multifilters. Graphs of* $|\hat{h}^{0,new}| + |\hat{h}^{1,new}|$ *with, from the left to the right,* $p = 1/5, 1/3, 2/5, 1/2$

7.1. An application to image compression. Multiwavelets have proved to be very effective in image compression problems ([27, 10, 11]). In particular, in [10] an efficient algorithm has been described which, taking into account the nature of a multiwavelet decomposition, realizes an efficient embedded coding of a an image. As pointed out in the same paper, this multiwavelet-based algorithm in most case outperforms a similar scalar wavelet-based one. However, due to the necessity of using a prefilter in conjunction with each multiwavelet basis, its complexity is much higher.

In this section we show some results that we have obtained in a image compression test on the classicla Lena test image with the same multiwavelet-based compression algorithm and the use of the p-depending balanced multifilters described above. We remark that the prefiltering step is no longer needed in this case.

In table 7.1 the results are shown in terms of the PSNR (Peak Signal to Noise Ratio) values computed after the reconstruction of the Lena image compressed at a compression ratio equal to 1:16.

According to what has been evidenced previously, the best results with the 2 and 4-balanced Lazy multifilters are obtained, respectively, with $p = 1/4$ and $p = 2/5$. Reconstructions using these bases are shown in figs 7.3:(a)-(b).

In comparison the CL basis without prefilter, as expected, give the worst result, while the absolute best result is achieved with the CL basis used with its proper prefilter (see figs. 7.3:(c)-(d))

Multifilter	**PSNR**
2-bal. Lazy ($p = -11/15$)	26.29
2-bal. Lazy ($p = -2/3$)	25.75
2-bal. Lazy ($p = -1/4$)	23.73
2-bal. Lazy ($p = 1/4$)	26.61
4-bal. Lazy ($p = 1/5$)	26.45
4-bal. Lazy ($p = 1/3$)	27.67
4-bal. Lazy ($p = 2/5$)	28.12
4-bal. Lazy ($p = 1/2$)	27.10
CL, no prefilter	11.79
CL, Haar prefilter	35.13

TABLE 7.1
PSNR's values

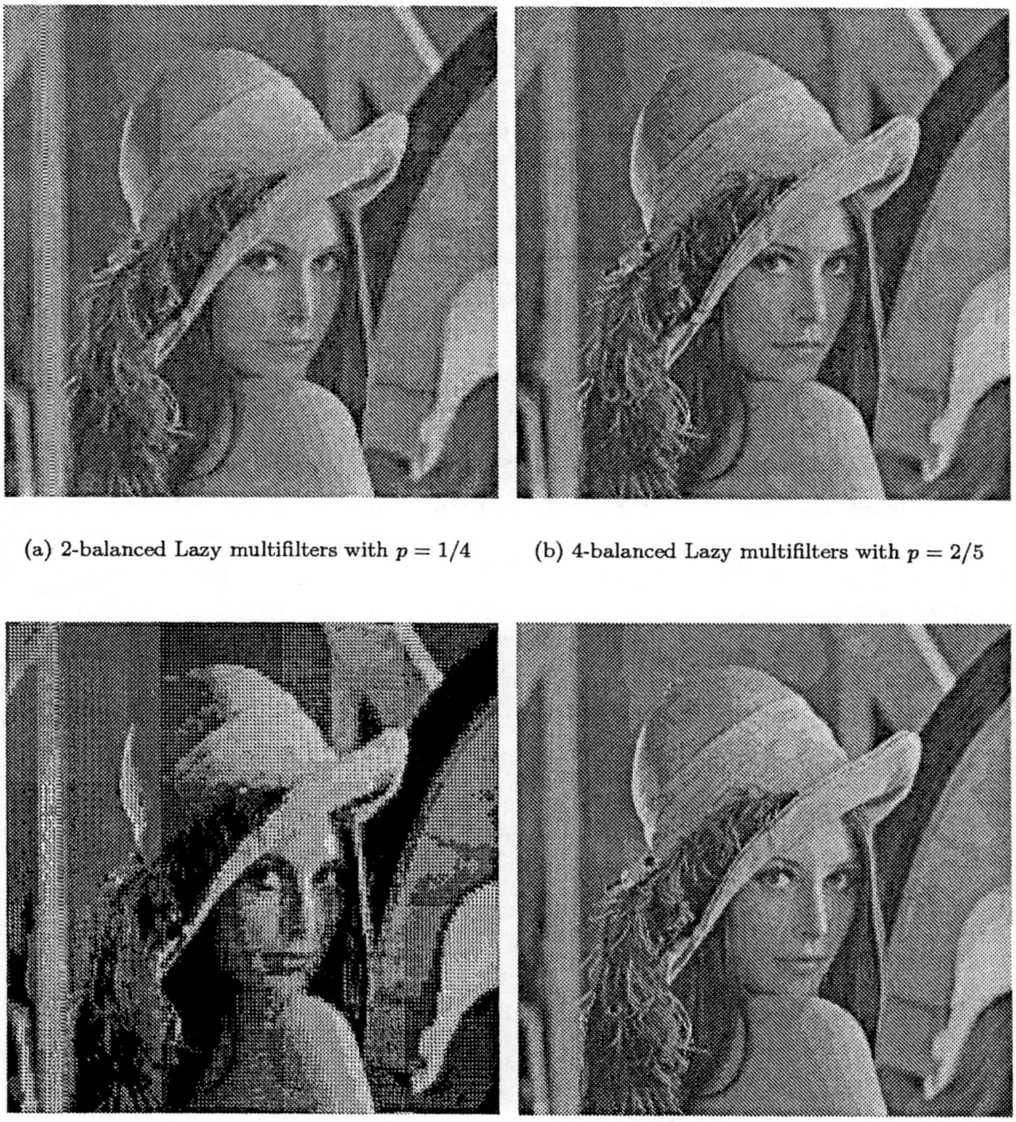

(a) 2-balanced Lazy multifilters with $p = 1/4$

(b) 4-balanced Lazy multifilters with $p = 2/5$

(c) Chui-Lian with no prefilter

(d) Chui-Lian with the Haar prefilter

FIG. 7.3. *Reconstructions of Lena image*

Acknowledgments Our researches have been developed within the activities of GNIM and supported by: MURST 40% and 60% projects; COFIN 97 project: "Analisi Numerica: Metodi e Software Matematico"; CNR project n. 98.01017.CT01.

REFERENCES

[1] S.BACCHELLI, *Block Toeplitz and Hurwitz matrices: a recursive approach*, Advances in Appl. Math., to appear

[2] S. BACCHELLI AND D. LAZZARO, *Some practical applications of block recursive matrices*, International Journal of Computers and Mathematics with Applications, to appear.

[3] S. BACCHELLI, M.COTRONEI AND D. LAZZARO, *An algebraic construction of k-balanced multiwavelets via the lifting scheme*, submitted to Numerical Algorithms.

[4] ———*A recursive approach to the construction of k-balanced biorthogonal multifilters*, submitted to the Proceedings of V International Conference on Mathematical Methods for Curves and Surfaces,1999.

[5] M. BARNABEI AND L. B. MONTEFUSCO, *Recursive Properties of Toeplitz and Hurwitz Matrices*, Linear Algebra and its Applications, vol. 274, 367–388, 1998.

[6] M. BARNABEI, L. B. MONTEFUSCO, AND C. GUERRINI, *Some algebraic aspects on signal processing*, Linear Algebra and its Applications, 1997.

[7] C.K. CHUI AND J. LIAN, *A study of orthonormal multi-wavelets*, Appl. Numer. Math., vol. 20, 273–298, 1996.

[8] C.K. CHUI, *An introduction to wavelets*, Academic Press, 1992.

[9] M. COTRONEI, *Multiwavelets: analisi teorica ed algoritmi*, Ph.D. Dissertation, 1996.

[10] M. COTRONEI, D. LAZZARO, L.B. MONTEFUSCO, AND L. PUCCIO, *Image compression through embedded multiwavelet transform coding*, IEEE Trans. on Image Process., to appear.

[11] ———, *Some experiments on image compression by means of multiwavelet transform*, Atti Accademia Peloritana dei Pericolanti, Classe I di Sc. Mat. Fis. e Natur., vol. LXXVI, to appear.

[12] M. COTRONEI, L.B. MONTEFUSCO, AND L. PUCCIO, *Multiwavelet analysis and signal processing*, IEEE Trans. on Circuits and Systems II: Analog and Digital Signal Processing, vol. 45, 970–987, 1998.

[13] I. DAUBECHIES AND W. SWELDENS,*Factoring wavelet transform into lifting steps*, Tech. Report, Bell Labs, 1996.

[14] J.S. GERONIMO, D. P. HARDIN, AND P.R. MASSOPUST, *Fractal functions and wavelet expansions based on several scaling functions*, J. Approx. Theory, vol. 78 n. 3, 373–401, 1994.

[15] S.S. GOH, Q. JIANG, AND T. XIA, *Construction of biorthogonal multiwavelets using the lifting scheme*, preprint.

[16] T. N. T. GOODMAN AND S. L. LEE, *Wavelets of multiplicity r*, Trans. Amer. Math. Soc., vol. 342(1):307–324, 1994.

[17] D.P. HARDIN AND D.W. ROACH, *Multiwavelet prefilters I: orthogonal prefilters preserving approximation order $p \leq 2$*, IEEE Trans. on Circuits and Systems II: Analog and Digital Signal Processing, vol. 45, 1106–1112, 1998.

[18] Q. JIANG, *On the construction of biorthogonal multiwavelet bases*, preprint.

[19] D. LAZZARO, *Biorthogonal M-band filter construction using the lifting scheme*, Num. Algorithms, to appear.

[20] J. LEBRUN AND M. VETTERLI, *Balanced multiwavelets: theory and design*, IEEE Trans. on Signal Process., vol. 46, 1998.

[21] ———*High order balanced multiwavelets: theory, factorization and design*, preprint.

[22] C.MICCHELLI AND T.SAUER, *Regularity of multiwavelets*, Adv. Comput. Math. 7, No.4, 455-545, 1997.

[23] G. PLONKA, *Approximation order provided by refinable function vectors*, Constructive Approximation, 1995.

[24] G. PLONKA AND V. STRELA, *Construction of multiscaling functions with approximation and symmetry*, SIAM J. Math. Anal., 1995.

[25] I. SELESNICK, *Multiwavelet bases with extra approximation properties*, IEEE Trans. on Signal Process., vol. 46, 2898–2908, 1998.

[26] G. STRANG AND V. STRELA, *Short wavelets and matrix dilation equations*, IEEE Trans. on Signal Process., vol. 43., 108–115, 1995.

[27] V. STRELA, P.N. HELLER, G. STRANG, P. TOPIWALA, AND C. HEIL, *The application of multiwavelet filter banks to signal and image processing*, IEEE Trans. on Image Process., to appear.

[28] W. SWELDENS, *The lifting scheme: a custom-design construction of biorthogonal wavelets*, ACHA, vol. 3, 186–200, 1996.

[29] P.P. VAIDYANATHAN, *Multirate Systems and Filter Banks*, Prentice-Hall, 1995.
[30] X.G. XIA, J.S. GERONIMO, D.P. HARDIN, AND B.W. SUTER, *Design of prefilters for discrete multiwavelet transforms*, IEEE Trans. on Signal Process., vol. 44, 25–35, 1996.

AN ALGEBRAIC FRAMEWORK FOR BIORTHOGONAL M-BAND FILTERS*

MARILENA BARNABEI†, CARLA GUERRINI‡, AND LAURA B. MONTEFUSCO§

Abstract. The building of M-band perfect reconstruction filter banks with finite filters involves a large number of conditions in comparison to the two-band case, and design problems grow as soon as the number of bands increases and filters are required to present better frequency separation characteristics. In this paper we consider a strategy to partially overcome such difficulties, by stating the problem in the algebraic context of banded recursive matrices and taking advantage of the mathematical tools that it allows us to use. In particular we consider the extension of the lifting scheme to the M-band case and we succeed in analyzing both its building and decomposition aspects and in characterizing its capabilities. Our analysis proves that the M-band lifting preserves most of the advantages of the two-band case, such as flexibility, simplicity, and speed.

Key words. M-band filter bank, recursive matrices, Laurent polynomial matrix, lifting scheme, computational complexity.

AMS subject classifications. 42C15, 42A38, 13F07, 65Y99.

1. Introduction. The design of M-band perfect reconstruction systems with finite filters represents a challenging problem, due to the great number of parameters involved, and the difficulties grow as soon as the number of bands taken into consideration increases and filters are required to meet a general set of pre-specified properties.

In the two-band case, a powerful tool, yielding an easy custom-design construction of biorthogonal filters, is the well known lifting scheme, which on the one hand can be seen as a simple tool for the construction of biorthogonal filters with pre-assigned features (building property), and on the other hand yields a factorization procedure that leads to faster analysis-synthesis algorithms (decomposition property).

In the M-band case, these two aspects of lifting have not yet been thoroughly examined, as far as we know, even if the decomposition aspect has been sketched in different settings in the filter bank literature (see, e.g., [6], [12]), while the building aspect has been considered in [7], where particular attention was paid to the computational aspects.

The aim of this paper is to exploit the mathematical tools offered by the algebraic context introduced in [1] and [2] in order to carefully analyze both aspects of the lifting scheme in the M-band setting and to show and characterize its capabilities. In fact, in this algebraic setting, it is natural to switch from the operational formulation of the problem to an equivalent formulation that makes use of particular bi-infinite matrices whose properties have been thoroughly investigated in the previously mentioned papers. Such properties lead us to simply restate the problems making use of the notion of the M-decimated matrix - an instrument closely connected with the widely-used polyphase representation - without going through the Fourier domain.

As far as the constructive aspect is concerned, our main result (Theorems 4.5 and 4.6) is a characterization of the algebraic form of all biorthogonal pairs of finite filter M-

*This work was supported by MURST ex 40% and 60% projects, by C.N.R. Project n. 98.01017.CT01, and by the University of Bologna, funds for selected topics.

† Department of Mathematics - University of Bologna, Piazza di Porta S. Donato 5, 40127 Bologna, Italy (barnabei@dm.unibo.it)

‡Department of Mathematics - University of Bologna,(guerrini@dm.unibo.it)

§Department of Mathematics - University of Bologna,(montelau@dm.unibo.it) tuples that can be obtained from a given pair, either by modifying the analysis low-pass filter and consequently updating the synthesis high-pass filters (lifting), or by modifying the analysis high-pass filters and consequently updating the synthesis low-pass filter (dual lifting). This characterization makes use only of linear combinations of Laurent polynomials, therefore reducing dramatically the computational complexity of designing M-band systems with good filtering properties. In fact, the perfect reconstruction is automatically enforced by the lifting scheme, and any pre-defined requirement can be fulfilled by simply imposing the appropriate constraints on the coefficients of the Laurent polynomials.

As regards the decomposition aspect, we succeed in proving that every biorthogonal pair of M-tuples of finite filters can be obtained from a suitably scaled trivial pair by a finite sequence of lifting and dual lifting steps, alternated with permutations. As already seen for the two-band case, lifting thus provides a framework that allows the construction - and hence the factorization - of all biorthogonal filter banks, with a consequent reduction of the computational costs in practical applications.

In conclusion, the present paper can be seen as an improvement of the existing results on this topic, and at the same time it provides a simple tool for the construction of M-band biorthogonal filters.

It is worthwhile recalling that the lifting scheme was originally introduced in [10] in the context of wavelet theory. However, quoted from [11], "The lifting scheme ... provides an answer to the algebraic stage of a wavelet construction. Whether these filters actually generate functions which form a stable basis (analytic stage) or have a smoothness (geometric stage), remains to be checked in each particular case". For this reason, since at the moment it does not seem possible to translate the analytic and geometric stages into some of the algebraic properties of filters, we prefer to state our results in terms of biorthogonal filters, without mentioning wavelets.

The paper is organized as follows. We start out by recalling in Section 2 the main notions and results about Laurent polynomials and the related bi-infinite matrices. In Section 3 we describe a multichannel filter bank system in algebraic notation, both in terms of M-decimated matrices and in terms of Hurwitz matrices. Finally, in Sections 4 and 5, with the notation introduced in Section 3, we are able to state and prove our main results, namely, the extension of the lifting scheme to the M-band case, in both its aspects.

2. Background Material. In this paper we will be concerned with the theory of FIR systems, and consequently we will deal with finite sequences of real numbers.
Such sequences can be identified with Laurent polynomials, as follows: if $(a_h, a_{h+1}, \dots, a_k)$, $h, k \in \mathbb{Z}$, is a finite sequence, its *generating function* is the (formal) Laurent polynomial

$$\alpha = \sum_{i=h}^{k} a_i t^i. \tag{1}$$

Moreover, we will denote by $[\alpha]$ the bi-infinite vector, with finite support,

$$[\alpha] = [\dots 0 \dots 0 \; a_h \; a_{h+1} \dots a_k \; 0 \dots 0 \dots]^T.$$

The degree of the Laurent polynomial α is defined as $deg\ \alpha := k-h$, provided that $a_h, a_k \neq 0$. The degree of the zero polynomial is conventionally set to be $-\infty$. It is well known that the set $\mathcal{P}$ of real Laurent polynomials under the usual sum and convolution product is a Euclidean domain, whose quotient field is the field $\mathcal{L}^+$ of Laurent series, namely, formal series of the kind

$$\sum_{i\geq d} a_i t^i$$

where d is an integer.

If $\alpha = \sum_{i=h}^{k} a_i t^i$ is a Laurent polynomial, we denote by α^* the Laurent polynomial

$$\alpha^* = \sum_{i=h}^{k} a_i t^{-i}.$$

For every fixed integer $M \geq 2$, the *M-decimated polynomials* of α are the M Laurent polynomials defined as

$$\mathbf{D}_r^M(\alpha) = \alpha_r := \sum_i a_{Mi+r} t^i \qquad r = 0, 1, \ldots, M-1.$$

The original polynomial can be recovered from its decimated polynomials, as follows:

(2) $$\alpha = \alpha_0 \circ t^M + t(\alpha_1 \circ t^M) + \ldots + t^{M-1}(\alpha_{M-1} \circ t^M)$$

where, if $\beta = \sum_i b_i t^i$ is any Laurent polynomial, then

$$\beta \circ t^M = \sum_i b_i t^{Mi}.$$

In the sequel, we will make use of the following identities concerning the interplay between decimation and "star" operators, which have been proved in [1]:

$$(\alpha_0)^* = (\alpha^*)_0$$

(3) $$(\alpha_r)^* = t(\alpha^*)_{M-r} \quad for \quad r = 1, 2, \ldots, M-1.$$

For the sake of simplicity, we will often write α_r^* for $(\alpha_r)^*$.

If α is a Laurent polynomial, the *M-decimated vector* of α will be the column vector

(4) $$\mathbf{\Delta}^M(\alpha) := \begin{bmatrix} \alpha_0 \\ \alpha_1 \\ \vdots \\ \alpha_{M-1} \end{bmatrix},$$

whose entries are the M-decimated polynomials of α. Moreover, the *starred M-decimated vector* of α will be the column vector

(5) $$\mathbf{\Delta}^M(\alpha)^* := \begin{bmatrix} \alpha_0^* \\ \alpha_1^* \vdots \\ \alpha_{M-1}^* \end{bmatrix}.$$

Similarly, if $\boldsymbol{\alpha} = (\alpha^{(0)}, \alpha^{(1)}, \ldots, \alpha^{(M-1)})$ is an ordered M-tuple of Laurent polynomials, the matrix

$$\mathbf{\Delta}^M(\boldsymbol{\alpha}) = \begin{bmatrix} \alpha_0^{(0)} & \alpha_0^{(1)} & \cdots & \alpha_0^{(M-1)} \\ \alpha_1^{(0)} & \alpha_1^{(1)} & \cdots & \alpha_1^{(M-1)} \\ \vdots & \vdots & \vdots & \vdots \\ \alpha_{(M-1)}^{(0)} & \alpha_{(M-1)}^{(1)} & \cdots & \alpha_{(M-1)}^{(M-1)} \end{bmatrix}. \tag{6}$$

will be called the *M-decimated matrix* of $\boldsymbol{\alpha}$.

Finally, we recall that, if $\alpha = \sum_i a_i t^i$ is a Laurent polynomial, the *M-upsampled polynomial* $(\uparrow M)\alpha$ relative to α is defined as:

$$(\uparrow M)\alpha := \alpha \circ t^M = \sum_i a_i t^{Mi}.$$

Laurent polynomials can be used to describe some banded bi-infinite matrices, widely used in practical applications. In fact, a banded Toeplitz matrix whose generating function is the Laurent polynomial α will be denoted by the symbol $R(t, \alpha)$, and hence its transpose is the matrix $R(t, \alpha^*)$.

Similarly, if M is a fixed integer $M \geq 1$, the symbol $R(t^M, \alpha)$ will denote the banded Hurwitz matrix with step M and generating function α.

Both banded Toeplitz and Hurwitz matrices can be recovered under the general theory of recursive matrices, and many of their algebraic properties have been derived in [1] from this combinatorial point of view. In order to make the present paper self-contained we recall here two key results for Hurwitz matrices, namely, the fact that the product of two Hurwitz matrices is again a Hurwitz matrix, with a different step, and that the product of the Hurwitz matrix $R(t^M, \alpha)$ and the transpose of the Hurwitz matrix $R(t^M, \beta)$ is a Toeplitz matrix, whose generating function can be explicitly described in terms of the M-decimated polynomials of α and β.

THEOREM 2.1. *Let α, β be non-zero Laurent polynomials and M, N fixed integers, $M, N \geq 1$. Then we have*

$$R(t^M, \alpha) \times R(t^N, \beta) = R(t^{MN}, (\alpha \circ t^N)\beta).$$

Moreover, the Hurwitz matrix $R(t^M, \alpha)$ can be factorized as follows:

$$R(t^M, \alpha) = R(t^M, 1) \times R(t, \alpha),$$

where 1 denotes the unity of the ring $\mathcal{P}$.

THEOREM 2.2. *Let α, β be non-zero Laurent polynomials, and M a fixed integer, $M \geq 2$. Set $H = R(t^M, \alpha)$, $K = R(t^M, \beta)$. The product $H \times K^T$ is a Toeplitz matrix, whose generating function λ is given by*

$$\lambda = \mathbf{\Delta}^M(\alpha)^T \times \mathbf{\Delta}^M(\beta)^*.$$

3. Basic Operators and Multichannel Filter Banks. Banded Toeplitz and Hurwitz matrices are associated with the basic operators of FIR filter theory, as shown in [2]. An M-channel maximally decimated filter bank system, ($M \geq 2$) is represented by a scheme of the form

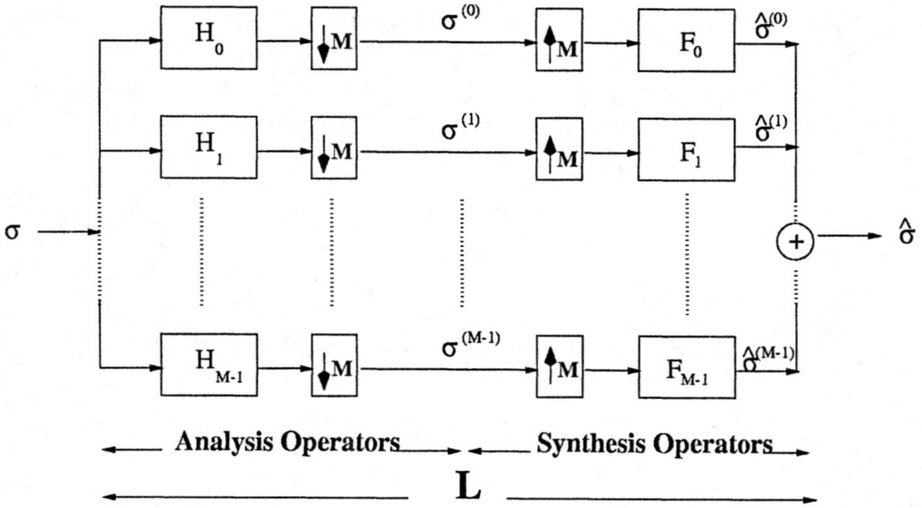

where, for every $i = 0, \ldots, M-1$, we denote by $C_i := R(t^M, \gamma^{(i)})$ the Hurwitz matrix associated with the analysis operator

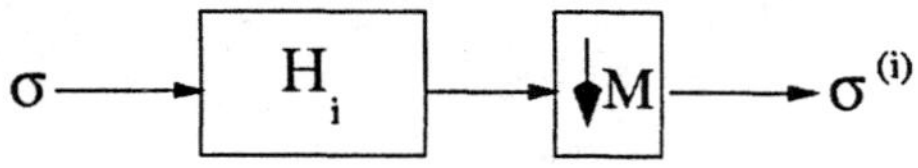

namely, $[\sigma^{(i)}] = C_i \times [\sigma]$, and by $A_i^T = R(t^M, \alpha^{(i)})^T$ the matrix associated with the synthesis operator

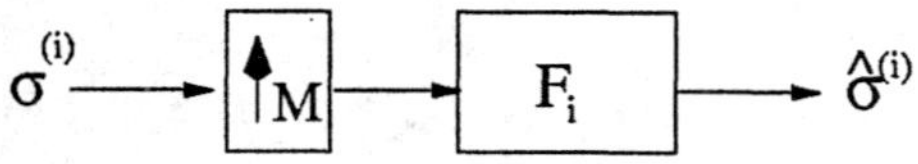

that is, $[\hat{\sigma}^{(i)}] = A_i^T \times [\sigma^{(i)}]$. Hence, setting

$$\boldsymbol{\alpha} := (\alpha^{(0)}, \alpha^{(1)}, \ldots, \alpha^{(M-1)}), \quad \boldsymbol{\gamma} := (\gamma^{(0)}, \gamma^{(1)}, \ldots, \gamma^{(M-1)})$$

the M-channel filter bank is uniquely represented by the pair of polynomial vectors $(\boldsymbol{\alpha}, \boldsymbol{\gamma})$. The 0–th components $\alpha^{(0)}, \gamma^{(0)}$ of $\boldsymbol{\alpha}$ and $\boldsymbol{\gamma}$, respectively, are called the *low-pass filters*, while the components $\alpha^{(1)}, \alpha^{(2)}, \ldots, \alpha^{(M-1)}$ and $\gamma^{(1)}, \gamma^{(2)}, \ldots, \gamma^{(M-1)}$ are called the *high-pass filters.*

The linear operator $\mathbf{L} : \sigma \to \hat{\sigma}$ corresponds to the matrix

$$A_0^T \times C_0 + A_1^T \times C_1 + \ldots + A_{M-1}^T \times C_{M-1},$$

which can be formally written as

$$[A_0^T A_1^T \dots A_{M-1}^T] \times \begin{bmatrix} C_0 \\ C_1 \\ \vdots \\ C_{M-1} \end{bmatrix}.$$

It has been shown in [1] that this is an $(M \times M)$-block Toeplitz matrix, whose generating function can be described in terms of the two polynomial matrices $\boldsymbol{\Delta}^M(\boldsymbol{\alpha})$ and $\boldsymbol{\Delta}^M(\boldsymbol{\gamma})$.

More precisely we have:

THEOREM 3.1. *Let $A_i = R(t^M, \alpha^{(i)})$, $C_i = R(t^M, \gamma^{(i)})$, $i = 0, 1, \dots, M-1$, be banded Hurwitz matrices. The matrix*

$$A_0^T \times C_0 + A_1^T \times C_1 + \dots + A_{M-1}^T \times C_{M-1}$$

is an $(M \times M)$-block Toeplitz matrix, whose generating function is the polynomial matrix given by

$$\boldsymbol{\Pi} = \boldsymbol{\Delta}^M(\boldsymbol{\alpha})^* \times \boldsymbol{\Delta}^M(\boldsymbol{\gamma})^T. \tag{7}$$

It is now worthwhile remarking that, if $\mathbf{T} : \mathcal{P} \to \mathcal{P}$ is a linear operator associated with the block Toeplitz matrix $R(t, \boldsymbol{\Omega})$, where $\boldsymbol{\Omega}$ is the square polynomial matrix

$$\boldsymbol{\Omega} = \begin{bmatrix} \omega_{0,0} & \omega_{0,1} & \dots & \omega_{0,M-1} \\ \omega_{1,0} & \omega_{1,1} & \dots & \omega_{1,M-1} \\ \vdots & \vdots & \vdots & \vdots \\ \omega_{M-1,0} & \omega_{1,M-1} & \dots & \omega_{M-1,M-1} \end{bmatrix} \qquad \omega_{i,j} \in \mathcal{P},$$

then the image $\hat{\sigma} = \mathbf{T}\sigma$ of the Laurent polynomial σ under the operator $\mathbf{T}$ can be described in terms of its M-decimated polynomials as follows:

$$\boldsymbol{\Delta}^M(\hat{\sigma}) = \boldsymbol{\Omega}^* \times \boldsymbol{\Delta}^M(\sigma), \tag{8}$$

where

$$\boldsymbol{\Omega}^* = \begin{bmatrix} \omega_{0,0}^* & \omega_{0,1}^* & \dots & \omega_{0,M-1}^* \\ \omega_{1,0}^* & \omega_{1,1}^* & \dots & \omega_{1,M-1}^* \\ \vdots & \vdots & \vdots & \vdots \\ \omega_{M-1,0}^* & \omega_{1,M-1}^* & \dots & \omega_{M-1,M-1}^* \end{bmatrix}.$$

Let $\mathbf{L}$ be the linear operator describing the action of an M-channel filter bank system relative to the pair $(\boldsymbol{\alpha}, \boldsymbol{\gamma})$. From the above considerations, we get $\mathbf{L}\sigma = \hat{\sigma}$, where $\hat{\sigma}$ is the unique Laurent polynomial such that

$$\boldsymbol{\Delta}^M(\hat{\sigma}) = \boldsymbol{\Delta}^M(\boldsymbol{\alpha}) \times \boldsymbol{\Delta}^M(\boldsymbol{\gamma})^{*^T} \times \boldsymbol{\Delta}^M(\sigma). \tag{9}$$

Formula (9) allows us to give a computationally efficient form for the analysis and synthesis algorithms, namely:

$$[\sigma^{(0)}\sigma^{(1)}\dots\sigma^{(M-1)}]^T = \mathbf{\Delta}^M(\gamma)^{*^T} \times \mathbf{\Delta}^M(\sigma) \quad (analysis)$$

$$\mathbf{\Delta}^M(\hat{\sigma}) = \mathbf{\Delta}^M(\boldsymbol{\alpha}) \times [\sigma^{(0)}\sigma^{(1)}\dots\sigma^{(M-1)}]^T \quad (synthesis).$$

We recall that the operator $\mathbf{L}$ has the *perfect reconstruction property* whenever $\mathbf{L}\sigma = \sigma$. Theorem 3.1 allows us to translate the perfect reconstruction condition in terms of Laurent polynomials as follows:

PROPOSITION 3.2. *The M-channel filter bank described by the pair of polynomial vectors* $(\boldsymbol{\alpha},\gamma)$ *corresponds to a perfect reconstruction operator if and only if the following condition is satisfied:*

$$\mathbf{\Delta}^M(\boldsymbol{\alpha}) \times \mathbf{\Delta}^M(\gamma)^{*^T} = I. \tag{10}$$

The perfect reconstruction condition can also be stated in terms of Hurwitz matrices:

PROPOSITION 3.3. *The M-channel filter bank described by the pair of polynomial vectors* $(\boldsymbol{\alpha},\gamma)$ *corresponds to a perfect reconstruction operator if and only if the Hurwitz matrices*

$$A_i = R(t^M, \alpha^{(i)}), \quad C_i = R(t^M, \gamma^{(i)}), \quad i = 0, 1, \dots, M-1$$

satisfy the following conditions:

$$C_i \times A_j^T = \delta_{i,j} I, \quad i,j = 0,1,\dots,M-1, \tag{11}$$

or, in matrix notation,

$$\begin{bmatrix} C_0 \\ C_1 \\ \vdots \\ C_{M-1} \end{bmatrix} \times [A_0^T A_1^T \dots A_{M-1}^T] = \begin{bmatrix} \mathrm{I} & 0 & \dots & 0 \\ 0 & \mathrm{I} & \dots & 0 \\ \dots & \dots & \dots & \dots \\ 0 & 0 & \dots & \mathrm{I} \end{bmatrix}.$$

Proof. Recall that the ring $\mathcal{P}$ of Laurent polynomials is an integral domain. Hence, condition (10) can be equivalently written as

$$\mathbf{\Delta}^M(\gamma)^{*^T} \times \mathbf{\Delta}^M(\boldsymbol{\alpha}) = I, \tag{12}$$

that is,

$$\mathbf{\Delta}^M(\gamma^{(i)})^{*^T} \times \mathbf{\Delta}^M(\alpha^{(j)}) = \delta_{ij}, \quad i,j = 0,1\dots,M-1. \tag{13}$$

The assertion now follows from Theorem 2.2. □

A pair $(\boldsymbol{\alpha},\gamma)$ satisfying one of the above equivalent conditions (10) or (11) will be called a pair of *dual* (or *biorthogonal*) *polynomial vectors.* The simplest pair of dual polynomial vectors is the "trivial" pair $(\boldsymbol{\delta}, \boldsymbol{\delta})$, where $\boldsymbol{\delta} = (1, t, t^2, \dots, t^{M-1})$.

It is well known that a square polynomial matrix admits an inverse with polynomial entries whenever its determinant is a non-zero monomial. Hence, an M-tuple $\boldsymbol{\alpha}$ of Laurent polynomials admits a dual M-tuple if and only if $det(\mathbf{\Delta}^M(\boldsymbol{\alpha}))$ is a non-zero monomial. If this is the case, the dual M-tuple γ of $\boldsymbol{\alpha}$ is uniquely determined by condition (10). Obviously, if $(\boldsymbol{\alpha}, \gamma)$ is a pair of dual polynomial vectors, the same holds for the pair $(\gamma, \boldsymbol{\alpha})$.

4. The Lifting Scheme. The algebraic results previously presented allow us to state in this section the generalization of the lifting scheme to the M-band case.

As we have already remarked, the design of FIR M-band perfect reconstruction systems leads, in general, to cumbersome numerical problems, due to the large number of parameters involved in the optimization procedures that must be used. This is not the case with lifting. In fact, every filter bank built with this method automatically satisfies the perfect reconstruction condition, while the other constraints necessary to obtain filter with pre-specified properties can be imposed directly on the filter coefficients. Moreover, the lifting scheme offers an implementation structure that makes the realization of the new filters very efficient.

The algebraic setting in which we have developed our considerations allows us to state our results both in terms of decimated matrices and of bi-infinite matrices, easily switching from one setting to the other, according to the context.

First of all, we give two preliminary results, that represent the theoretical support of the lifting scheme. The first one provides a simple characterization of all the polynomial vectors γ which admit a dual vector and share the same 0-th component $\gamma^{(0)}$.

THEOREM 4.1. *Let $(\boldsymbol{\alpha},\boldsymbol{\gamma})$ be a pair of dual polynomial vectors. For any fixed index $i = 1, 2, \ldots, M-1$, every Laurent polynomial $\hat{\gamma}^{(i)}$ such that*

$$\boldsymbol{\Delta}^M(\hat{\gamma}^{(i)})^{*^T} \times \boldsymbol{\Delta}^M(\alpha^{(j)}) = \delta_{ij} \quad j = 1, 2, \ldots, M-1 \tag{14}$$

is given by

$$\hat{\gamma}^{(i)} = \gamma^{(i)} + (\lambda \circ t^M)\gamma^{(0)}$$

as λ ranges over the set of all Laurent polynomials.

Proof. Conditions (14) yield a system of M-1 linear equations in the M unknowns $(\hat{\gamma}_0^{(i)})^*, (\hat{\gamma}_1^{(i)})^*, \ldots, (\hat{\gamma}_{M-1}^{(i)})^*$, namely,

$$\begin{cases} \alpha_0^{(1)}(\hat{\gamma}_0^{(i)})^* + \ldots + \alpha_{M-1}^{(1)}(\hat{\gamma}_{M-1}^{(i)})^* & = 0 \\ \ldots & \\ \alpha_0^{(i)}(\hat{\gamma}_0^{(i)})^* + \ldots + \alpha_{M-1}^{(i)}(\hat{\gamma}_{M-1}^{(i)})^* & = 1 \\ \ldots & \\ \alpha_0^{(M-1)}(\hat{\gamma}_0^{(i)})^* + \ldots + \alpha_{M-1}^{(M-1)}(\hat{\gamma}_{M-1}^{(i)})^* & = 0. \end{cases} \tag{15}$$

The M-tuple $\boldsymbol{\Delta}^M(\gamma^{(i)})^*$ is a solution of the system (15), while $\boldsymbol{\Delta}^M(\gamma^{(0)})^*$ is a solution of the associated homogeneous linear system. Hence, in the quotient field $\mathcal{L}^+$ of Laurent series, the general solution of (15) is given by

$$\chi = \boldsymbol{\Delta}^M(\gamma^{(i)})^* + \tau\boldsymbol{\Delta}^M(\gamma^{(0)})^*, \tag{16}$$

where τ is a Laurent series. We now need to characterize the polynomial solutions of system (15). We have assumed that the polynomial vector $\boldsymbol{\gamma}$ admits a dual vector $\boldsymbol{\alpha}$, which implies that $det\boldsymbol{\Delta}^M(\boldsymbol{\gamma})^*$ is a non-zero monomial. By using the Laplace expansion formula for the determinant, we find that the g.c.d. of the M-tuple $((\gamma_0^{(0)})^*, (\gamma_1^{(0)})^*, \ldots, (\gamma_{M-1}^{(0)})^*)$ is a monomial. This implies that the solution χ given by (16) is a polynomial vector if and only if τ is a Laurent polynomial.

At this point we note that every polynomial solution $\chi = (\chi_0, \chi_1, \ldots, \chi_{M-1})$ of (16), can be seen as the starred M-decimated vector of the Laurent polynomial

$$\hat{\gamma}^{(i)} = \chi_0^* \circ t^M + t(\chi_1^* \circ t^M) + \ldots + t^{M-1}(\chi_{M-1}^* \circ t^M)$$

$$= \gamma_0^{(i)} \circ t^M + (\tau^* \circ t^M)(\gamma_0^{(0)} \circ t^M) + t(\gamma_1^{(i)} \circ t^M + (\tau^* \circ t^M)(\gamma_1^{(0)} \circ t^M)) + \ldots$$

$$+ t^{M-1}(\gamma_{M-1}^{(i)} \circ t^M + (\tau^* \circ t^M)(\gamma_{M-1}^{(0)} \circ t^M))$$

$$= \gamma^{(i)} + (\tau^* \circ t^M)\gamma^{(0)};$$

the thesis now follows by setting $\lambda = \tau^*$. □

The second basic result is a characterization of all polynomial vectors γ which admit a dual vector and share the same components $\gamma^{(1)}, \gamma^{(2)}, \ldots, \gamma^{(M-1)}$.

THEOREM 4.2. *Given a pair of dual M-tuples of Laurent polynomials (α,γ), every Laurent polynomial $\hat{\gamma}^{(0)}$ such that*

$$\Delta^M(\hat{\gamma}^{(0)})^{*^T} \times \Delta^M(\alpha^{(0)}) = 1 \tag{17}$$

is given by

$$\hat{\gamma}^{(0)} = \gamma^{(0)} + (\mu^{(1)} \circ t^M)\gamma^{(1)} + \ldots + (\mu^{(M-1)} \circ t^M)\gamma^{(M-1)},$$

as $\mu^{(1)}, \mu^{(2)}, \ldots, \mu^{(M-1)}$ range over the set of all Laurent polynomials.

Proof. Condition (17) yields a linear equation in the M unknowns $(\hat{\gamma}_0^{(i)})^*$, $(\hat{\gamma}_0^{(i)})^*$, …, $(\hat{\gamma}_{M-1}^{(i)})^*$, namely,

$$\alpha_0^{(0)}(\hat{\gamma}_0^{(0)})^* + \alpha_0^{(1)}(\hat{\gamma}_0^{(1)})^* + \ldots + \alpha_{M-1}^{(0)}(\hat{\gamma}_{M-1}^{(0)})^* = 1. \tag{18}$$

The M-tuple $\Delta^M(\gamma^{(0)})^*$ is a solution of equation (18), while the M-tuples

$$\Delta^M(\gamma^{(1)})^*, \Delta^M(\gamma^{(2)})^*, \ldots, \Delta^M(\gamma^{(M-1)})^*$$

are M-1 linearly independent solutions of the associated homogeneous equation. Hence, the general solution of (18) is given by

$$\chi = \Delta^M(\gamma^{(0)})^* + \mu_1\Delta^M(\gamma^{(1)})^* + \ldots + \mu_{M-1}\Delta^M(\gamma^{(M-1)})^*, \tag{19}$$

where $\mu_1, \mu_2, \ldots, \mu_{M-1}$ belong to the field of Laurent series. We now prove that the solution χ is a polynomial vector if and only if $\mu_1, \mu_2 \ldots, \mu_{M-1}$ are Laurent polynomials. Indeed, suppose that χ is a polynomial solution of equation (18). Without loss of generality, we can set $\mu^{(i)} = \tau^{(i)}/\rho$, with $\tau^{(i)}, \rho \in \mathcal{P}$, $i = 1, 2, \ldots, M-1$.

Then, setting $\chi' := \rho(\chi - \mathbf{\Delta}^M \gamma^{(0)*})$, the ($M$-1)-tuple $(\tau^{(1)}, \dots, \tau^{(M-1)})$ is a polynomial solution of the system of M linear equations in the M-1 unknowns $\phi_1, \dots, \phi_{M-1}$:

$$(20) \qquad \begin{bmatrix} (\gamma_0^{(1)})^* & (\gamma_0^{(2)})^* & \dots & (\gamma_0^{(M-1)})^* \\ (\gamma_1^{(1)})^* & (\gamma_1^{(2)})^* & \dots & (\gamma_1^{(M-1)})^* \\ \dots & \dots & \dots & \dots \\ (\gamma_{M-1}^{(1)})^* & (\gamma_{M-1}^{(2)})^* & \dots & (\gamma_{M-1}^{(M-1)})^* \end{bmatrix} \times \begin{bmatrix} \phi_1 \\ \phi_2 \\ \vdots \\ \phi_{n-1} \end{bmatrix} = \begin{bmatrix} \chi_0' \\ \chi_1' \\ \vdots \\ \chi_{M-1}' \end{bmatrix}.$$

It is known ([5], p. 111) that a linear system whose coefficients belong to a Euclidean domain E admits a solution in E whenever its matrix of coefficients has the same rank k, say, as the augmented matrix, and the g.c.d. of the non-zero minors of order k of the augmented matrix divides the g.c.d. of the minors of the same order of the matrix of coefficients. In our case, the M minors of order M-1 of the matrix of coefficients are the cofactors of the elements of the 0–th row of the matrix $\mathbf{\Delta}^M(\gamma)^*$, and therefore their g.c.d. is a monomial. On the other hand, the g.c.d. of the minors of order M-1 of the augmented matrix is divisible by ρ, and this implies that ρ is a monomial. □

Theorems 4.1 and 4.2 can be equivalently formulated in terms of Hurwitz matrices. In fact, Theorem 2.1 and Proposition 3.3 allow us to state the following results:

THEOREM 4.3. *Let $(\boldsymbol{\alpha}, \boldsymbol{\gamma})$ be a pair of dual polynomial vectors. Set*

$$A_i = R(t^M, \alpha^{(i)}), \quad C_i = R(t^M, \gamma^{(i)}), \quad i = 0, 1, \dots, M-1.$$

For any fixed index $i = 1, 2, \dots, M-1$, every Hurwitz matrix $\hat{C}_i = R(t^M, \hat{\gamma}^{(i)})$ such that

$$\hat{C}_i \times A_j^T = \delta_{ij} I, \quad j = 1, 2, \dots, M-1$$

is given by

$$(21) \qquad \hat{C}_i = C_i + T_i \times C_0,$$

as $T_i = R(t, \mu^{(i)})$ ranges over the set of banded Toeplitz matrices.

THEOREM 4.4. *Let $(\boldsymbol{\alpha}, \boldsymbol{\gamma})$ be a pair of dual polynomial vectors. Set*

$$A_i = R(t^M, \alpha^{(i)}), \quad C_i = R(t^M, \gamma^{(i)}), \quad i = 0, 1, \dots, M-1.$$

Every Hurwitz matrix $\hat{C}_0 = R(t^M, \hat{\gamma}^{(0)})$ such that

$$\hat{C}_0 \times A_0^T = I$$

is given by

$$(22) \qquad \hat{C}_0 = C_0 + \sum_{j=1}^{M-1} Q_j \times C_j,$$

as

$$Q_j = R(t, \lambda^{(j)}), \quad j = 1, 2, \dots, M-1$$

range over the set of all banded Toeplitz matrices.

Making use of the preceding results, we can now extend the lifting scheme to the M-band case, namely, given a dual pair $(\boldsymbol{\alpha}, \boldsymbol{\gamma})$, we are able to characterize in algebraic form all dual pairs $(\overline{\boldsymbol{\alpha}}, \overline{\boldsymbol{\gamma}})$ such that $\overline{\boldsymbol{\alpha}}$ has the same low pass filter as $\boldsymbol{\alpha}$, while $\overline{\boldsymbol{\gamma}}$ has the same (M-1)-tuple of high-pass filters as $\boldsymbol{\gamma}$.

THEOREM 4.5 (Lifting Scheme). *Let $(\boldsymbol{\alpha}, \boldsymbol{\gamma})$ be a dual pair of Laurent polynomials, with*

$$\boldsymbol{\alpha} = (\alpha^{(0)}, \alpha^{(1)}, \ldots, \alpha^{(M-1)}), \quad \boldsymbol{\gamma} = (\gamma^{(0)}, \gamma^{(1)}, \ldots, \gamma^{(M-1)}).$$

Every dual pair $(\overline{\boldsymbol{\alpha}}, \overline{\boldsymbol{\gamma}})$ of the form

$$\overline{\boldsymbol{\alpha}} = (\alpha^{(0)}, \hat{\alpha}^{(1)}, \ldots, \hat{\alpha}^{(M-1)}), \quad \overline{\boldsymbol{\gamma}} = (\hat{\gamma}^{(0)}, \gamma^{(1)}, \ldots, \gamma^{(M-1)}). \tag{23}$$

can be obtained by choosing

$$\hat{\gamma}^{(0)} = \gamma^{(0)} + \sum_{j=1}^{M-1} (\lambda^{(j)} \circ t^M)\gamma^{(j)}, \tag{24a}$$

$$\hat{\alpha}^{(i)} = \alpha^{(i)} - ((\lambda^{(i)})^* \circ t^M)\alpha^{(0)}, \quad i = 1, 2, \ldots, M-1, \tag{24b}$$

where $\lambda^{(1)}, \lambda^{(2)}, \ldots, \lambda^{(M-1)}$ are arbitrary Laurent polynomials.

Proof. Set $A_i = R(t^M, \alpha^{(i)})$, $C_i = R(t^M, \gamma^{(i)})$, $i = 0, 1, \ldots, M\text{-}1$, $\hat{A}_i = R(t^M, \hat{\alpha}^{(i)})$, $i = 1, 2, \ldots, M-1$, $\hat{C}_0 = R(t^M, \hat{\gamma}^{(0)})$. By Proposition 3.3, the condition of $(\overline{\boldsymbol{\alpha}}, \overline{\boldsymbol{\gamma}})$ being a dual pair is equivalent to the following three relations involving the associated Hurwitz matrices:

i) $\hat{C}_0 \times A_0^T = I$,
ii) $C_j \times \hat{A}_i^T = \delta_{ij} I$, $i, j = 1, 2, \ldots, M-1$, and
iii) $\hat{C}_0 \times \hat{A}_i^T = 0$, $i = 1, 2, \ldots, M-1$.

Making use of the symmetry of the duality relation, Theorems 4.1 and 4.2 imply that conditions i) and ii) are satisfied if and only if

$$\hat{C}_0 = C_0 + \sum_{j=1}^{M-1} Q_j \times C_j$$

and

$$\hat{A}_i = A_i + T_i \times A_0, \quad i = 1, 2, \ldots, M-1,$$

where $Q_j = R(t, \lambda^{(j)})$, $T_i = R(t, \mu^{(i)})$ are arbitrary banded Toeplitz matrices.

We now get, for every $i = 1, 2, \ldots, M-1$:

$$\hat{C}_0 \times \hat{A}_i^T = \left(C_0 + \sum_{j=1}^{M-1} Q_j \times C_j \right) \times (A_i^T + A_0^T \times T_i^T) =$$

$$C_0 \times A_i^T + C_0 \times A_0^T \times T_i^T + \sum_{j=1}^{M-1} Q_j \times C_j \times A_i^T + \sum_{j=1}^{M-1} Q_j \times C_j \times A_0^T \times T_i^T =$$

$$\mathbf{0} + T_i^T + Q_i + \mathbf{0} = R(t, (\mu^{(i)})^*) + R(t, \lambda^{(i)}).$$

Hence, condition iii) holds if and only if

$$-R(t, (\mu^{(i)})^*) = R(t, \lambda^{(i)}),$$

namely,

$$\mu^{(i)} = -(\lambda^{(i)})^*.$$

□

The above result shows how, starting from a given perfect reconstruction M-channel filter bank, it is possible to modify the analysis low-pass filter, and consequently update the synthesis high-pass filters, while still maintaining the perfect reconstruction property. The polynomials $\lambda^{(i)}$ in equations (24) above represent the degrees of freedom that are left after imposing the biorthogonality conditions, and hence they give us the full control over all dual pairs $(\overline{\boldsymbol{\alpha}}, \overline{\boldsymbol{\gamma}})$ of the form (23).

In practical situations, especially in the context of image compression, it is sometimes more useful to modify the analysis high-pass filters by adding some desirable properties, such as vanishing moments, prefixed shape, etc., and consequently updating the low-pass synthesis filter. This second construction, known as the *dual lifting scheme*, is simply obtained by applying the lifting scheme to the pair $(\boldsymbol{\gamma}, \boldsymbol{\alpha})$:

THEOREM 4.6 (Dual Lifting Scheme). *Let $(\boldsymbol{\alpha}, \boldsymbol{\gamma})$ be a dual pair of Laurent polynomials, with*

$$\boldsymbol{\alpha} = (\alpha^{(0)}, \alpha^{(1)}, \ldots, \alpha^{(M-1)}), \quad \boldsymbol{\gamma} = (\gamma^{(0)}, \gamma^{(1)}, \ldots, \gamma^{(M-1)}).$$

Every dual pair $(\overline{\boldsymbol{\alpha}}, \overline{\boldsymbol{\gamma}})$ of the form

$$\overline{\boldsymbol{\alpha}} = (\hat{\alpha}^{(0)}, \alpha^{(1)}, \ldots, \alpha^{(M-1)}), \quad \overline{\boldsymbol{\gamma}} = (\gamma^{(0)}, \hat{\gamma}^{(1)}, \ldots, \hat{\gamma}^{(M-1)}) \tag{25}$$

can be obtained by choosing

$$\hat{\gamma}^{(i)} = \gamma^{(i)} + (\lambda^{(i)} \circ t^M)\gamma^{(0)}, \quad i = 1, 2, \ldots, M-1, \tag{26a}$$

$$\hat{\alpha}^{(0)} = \alpha^{(0)} - \sum_{j=1}^{M-1} (\lambda^{(j)*} \circ t^M)\alpha^{(j)}, \tag{26b}$$

where $\lambda^{(1)}, \lambda^{(2)}, \ldots, \lambda^{(M-1)}$ are arbitrary Laurent polynomials.

In order to fully exploit the capabilities offered by the algebraic framework introduced in the previous section, we now give an explicit formulation of the lifting scheme in terms of Hurwitz matrices.

THEOREM 4.7. *Let $(C_0, C_1, \ldots, C_{M-1})$, $(A_0, A_1, \ldots, A_{M-1})$ be two M-tuples of banded Hurwitz matrices corresponding to a perfect reconstruction M-channel filter bank, namely,*

$$C_i \times A_j^T = \delta_{i,j} I, \quad i,j = 0, 1, \ldots, M-1.$$

Every pair of M-tuples of banded Hurwitz matrices of the form

$$(\hat{C}_0, C_1, \ldots, C_{M-1}), \quad (A_0, \hat{A}_1, \ldots, \hat{A}_{M-1})$$

satisfying

$$\begin{bmatrix} \hat{C}_0 \\ C_1 \\ \vdots \\ C_{M-1} \end{bmatrix} \times \left[A_0^T \hat{A}_1^T \ldots \hat{A}_{M-1}^T \right] = \begin{bmatrix} \mathbf{I} & \mathbf{0} & \ldots & \mathbf{0} \\ \mathbf{0} & \mathbf{I} & \ldots & \mathbf{0} \\ \ldots & \ldots & \ldots & \ldots \\ \mathbf{0} & \mathbf{0} & \ldots & \mathbf{I} \end{bmatrix}$$

can be obtained by choosing

$$\hat{C}_0 = C_0 + \sum_{i=1}^{M-1} Q_i \times C_i \tag{27}$$

and

$$\hat{A}_j = A_j - Q_j^T \times A_0, \quad j = 1, 2, \ldots, M-1, \tag{28}$$

where $Q_1, Q_2, \ldots, Q_{M-1}$ are arbitrary banded Toeplitz matrices.

This last formulation of lifting suggests a computationally efficient strategy for the realization of the lifted filters. In fact, it is not necessary to explicitly evaluate the new filters given by Formulas (24), since, using (27) and (28), we find that the output of the new filters can be computed starting from the output of the old system, simply by knowing the Toeplitz matrices Q_i. More precisely, we can sketch the following optimized analysis-synthesis algorithm.

Analysis:

$$[\overline{\sigma}^{(0)}] = \hat{C}_0 \times [\sigma] = C_0 \times [\sigma] + \sum_{i=1}^{M-1} Q_i \times C_i \times [\sigma] \tag{29}$$

$$= [\sigma^{(0)}] + \sum_{i=1}^{M-1} Q_i \times [\sigma^{(i)}],$$

where

$$[\sigma^{(i)}] = C_i \times [\sigma] \quad i = 1, 2, \ldots, M-1.$$

Synthesis:

$$[\overline{\sigma}] = A_0^T \times [\overline{\sigma}^{(0)}] + \sum_{i=1}^{M-1} \hat{A}_i^T \times [\overline{\sigma}^{(i)}] = A_0^T \times [\sigma^{(0)}] + \sum_{i=1}^{M-1} A_i^T \times [\sigma^{(i)}],$$

where we have used (29) to get

$$[\sigma^{(0)}] = [\overline{\sigma}^{(0)}] - \sum_{i=1}^{M-1} Q_i \times [\sigma^{(i)}].$$

We refer to [7] for a detailed analysis of the computational aspects of this algorithm.

5. Factorization into Lifting Steps. The lifting scheme represents a powerful tool for M-band biorthogonal filter construction, particularly due to the fact that it can be iterated. In fact, after adding suitable properties to the analysis low-pass filter, one can use the dual lifting scheme to add other properties to the analysis high-pass filters, thereby obtaining filters with any desired properties after a finite number of lifting steps.

The flexibility of the lifting scheme prompts us to ask whether any M-channel biorthogonal filter bank can be obtained using lifting. In the two-band case, it has been shown [3] that every dual pair of finite filters can be obtained with a finite number of lifting steps. In the following we show that a similar result also holds in the M-channel case, even if this case needs some additional work, due to the higher complexity of handling $M \times M$ polynomial matrices.

In order to prove this result, we need some preliminary considerations.

First of all, we remark that the lifting step (24a) and (24b) can be represented in matrix notation as follows:

(30a)
$$\boldsymbol{\Delta}^M(\overline{\boldsymbol{\gamma}}) = \boldsymbol{\Delta}^M(\boldsymbol{\gamma}) \times \boldsymbol{\Lambda}^M(\boldsymbol{\lambda}),$$

(30b)
$$\boldsymbol{\Delta}^M(\overline{\boldsymbol{\alpha}}) = \boldsymbol{\Delta}^M(\boldsymbol{\alpha}) \times \boldsymbol{\Lambda}^M(-\boldsymbol{\lambda}^*)^T,$$

where $\boldsymbol{\lambda}$ is the vector whose components are the M-1 polynomials $\lambda^{(1)}, \lambda^{(2)}, \ldots, \lambda^{(M-1)}$, and $\boldsymbol{\Lambda}^M(\boldsymbol{\lambda})$ is the matrix

(31)
$$\boldsymbol{\Lambda}^M(\boldsymbol{\lambda}) = \begin{bmatrix} 1 & 0 & \ldots & 0 \\ \lambda^{(1)} & 1 & \ldots & 0 \\ \vdots & \vdots & \vdots & \vdots \\ \lambda^{(M-1)} & 0 & \ldots & 1 \end{bmatrix},$$

which will be called the *lifting matrix* relative to $\boldsymbol{\lambda}$. Consequently,

(32)
$$\boldsymbol{\Lambda}^M(-\boldsymbol{\lambda}^*)^T = \begin{bmatrix} 1 & -\lambda^{(1)*} & \ldots & -\lambda^{(M-1)*} \\ 0 & 1 & \ldots & 0 \\ \vdots & \vdots & \vdots & \vdots \\ 0 & 0 & \ldots & 1 \end{bmatrix} = (\boldsymbol{\Lambda}^M(\boldsymbol{\lambda}^*)^{-1})^T.$$

The dual lifting step (26a) and (26b) can be similarly represented as

(33a)
$$\boldsymbol{\Delta}^M(\overline{\boldsymbol{\gamma}}) = \boldsymbol{\Delta}^M(\boldsymbol{\gamma}) \times \boldsymbol{\Lambda}^M(\boldsymbol{\lambda})^T,$$

(33b)
$$\boldsymbol{\Delta}^M(\overline{\boldsymbol{\alpha}}) = \boldsymbol{\Delta}^M(\boldsymbol{\alpha}) \times \boldsymbol{\Lambda}^M(-\boldsymbol{\lambda}^*).$$

Moreover, we define the notion of partial lifting matrices as follows: if $N < M$, let $\boldsymbol{\mu} = (\mu^{(1)}, \mu^{(2)}, \dots, \mu^{(N-1)})$ be an (N-1)-tuple of Laurent polynomials. The *N-order partial lifting matrix* relative to $\boldsymbol{\mu}$ will be the $M \times M$ matrix

$$\Lambda_M^N(\boldsymbol{\mu}) = \left[\begin{array}{c|c} \boldsymbol{I}_{M-N} & \boldsymbol{0} \\ \hline \boldsymbol{0} & \Lambda^N(\boldsymbol{\mu}) \end{array}\right]. \tag{34}$$

We are now in position to prove that every dual pair of Laurent polynomial vectors can be obtained with a finite number of lifting and dual lifting steps, alternated with permutations, starting from a suitably scaled trivial pair.

THEOREM 5.1. *Let $(\boldsymbol{\alpha}, \boldsymbol{\gamma})$ be a dual pair of Laurent polynomial vectors. There are always Laurent polynomial vectors $\boldsymbol{\lambda}_k$, $k = 0, 1, \dots, M$-2, $\boldsymbol{\mu}_h^j$, $j = 0, 1, \dots, n_h$, $h = 2, 3, \dots, M$, with (M-k-1) and (h-1) components respectively, such that*

$$\Delta^M(\boldsymbol{\alpha}) = \boldsymbol{D} \prod_{k=0}^{M-2} \Lambda_M^{M-k}(\boldsymbol{\lambda}_k) \prod_{h=2}^{M} \boldsymbol{\Omega}_h, \tag{35}$$

where $\boldsymbol{D}$ is a diagonal matrix whose determinant is a non-zero monomial, and, for every $h = 2, 3, \dots, M$,

$$\boldsymbol{\Omega}_h = \prod_{j=0}^{n_h} \Lambda_M^h(\boldsymbol{\mu}_h^j)^T \boldsymbol{P}_j^{(h)},$$

$\boldsymbol{P}_j^{(h)}$ being suitable permutation matrices acting on the last h columns, and n_h an integer bounded by the maximum degree of the non-zero elements of the h-th row of $\Delta^M(\boldsymbol{\alpha})$. Moreover,

$$\Delta^M(\boldsymbol{\gamma})^{*^T} = \prod_{h=M}^{2} \boldsymbol{\Omega}_h^{-1} \prod_{k=M-2}^{0} \Lambda_M^{M-k}(-\boldsymbol{\lambda}_k)\boldsymbol{D}^{-1}. \tag{36}$$

Proof. The proof is essentially based on the fact that any square matrix $\boldsymbol{\Sigma}$ with Laurent polynomial elements whose determinant is a non-zero monomial can be diagonalized by means of right elementary operations, as stated in ([4], ch.VI, §2). We recall here the main steps of the proof, in order to highlight the connections between the elementary operations involved and the matricial formulation of lifting and dual lifting steps.

The diagonalization process consists of two phases, the first of which reduces the matrix $\boldsymbol{\Sigma}$ to a lower triangular form by right multiplication by a finite number of permutation matrices and by matrices, with polynomial entries, of the form

$$
\left[\begin{array}{ccccc|ccccc}
1 & 0 & 0 & \dots & 0 & & & & & \\
0 & 1 & 0 & \dots & 0 & & & \mathbf{0} & & \\
\dots & \dots & \dots & \dots & \dots & & & & & \\
0 & 0 & \dots & 0 & 1 & & & & & \\
\hline
 & & & & & 1 & \xi_1 & \xi_2 & \dots & \xi_h \\
 & & & & & 0 & 1 & 0 & \dots & 0 \\
 & & \mathbf{0} & & & \dots & \dots & \dots & \dots & \dots \\
 & & & & & 0 & 0 & \dots & 0 & 1
\end{array}\right]. \tag{37}
$$

Note that these can be seen as dual partial lifting matrices. The second phase reduces the lower triangular matrix previously obtained to a diagonal matrix $\boldsymbol{D}$ by right multiplication by a finite number of matrices of the form

$$
\left[\begin{array}{ccccc|ccccc}
1 & 0 & 0 & \dots & 0 & & & & & \\
0 & 1 & 0 & \dots & 0 & & & \mathbf{0} & & \\
\dots & \dots & \dots & \dots & \dots & & & & & \\
0 & 0 & \dots & 0 & 1 & & & & & \\
\hline
 & & & & & 1 & 0 & 0 & \dots & 0 \\
 & & & & & \psi_1 & 1 & 0 & \dots & 0 \\
 & & \mathbf{0} & & & \dots & \dots & \dots & \dots & \dots \\
 & & & & & \psi_k & 0 & \dots & 0 & 1
\end{array}\right], \tag{38}
$$

which correspond to partial lifting steps. We point out that the elementary operations involved in the whole process do not change the determinant of the original matrix, up to a sign. Therefore, the diagonal elements of the matrix $\boldsymbol{D}$ are non-zero monomials. Formula (35) can now be easily deduced from the above factorization result, while (36) is an immediate consequence of (35) and (10). □

We remark that the proof of Theorem 5.1 is based upon a factorization theorem which deeply differs from the well-known Smith factorization result, since it makes use only of right multiplication by elementary matrices. In fact, in this context, these are the only operations which can be seen as lifting steps.

We give in the following an example of the factorization of the 4-decimated matrix $\boldsymbol{\Delta}^4(\boldsymbol{\alpha})$ relative to the 4-band filters, with two vanishing moments, constructed in [8], namely,

$$
\begin{aligned}
\alpha^{(0)} &= .269 + .394t + .519t^2 + .644t^3 + .230t^4 + .105t^5 - .019t^6 - .144t^7, \\
\alpha^{(1)} &= -.098 - .072t - .206t^2 - .180t^3 + .751t^4 + .343t^5 - .064t^6 - .472t^7, \\
\alpha^{(2)} &= .500 - .500t - .500t^2 + .500t^3, \\
\alpha^{(3)} &= .217 - .677t + .657t^2 - .237t^3 + .053t^4 + .024t^5 - .004t^6 - .033t^7,
\end{aligned}
$$

where, for the sake of simplicity, we have written only the first three digits. We have

$$
\boldsymbol{\Delta}^4(\boldsymbol{\alpha}) = \begin{bmatrix}
.269 + .230t & -.098 + .751t & .500 & .217 + .053t \\
.394 + .105t & -.072 + .343t & -.500 & -.677 + .024t \\
.519 - .019t & -.206 - .064t & -.500 & .657 - .004t \\
.644 - .144t & -.180 - .472t & .500 & -.237 - .033t
\end{bmatrix},
$$

and the algorithm described in the proof of Theorem 5.1 yields

$$\Delta^4(\alpha) := \mathbf{D}\Lambda^4(\lambda_0)\Lambda^3(\lambda_1)\Lambda^2(\lambda_2)\Omega_2\Omega_3\Omega_4,$$

where

$$\mathbf{D} := \begin{bmatrix} .5 & 0 & 0 & 0 \\ 0 & .7169 & 0 & 0 \\ 0 & 0 & 1.448 & 0 \\ 0 & 0 & 0 & -1.9255t \end{bmatrix}, \quad \Lambda^4(\lambda_0) := \begin{bmatrix} 1 & 0 & 0 & 0 \\ -.697t^{-1} & 1 & 0 & 0 \\ -.345t^{-1} & 0 & 1 & 0 \\ -.259t^{-1} & 0 & 0 & 1 \end{bmatrix},$$

$$\Lambda^3(\lambda_1) := \begin{bmatrix} 1 & 0 & 0 & 0 \\ 0 & 1 & 0 & 0 \\ 0 & .609t^{-1} & 1 & 0 \\ 0 & -.207t^{-1} & 0 & 1 \end{bmatrix}, \quad \Lambda^2(\lambda_2) := \begin{bmatrix} 1 & 0 & 0 & 0 \\ 0 & 1 & 0 & 0 \\ 0 & 0 & 1 & 0 \\ 0 & 0 & .102t^{-1} & 1 \end{bmatrix},$$

$$\Omega_2 = \Lambda^2(\mu_2^0)^T P_0^{(2)} = \begin{bmatrix} 1 & 0 & 0 & 0 \\ 0 & 1 & 0 & 0 \\ 0 & 0 & 1 & -.064 - .456t \\ 0 & 0 & 0 & 1 \end{bmatrix} \times \begin{bmatrix} 1 & 0 & 0 & 0 \\ 0 & 1 & 0 & 0 \\ 0 & 0 & 0 & 1 \\ 0 & 0 & 1 & 0 \end{bmatrix},$$

$$\Omega_3 = \Lambda^3(\mu_3^0)^T P_0^{(3)} \Lambda^3(\mu_3^1)^T$$

$$= \begin{bmatrix} 1 & 0 & 0 & 0 \\ 0 & 1 & -.238 + 1.527t & -.625 \\ 0 & 0 & 1 & 0 \\ 0 & 0 & 0 & 1 \end{bmatrix} \times \begin{bmatrix} 1 & 0 & 0 & 0 \\ 0 & 0 & 1 & 0 \\ 0 & 1 & 0 & 0 \\ 0 & 0 & 0 & 1 \end{bmatrix} \times \begin{bmatrix} 1 & 0 & 0 & 0 \\ 0 & 1 & .306 & .071 \\ 0 & 0 & 1 & 0 \\ 0 & 0 & 0 & 1 \end{bmatrix}$$

$$\Omega_4 = \Lambda^4(\mu_4^0)^T P_0^{(4)}$$

$$= \begin{bmatrix} 1 & -.196 + 1.502t & +.539 + .460t & +.4343 + .106t \\ 0 & 1 & 0 & 0 \\ 0 & 0 & 1 & 0 \\ 0 & 0 & 0 & 1 \end{bmatrix} \times \begin{bmatrix} 0 & 0 & 1 & 0 \\ 0 & 1 & 0 & 0 \\ 1 & 0 & 0 & 0 \\ 0 & 0 & 0 & 1 \end{bmatrix}.$$

It is easy to check that the above factorization reduces the computational complexity of the analysis-synthesis algorithm.

REFERENCES

[1] M. Barnabei and L.B. Montefusco, Recursive Properties of Toeplitz and Hurwitz Matrices, *Linear Algebra and Applications* , 274, pp. 367-388, 1998.
[2] M. Barnabei, C. Guerrini, and L.B. Montefusco, Some Algebraic Aspects of Signal Processing, *Linear Algebra and its Applications*, vol.284, 1-3, pp.3-17, 1998.
[3] I. Daubechies, W. Sweldens, Factoring Wavelet Transforms into Lifting Steps, *J. Fourier Anal. Appl.*, Vol. 4, Nr. 3, pp. 247-269, 1998.
[4] F.R. Gantmacher, *The Theory of Matrices*, Chelsea Publishing Company, New York, N. Y., 1959.

[5] I. Heger, Über die Auflösung eines Systemes von mehreren unbest immtem Gleichungen des ersten Grades in ganzen Zahlen, *Denkschriften der Königlichen Akademie der Wissenschaften (Wien), Mathematisch-naturwissenschaftliche Klasse* 14 (2), pp. 1-122, (1858)

[6] J. Kovacevic and W. Sweldens, Wavelet Families of Increasing Order in Arbitrary Dimensions, *Preprint, Bell Laboratories, Lucent Technologies*, December 1997.

[7] D. Lazzaro, Biorthogonal M-band Filter Construction Using the Lifting Scheme, *Numerical Algorithm (1999)* n.22 pp. 53-72.

[8] L.B. Montefusco and D. Lazzaro, Discrete Orthogonal Transform and M-band Wavelets for Image Compression, *Surface Fitting and Multiresolution Methods*, L.Schumacker, A. Le Mehaute, C. Rabut, editors, pp.261-270, 1997.

[9] G. Strang and T. Nguyen, *Wavelets and Filter Banks*, Wellesley-Cambridge Press, 1996.

[10] W. Sweldens, The lifting scheme: A costum-design construction of biorthogonal wavelets, *Appl. Comput. Harmon. Anal.* 3(2), pp.186-200, 1996.

[11] W. Sweldens, The lifting scheme: A construction of second generation wavelets, *Siam J. Math. Anal*, 29(2) pp. 511-546, 1997.

[12] L.M.G. Tolhuizen, H.D. Hollmann, and A.C.M. Kalker, On the Realizability of bi-orthogonal M-dimensinal 2-band Filter Banks, *IEEE Trans. on Sign. Proc.* 43(3), pp.640-648, 1995.

[13] P.P. Vaidyanathan, *Multirate Systems and Filter Banks*, Prentice Hall Signal Processing Series, Englewood Cliffs, N. J., 1993.

CONTRACTIVE INITIALIZING METHODS FOR THE PANTOGRAPH EQUATION OF NEUTRAL TYPE

A. BELLEN , S. MASET , AND L. TORELLI *

Abstract. In this paper we propose a method for the first step integration of (1.1) on $[0,h]$ by an s–stage continuous RK method of uniform order q and we prove that the uniform accuracy order is $O\left(h^{\min\{q,s\}}\right)$. We analyze when the numerical solution preserves the contractivity property for any $h>0$. We find methods with $s=q=1,2$ for variable matrix coefficient and with $s=q=3$ for the special case of constant scalar real coefficients.

1. Introduction. Let us consider the pantograph equation

$$\begin{cases} y'(t) = L(t)\,y(t) + M(t)\,y(qt) + N(t)\,y'(qt)\,,\ t\geq 0,\ 0<q<1, \\ y(0) = y_0, \end{cases} \tag{1.1}$$

where $|N(t)| \leq \gamma < 1$ for all $t\geq 0$, and observe that it can be viewed as a delay differential equation

$$\begin{cases} y'(t) = L(t)\,y(t) + M(t)\,y(t-\tau(t)) + N(t)\,y'(t-\tau(t))\,,\ t\geq 0 \\ y(0) = y_0 \end{cases}$$

with non negative delay $\tau(t) = (1-q)\,t$.

In solving (1.1) by the standard approach based on a discrete ODE-scheme endowed by a suitable interpolation, one must take care of two facts on the delay $\tau(t)$: it vanishes at $t=0$ and is unbounded as $t\to\infty$.

The former leads to some complicacy in carrying out the integration on the first step $[0,h]$ where the equation is "intrinsically functional", i.e. it can't be reduced to an ODE, and in preserving the contractivity of (1.1) when it occurs. The latter leads to difficulties in storing the larger and larger amount of approximated values needed in the long term integration and in preserving the asymptotic behavior of the solution. Here we are concerned with the first of the two problems, that is the integration of the first step-interval $[0,h]$. In particular we look for "unconditionally contractive" methods, that is methods such that the continuous approximation $\eta(t)\simeq y(t)$ satisfies the condition

$$\max_{\theta\in[0,1]} |\eta(\theta h)| \leq |y_0| \quad \forall h>0,\ \forall q\in(0,1)$$

whenever

$$\max_{\theta\in[0,1]} |y(\theta h)| \leq |y_0| \quad \forall h>0,\ \forall q\in(0,1)\,.$$

It is known that B–stability (or A–stability for the constant coefficient case) is not sufficient for our goals. In fact even in the simplest case of non neutral real constant coefficient scalar test equation

$$\begin{cases} y'(t) = ay(t) + by(qt)\,,\ t\geq 0,\ 0<q<1 \\ y(0) = y_0, \end{cases}$$

* Dipartimento di Scienze Matematiche, Universita' di Trieste, 34100 Trieste, Italy. Work supported by MURST and CNR.

which is contractive for $a \leq -|b|$, the trapezoidal rule with linear interpolation

$$y_1 = y_0 + \frac{h}{2}\left[ay_0 + by_0 + ay_1 + b\left((1-q)\,y_0 + qy_1\right)\right]$$

is not contractive for all h and q. In fact contractivity condition $|y_1| \leq |y_0|$ holds only for h and q satisfying

$$bh \leq -ah \quad \text{and} \quad bh > \frac{2}{q-1}.$$

2. The method. The method we analyze is a modification of the method introduced by Bellen, Guglielmi and Zennaro in [1] for non-vanishing delay and is based on the change of variable

$$z(t) = y'(t) - L(t)\,y(t)\,.$$

It leads to the algebraic differential system

$$\begin{cases} y'(t) = L(t)\,y(t) + z(t)\,, \ t \geq 0 \\ y(0) = y_0 \\ z(t) = N(t)\,z(qt) + \left[N(t)\,L(qt) + M(t)\right]y(qt)\,, \ t \geq 0 \end{cases} \tag{2.1}$$

for which we provide an approximation in $[0,h]$ by the following numerical method. Let $(A, b(\cdot), c)$ be a nonconfluent continuous Runge-Kutta method. An approximation $\eta(t)$, $t \in [0,h]$, to $y(t)$ is obtained by

$$\eta(\theta h) = y_0 + h\sum_{i=1}^{s} b_i(\theta)\,K_i, \tag{2.2}$$

$$K_i = L(c_i h)\left(y_0 + h\sum_{j=1}^{s} a_{ij}K_j\right) + \alpha(c_i h)\,, \quad i = 1, \ldots, s, \tag{2.3}$$

where $\alpha(t)$, $t \in [0,h]$, is an approximation to $z(t)$ defined by

$$\alpha(t) = N(t)\,P(\alpha; qt) + \left[N(t)\,L(qt) + M(t)\right]\eta(qt) \tag{2.4}$$

and

$$P(\alpha; \cdot) = \sum_{j=1}^{s} l_j(\cdot)\,\alpha(c_j h)$$

is the polynomial interpolating α at the nodes $c_i h$, $i = 1, \ldots, s$. Therefore the function α is defined in (2.4) by means of the values $Z_i = \alpha(c_i h)$, $i = 1, \ldots, s$, which are needed in (2.3). Such values satisfies the equations

$$Z_i = N(c_i h)\sum_{j=1}^{s} l_j(qc_i h)\,Z_j + \left[N(c_i h)\,L(qc_i h) + M(c_i h)\right]\eta(qc_i h)\,, \quad i = 1, \ldots, s. \tag{2.5}$$

In a compact form we can rewrite (2.2), (2.3) and (2.5) as

$$\eta(\theta h) = y_0 + h\left(b^T(\theta) \otimes I_m\right) K,$$

$$K = \overline{L}\left[1_s y_0 + h\left(A \otimes I_m\right) K\right] + Z,$$

$$DZ = \overline{Q}\overline{\eta},$$

where

$$\overline{L} = diag\left(L\left(c_i h\right)\right)_{i=1,\ldots,s},$$

$$\overline{Q} = diag\left(N\left(c_i h\right) L\left(q c_i h\right) + M\left(c_i h\right)\right)_{i=1,\ldots,s},$$

$$K = \left(K_i\right)_{i=1,\ldots,s}, \; Z = \left(Z_i\right)_{i=1,\ldots,s}, \; \overline{\eta} = \left(\eta\left(q c_i h\right)\right)_{i=1,\ldots,s},$$

$$D = I - \overline{N}\left(\mathcal{L} \otimes I_m\right),$$

$$\overline{N} = diag\left(N\left(c_i h\right)\right)_{i=1,\ldots,s},$$

and

$$[\mathcal{L}]_{ij} = l_j\left(q c_i h\right), \quad i, j = 1, \ldots, s.$$

Now we show that there exists a unique function η such that (2.2), (2.3) and (2.5) hold. For $h = 0$ we have

$$D = I - \left(I_s \otimes N(0)\right)\left(\mathcal{L}_0 \otimes I_m\right)$$

where $\mathcal{L}_0$ is the rank-1 matrix

$$\mathcal{L}_0 = \left[l_1(0) \; \ldots \; l_s(0)\right]^T 1_s^T.$$

By the Shermann-Morrison formula, the eigenvalues of $\mathcal{L}_0$ are 0 and

$$\left[l_1(0) \; \ldots \; l_s(0)\right] 1_s = 1$$

Moreover, being $\|N\| < 1$, we have that D is nonsingular for $h = 0$ and then D is nonsingular for sufficiently small h. Hence, by setting $B = \left(b_i\left(q c_j\right)\right)_{i,j=1,\ldots,s}$,

$$K = \overline{L}\left[1_s y_0 + h\left(A \otimes I_m\right) K\right] + D^{-1}\overline{Q}\left[1_s y_0 + h\left(B \otimes I_m\right) K\right]$$

$$= \left(\overline{L} + D^{-1}\overline{Q}\right) 1_s y_0 + h\left[\overline{L}\left(A \otimes I_m\right) + D^{-1}\overline{Q}\left(B \otimes I_m\right)\right] K$$

and an unique vector K is defined for sufficiently small h.

In the next theorem we estimate the error of the previous numerical process by using the notation $\|x\|_h = \max_{t \in [0,h]} |x(t)|$.

THEOREM 2.1. *If the continuous RK method has uniform order q and has s stages, then*

$$\|\eta - y\|_h = O\left(h^{\min\{q,s\}}\right).$$

Proof. We write

$$\|\eta - y\|_h \leq \|\eta - w\|_h + \|w - y\|_h \tag{2.6}$$

where w is the solution of

$$\begin{cases} w'(t) = L(t)\, w(t) + \alpha(t), & t \in [0,h] \\ w(0) = y_0 \end{cases}$$

and α is defined in (2.4). On the other hand, since

$$\begin{cases} \frac{d}{dt}[w(t) - y(t)] = L(t)[w(t) - y(t)] + \alpha(t) - z(t), & t \in [0,h] \\ w(0) - y(0) = 0 \end{cases}$$

we obtain

$$\|w - y\|_h \leq \frac{e^{h\|L\|_h} - 1}{\|L\|_h} \|\alpha - z\|_h. \tag{2.7}$$

Now, by the last of (2.1) and the following modification of (2.4)

$$\alpha(t) = N(t)\,\alpha(qt) + [N(t)\,L(qt) + M(t)]\,\eta(qt) + N(t)\,[P(\alpha; qt) - \alpha(qt)], \quad t \in [0,h]$$

we get

$$(\alpha - z)(t) = N(t)\,[(\alpha - z)(qt)] + \\ [N(t)\,L(qt) + M(t)]\,(\eta(qt) - y(qt)) + N(t)\,\delta(t), \quad t \in [0,h] \tag{2.8}$$

where

$$\delta(t) = P(\alpha; qt) - \alpha(qt), \quad t \in [0,h].$$

Since $\|N\|_h < 1$, by (2.8) we obtain

$$\|\alpha - z\|_h \leq \frac{1}{1 - \|N\|_h} [\|NL(q\cdot) + M\|_h \|\eta - y\|_h + \|N\|_h \|\delta\|_h]. \tag{2.9}$$

The inequalities (2.6), (2.7) and (2.9), yield, for a sufficiently small h,

$$\|\eta - y\|_h \leq C_1 \|\eta - w\|_h + C_2 \|\delta\|_h$$

where C_1, C_2 are independent of h. Since the continuous RK method has uniform order q, we have $\|\eta - w\|_h = O(h^q)$, and since $P(\alpha; \cdot)$ is the polynomial interpolating α at the nodes $c_i h$, $i = 1, \ldots, s$, we have $\|\delta\|_h = O(h^s)$ being the derivative $\alpha^{(s+1)}$ uniformly bounded with respect to h. The assert follows. □

3. Contractivity of the first step. It is known (see [3]) that if

$$\mu\left[L\left(t\right)\right]<0,\quad \frac{\left\|NL\left(q\cdot\right)+M\right\|_{t}}{1-\left\|N\right\|_{t}}+\mu\left[L\left(t\right)\right]\leq 0,\quad t\geq 0 \tag{3.1}$$

then, for the solution y of (2.1), the contractivity property

$$\left|y\left(t\right)\right|\leq\left|y_{0}\right|,\quad t\geq 0$$

holds.

As for the preservation of the contractivity property for the numerical scheme introduced above we will prove a theorem based on the following definitions of BN_f-, A_f- and $A_f(0)-$stability.

DEFINITION 3.1. *(see [2]) Consider the system*

$$\begin{cases} y'\left(t\right)=f\left(t,y\left(t\right),z\left(t\right)\right)\\ y\left(t_{0}\right)=y_{0}\end{cases}$$

with $f\left(t,0,0\right)\equiv 0$, *and the functions*

$$Y\left(t\right)=\sup_{z,y_{1}\neq y_{2}}\frac{\Re\left\langle f\left(t,y_{1},z\right)-f\left(t,y_{2},z\right),y_{1}-y_{2}\right\rangle}{\left\|y_{1}-y_{2}\right\|^{2}}$$

$$X\left(t\right)=\sup_{y,z_{1}\neq z_{2}}\frac{\left\|f\left(t,y,z_{1}\right)-f\left(t,y,z_{2}\right)\right\|}{\left\|z_{1}-z_{2}\right\|}$$

for some inner product norm $\left\|\cdot\right\|=\left\langle\cdot,\cdot\right\rangle^{\frac{1}{2}}$. *An* $s-$*stage continuous RK method* $\left(A,b\left(\cdot\right),c\right)$ *is called* BN_f-*stable if for any system such that* $Y\left(t\right)\leq 0$ *and* $X\left(t\right)=r\left(t\right)Y\left(t\right)$ *for a locally integrable function* $r\left(t\right)\leq 0$, *then the continuous numerical solution* η *satisfies*

$$\left\|\eta\left(t_{0}+\theta h\right)\right\|=\max\left\{\left\|y_{0}\right\|,\max_{i=1,\ldots,s}\left|r\left(t_{n}+c_{i}h\right)\right|\left\|z\left(t_{n}+c_{i}h\right)\right\|\right\}$$

for all $\theta\in\left[0,1\right]$ *and* $h>0$. *In particular, if* f *is scalar and* $f\left(t,y,z\right)\equiv Ly+z$, *we get the definition of* A_f-*stability and* $A_f\left(0\right)-$*stability for the real and complex case respectively.*

THEOREM 3.2. *Let (3.1) hold for the problem (2.1). If the continuous RK method is* BN_f-*stable and the Lagrange projection* P *has norm* 1 *(as operator acting on* $C\left(\left[0,h\right],\mathbb{C}^{m}\right)$ *equipped by the maximum norm), then* $\left|\eta\left(t\right)\right|\leq\left|y_{0}\right|$ *for all* $t\in\left[0,h\right]$.

Proof. If (3.1) holds, then

$$\frac{\left\|NL\left(q\cdot\right)+M\right\|_{t}}{-\mu\left[L\left(t\right)\right]\left(1-\left\|N\right\|_{t}\right)}\leq 1\quad for\; t\geq 0.$$

Let us suppose that

$$\max_{t\leq h}\frac{\left\|NL\left(q\cdot\right)+M\right\|_{t}}{-\mu\left[L\left(t\right)\right]\left(1-\left\|N\right\|_{t}\right)}<1. \tag{3.2}$$

and prove that

$$\left\|\eta\right\|_{h}\leq\left|y_{0}\right|. \tag{3.3}$$

Since in our case

$$Y(t) = \mu[L(t)] < 0, \ X(t) = 1$$

and the continuous RK method is BN_f−stable we have $r(t) = \frac{1}{\mu[L(t)]}$ locally integrable and hence

$$\|\eta\|_h \leq \max\left\{|y_0|, \max_{\xi\leq h} \frac{|\alpha(\xi)|}{-\mu[L(\xi)]}\right\}.$$

If $\max\left\{|y_0|, \max_{\xi\leq h} \frac{|\alpha(\xi)|}{-\mu[L(\xi)]}\right\} = |y_0|$, (3.3) is proved, otherwise

$$\|\eta\|_h \leq \max_{\xi\leq h} \frac{|\alpha(\xi)|}{-\mu[L(\xi)]} \leq \max_{\xi\leq h} \frac{\|\alpha\|_\xi}{-\mu[L(\xi)]}$$

and, by (2.4) and by the hypothesis on P we have, for $\xi \leq h$,

$$\|\alpha\|_\xi \leq \frac{\|NL(q\cdot) + M\|_\xi}{1 - \|N\|_\xi} \|\eta\|_\xi.$$

Therefore

$$\|\eta\|_h \leq \max_{\xi\leq h} \frac{\|NL(q\cdot) + M\|_\xi}{-\mu[L(\xi)]\left(1 - \|N\|_\xi\right)} \|\eta\|_\xi$$

and, by the assumption (3.2), we have $\|\eta\|_h = 0$. Hence (3.3) is proved too. When

$$\max_{t\leq h} \frac{\|NL(q\cdot) + M\|_t}{-\mu[L(t)](1 - \|N\|_t)} = 1$$

a continuity argument is used to prove (3.3). □

When L, M and N are scalar constants the requests on the continuous RK method are weaker (A_f− or $A_f(0)$− instead of BN_f−stability), and the restriction on P drops according to the following theorem.

THEOREM 3.3. *Let L, M and N in (2.1) be scalar constants and assume that (3.1) holds. If the s−stage continuous RK method is A_f−stable ($A_f(0)$−stable for real L, M and N) with interpolant η of degree $\leq s-1$, then $|\eta(t)| \leq |y_0|$ for all $t \in [0,h]$.*

Proof. Since L, M and N are real constants and η is a polynomial of degree $\leq s-1$, then α turns out to be a polynomial of degree $s-1$ and so $\alpha = P(\alpha,\cdot)$. Thus

$$\alpha(t) = N\alpha(qt) + (NL + M)\eta(qt), \quad t \in [0,h]$$

and

$$\|\alpha\|_h \leq \frac{|NL + M|}{1 - |N|} \|\eta\|_h.$$

Now the proof proceeds as in Theorem 3.2 by using A_f− or $A_f(0)$−stability instead of BN_f−stability. □

In dealing with variable coefficient systems, the known continuous RK methods which satisfy the hypothesis of Theorem 3.2 are Backward Euler with constant interpolant ($s =$

$q = 1$) and two-stage Lobatto IIIC with linear interpolant ($s = q = 2$) (see [2]). In dealing with scalar equations with complex constant coefficients, the sole continuous RK method with $s = q = 1$ which satisfies the A_f–stability required in Theorem 3.3 is Backward Euler with constant interpolation, while for $s = q = 2$ and linear interpolation the class is characterized by

$$A = \begin{pmatrix} b_1 & c_1 - b_1 \\ b_1 & b_2 \end{pmatrix}, \; b^T = \begin{pmatrix} b_1 & b_2 \end{pmatrix}, \; c = \begin{pmatrix} c_1 \\ 1 \end{pmatrix}, \; \frac{1}{2} \le b_1 \le \frac{3}{4}, \; 0 \le c_1 \le \frac{1}{3}.$$

Finally, in the case of scalar equations with real constant coefficients, the class of $A_f\,(0)$ –stable methods with $s = q = 2$ and linear interpolation is larger and is characterized by

$$A = \begin{pmatrix} a_{11} & c_1 - a_{11} \\ b_1 & b_2 \end{pmatrix}, \; b^T = \begin{pmatrix} b_1 & b_2 \end{pmatrix}, \; c = \begin{pmatrix} c_1 \\ 1 \end{pmatrix}, \; b_1 \le a_{11}, \; 0 \le c_1 \le \frac{1}{2}.$$

Moreover we have some $A_f\,(0)$ –stable methods with $s = q = 3$ and quadratic interpolation, for example (see [4])

$$A = \begin{pmatrix} \frac{5}{2} & -2 & -\frac{1}{2} \\ -1 & 2 & -\frac{1}{2} \\ \frac{1}{6} & \frac{2}{3} & \frac{1}{6} \end{pmatrix}, \; b^T(\theta) = \begin{pmatrix} -\frac{5}{6}\theta^2 + \theta & \frac{2}{3}\theta^2 & \frac{1}{6}\theta^2 \end{pmatrix}, \; c = \begin{pmatrix} 0 \\ \frac{1}{2} \\ 1 \end{pmatrix}.$$

REFERENCES

[1] A. Bellen, N. Guglielmi, M. Zennaro. On the contractivity and asymptotic stability of systems of delay differential equations of neutral type. BIT 1999, Vol. 39, No. 1, pp. 1-24.

[2] A. Bellen, M. Zennaro. Strong contractivity properties of numerical methods for ordinary and delay differential equations. Appl. Num. Math. 9 (1992) 321-346.

[3] A. Iserles, J. Terjèki. Stability and asymptotic stability of functional-differential equations. J. London Math. Soc. (2) 51 (1995) 559-572.

[4] M. Zennaro. Contractivity of the Runge-Kutta methods with respect to forcing terms. Appl. Num. Math. 10 (1993) 321-345.

REGULARIZATION OF LARGE DISCRETE ILL–POSED PROBLEMS IN IMAGE PROCESSING

M. BERTAIA *, S. MORIGI †, E. LOLI PICCOLOMINI ‡, F. SGALLARI §, AND F. ZAMA ¶

Abstract. In this work we review some different approaches that can be used to compute regularized solutions of discretized Fredholm integral equations that occur in many imaging problems. In particular the truncated Conjugate Gradient iterations and the Tikhonov method are considered. In the presence of linear constraints we propose the Hestenes method. In the case of underdetermined systems a method based on the Lanczos iterations is proposed together with a new strategy for the computation of the regularization parameter based on the L–curve. Some numerical results are presented both on scalar and parallel computational environments.

Key words. regularization methods, L-curve, Hestenes method.

1. Introduction. Many problems in the area of image processing are modeled by a Fredholm integral equation of the first kind such as:

$$g(s) = \int h(s,r)f(r)dr + e \tag{1.1}$$

where f and g are the functions describing respectively the object and the recorded image, h is the response of the image recording instrument and e represents the noise. The knowledge of e is usually limited to statistical information. In image restoration and reconstruction, the degradation function h is known a priori and it represents the system response to a two–dimensional unit impulse. Since in optics the impulse is a point of light, the function h is also called Point Spread Function (PSF). The analytical form of the PSF plays a fundamental role in the image restoration process and in the choice of suitable numerical methods for the solution of equation (1.1). For instance, some image systems can be described by space–invariant PSFs and the solution of this problem can be efficiently computed in the frequency domain by means of the FFT transforms with a low computational complexity. However, the above simplification is not possible in many real systems.

The problem (1.1) is an ill–posed problem [12, 9] in the sense that the solution does not depend continuously on the data, or that bounded perturbations of the data result in an unbounded perturbation of the solution.

The discretization of the integral equation (1.1), by means of a quadrature formula, leads to the system:

$$\mathbf{b} = H\mathbf{f} \quad \text{where} \quad \mathbf{b} = \mathbf{g} + \mathbf{e} \tag{1.2}$$

where the matrix $H \in I\!R^{m\times n}$ is usually very *ill–conditioned*, $\mathbf{b},\mathbf{g},\mathbf{e} \in I\!R^m$, $\mathbf{f} \in I\!R^n$. The unique least–squares solution of minimal norm

$$\mathbf{f} = \sum_{i=1}^{r=rank(H)} \frac{1}{z_i}\mathbf{u_i}\mathbf{v_i}^T\mathbf{b}$$

*Dept. of Mathematics, University of Bologna, Italy; E-mail: bertaia@dm.unibo.it.
†Dept. of Mathematics, University of Bologna, Italy; E-mail: morigi@dm.unibo.it.
‡Dept. of Mathematics, University of Bologna, Italy; E-mail: piccolom@dm.unibo.it.
§Dept. of Mathematics, University of Bologna, Italy; E-mail: sgallari@dm.unibo.it.
¶Dept. of Mathematics, University of Bologna, Italy; E-mail: zama@dm.unibo.it.

(where $H = UZV^T$ and $Z = diag(z_1, z_2, \ldots z_r)$) may be too sensitive to the presence of noise. For this reason, an approximating regularized solution

$$\mathbf{f}_{reg} = \sum_{i=1}^{r=rank(H)} \phi_i(\mu)\frac{1}{z_i}\mathbf{u_i}\mathbf{v_i}^T\mathbf{b}$$

is searched using a regularization method ($\phi_i(\mu)$ are the filter factors and $\mu > 0$ is the regularization parameter [9]).

In this work we report some different approaches that can be applied to find a regularized solution of (1.2) when deriving from the discretization of imaging problems such as image restoration or image reconstruction problem.

In sections 2.1 and 2.2 we introduce methods which can be suitably applied to find a least squares solution in the case of a square or overdetermined linear system. The method reported in section 3 is designed to be efficiently applied in the case of an underdetermined linear system. In section 4 some numerical results are reported both for the image restoration and reconstruction problem.

2. Regularization of square and overdetermined systems. In this section we analyse two different approaches to compute the least-squares regularized solution of (1.2) in the normal and overdetermined cases.

2.1. Conjugate Gradients Regularization. The Conjugate Gradients (CG) method can be used as a regularization method [9, 7] for large size problems when the regularity properties of the solution are not known and in absence of constraints. The approximated solution $\mathbf{f}_k$ is obtained after k CG iterations applied to the normal equations:

$$H^T H\mathbf{f} = H^T\mathbf{b}.$$

The process is non–convergent because of the critical ill–conditioning of the matrix H, but it has been observed that in the first steps the method converges to a regularized solution and then it starts to diverge. This phenomenon is known as *semiconvergence* [9]. The iteration at which we stop the CG plays the role of the regularization parameter. In [19] we tested three heuristic criteria which do not require any additional knowledge about the noise and the solution.

2.2. Constrained Regularization. In some cases it is necessary to add constraints that are usually derived from the physical nature of the problem. The numerical solution can then be obtained by adding the constraints to the Tikhonov regularization method. The Tikhonov regularization method [23] solves the following quadratic programming problem:

$$\min_{\mathbf{f}\in \mathbb{R}^m} \{\|H\mathbf{f} - \mathbf{b}\|^2 + \mu\|C\mathbf{f}\|^2\} \tag{2.1}$$

where $\|.\|$ denotes here and in the following the euclidean norm, μ is the regularization parameter ($\mu > 0$) and represents a trade–off between the accuracy of the solution and the smoothness defined by the regularization matrix C. Several methods exist to choose a good value of μ, but in this section we used fixed values heuristically determined. A common choice of the matrix C consists in the identity matrix I or in a discrete differential operator as smoothing condition. The problem (2.1) can be written as:

$$\min_{\mathbf{f}\in \mathbb{R}^m} \{\frac{1}{2}\mathbf{f}^{\mathbf{T}} K\mathbf{f} - \mathbf{f}^{\mathbf{T}} H^T\mathbf{b}\} \tag{2.2}$$

where $K = H^T H + \mu C^T C$ is a symmetric $n \times n$ matrix. The system deriving from (2.2) is:

$$Kf = H^T b \tag{2.3}$$

whose solution can be computed either by direct or iterative methods. If the dimensions are large and H is sparse the Conjugate Gradient method can be used to exploit the sparsity of the matrix H. When linear equality constraints such as:

$$B\mathbf{f} = \mathbf{s} \tag{2.4}$$

are present, the constrained quadratic programming problem defined by the equations (2.2)–(2.4) must be solved. A suitable method is the Hestenes method of multipliers [11], given by the following iterations:

$$\lambda_{i+1} = \lambda_i + c(B\mathbf{f}_i - \mathbf{s}), \quad i = 0, 1, \ldots, \qquad \lambda_0 = 0 \tag{2.5}$$

where $\mathbf{f}_i$ is the solution of the system:

$$(K + cB^T B)\mathbf{f}_i = -B^T \lambda_i + \mathbf{r} + cB^T \mathbf{s}; \quad \mathbf{r} = H^T \mathbf{b} \tag{2.6}$$

and c is a suitable penalty parameter ($c > 0$). In [3], it is shown that the method converges for moderately large values of c in a few iterations.

3. Regularization of underdetermined systems. Now we focus on regularizing methods that are well suited for underdetermined large-scale systems (1.2) arising from image reconstruction problems.

The solution of minimal norm of (1.2) can be written as $\mathbf{f} := H^T \mathbf{y}$, where $\mathbf{y} \in I\!R^m$ solves the linear system

$$A\mathbf{y} = \mathbf{b}, \tag{3.1}$$

with $A := HH^T$, and the associated regularized system can be given by

$$(A + \mu I)\mathbf{y} = \mathbf{b}. \tag{3.2}$$

We denote the regularized solution of (3.2) by $\mathbf{y}^\mu$. It can be easily shown that if the solution $\mathbf{y}^\mu$ solves the minimization problem

$$\min_{\mathbf{y} \in I\!R^m} \{\|A\mathbf{y} - \mathbf{b}\|^2 + \mu \|H^T \mathbf{y}\|^2\}, \tag{3.3}$$

then $\mathbf{f}^\mu = H^T \mathbf{y}^\mu$ solves the Tikhonov system (2.1) with $C = I$. We should notice that this approach allows for a computer storage and arithmetic work reduction when the linear system has many more columns than rows [2]. The regularized solution depends on a parameter μ, which suitable value determines the convergence (in norm) to the exact solution as the perturbation in the right-hand side tends to zero. We consider some techniques to determine an approximation of the solution vector $\mathbf{f}^\mu$ and a suitable value of the regularization parameter μ.

If the dimensions of the problem are very large iterative methods must be used to compute $\mathbf{f}^\mu$. The iterative methods we propose for the computation of approximations of μ and $\mathbf{f}^\mu$ are based on the Lanczos bidiagonalization algorithm by Paige and Saunders [21]. The algorithm carries out k steps of the Lanczos bidiagonalization process applied to the matrix H, determines the matrices $V_k = [\boldsymbol{v}_1, \boldsymbol{v}_2, \ldots, \boldsymbol{v}_k] \in I\!R^{m\times k}$, $V_{k+1} = [V_k, \boldsymbol{v}_{k+1}] \in I\!R^{m\times(k+1)}$ and $U_k = [\boldsymbol{u}_1, \boldsymbol{u}_2, \ldots, \boldsymbol{u}_k] \in I\!R^{n\times k}$ with orthonormal columns, as well as the lower $k \times k$ bidiagonal matrix

$$L_k = \begin{bmatrix} \rho_1 & & & & \\ \sigma_2 & \rho_2 & & & \\ & \ddots & \ddots & & \\ & & \sigma_{k-1} & \rho_{k-1} & \\ & & & \sigma_k & \rho_k \end{bmatrix}. \tag{3.4}$$

It follows that

$$HU_k = V_k L_k + \sigma_{k+1}\boldsymbol{v}_{k+1}\boldsymbol{e}_k^T, \qquad H^T V_k = U_k L_k^T, \qquad \boldsymbol{b} = \|\boldsymbol{b}\| V_k \boldsymbol{e}_1, \tag{3.5}$$

where $\boldsymbol{e}_k = [0, \ldots, 0, 1, 0, \ldots, 0]^T$ denotes the kth axis vector. Combining these equations shows that

$$HH^T V_k = V_k L_k L_k^T + \sigma_{k+1}\rho_k \boldsymbol{v}_{k+1}\boldsymbol{e}_k^T. \tag{3.6}$$

It follows that $\{\boldsymbol{v}_j\}_{j=1}^{k+1}$ are Lanczos vectors and

$$T_k = L_k L_k^T \tag{3.7}$$

is the symmetric tridiagonal Lanczos matrix obtained when applying k steps of the Lanczos algorithm for symmetric matrices to the matrix HH^T with initial vector $\boldsymbol{b}$; see, e.g., [5, Chapter 9.1] for a discussion of the Lanczos algorithm.

3.1. L-curve for the choice of the regularization parameter. If we assume the knowledge of the norm of the noise in $\mathbf{b}$, we can consider a method presented in [2] to choose the value of μ so that the associated solution satisfies the Morozov discrepancy principle. When the norm of the error $\mathbf{e}$ is not explicitly known, the Morozov discrepancy principle cannot be applied. Among the available criteria for the choice of a suitable regularization parameter in this case, we consider the L-curve method.

This criterion was first proposed by Lawson and Hanson, see [15, Chapter 26], and has recently been considered by Hansen and O'Leary [8, 10]. It is based on the observation that the effect of the regularization on the norm of the solution $\mathbf{f}^\mu$ of (2.1) and on the norm of the discrepancy

$$\boldsymbol{d}^\mu = H\mathbf{f}^\mu - \boldsymbol{b} \tag{3.8}$$

can be conveniently illustrated by plotting the curve

$$\mathbf{L} = \{(\log(\|\mathbf{f}^\mu\|^2), \log(\|\boldsymbol{d}^\mu\|^2)) : \mu > 0\}. \tag{3.9}$$

Recent work by Engl and Grever [20], Vogel [24], points out limitations of the L-curve criterion. Nevertheless, computational experience shows that the L-curve criterion gives suitable regularization parameter values for many real problems.

As the computation of each point on the L-curve is quite expensive, if the dimensions of the problem are very large, only a few points on the L-curve can be determined. This can create difficulties in the identification of the vertex of the L-shaped plot.

If we define the L-curve coordinates as $(\log(\rho(\mu)), \log(\eta(\mu)))$ where

$$\rho(\mu) := \|H^T \mathbf{y}^\mu\|, \qquad \eta(\mu) := \|A\mathbf{y}^\mu - \mathbf{b}\|,$$

it is straightforward to verify that $\rho(\mu)$ is a monotonically decreasing function of $\eta(\mu)$ and if a few reasonable assumptions are made about the ill-posed problem, then it is possible to characterize the L-curve behaviour analogously to that done in [10] and in [8] for the Tikhonov regularization and the L-curve is shown to exhibit a corner behaviour as a function of μ [13].

In actual computations, the values $\rho(\mu)$ and $\eta(\mu)$ are usually not available, hence we search for good approximations. Let $\mathbf{y}_k^\mu$ be an approximation of the solution $\mathbf{y}^\mu$ of (3.2), we introduce the residual vector

$$\mathbf{r}_k^\mu := \mathbf{b} - (A + \mu I)\mathbf{y}_k^\mu, \tag{3.10}$$

and the quantities

$$\rho_k(\mu) := \|A\mathbf{y}_k^\mu - \mathbf{b}\|, \tag{3.11}$$

$$\eta_k(\mu) := \|H^T \mathbf{y}_k^\mu\|. \tag{3.12}$$

We present an iterative procedure (see Algorithm 1) to determine successive encapsulated intervals containing the optimal regularization parameter μ^{opt} corresponding to the vertex of the L-curve. At each step some new points on the discrete L-curve are computed by a few steps of the Lanczos algorithm, while the extreme points of a new interval for μ are determined as the points of minimum distance from a suitable reference point (*local origin* [13]). The final value obtained for the regularization parameter is used to compute an approximation of $\mathbf{f}^\mu$, solution of the regularized system (3.2).

The initialization step needs the definition of an initial interval $[\mu_i, \mu_f]$ for the regularizing parameter, the number of required points (N) on the discrete L-curve, and the initial dimension *kval* of the Krylov space. We will refer to each step j of the repeat loop as refinement step j. Each of these steps consists of a L-curve computation phase, which calculates N even spaced points on the discrete L-curve belonging to the interval $[\mu_i^j, \mu_f^j]$, and of a detection phase for determining the corner of the L-curve at step j. More details about this algorithm are given in [13].

In the next paragraphs we present two different approaches to the L-curve computation phase, named METHOD 1 and METHOD 2 in Algorithm 1.

3.2. Method 1. A good approximation of the discrete L-curve can be obtained by computing $(\rho_k(\mu), \eta_k(\mu))$ using (3.11) and (3.12), and by quantifying the errors by the length of the residual vector.

The following proposition, whose proof is in [13], is concerned with how well the functions ρ_k and η_k approximate ρ and η, respectively.

PROPOSITION 3.1. *Let $\rho_k(\mu)$ and $\eta_k(\mu)$ be defined by (3.11) and (3.12), respectively. Then*

$$|\rho_k(\mu) - \rho(\mu)| \le \|\mathbf{r}_k^\mu\|, \tag{3.13}$$

$$|\eta_k(\mu) - \eta(\mu)| \le \|\mathbf{r}_k^\mu\|. \tag{3.14}$$

Input: $H \in I\!R^{m\times n}$, $\mathbf{b} \in I\!R^m$, ϵ_d, $\epsilon_r \in I\!R^+$, $[\mu_i,\mu_f]$,$\mu_i, \mu_f \in I\!R^+$, $N, kval \in I\!N$;
Output: μ_{opt}, regularized approx. solution $\mathbf{f}_k^{\mu_{opt}} \in I\!R^n$;
% Initialization %
. k:=kval; j=0; $\mu_i^1 = \mu_i$,$\mu_f^1 = \mu_f$;$\mu_{opt}^0 := 0$;
repeat
% Refinement step j on the interval $[\mu_i^j, \mu_f^j]$ %
% [1.] **L-curve computation phase** %
. Compute N even spaced points on the L-curve on $[\mu_i^j, \mu_f^j]$;
. For each point $P_\ell(\rho_k(\mu_\ell), \eta_k(\mu_\ell))$ apply k Lanczos steps
.Estimate $\rho_k(\mu_\ell)$ by METHOD 1 or METHOD 2
.Estimate $\eta_k(\mu_\ell)$ by METHOD 1 or METHOD 2
% [2.] **Detection corner phase** %
. Detect the smallest subinterval $[\mu_{inew}^j, \mu_{fnew}^j] \subset [\mu_i^j, \mu_f^j]$ that contains the closest point P_j to the origin;
. Compute the associated μ_{opt}^j;
until $(|\mu_{opt}^{j-1} - \mu_{opt}^j| \leq \epsilon_d|\mu_{opt}^{j-1}|)$;
% Resolution phase %
. Evaluate the final solution of the linear system $H\mathbf{f} = \mathbf{b}$ by solving $\mathbf{f}_k^{\mu_{opt}} = H^T\mathbf{y}_k^{\mu_{opt}}$.

FIG. 3.1. *Algorithm 1*

Thus $\|\mathbf{r}_k^\mu\|$ quantifies the error introduced when approximating. We determine an approximate solution of (3.2) of the form $\mathbf{y}_k^\mu = V_k\mathbf{z}$ by solving the minimization problem

$$\begin{aligned}\min_{\mathbf{z}\in I\!R^k} \|(A+\mu I)V_k\mathbf{z} - \mathbf{b}\| &= \min_{\mathbf{z}\in I\!R^k} \|V_{k+1}(T_{k+1,k} + \mu I_{k+1,k})\mathbf{z} - V_{k+1}\mathbf{e}_1\|\mathbf{b}\|\| \\ &= \min_{\mathbf{z}\in I\!R^k} \|(T_{k+1,k} + \mu I_{k+1,k})\mathbf{z} - \mathbf{e}_1\|\mathbf{b}\|\|,\end{aligned} \tag{3.15}$$

where

$$T_{k+1,k} = \begin{bmatrix} T_k \\ \sigma_{k+1}\rho_k\mathbf{e}_k^T \end{bmatrix} \in I\!R^{(k+1)\times k}, \tag{3.16}$$

and $I_{k+1,k}$ is the leading principal $(k+1)\times k$ submatrix of the identity matrix of order $k+1$. The quantities ρ_k and η_k defined by (3.11) and (3.12), respectively, are evaluated as

$$\rho_k(\mu) = \|T_{k+1,k}\mathbf{z}_k^\mu - \mathbf{e}_1\|\mathbf{b}\|\|, \tag{3.17}$$

$$\eta_k(\mu) = \|L_k^T\mathbf{z}_k^\mu\|, \tag{3.18}$$

where $\mathbf{z}_k^\mu$ is derived from (3.15) using a QR factorization of $(T_{k+1,k} + \mu I_{k+1,k})$. We should observe that $\rho_k(\mu)$ and $\eta_k(\mu)$, for fixed k, can be evaluated for different values of μ without recomputing the Lanczos decomposition. Each new value of μ yields a new tridiagonal matrix, whose QR-factorization, has to be computed. The computation of this factorization is inexpensive; it requires only $\mathcal{O}(k)$ arithmetic operations, since it can be carried with k Givens rotations. All the details are reported in [13].

% [1.] L-curve computation phase %
. Compute N even spaced points on the L-curve on $[\mu_i^j, \mu_f^j]$;
. For each point $P_\ell(\rho_k(\mu_\ell), \eta_k(\mu_\ell))$ apply k Lanczos steps
    .Estimate $\rho_k(\mu_\ell)$ by (3.17) and $\eta_k(\mu_\ell)$ by (3.18) if ($\|\mathbf{r}_k\| \leq \epsilon_r$) else
    .Apply one more Lancsoz step;

FIG. 3.2. *METHOD 1*

In Figure 3.2 we give some details of the L-curve computation phase in Algorithm 1 by applying METHOD 1.

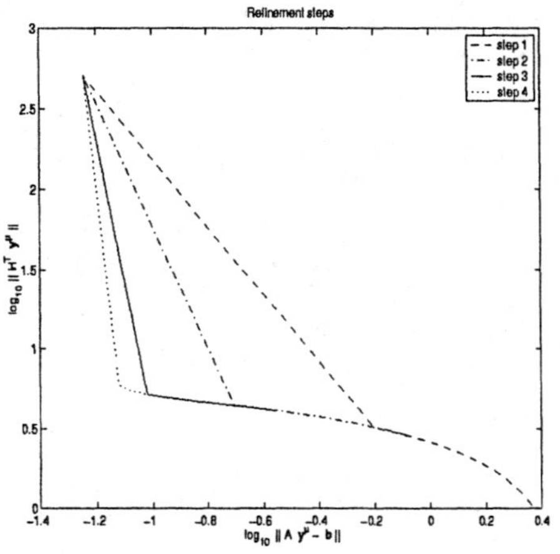

FIG. 3.3. *L-curve for 4 refinement steps* (0.01 *noise level), METHOD 1.*

Figure 3.3 shows the approximated L-curve for the first four successive refinement steps of Algorithm 1 using METHOD 1, at which correspond the nested intervals $[\mu_i^j, \mu_f^j]$, $j = 1, \ldots, 4$, for a test problem using a matrix $H \in \mathbb{R}^{100\times 200}$ defined by its singular value decomposition ($z_i = e^{-0.2(i-m)}, i = 1, \ldots, 100$) with noise level 0.01 ($\|\mathbf{e}\|/\|\mathbf{b}\|$). The curve moves from the upper left to the lower right as μ increases. One can clearly see that for smaller intervals, the L-curves have well defined shapes and corners at which the distance from the local origin is minimized.

For all the reported results we used for the stopping criterion ϵ_d of Algorithm 1, the value 10^{-6}, and for ϵ_r a value greater than 0.001 because experimental results showed us that there was no significant difference in the L-shape to justify greater and expensive precisions.

3.3. Method 2. This section considers a different estimation of the L-curve points by using upper and lower bounds for matrix functionals of the form

$$F(A) = u^T f(A) u, \qquad A \in \mathbb{R}^{m\times m}, \quad u \in \mathbb{R}^m, \tag{3.19}$$

where A is a large sparse symmetric matrix, and u is a given vector.

In several papers Golub and collaborators have exploited the connection between the matrix functional (3.19), Stieltjes integrals, Gauss-type quadrature rules and the Lanczos process to derive powerful inexpensive algorithms for computing upper and lower bounds for $F(A)$; see, e.g., [4, 6]. Such bounds can be computed when the derivatives $d^j f/dt^j$, $j = 1, 2, \dots$, are of constant sign on the spectrum of A and bounds for the largest or smallest eigenvalues of A are known. However, in several applications functions f arise that do not have derivates of constant sign on the spectrum of A. In [1] a new method for inexpensively computing estimates of upper and lower bounds of $F(A)$ is presented. This method is based on applying pairs of Gauss and anti-Gauss quadrature rules, and neither imposes requirements on the sign of the derivatives of f nor requires bounds for an extreme eigenvalue of A. Anti-Gauss rules are obtained by a modification of Gauss rules, and were recently described by Laurie [14].

In [1] a brief review about Gauss and anti-Gauss quadrature rules is given, and in this section we present the application of these estimations to the computation of points on the L-curve. Introduce the spectral factorization

$$A = S\Lambda S^T, \quad \Lambda = \mathrm{diag}[\lambda_1, \lambda_2, \dots, \lambda_m], \quad S^T S = I, \quad \lambda_1 \leq \lambda_2 \leq \dots \leq \lambda_m.$$

Define

$$f(A) := Sf(\Lambda)S^T, \quad and \quad u^T S := [\beta_1, \beta_2, \dots, \beta_m]$$

then

$$F(A) := u^T S f(\Lambda) S^T u = \sum_{j=1}^{m} f(\lambda_j)\beta_j^2.$$

The right-hand side is a Stieltjes integral,

$$F(A) = \mathcal{I}f := \int_{-\infty}^{\infty} f(t)d\beta(t),$$

where the non-negative measure $d\beta(t)$ is such that $\beta(t)$ is a non-decreasing step function with jumps of height β_j^2 at $t = \lambda_j$. The k-point Gauss quadrature rule associated with the measure $d\beta(t)$ is of the form

$$\mathcal{G}_k f = \sum_{j=1}^{k} f(\theta_j)\gamma_j^2.$$

$\mathcal{G}_k f$ is characterized by the fact that

$$\mathcal{I}f = \mathcal{G}_k f, \qquad \forall f \in \mathbf{P}_{2k-1},$$

where $\mathbf{P}_{2k-1}$ denotes the set of polynomials of degree at most $2k-1$. Let

$$\mathcal{H}_{k+1} f := \sum_{j=1}^{k+1} f(\tilde{\theta}_j)\tilde{\gamma}_j^2 \tag{3.20}$$

be the $(k+1)$-point anti-Gauss quadrature rule associated with the measure $d\beta(t)$. This rule is determined by the fact that the error for polynomials of degree $2k+1$ has the same modulus but opposite sign of the error in $\mathcal{G}_k$. Evaluations of pairs of Gauss and anti-Gauss rules provide accurate estimations of upper and lower bounds of $F(A)$. These estimates do not require that the derivatives $f^{(k)}(t)$ and $f^{(k+1)}(t)$ be of constant sign. A good approximation of $F(A)$ is then given by

$$\mathcal{L}_{2k+1}f := 1/2(\mathcal{G}_k + \mathcal{H}_{k+1})f. \tag{3.21}$$

which is a quadrature rule of degree $2k+1$ [14, 1].

Gauss quadrature rules can be determined by the Lanczos process applied to matrix $A = HH^T$ with initial vector $v_1 = u$, where u is of unit norm.

Applications of the Lanczos process yields the decomposition

$$AV_k = V_kT_k + \sigma_{k+1}\rho_k \mathbf{v}_{k+1}\mathbf{e}_k^T \tag{3.22}$$

where T_k is the symmetric tridiagonal matrix defined in (3.7). Introduce the spectral factorization of T_k

$$T_k = Q_kD_kQ_k^T$$

where $D_k = diag[\theta_1, \theta_2, .., \theta_m]$ and $Q_k^TQ_k = I_k$. It is well known that the weights of the Gauss rule are given by $\gamma_j^2 = (e_1^TQ_ke_j)^2, \quad 1 \le j \le k$, and the nodes are the eigenvalues of T_k. Then the Gauss rule can be written in the form

$$\mathcal{G}_k(f) = e_1^TQ_kf(D_k)Q_k^Te_1 = e_1^Tf(T_k)e_1. \tag{3.23}$$

Thus the value $\mathcal{G}_k(f)$ can be determined by first computing the Lanczos decomposition (3.22) and then evaluating (3.23).

Anti-Gaussian quadrature formulas can be trivially computed from Gaussian rules [1]. In particular, the symmetric tridiagonal matrix associated with the quadrature rules $\mathcal{H}_{k+1}(f)$ is given by

$$\tilde{T}_{k+1} := \begin{bmatrix} \alpha_1 & \gamma_1 & & & \\ \gamma_1 & \alpha_2 & \gamma_2 & & \\ & \ddots & \ddots & \ddots & \\ & & \gamma_{k-1} & \alpha_k & \sqrt{2}\gamma_k \\ & & & \sqrt{2}\gamma_k & \alpha_{k+1} \end{bmatrix} \in I\!R^{(k+1)\times(k+1)}$$

The entries of $\tilde{T}_{k+1}$ can be computed by $k+1$ steps of the Lanczos process (see (3.4) and (3.7)). Similarly to the formula (3.23), the anti-Gauss quadrature rule (3.20) can be evaluated according to

$$\mathcal{H}_{k+1}(f) = e_1^Tf(\tilde{T}_{k+1})e_1.$$

Then

$$\begin{aligned} \rho(\mu)^2 &= \|(\mathbf{b} - A\mathbf{y}^{\mu})\|^2 \\ &= \mathbf{b}^T\mathbf{b} - 2\mathbf{b}^T(A+\mu I)^{-1}A\mathbf{b} + \mathbf{b}^T(A+\mu I)^{-1}A^2(A+\mu I)^{-1}\mathbf{b}, \end{aligned}$$

$$\eta^2(\mu) = \|H^T\mathbf{y}\|^2 = \mathbf{b}^T(A+\mu I)^{-1}A(A+\mu I)^{-1}\mathbf{b}.$$

Introduce the functions

$$\phi_1(t) := -2(t+\mu)^{-1}t + (t+\mu)^{-1}t^2(t+\mu)^{-1}$$

% [1.] L-curve computation phase %
. Compute N even spaced points on the L-curve on $[\mu_i^j, \mu_f^j]$;
. For each point $P_\ell(\rho_k(\mu_\ell), \eta_k(\mu_\ell))$ apply k Lanczos steps
.Estimate $\rho_k(\mu_\ell)$ by (3.25) using $\mathcal{L}_{2k+1}(\phi_2) = \frac{1}{2}(\mathcal{G}_k + \mathcal{H}_{k+1})(\phi_2)$
.Estimate $\eta_k(\mu_\ell)$ by (3.24) using $\mathcal{L}_{2k+1}(\phi_1) = \frac{1}{2}(\mathcal{G}_k + \mathcal{H}_{k+1})(\phi_1)$

FIG. 3.4. *METHOD 2*

$$\phi_2(t) := (t+\mu)^{-1} t (t+\mu)^{-1}$$

and define the quantities

$$s_i := \mathbf{b}^T \phi_i(A)\mathbf{b}, \quad i = 1, 2.$$

Then the coordinates of the point P_μ on the L-curve can be expressed in terms of s_i

$$\rho(\mu) = (\mathbf{b}^T\mathbf{b} + s_1)^{1/2}, \tag{3.24}$$

$$\eta(\mu) = s_2^{1/2}. \tag{3.25}$$

The s_i can be represented as Stieltjes integrals which can be approximated by Gauss quadrature rules. We can improve the approximation of these integrals $\mathcal{I}f$ by using the $(2k+1)$ quadrature rule (3.21).

In Figure 3.4 we specify the L-curve computation phase given by METHOD 2 by means of estimations given by (3.21).

Figure 3.5 shows the convergence of the approximated L-curves to the exact L-curve for increasing μ values, when we apply Algorithm 1 with both METHOD 2 and METHOD 1, for the same test problem considered in Figure 3.3.

The different behaviour of the three different L-curves is more evident around the corner, as illustrated in Figure 3.6, where we considered a subinterval of the interval considered in Figure 3.5.

From the figures, we observe a better performance of METHOD 2 rather than METHOD 1, but we should say that in the first case the number of Lanczos steps k is fixed in advance and it may be bigger than the necessary, while for METHOD 1 we use for k the smallest value required to satisfy the residual tolerance.

4. Numerical Results. In this section we report some results obtained in the treatment of two problems modelled by the equation (1.1), i.e. the image restoration and image reconstruction problems, using the regularization methods presented in section 2. In both cases parallel algorithms are developed: in the case of image restoration the domain decomposition strategy is used while in the case of image reconstruction the functional parallelism is exploited.

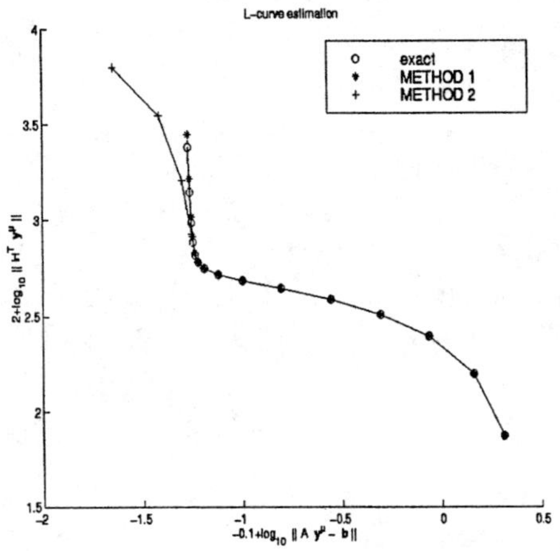

FIG. 3.5. *Exact and approximated points on the L-curve; Noise level 0.01, Number of points 15,* $\mu \in [1.0 \times 10^{-7}, 2.0]$ *Lanczos steps: Method 1. max k=30* ($\|\mathbf{r}_k\| \leq 0.01$), *Method 2. k=30.*

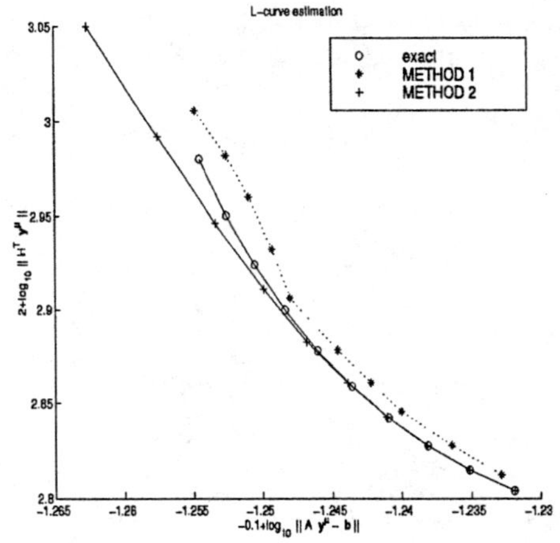

FIG. 3.6. *Exact and approximated points on the L-curve; Noise level 0.01, Number of points 10* $\mu \in [3.0 \times 10^{-6}, 4.4 \times 10^{-5}]$

4.1. Image Restoration. A relevant problem modeled by (1.1) is the image restoration problem. The PSF is the system response function and can be assumed to have a Gaussian shape:

$$h(x, \xi, y, \eta) = \frac{exp\left(-\frac{(\xi-x)^2+(\eta-y)^2}{2\sigma^2(x,y)}\right)}{\sqrt{2\pi}\sigma(x,y)} \tag{4.1}$$

We considered two PSFs, PSF1 and PSF2, characterized by two different expressions of the variance, $\sigma_1(x, y)$ and $\sigma_2(x, y)$ respectively:

$$\sigma_1(x, y) = \begin{cases} |ksin(x+y)|, & \text{if } \sin(\text{x+y}) \not\equiv 0 \quad k \in [1, 2] \\ k, & \text{if } \sin(\text{x+y}) = 0 \end{cases} \tag{4.2}$$

$$\sigma_2(x, y) = k(x + y); \quad k \in [0.01, 0.05]. \tag{4.3}$$

1. for $i = 1, \dots, \tau$ in parallel
1.2 Compute the approximate solution in the sets S_i.
$\min_{f_i \in S_i} \{\frac{1}{2}\mathbf{f_i^T} K_i \mathbf{f_i} - \mathbf{f^T}_i H_i^T \mathbf{b}_i \}$
1.3 solve $\min_{f_l \in D_l} \{\frac{1}{2}\mathbf{f_l^T} K_l \mathbf{f_l} - \mathbf{f^T}_l H_l^T \mathbf{b}_l \}$
$B\mathbf{f_l} = \mathbf{s_l}$
2. collect $\mathbf{f}$ from $\mathbf{f_l}$

FIG. 4.1. *Parallel algorithm for image restoration*

The discretization of the integral equation gives raise to a linear system (1.2) of very large size that requires a lot of computational time to be solved. We have applied a domain decomposition technique. The image domain has been partitioned as in fig. 4.1 and solutions $\mathbf{s_i}$ of the least squares problem (1.2) are computed on small stripes centered on the borders S_i of the rectangles. Then, constrained quadratic programming problems of the form (2.2)–(2.4) are solved in the rectangles, after assigning the computed vectors $\mathbf{s_i}$ as constraints on the borders, $B\mathbf{f_i} = \mathbf{s_i}$, $i = 1, \dots, \tau$. This algorithm has a high degree of parallelism and it has been efficienly implemented both on shared and distributed memory architectures [16].

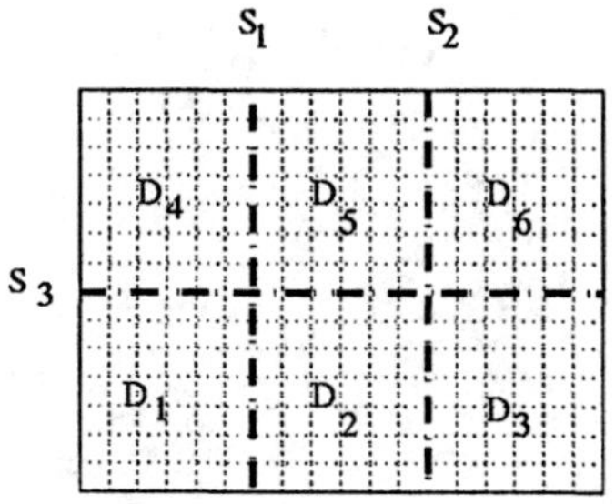

FIG. 4.2. *Domain partitioning.*

Two examples of blurred and restored images are shown in fig. 4.3 and 4.4. From many tests executed on different images and reported in [18, 25, 16], we can conclude that the quality of the restorations is good and only slightly depends on the subdivision grain size. Evaluating the performance of the applications on parallel computers, we can state that the domain decomposition is a very good technique to realize parallel programs in terms of speed–up and efficiency when applied to constrained regularization algorithms such as the Hestens method defined in (2.5) and (2.6).

4.2. Image reconstruction. A relevant problem in the area of medical imaging is the reconstruction of human brain sections from experimental data by means of Single Photon Emission Tomography (SPECT). It consists in the detection of the gamma rays emitted singly from radiactive atoms (radionuclides) contained in the radiopharmaceutical injected in a particular organ to determine its functional activity. The object is reconstructed as a set of N slices. Each slice requires the solution of an integral equation (1.1), where

$r = (x, y)$ are the object spatial coordinates and $s = (\theta, t)$ the projection coordinates. The geometrical system response is described by a gaussian function whose variance σ depends on the distance q of the point (x, y) from the Gamma camera collimator surface [22]. We then obtain N linear systems,

$$\mathbf{H}\,\mathbf{f}_i = \mathbf{b}_i, \qquad i = 1, \ldots, N,$$

which have been solved with the Conjugate Gradients regularization method applied to the Normal Equations.

In real applications the total number of equations is large (e.g. 10^5) and in order to have real time solutions we have implemented the CG method also on massively parallel architectures [17]. In figure 4.5 (a) and (b) we report the reconstruction of four sections of human brain obtained by a real acquisition from phantom data and from patient data, respectively.

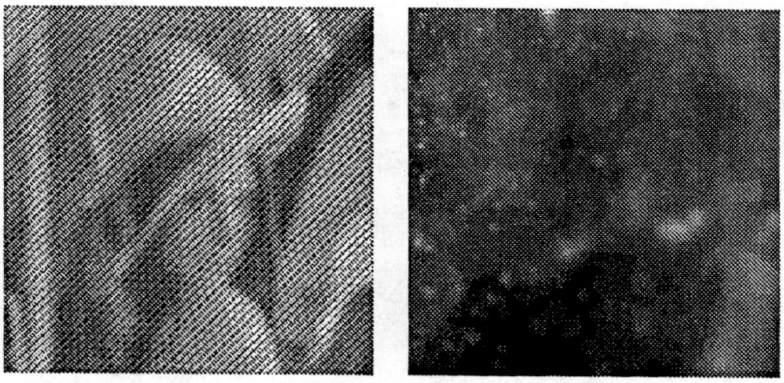

256 × 256 blurred image with PSF1. 200 × 200 blurred image with PSF2.

FIG. 4.3. *Blurred images*

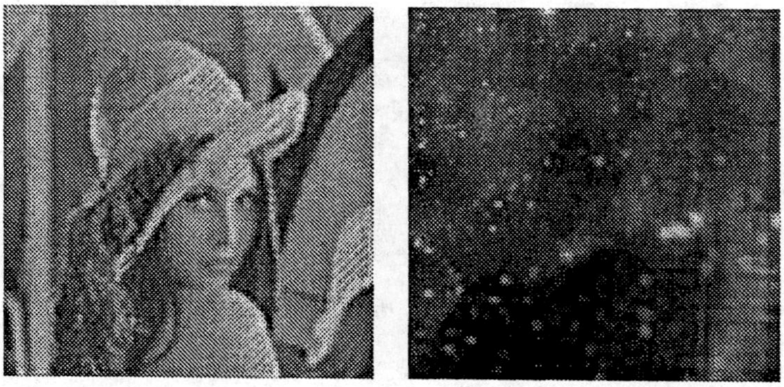

256 × 256 restored image with algorithm of fig. 4.1. 200 × 200 restored image with algorithm of fig. 4.1.

FIG. 4.4. *Restored images*

Acknowledgments. The authors are really greatful to Professor Ilio Galligani who has always given valuable suggestions in emerging research fields ranging from ill posed

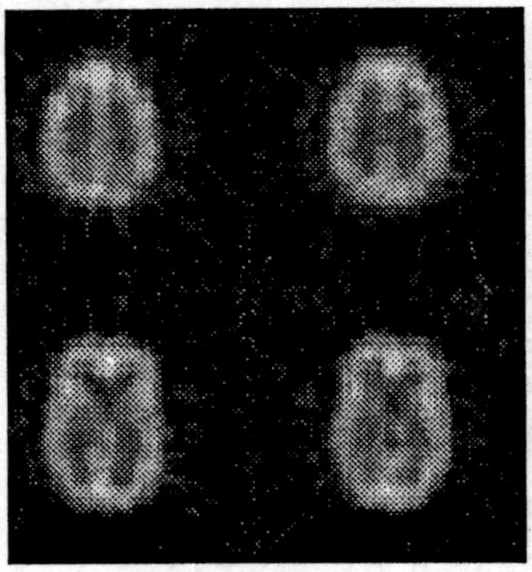

Reconstruction Phantom data with 8 CG iterations.

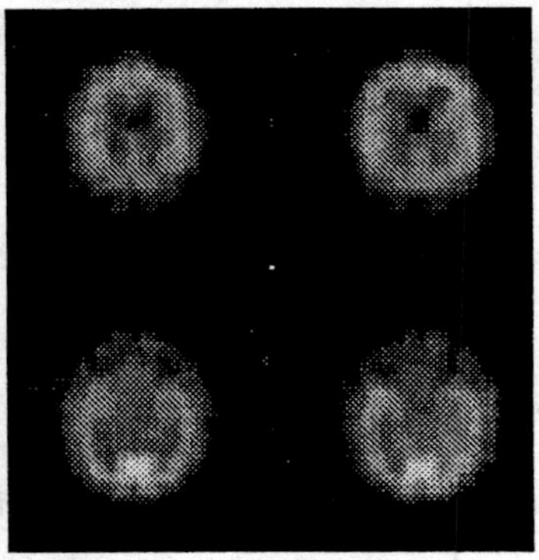

Reconstruction Patient data with 6 CG iterations.

FIG. 4.5. *Reconstructed SPECT images*

problems, image processing to parallel computation. In this paper we have tried to briefly review and collect all of these topics.

REFERENCES

[1] D.Calvetti, L.Reichel, F.Sgallari, *Application of Anti-Gauss quadrature rules in Linear algebra*, in Applications and Computation of Orthogonal Polynomials, eds. W. Gautschi, G.H.Golub, and G.Opfer, Birkhauser, Basel, 1999, pp. 41–56.

[2] D. Calvetti, L. Reichel, F. Sgallari and G. Spaletta, *A regularizing Lanczos iteration method for underdetermined linear systems*, J. Comput. Appl. Math., in press.

[3] I. Galligani V. Ruggiero. *Numerical solution of equality–constrained quadratic programming problems on vector–parallel computers*, Optimization Methods: Software and Algorithms, 2 (1993), pp. 233–247.

[4] G. H. Golub and G. Meurant, *Matrices, moments and quadrature*, in Numerical Analysis 1993, eds. D. F. Griffiths and G. A. Watson, Longman, Essex, England, 1994, pp. 105–156.

[5] G. H. Golub and C. F. Van Loan, *Matrix Computations*, 3rd ed., Johns Hopkins University Press, Baltimore, 1996.

[6] G. H. Golub and Z. Strakos, *Estimates in quadratic formulas*, Numer. Algorithms, 8 (1994), pp. 241–268.

[7] M. Hanke. *Conjugate gradient type methods for ill posed problems*, In Pitman Reserch Notes in Mathematics Series. Longman, 1995.

[8] P. C. Hansen, *Analysis of discrete ill-posed problems by means of the L-curve*, SIAM Review, 34 (1992), pp. 561–680.

[9] P. C. Hansen, *Rank-Deficient and Discrete Ill-Posed Problems*, SIAM, Philadelphia, 1998.

[10] P. C. Hansen and D. P. O'Leary, *The use of the L-curve in the regularization of discrete ill-posed problems*, SIAM J. Sci. Comput., 14 (1993), pp. 1487–1503.

[11] M. R. Hestenes, *Optimization Theory. The finite dimensional case*. J. Wiley and Sons, 1975.

[12] B. R. Hunt H. C. Andrews. *Digital Image Restoration*. Prentice Hall, INC., Englewood Press, 1977.

[13] S.Morigi, F.Sgallari, *A regularizing L-curve Lanczos method for underdetermined linear systems*, Appl. Math. Comp., in press.

[14] D. P. Laurie, *Anti-Gaussian quadrature formulas*, Math. Comp., 65 (1996), pp. 739–747.

[15] C.L.Lawson, R.J. Hanson, *Solving Least Square Problems*, SIAM Philadelphia, 1995.

[16] E.Loli Piccolomini F. Zama. *Parallel image restoration with domain decomposition*, in Special issue on Fast–Energy–Minimization–Based Imaging and Vision Techniques on Real–Time Imaging, in press.

[17] E. Loli Piccolomini F. Zama *Parallel Application on SPECT Data Reconstruction*, Proceedings of the Conference ECMI98, Goteborg, 1998.

[18] E. Loli Piccolomini F. Zama. *An experiment in image restoration using transputer networks*, Parallel

Algorithms and Applications, 6, (1994).

[19] E.Loli Piccolomini and F. Zama. *The conjugate gradient regularization method in computed tomography problems*, Applied Mathematics and Computation, 102 (1999), pp. 87–99.

[20] H.W.Engl, W.Grever, *Using the L-curve for determining optimal regularization parameters*, Numer. Math., 69 (1994), pp. 25–31.

[21] C. C. Paige and M. A. Saunders, *LSQR: An algorithm for sparse linear equations and sparse least squares*, ACM Trans. Math. Software, 8 (1982), pp. 43–71.

[22] A. Passeri, A.R. Formiconi, and U. Meldolesi. *Physical modelling (geometrical system response, Compton scattering and attenuation) in brain SPECT using the conjugate gradients reconstruction method*, Phys. Med. Biol., 37 (1992), pp. 172–174.

[23] V. Y. Arsenin A. N. Tikhonov. *Solutions of Ill–Posed Problems*, John Wiley & Sons, 1977.

[24] C. R. Vogel, *Non-convergence of the L-curve regularization parameter selection method*, Tech.Report, Department of Mathematics, Montana State University, Bozeman, MT, 1996.

[25] F. Zama I. Galligani, E. Loli Piccolomini V. Ruggiero. *A substructuring method for solving the image restoration problem on a multiprocessor system*, SMS TPE 94 (Software for Multiprocessrs and Supercomputers), IOS Press, 1994.

IMPLEMENTATION OF THE STABILIZED THREE-FIELD FORMULATION

FRANCO BREZZI* AND L. DONATELLA MARINI*

Abstract. This paper deals with the computational aspects of the so-called three-field formulation, introduced first in [6] and analysed in [8]. In particular, we introduce the possible use of *virtual bubbles* (in the basic framework of [1]) and we show that their use can simplify the implementation of the method. A variety of numerical experiments is also presented. They show that the use of virtual bubbles instead of conventional ones does not affect the quality of the final accuracy. More generally, they show clearly the general robustness of the three-field formulation, together with the possibility of using powerful preconditioners when the interface grid is chosen to be uniform.

1. Introduction. Let us briefly recall, for the convenience of the reader, the main idea of the three-field formulation. Assume that we have to solve a linear elliptic problem whose variational formulation is given by:

$$\text{find } w \in V \text{ such that } \qquad a(w,v) = (f,v) \qquad \forall v \in V, \tag{1.1}$$

on a domain Ω. We assume that the problem is associated to a second-order differential operator, so that the space V will be a subspace of $H^1(\Omega)$. By splitting Ω into subdomains Ω^s one introduces suitable subspaces V^s of $H^1(\Omega^s)$, and defines $M^s = H^{-1/2}(\partial\Omega^s)$, $\Sigma = \cup_s \partial\Omega^s$, and $\Phi = \{\text{traces on } \Sigma \text{ of the functions of } V\}$. Setting $V^* := \Pi_s V^s$ and $M^* := \Pi_s M^s$, the three-field formulation of (1.1) then reads

$$\begin{cases} \text{find } u \in V^*,\ \lambda \in M^*, & \text{and } \psi \in \Phi \text{ such that} & & \\ \text{i)}\ \ a_s(u^s,v) - <\lambda^s, v>_s & = (f,v)_s & \forall v \in V^s & \forall s \\ \text{ii)}\ \ <\mu^s, u_s>_s & = <\mu^s,\psi>_s & \forall \mu^s \in M^s & \forall s \\ \text{iii)}\ \ \sum_s <\lambda^s,\varphi>_s & = 0 & \forall \varphi \in \Phi & \end{cases} \tag{1.2}$$

with obvious meaning for a_s, $<.,.>_s$ and $(.,.)_s$ (see also (2.2)- (2.3) in the next section). This formulation was originally introduced in [6, 7], where it was proved that, under reasonable assumptions, problem (1.2) has a unique solution related to the solution of (1.1) by

$$\begin{cases} \text{i)}\ \ u^s & = w_{|\Omega^s} & \text{for each } s \\ \text{ii)}\ \ \lambda^s & = \dfrac{\partial w}{\partial n_a}_{|\partial\Omega^s} & \text{for each } s \\ \text{iii)}\ \ \psi & = w_{|\Sigma}, & \end{cases} \tag{1.3}$$

with $\frac{\partial}{\partial n_a}_{|\partial\Omega^s}$ = outward conormal derivative to $\partial\Omega^s$. For other interesting variants of (1.2) see [14] In order to approximate (1.2) one has to choose, for every s, finite dimensional subspaces V_h^s, M_h^s and Φ_h of V^s, M^s and Φ respectively, to construct $V_h^* := \Pi_s V_h^s$ and $M_h^* := \Pi_s M_h^s$, and to consider the following discretized problem:

$$\begin{cases} \text{find } u_h \in V_h^*,\ \lambda_h \in M_h^*, & \text{and } \psi_h \in \Phi_h \text{ such that} & & \\ \text{i)}\ \ a_s(u_h^s,v) - <\lambda_h^s, v>_s & = (f,v)_s & \forall v \in V_h^s & \forall s \\ \text{ii)}\ \ <\mu^s, u_h^s>_s & = <\mu^s,\psi_h>_s & \forall \mu^s \in M_h^s & \forall s \\ \text{iii)}\ \ \sum_s <\lambda_h^s,\varphi>_s & = 0 & \forall \varphi \in \Phi. & \end{cases} \tag{1.4}$$

*Dipartimento di Matematica, Università di Pavia, via Ferrata 1, 27100 Pavia, Italy and Istituto di Analisi Numerica del CNR, via Ferrata 1, 27100 Pavia, Italy

If the three families of subspaces are suitably chosen, problem (1.4) will be well posed, and its solution will converge to the exact solution of (1.2) with optimal order. In general, the finite element spaces Φ_h, V_h^s and M_h^s will be based on three corresponding decompositions, that cannot be chosen arbitrarily. In [8], we considered the case where two independent choices are given, a priori, for Φ_h and V_h^s. It seems particularly reasonable to assume the grid for Φ_h as given, since in many cases this will be the space where multidomain preconditioners (for the Schur complement) will be applied: to have a convenient grid might then allow the use of more powerful preconditioners (see for instance [11, 13, 2, 4, 17, 12, 15, 16].) Similarly, it is also reasonable to assume the grid for V_h^s as given, as it is often the case for real applications, where the nature of the original problem (1.1) dictates different grids in different parts of the domain. Let us assume, for simplicity, that both Φ_h and V_h^s are made of piecewise linear continuous functions. The task is then to start from the two given grids, and to reach, with minor modifications, a problem of the type (1.4) which is well posed and optimally convergent. The main idea of [5], made precise and perfectioned in [9] and then in [8], was to chose first a suitable grid for the multipliers M_h^s, in each subdomain, starting from the given grids T_ψ (for ψ at the interfaces) and T_u^s (for u within each subdomain). Essentially, the choice considered was to take, in each subdomain,

$$\mathrm{T}_\lambda^s = \text{ merge } \{(\mathrm{T}_u^s)_{|\Gamma^s}, (\mathrm{T}_\psi)_{|\Gamma^s}\} \qquad \forall s. \tag{1.5}$$

Once the grids T_λ^s were given, the spaces M_h^s were chosen as piecewise constants on T_λ^s. In general, this choice cannot actually guarantee stability. However, the idea was to increase artificially the original spaces V_h^s with the addition of *boundary bubbles* (essentially: functions having support in a single boundary triangle.) If, for every s, B_h^s is the space of bubbles to be used in order to augment the space V_h^s, we replace then V_h^s by the augmented space

$$\widetilde{V}_h^s = V_h^s \oplus B_h^s \qquad \forall s. \tag{1.6}$$

Setting finally $\widetilde{V}_h^* = \mathbf{\Pi}_s \widetilde{V}_h^s$, the stabilized discrete problem becomes

$$\begin{cases} \text{find } u_h \in \widetilde{V}_h^*,\ \lambda_h \in M_h^* \text{ and } \psi_h \in \Phi_h & \text{such that} & & \\ \text{i)}\ \ a_s(u_h^s, v) - <\lambda_h^s, v>_s & = (f, v)_s & \forall v \in \widetilde{V}_h^s & \forall s \\ \text{ii)}\ \ <\mu^s, u_h^s>_s & = \ <\mu^s, \psi_h>_s & \forall \mu^s \in M_h^s & \forall s \\ \text{iii)}\ \ \sum_s <\lambda_h^s, \varphi>_s & = 0 & \forall \varphi \in \Phi_h. & \end{cases} \tag{1.7}$$

All this will become clearer, and more precise, in the next sections, where a particular case will be considered. For the time being, we just point out that, under reasonable assumptions on the original given grids T_ψ and T_u^s, we were able to indicate a suitable choice of piecewise quadratic boundary bubbles, whose addition makes (1.7) stable and optimally convergent. Moreover, (see also [9]), it is possible to eliminate a priori *both* the bubbles and the multipliers in such a way that, in each subdomain, one is left with a problem in the only (original) unknown u_h.

In the present paper, we reconsider the choice of the bubble functions, that we now allow to be much more general. In particular, we show that the elimination process can be simplified with a nonstandard choice of the bubble shape. We also indicate explicitly a possible very effective preconditioner that can be used if the grid T_ψ is uniform, and we test it on a model problem. More generally, we performed several series of tests, always on a simple model problem, in order to assess the robustness of the method.

An outline of the paper is as follows. In section 2, we choose the model problem, and we present, in this particular case, the spaces to be used, the type of bubbles that we are going to use, and the corresponding elimination procedure in more detail. Always on the model problem, we describe the preconditioner as well. In section 3 we present the numerical results that we have obtained. The possible parameters that we might have taken into account are several: the ratio between the two abutting grids for u, the ratio among the grids for u and for ψ, the shape of the bubbles, and the general behaviour when all the meshes are refined (say, halved) at the same time. We tried to point out the most meaningful among those relationships. In particular, the theoretical error bounds proved in [8] are fully verified, even with the present more general bubble shape.

2. The problem, the spaces and the bubbles. In order to simplify the presentation, we confine ourselves on a particularly simple toy-problem. It should be clear, however, that all the contents of this section can be very easily extended to much more complex situations.

We assume therefore that our bilinear form $a(u, v)$ is associated with the Laplace operator:

$$a(u,v) = \int_{\Omega} \nabla u \cdot \nabla v \mathrm{d}x\, \mathrm{d}y. \tag{2.1}$$

Always to make life simple, assume that Ω is the rectangle $]-1,1[\times]0,1[$, split into $\Omega^1 =]-1,0[\times]0,1[$ and $\Omega^2 =]0,1[\times]0,1[$. Accordingly, we set

$$a_s(u,v) = \int_{\Omega_s} \nabla u \cdot \nabla v \, \mathrm{d}x\, \mathrm{d}y, \tag{2.2}$$

for $s = 1, 2$. Similarly we set

$$< u, v >_s = \int_{\partial\Omega^s} u\, v\, \mathrm{d}\Gamma, \qquad (u,v)_s = \int_{\Omega^s} u\, v\, \mathrm{d}x\, \mathrm{d}y. \tag{2.3}$$

For the time being, we consider homogeneous Dirichlet boundary conditions on $\partial\Omega$, although the numerical experiments have been done with nonhomogeneous ones. It is true that, in the finite element context, nonhomogeneous Dirichlet boundary conditions are easier to implement than to explain. Accordingly, we then have

$$V = H_0^1(\Omega), \qquad V^s = \{v \in H^1(\Omega^s), \quad v = 0 \quad on \quad \partial\Omega\}, \quad s = 1,2. \tag{2.4}$$

We assume that we are given a (uniform) decomposition T_ψ of the interface, and two given decompositions $\mathcal{T}_u^1$ and $\mathcal{T}_u^2$ of Ω^1 and Ω^2 respectively. For simplicity, we might also assume that they are both uniform (right triangles), as those used in the numerical experiments, although this is not really necessary. What is important, in order to have a meaningful case, is that the two decompositions *do not* agree on the interface $\Sigma = \{0\}\times]0,1[$. We consider then, as we announced in the introduction, the spaces Φ_h, V_h^1 and V_h^2 as made of piecewise linear continuous functions on the given decompositions. We then use the basic idea (1.5) to define the grids for the multiplier, which are given here by

$$\mathrm{T}_\lambda^s = \text{ merge } \{(\mathrm{T}_u^s)_{|\Sigma}, (\mathrm{T}_\psi)\} \qquad s = 1,2, \tag{2.5}$$

and we take M_h^1 and M_h^2 made of piecewise constants on their respective grids. Then $M_h^* = M_h^1 \times M_h^2$.

We are now ready for the introduction of the bubbles. We restrict our attention to Ω^1. The procedure on Ω^2 will be identical. Let T be a triangle in $\mathcal{T}_u^1$ having an edge L on Σ, and let P be the vertex of T opposite to L. Let now $I_1, ..., I_K$ be the intervals of $\mathcal{T}_\lambda^1$ contained in L. For every $k = 1, .., K$ we consider the triangle T_k having I_k as one edge, and P as opposite vertex. Applying this procedure to every triangle having an edge on Σ, we obtain as many subtriangles T_k as there are intervals in $\mathcal{T}_\lambda^1$. For every subtriangle T_k we choose now *one* function b_k, with the following properties: we assume that $\int_{I_k} b_k$ is different from zero, so that we can normalise it to one. We also assume that b_k vanishes identically on the other two edges of T_k. When convenient, we shall identify b_k with its extension by zero on Ω^1, that will clearly belong to V^1. We now call B^1 the subspace of V^1 spanned by all the *bubbles* b_k, and we set, as in (1.6)

$$\widetilde{V}^1 = V^1 \oplus B^1. \tag{2.6}$$

By applying the same procedure in Ω^2 we will obtain a space $\widetilde{V}^2$ that, together with $\widetilde{V}^1$, gives then rise to $\widetilde{V}_h^* = \widetilde{V}^1 \times \widetilde{V}^2$. With all the spaces made precise we can now reconsider the stabilized problem (1.7), that we report here for the convenience of the reader:

$$\begin{cases} \text{find } u_h \in \widetilde{V}_h^*,\ \lambda_h \in M_h^* \text{ and } \psi_h \in \Phi_h & \text{such that} & & \\ \text{i)}\ \ a_s(u_h^s, v) - < \lambda_h^s, v >_s & = (f, v)_s & \forall v \in \widetilde{V}_h^s & s = 1,2 \\ \text{ii)}\ \ < \mu^s, u_h^s >_s & = \ < \mu^s, \psi_h >_s & \forall \mu^s \in M_h^s & s = 1,2 \\ \text{iii)}\ \sum_s < \lambda_h^s, \varphi >_s & = 0 & \forall \varphi \in \Phi_h. & \end{cases} \tag{2.7}$$

Looking at (2.7) it is clear that, for fixed ψ_h, the first two equations can be solved independently and *in parallel*. We can therefore consider the mapping $\mathrm{S}_h = (\mathrm{S}_u, \mathrm{S}_\lambda)$ that associates to the pair (f, ψ_h) the solution (u_h, λ_h) of the first two equations of (2.7). With this notation, problem (2.7) can be written as:

$$\text{find } \psi_h \in \Phi_h \text{ such that} \qquad \sum_{s=1}^{2} < \mathrm{S}_\lambda(f, \psi_h), \varphi >_s = 0, \qquad \forall \varphi \in \Phi_h. \tag{2.8}$$

It is clear that in (2.8) a crucial role is played by the linear operator S_h, from Φ_h to its dual space, defined by:

$$< S_h(\psi_h), \varphi > := \sum_{s=1}^{2} < \mathrm{S}_\lambda(0, \psi_h), \varphi >_s, \tag{2.9}$$

which is commonly called Schur complement, and whose spectral properties have a paramount relevance in solving (2.8) by iterative methods. We shall come back to this very important point when discussing the preconditioner.

For the time being, we concentrate first on the procedure of eliminating bubbles and multipliers. As we shall see, the procedure can be implemented, separately and in parallel, in each subdomain Ω^s. Hence, to describe it, we concentrate on a single domain. We assume ψ_h to be given, and we see how to compute the corresponding u_h^s and λ_h^s. For the sake of simplicity, we might drop at this point the superscript s, as the same identical procedure will be applied, in parallel, in each subdomain. With an abuse of notation, we are also going to call Ω (instead of Ω^s) the current subdomain, and so on. No confusion should arise. With this simplified notation the local problem becomes:

$$\begin{cases} \text{given } \psi_h \text{ on } \partial\Omega, \text{ find } u_h \in \widetilde{V}_h & \text{and } \lambda_h \in M_h \quad \text{such that} \\ \text{i)} \quad a(u_h, v) - <\lambda_h, v> & = (f, v) \qquad \forall v \in \widetilde{V}_h \\ \text{ii)} \quad <\mu, u_h> & = <\mu, \psi_h> \quad \forall \mu \in M_h, \end{cases} \tag{2.10}$$

where M_h is made of piecewise constants on T_λ, and $\widetilde{V}_h$ is obtained as the sum of piecewise linear functions on T_u (that we will denote by V_L) and the space of boundary bubbles B_h. Accordingly, u_h will be written as

$$u_h(x) = u_L(x) + \sum_k u_B^k \, b_k(x), \tag{2.11}$$

where k ranges now on the set of all the bubbles.

From the second equation of (2.10), by taking μ as the characteristic function of an interval I_k of T_λ, we obtain:

$$u_B^k = \left(\int_{I_k} \psi_h \mathrm{d}\Gamma - \int_{I_k} u_L \mathrm{d}\Gamma \right) / \int_{I_k} b_k \mathrm{d}\Gamma. \tag{2.12}$$

We remark that both u_L and ψ_h are linear in each I_k. We also remind that the bubbles were normalized in such a way that $\int_{I_k} b_k \mathrm{d}\Gamma = 1$ for all k. Therefore, indicating by m_k the midpoint of I_k, we immediately obtain from (2.12) that

$$u_B^k = (\psi_h(m_k) - u_L(m_k))|I_k|. \tag{2.13}$$

We can now use the first equation of (2.10), with $v = b_k$, in order to express λ_k as a function of the other variables:

$$\lambda_k = a(u_L, b_k) + u_B^k \, a(b_k, b_k) - (f, b_k). \tag{2.14}$$

Assume now, for simplicity, that the right-hand side f is piecewise constant. This is not really restrictive, as in most cases f is approximated anyhow by a piecewise constant in the actual implementation. We remark now that, integrating by parts, for every $v_L \in V_L$ and for every k we have:

$$a(v_L, b_k) = a(b_k, v_L) = \int_{I_k} \frac{\partial v_L}{\partial n} b_k \mathrm{d}\Gamma = \frac{\partial v_L}{\partial n}, \tag{2.15}$$

where clearly $\partial v_L / \partial n$ is the (constant) outward normal derivative of v_L in the triangle (of $\mathcal{T}_u$) containing T_k. We also set:

$$\gamma_k^T = \int_{T_k} b_k \mathrm{d}x \, \mathrm{d}y, \qquad \gamma_k^a = a(b_k, b_k), \tag{2.16}$$

so that (2.14) becomes

$$\lambda_k = \frac{\partial u_L}{\partial n} + u_B^k \gamma_k^a - f_k \gamma_k^T. \tag{2.17}$$

With some manipulations, using (2.13) and (2.15), we obtain

$$a(\sum_k u_B^k b_k, v_L) = \int_{\partial\Omega} (\psi_h - u_L)\frac{\partial v_L}{\partial n} d\Gamma. \tag{2.18}$$

Taking now the first equation of (2.10) for $v = v_L$, using (2.13) (2.17) and (2.15), we obtain with easy computations that

$$\begin{aligned} &a(u_L, v_L) - < u_L, \frac{\partial v_L}{\partial n} > - < v_L, \frac{\partial u_L}{\partial n} > + \textstyle\sum_k |I_k|^2 \gamma_k^a u_L(m_k) v_L(m_k) = \\ &(f, v_L) - \textstyle\sum_k f_k \gamma_k^T |I_k| v_L(m_k) - < \psi_h, \frac{\partial v_L}{\partial n} > + \textstyle\sum_k |I_k|^2 \gamma_k^a \psi_h(m_k) v_L(m_k). \end{aligned} \tag{2.19}$$

At this point we have to discuss the actual shape of the bubble functions. Neglecting for a while the problem related to the coefficient γ_k^T in (2.16) we remark that, having fixed the integral of b_k on I_k to be one, the coefficient γ_k^a has a *lower bound* which depends on the shape of T_k. In particular we have the following result.

THEOREM 2.1. *Let T be a triangle, and ℓ one of its edges. Let v be an element of $H^1(T)$ vanishing on the other two edges of T. Assume that $\int_\ell v\, d\Gamma = 1$. Then $|v|_{1,T} \geq |w|_{1,T}$, where w is the unique harmonic function on T such that $\partial w/\partial n =$ constant on ℓ and $w = 0$ at the other two edges of T, scaled in such a way that $\int_\ell w\, d\Gamma = 1$.*

Proof. The result comes out easily by minimising the quantity $|v|_{1,T}^2$ under the constraint $\int_\ell v\, d\Gamma = 1$. Let w be the minimizer. Then, by the usual technique of Lagrange multipliers, we have that

$$\int_T \nabla w \cdot \nabla \zeta dx\, dy = c \int_\ell \zeta d\Gamma. \tag{2.20}$$

for all $\zeta \in H^1(T)$ vanishing on $\partial T \setminus \ell$, where c is a suitable constant to be determined. The result follows immediately. ∎

We consider first a fixed triangle, say $\hat{T}$, having vertices in $(0,0)$, $(1,1)$, and $(1,-1)$. Here a harmonic function $\hat{w}$ having constant normal derivative on the edge $\hat{\ell}$ of equation $x = 1$ and vanishing on the other two edges is easily found: actually we have $\hat{w} = x^2 - y^2$. For this function we have

$$|\hat{w}|_{1,\hat{T}}^2 = 8/3, \qquad \int_{\hat{\ell}} \hat{w} d\Gamma = 4/3, \qquad \int_{\hat{T}} \hat{w} dx dy = 1/3. \tag{2.21}$$

For any positive t we can now scale the triangle $\hat{T}$ down to a new triangle T_t having vertices in (0,0), (t,t), and $(t,-t)$. If w_t is the image of $\hat{w}$ in T_t (that means that $w_t = (x/t)^2 - (y/t)^2$), we easily get

$$|w_t|_{1,T_t}^2 = 8/3, \qquad \int_{\ell_t} w_t d\Gamma = 4t/3, \qquad \int_{T_t} w_t dx dy = t^2/3. \tag{2.22}$$

Assume, for simplicity, that the two angles of T_k insisting on the edge I_k are both $\geq \pi/4$, as it was the case in our tests. Hence, we can always insert a triangle like T_t inside T_k, (see fig. 2.1) Notice that in this case we shall have $t = |I_k|/2$. In order to normalize the boundary integral to one, we set then

$$b_k = \frac{3}{4t} w_t \qquad in \quad T_t, \tag{2.23}$$

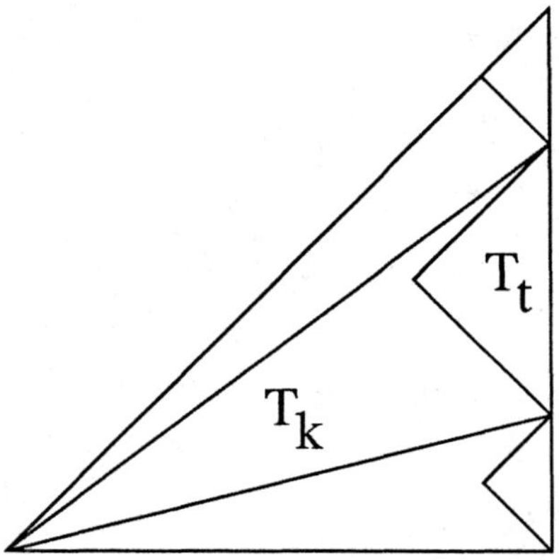

FIG. 2.1.

and $b_k = 0$ in the rest of T_k. Recalling (2.22) and $t = |I_k|/2$, we have therefore

$$\gamma_k^a = |b_k|_{1,T_k}^2 = 6|I_k|^{-2}, \qquad \int_{I_k} b_k \mathrm{d}\Gamma = 1, \qquad \gamma_k^T = \int_{T_k} b_k \mathrm{d}x\mathrm{d}y = |I_k|/8. \tag{2.24}$$

We see therefore, in particular, that the lower bound for γ_k^a will always be $\leq 6|I_k|^{-2}$. Substituting (2.24) in (2.19) yields:

$$\begin{aligned} &a(u_L, v_L) - < u_L, \frac{\partial v_L}{\partial n} > - < v_L, \frac{\partial u_L}{\partial n} > + \textstyle\sum_k 6\, u_L(m_k) v_L(m_k) = \\ &(f, v_L) - \textstyle\sum_k f_k \frac{1}{8} |I_k|^2 v_L(m_k) - < \psi_h, \frac{\partial v_L}{\partial n} > + \textstyle\sum_k 6\, \psi_h(m_k) v_L(m_k), \end{aligned} \tag{2.25}$$

whose implementation is reasonably easy.

We can also consider other possible choices of the bubble shape. In particular we could think of choosing a bubble function b_k such that its integral over T_k vanishes. This would make the second term in the right-hand side of (2.19) equal to zero. One possibility is to consider a new bubble still given by (2.23) in T_t, but which is now different from zero somewhere else in T_k. For instance, we can start by considering a square S_t with sidelength $2t$, and the function φ_t having value t at the center, vanishing at the boundary ∂S_t and linear in each of the four subtriangles obtained by a criss-cross splitting of S_t. An elementary computation shows that

$$|\varphi_t|_{1,S_t}^2 = 4t^2, \qquad \int_{S_t} \varphi_t \mathrm{d}x\mathrm{d}y = \frac{4t^3}{3}. \tag{2.26}$$

Now we choose b_k in S_t as a multiple of φ_t, in such a way that

$$\int_{S_t} b_k \mathrm{d}x\mathrm{d}y = -\frac{|I_k|}{8}. \tag{2.27}$$

This easily gives $b_k = -\dfrac{3|I_k|\varphi_t}{32t^3}$, and consequently, using (2.26)

$$|b_k|_{1,S_t}^2 = 9|I_k|^2/256t^4. \tag{2.28}$$

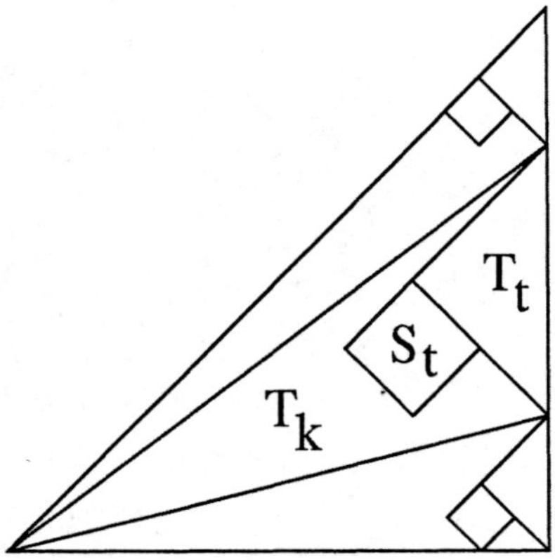

FIG. 2.2.

It is easy to check (see fig. 2.2) that we can always fit inside T_k a little square S_t that does not intersect T_t, provided that $t \le |I_k|/4\sqrt{2}$. Using the equality, and substituting the corresponding value of t in (2.28) we get

$$|b_k|^2_{1,S_t} = 36|I_k|^{-2}. \tag{2.29}$$

Finally, with this new choice of b_k, using (2.24) and (2.29), we have

$$\gamma_k^a = |b_k|^2_{1,T_k} = |b_k|^2_{1,T_t} + |b_k|^2_{1,S_t} = 42|I_k|^{-2}. \tag{2.30}$$

In this case the scheme (2.19) becomes

$$\begin{aligned} &a(u_L, v_L) - < u_L, \frac{\partial v_L}{\partial n} > - < v_L, \frac{\partial u_L}{\partial n} > + \textstyle\sum_k 42\, u_L(m_k) v_L(m_k) = \\ &(f, v_L) - < \psi_h, \frac{\partial v_L}{\partial n} > + \textstyle\sum_k 42\, \psi_h(m_k) v_L(m_k). \end{aligned} \tag{2.31}$$

We come back now, briefly, to the Schur complement operator S_h, as described in (2.9). It is easy to see that S_h is the discretization of a pseudo-differential operator S, of order 1, on Σ. More precisely, in this case S is the operator acting in the following way: given a function $\varphi \in H^{1/2}_{00}(\Sigma)$, we first construct the two harmonic functions u^1 and u^2 (in Ω^1 and Ω^2 respectively) having trace equal to φ on Σ and zero elsewhere on $\partial\Omega_i$ $(i = 1, 2)$. The operator S applied to φ is now given on Σ by

$$S\varphi = \frac{\partial u^1}{\partial n_1} + \frac{\partial u^2}{\partial n_2}. \tag{2.32}$$

In order to precondition (2.8) one has to find a cheaply computable operator that could be regarded as an approximation of S^{-1}. Several choices for that are found in the literature. For instance one could solve local Neumann problems in order to get an approximation of the local Steklov-Poincaré operators. Or one can define a sort of $H^{1/2}$ inner product on Φ_h, and invert the associated linear operator. We refer to the survey [10], and to the impressive set of proceedings of the various meetings on Domain Decomposition Methods, whose last volume is [3]. It is clear, however, that in doing that the task is made much easier if the grid on Σ is uniform, as it was the case in our experiments. For instance one could use fast solvers for the local Neumann problems, or find easy expressions for the $H^{1/2}$ inner product. In particular, we used an FFT algorithm for going from a function φ_h to his Fourier (sine) coefficients $\mathbf{c}^{\varphi}$. Then both the above choices become easy. For instance, the second one ($H^{1/2}$ inner product) corresponds to take, for each frequency n, the product of c_n^{φ} times n and then transform back. The first one (local Neumann problems) actually came out to be a minor variant of the first. If L_1 and L_2 are the widths of the two subdomains Ω^1 and Ω^2, respectively, the coefficient c_n^{φ} has to be multiplied by $n(\tanh(L_1)+\tanh(L_2))$ instead of n.

Both choices, as we shall see in the next section, gave very good results, in particular when the grids in the two subdomains Ω_i were not too coarse compared with the interface grid for ψ. This is very reasonable, as we used, somehow, something close to S to precondition S_h, which is clearly good only for h small enough.

3. Numerical experiments. The above formulation was tested on the academic problem

$$-\Delta u = f \quad \text{in } \Omega, \qquad u = g \quad \text{on } \partial\Omega, \tag{3.1}$$

with $\Omega = [-1,1]\times[0,1]$, $f=0$, $g=x^2-y^2$, so that the exact solution of (3.1) is $u(x,y) = x^2-y^2$. Notice that, with this choice, the use of piecewise linear finite elements on uniform grids for u should produce the exact solution at the nodes. Therefore, the only source of errors is the presence of the interface with non-matching grids. The domain Ω was split into two subdomains with interface at $x=0$, and in each subdomain a uniform decomposition into triangles was taken (see fig. 3.1).

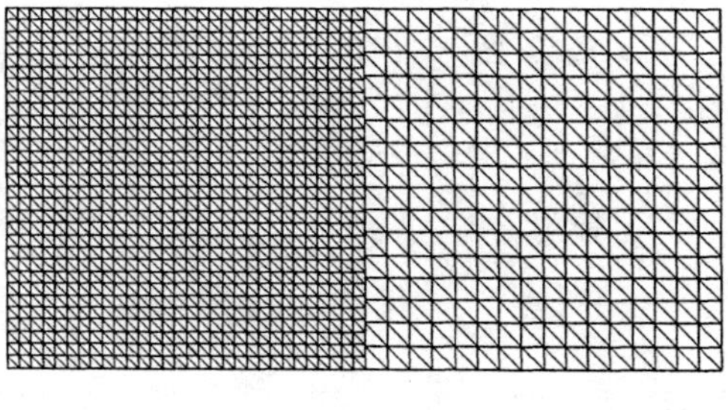

FIG. 3.1. *Example of mesh.*

For each coupling of non-matching grids for the u–variable, various meshes at the interface were taken for the ψ–variable, in order to see if and how the presence of very small intervals in T_λ^s can influence the performance of the method. Apparently, this fact does not in practice alter the convergence of the scheme. In [8] a crucial hypothesis for deriving error estimates for the ψ–variable was that each interval I_ψ must contain at least two regular intervals of T_λ^s. The experiments carried out so far showed that this is not necessary in practice.

In figs. 3.2-3.3 we report the behaviour of the error for u, λ, ψ respectively as T_ψ varies, while the meshes for u are kept fixed: $\mathcal{T}_u^1 = 30\times 30$, $\mathcal{T}_u^2 = 16\times 16$. From these results we see that the best values are reached when the number of nodes of T_ψ is in between those of $\mathcal{T}^1_{u\,|\Sigma}$ and $\mathcal{T}^2_{u\,|\Sigma}$ (i.e., in between 16 and 30 in our case). When T_ψ is too coarse we have a natural loss of accuracy. On the other hand, taking an unnecessary fine grid for ψ does not improve the accuracy (it would be unrealistic to expect that), but does not spoil it, and the results are pretty stable. This is particularly true for the most important unknowns, namely, u and ψ.

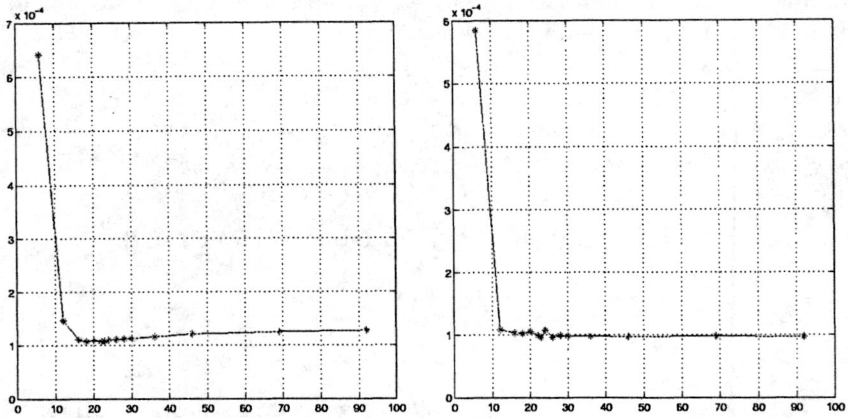

(a) (b)

FIG. 3.2. *(a) Error on u (left); (b) Error on u (right).*

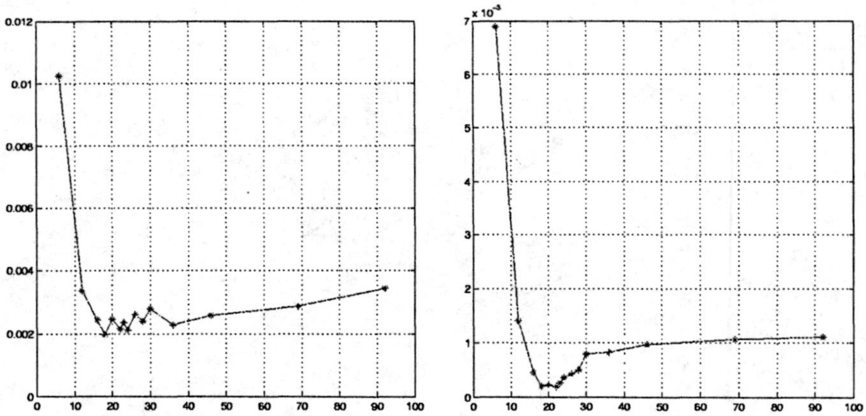

(a) (b)

FIG. 3.3. *(a) Error on λ (right); (b) Error on ψ.*

We also wanted to test the dependence of the quality of the scheme with respect to γ_k^a. For this, we chose different values of γ_k^a of the type $\gamma_k^a = \gamma|I_k|^{-2}$ as in (2.24) and (2.30), with γ varying from 0.75 up to 360. Fig. 3.4 shows the error on u (on the right) as a function of γ, for the same grids for u as before, and 22 internal nodes for ψ at the interface. We can clearly see that big changes in γ have very little impact on the results.

Another kind of test was meant to check the error estimates proved in [8]. For this,

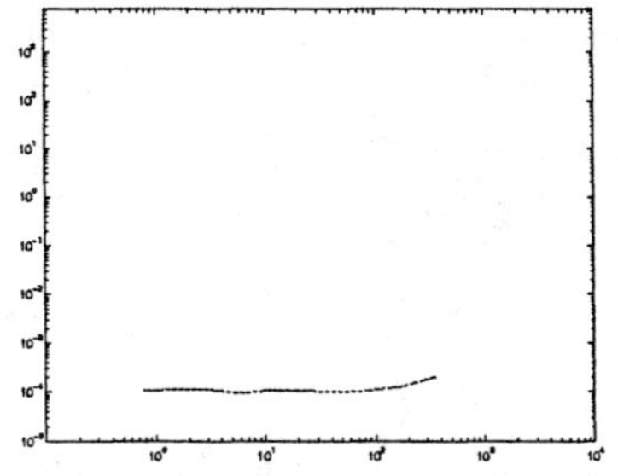

FIG. 3.4. *Dependence of u (right) on γ.*

various coupling of meshes in the two subdomains were taken:

$$\begin{aligned} \mathcal{T}_u^1 &= [10 \times 10];\ [20 \times 20];\ [30 \times 30];\ [40 \times 40];\ [50 \times 50], \\ \mathcal{T}_u^2 &= [8 \times 8];\ [16 \times 16];\ [24 \times 24];\ [32 \times 32];\ [40 \times 40]. \end{aligned} \tag{3.2}$$

According to the suggestions from the previous results, the corresponding T_ψ was

$$T_\psi = [9;\ 18\ ;\ 27;\ 36;\ 45]. \tag{3.3}$$

The test was also meant to assess the quality of the preconditioners proposed in the previous section. We solved (2.8) with preconditioned conjugate gradient, stopped when a reduction of 10^{-12} on the initial residual was achieved. Both preconditioners behaved similarly, and the number of iterations to achieve convergence was [5; 7; 8; 8; 8] for T_ψ as in (3.3). Figs. 3.5-3.6 show the corresponding errors on a log log scale. In each figure the line in the lower left corner represents the theoretical order of convergence. As it can be seen, the slope of the experimental results matches very well that predicted by the theory.

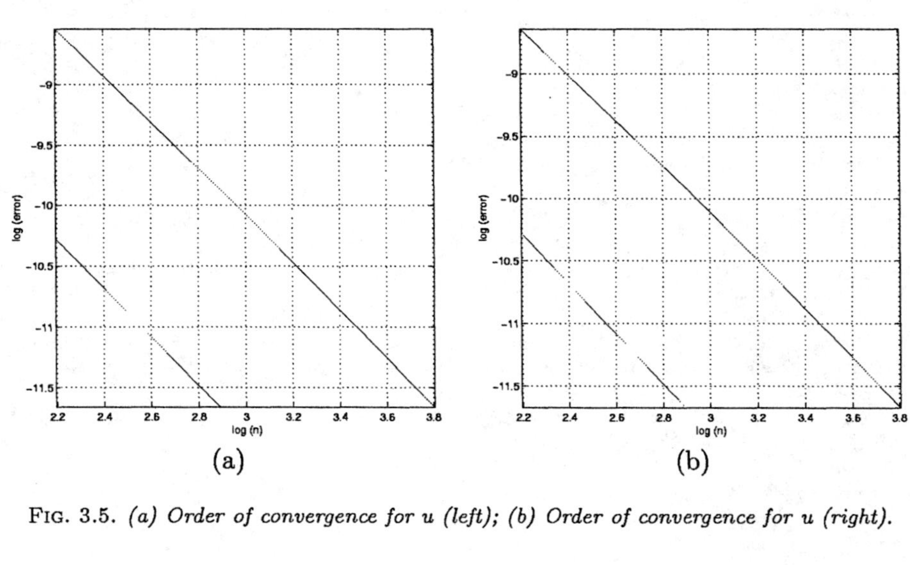

FIG. 3.5. *(a) Order of convergence for u (left); (b) Order of convergence for u (right).*

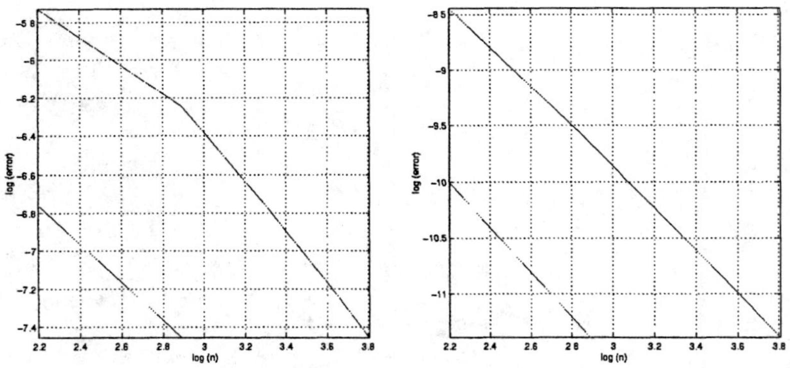

(a) (b)

FIG. 3.6. *(a) Order of convergence for λ (right); (b) Order of convergence for ψ.*

REFERENCES

[1] C. BAIOCCHI, F. BREZZI, AND L.P. FRANCA, *Virtual bubbles and Ga.L.S.*, Comp. Meth. in Appl. Mech. and Eng., 105 (1993), pp. 125–141.

[2] P.E. BJØRSTAD AND O.B. WIDLUND, *Iterative methods for the solution of elliptic problems on regions partitioned into substructures*, SIAM J. Numer. Anal., 23 (1986), pp. 1093–1120.

[3] P.E. BJØRSTAD, M.S. ESPEDAL AND D.E. KEYES (EDS.), *Domain Decomposition Methods in Science and Engineering*, Domain Decomposition Press, Bergen, 1998.

[4] J.H. BRAMBLE, J. PASCIAK AND A. SCHATZ, *The construction of preconditioners for elliptic problems by substructuring, IV*, Math. Comput., 53 (1989), pp. 1–24.

[5] F. BREZZI, L.P. FRANCA, L.D. MARINI, AND A. RUSSO, *Stabilization techniques for domain decomposition methods with non-matching grids.* In Domain Decomposition Methods in Science and Engineering, (9*th* International Conference, Bergen, Norway, 1996) P. Bjørstad et al., ed., Domain Decomposition Press, Bergen, 1998, pp. 1–11.

[6] F. BREZZI AND L. D. MARINI, *Macro hybrid elements and domain decomposition methods*, in Optimization et Controle, J. Desideri et al., ed., Toulouse, 1993, CÉPADUÈs-Editions, pp. 89–96.

[7] ———, *A three-field domain decomposition method*, in Domain Decomposition Methods in Science and Engineering, A. Quarteroni et al., ed., Providence, 1994, AMS, Series CONM 157, pp. 27–34.

[8] ———, *Error estimates for the three-field formulation with bubble stabilization.* Math. of Comp., to appear.

[9] A. BUFFA, *Error estimate for a stabilized domain decomposition method with nonmatching grid.* Numer. Math., submitted.

[10] T.F. Chan and T.P. Mathew , *Domain decomposition algorithms*, Acta Numerica 1994, pp. 61–144.

[11] M. DRYJA, *A finite element-capacitance method for elliptic problems on regions partitioned into subregions*, Numer. Math., 44 (1984), pp. 153–168.

[12] R. GLOWINSKI, W.K. KINTON AND M.F. WHEELER, *Acceleration of domain decomposition algorithms for mixed finite elements by multilevel methods* , in Third Int. Symp. on Domain decomposition methods for partial differential equations, T. Chan et al., ed., SIAM (Philadelphia), 1990.

[13] G. GOLUB AND D. MAYERS, *The use of preconditioning over irregular regions*, in Computing Methods in Applied Sciences and Engineering, VI, R. Glowinski and J.L. Lions, eds., North Holland, 1984, pp. 3–14.

[14] YU. A. KUZNETSOV AND M.F. WHEELER, *Optimal order substructuring preconditioners for mixed finite element methods on nonmatching grids*, East-West Journal of Numerical Mathematics, 3 (1995), pp. 127–144.

[15] B.F. SMITH AND O.B. WIDLUND, *A domain decomposition algorithm using a hierarchical basis*, SIAM J. Sci. Comput., 11 (1990), pp. 1212–1220.

[16] C.H. TONG, T.F. CHAN AND C.C.J. KUO, *A domain decomposition preconditioner based on a change to a multilevel nodal basis*, SIAM J. Sci. Comput., 12 (1991), pp. 1486–1495.

[17] J. XU, *Theory of multilevel methods*, PhD thesis, Cornell University, 1989.

THE CG METHODS - A BRIEF REVIEW

C. G. BROYDEN* AND M.T. VESPUCCI †

Abstract. In this paper a brief review of the CG methods is given. The direct methods for symmetric systems introduced by Morris in 1946 and by Fox, Huskey and Wilkinson in 1948 are first presented, which require an additional storage proportional to n^2. Versions of these methods with an additional storage proportional to n were then introduced in 1952 by Hestenes and Stiefel and by Lanczos. In section 2 the block form of the CG algorithm is discussed, which allows nonsymmetric systems to be solved. In the last section transpose-free variants are presented which only involve the matrix $\mathbf{A}$ (and not $\mathbf{A}^T$) in the computation.

1. Historical Introduction. The solution of a system of linear simultaneous algebraic equations, which may be expressed as

$$\mathbf{A}\mathbf{x} = \mathbf{b} \tag{1.1}$$

where $\mathbf{A} \in \mathbb{R}^{n \times n}$ and is nonsingular and $\mathbf{b} \in \mathbb{R}^n$, was one of the major problems of numerical analysis even before the advent of the electronic computer. Direct methods like Gaussian elimination and indirect ones like the various forms of relaxation method were well-known before 1945 (which we may take as being the date when electronic computers first began to make a serious impact on numerical mathematics), although the emphasis then was principally on the direct methods. This is because indirect methods come into their own with sparse matrices, sparsity only begins to be a factor when matrices become large, and prior to 1945 only small problems were normally tackled for practical reasons. With the coming of the computer came not only the desire and the ability to solve larger problems but also a spate of new algorithms with which to solve them, and the decade 1945-54 saw enormous advances in the field of numerical linear algebra in general and the solution of simultaneous algebraic equations in particular.

One of the first of the new methods was proposed in 1946 and was intended to solve equation (1.1) when $\mathbf{A}$ is symmetric. It was due to Morris [22] who called it the *escalator method*. It was rediscovered some two years later by Fox, Huskey and Wilkinson [10], who referred to it as the *method of orthogonal vectors*, and works as follows. Suppose that we want to solve

$$\mathbf{G}\mathbf{x} = \mathbf{h} \tag{1.2}$$

where $\mathbf{G} \in \mathbb{R}^{n \times n}$ is symmetric and nonsingular, and $\mathbf{h} \in \mathbb{R}^n$. Let $\mathbf{x}_i$ be an approximation to the solution and $\mathbf{p}_i$ an arbitrary vector. Then if $\mathbf{f}_i$ is defined by

$$\mathbf{f}_i = \mathbf{G}\mathbf{x}_i - \mathbf{h}, \tag{1.3}$$

i.e. $\mathbf{f}_i$ is the vector of residuals of equation (1.2) calculated at $\mathbf{x}_i$, and if $\mathbf{x}_{i+1}$ is given by

$$\mathbf{x}_{i+1} = \mathbf{x}_i - \mathbf{p}_i \frac{\mathbf{p}_i^T \mathbf{f}_i}{\mathbf{p}_i^T \mathbf{G} \mathbf{p}_i} \tag{1.4}$$

* Università di Bologna, Facoltà di Scienze MM.FF.NN., Via Sacchi 3, 47023 Cesena (FO), Italy
† Università di Bergamo, Facoltà di Ingegneria, Via G. Marconi 5, 24044 Dalmine (BG), Italy

where $\mathbf{p}_i$ is some arbitrary vector then, trivially,

$$\mathbf{f}_{i+1} = \mathbf{f}_i - \mathbf{G}\mathbf{p}_i \frac{\mathbf{p}_i^T \mathbf{f}_i}{\mathbf{p}_i^T \mathbf{G}\mathbf{p}_i} \tag{1.5}$$

so that

$$\mathbf{p}_i^T \mathbf{f}_{i+1} = 0. \tag{1.6}$$

Suppose now that $\mathbf{x}_i$ has been similarly calculated from $\mathbf{x}_{i-1}$ so that $\mathbf{p}_{i-1}^T \mathbf{f}_i = 0$. It is again straightforward to show that if $\mathbf{p}_{i-1}^T \mathbf{G}\mathbf{p}_i = 0$ then, in addition to equation (1.6) being satisfied, we also have $\mathbf{p}_{i-1}^T \mathbf{f}_{i+1} = 0$. This result may be generalised. If $\mathbf{x}_1$ is the arbitrary initial approximation to the solution of equation (1.2), $\mathbf{x}_{j+1}$, $j = 1, 2, \ldots, i$, is computed by equation (1.4) and if

$$\mathbf{p}_j^T \mathbf{G}\mathbf{p}_k = 0, \qquad 1 \le j < k \le i, \tag{1.7}$$

then

$$\mathbf{p}_j^T \mathbf{f}_{i+1} = 0, \qquad 1 \le j \le i. \tag{1.8}$$

Moreover, if

$$\mathbf{p}_j^T \mathbf{G}\mathbf{p}_j \ne 0 \qquad \text{for } 1 \le j \le i, \tag{1.9}$$

i.e. if the process does not break down for these values of j, the vectors $\mathbf{p}_1, \mathbf{p}_2, \ldots, \mathbf{p}_i$ are linearly independent (see e.g. [5]) and this implies that, as i increases, the vectors $\mathbf{f}_{i+1}$ are restricted to lie in subspaces of constantly reducing dimension. If i reaches n without breakdown (breakdown occurs when $\mathbf{p}_j^T \mathbf{G}\mathbf{p}_j = 0$, either because $\mathbf{p}_j = \mathbf{0}$ or $\mathbf{G}$ is indefinite) then $\mathbf{f}_{i+1}$ is orthogonal to n linearly independent $n-$vectors and is hence null. This method, therefore, terminates either with breakdown or with the exact solution of equation (1.2) and so may be regarded, as it was by Morris and by Fox *et al*, as a direct method. In this respect it is rather like solving equation (1.2) using triangular decomposition without interchanges, which similarly breaks down if a leading principal submatrix of $\mathbf{G}$ is singular.

So far so good. We have derived an algorithm that will, for most of the time, solve a symmetric system of equations providing we can construct a sequence of vectors $\{\mathbf{p}_i\}$ that satisfy equation (1.7). How can we do this?

Let us define the sequence of matrices $\{\mathbf{Q}_j\}$ by $\mathbf{Q}_0 = \mathbf{I}$ and, for $1 \le j \le i$,

$$\mathbf{Q}_j = \mathbf{I} - \sum_{k=1}^{j} \frac{\mathbf{G}\mathbf{p}_k \mathbf{p}_k^T}{\mathbf{p}_k^T \mathbf{G}\mathbf{p}_k}. \tag{1.10}$$

If the vectors $\mathbf{p}_k$, $1 \le k \le j$, are conjugate then it follows from equation (1.10) that

$$\mathbf{Q}_j \mathbf{G}\mathbf{p}_k = \mathbf{0}, \qquad 1 \le k \le j \tag{1.11}$$

so if $\mathbf{p}_{j+1}$ is defined by

$$\mathbf{p}_{j+1} = \mathbf{Q}_j^T \mathbf{w}_{j+1} \tag{1.12}$$

for *any* arbitrary vector $\mathbf{w}_{j+1}$ then, automatically $\mathbf{p}_{j+1}$ is conjugate to $\mathbf{p}_k$, $1 \le k \le j$. We thus have a device for computing our conjugate directions $\mathbf{p}_j$. Note that the matrices $\mathbf{Q}_j$ are idempotent so the calculation of $\mathbf{p}_{j+1}$ is a projection, and that this calculation is virtually the Gram-Schmidt method.

We may now define different algorithms in terms of the choice of the vectors $\mathbf{w}_j$. If $\mathbf{G} = \mathbf{A} = \mathbf{A}^T$ and these vectors are chosen to be the successive columns of the unit matrix then we have the escalator method [22] or, equivalently, the method of orthogonal vectors [10]. If $\mathbf{G} = \mathbf{A} \neq \mathbf{A}^T$,

$$\mathbf{Q}_{i+1} = \mathbf{Q}_i - \frac{\mathbf{Q}_i \mathbf{u}_i \mathbf{v}_i^T \mathbf{Q}_i}{\mathbf{v}_i^T \mathbf{Q}_i \mathbf{u}_i}$$

where $\mathbf{v}_i$ and $\mathbf{w}_i$ are arbitrary save that $\mathbf{v}_i^T \mathbf{Q}_i \mathbf{u}_i \neq 0$, and $\mathbf{u}_i$ is the *i-th* column of $\mathbf{A}^T$ then we have the original ABS method [1]. If $\mathbf{G} = \mathbf{A}^T\mathbf{A}$ and $\mathbf{h} = \mathbf{A}^T\mathbf{b}$ so that $\mathbf{f}_j = \mathbf{A}^T\mathbf{r}_j$, where

$$\mathbf{r}_j = \mathbf{A}\mathbf{x}_j - \mathbf{b}, \tag{1.13}$$

then the choice $\mathbf{w}_j = \mathbf{r}_j$ gives the GCR method [9] or Orthomin [28]. If, with the same choices of $\mathbf{G}$ and $\mathbf{h}$, $\mathbf{w}_j = \mathbf{A}\mathbf{p}_{j-1}$ then we obtain the method Orthodir [29]. Finally, if $\mathbf{G} = \mathbf{I}$ and $\mathbf{w}_j = \mathbf{A}\mathbf{p}_{j-1}$ where $\mathbf{A}$ is not symmetric, equations (1.10) and (1.12) may be used to construct a sequence of *orthogonal* vectors $\{\mathbf{p}_i\}$, but since they are orthogonal we cannot use equation (1.4) to compute $\mathbf{x}_{i+1}$. A more sophisticated method has to be devised, and it is this combination of ideas that forms the basis of GMRES [24].

All these methods are direct methods in the sense that, in the absence of both breakdown and rounding, they will eventually compute the exact solution of (1.2). Their disadvantage is that as i approaches n the amount of storage required to define $\mathbf{Q}_i$ approaches n^2 and the advantage of these methods over Gaussian elimination or triangular decomposition disappears. There is even the disadvantage that whereas breakdown in Gaussian elimination or triangular decomposition may be avoided by pivoting, the corresponding operation for the projection method as described here is more complicated. In fact it is likely that without two extraordinary contributions to the theory, one by Lanczos [21] in 1952 and the other by Hestenes and Stiefel [20], also in 1952, these methods (with the exception of GMRES) would be no better known today than the escalator method. In our notation, they showed that if the vectors $\mathbf{w}_j$ were chosen appropriately then the number of terms in the sum defining $\mathbf{Q}_i$ could be reduced to three, or even two. More precisely, if $\mathbf{w}_j = \mathbf{K}\mathbf{G}\mathbf{p}_{j-1}$ and $\mathbf{K}$ is symmetric, equation (1.12) becomes

$$\mathbf{p}_{i+1} = \left(\mathbf{I} - \frac{\mathbf{p}_{i-1}\mathbf{p}_{i-1}^T\mathbf{G}}{\mathbf{p}_{i-1}^T\mathbf{G}\mathbf{p}_{i-1}} - \frac{\mathbf{p}_i\mathbf{p}_i^T\mathbf{G}}{\mathbf{p}_i^T\mathbf{G}\mathbf{p}_i}\right)\mathbf{K}\mathbf{G}\mathbf{p}_i \tag{1.14}$$

and if $\mathbf{w}_j = \mathbf{K}\mathbf{f}_j$, again with $\mathbf{K}$ symmetric, equation (1.12) is reduced to the even simpler

$$\mathbf{p}_{i+1} = \left(\mathbf{I} - \frac{\mathbf{p}_i\mathbf{p}_i^T\mathbf{G}}{\mathbf{p}_i^T\mathbf{G}\mathbf{p}_i}\right)\mathbf{K}\mathbf{f}_{i+1}. \tag{1.15}$$

resulting in savings both in storage and computing time.

Further savings may be made by using equivalent alternative versions of equations (1.4), (1.5), (1.14) and (1.15). If δ_i and γ_i are defined by

$$\delta_i = \mathbf{p}_i^T\mathbf{G}\mathbf{p}_i \quad \text{and} \quad \gamma_i = \mathbf{f}_i^T\mathbf{K}\mathbf{f}_i, \tag{1.16}$$

premultiplying equation (1.15) by $\mathbf{f}_{i+1}^T$ gives, since $\mathbf{p}_i^T\mathbf{f}_{i+1} = 0$,

$$\mathbf{f}_{i+1}^T\mathbf{p}_{i+1} = \gamma_{i+1}. \tag{1.17}$$

Equations (1.4) and (1.5) may thus be written

$$\mathbf{x}_{i+1} = \mathbf{x}_i - \mathbf{p}_i \left(\frac{\gamma_i}{\delta_i} \right) \tag{1.18}$$

and

$$\mathbf{f}_{i+1} = \mathbf{f}_i - \mathbf{G}\mathbf{p}_i \left(\frac{\gamma_i}{\delta_i} \right). \tag{1.19}$$

A slightly more complicated analysis then shows that

$$\frac{\mathbf{p}_i^T \mathbf{G}\mathbf{K}\mathbf{f}_{i+1}}{\delta_i} = -\frac{\gamma_{i+1}}{\gamma_i}$$

so that equation (1.15) reduces to

$$\mathbf{p}_{i+1} = \mathbf{K}\mathbf{f}_{i+1} + \mathbf{p}_i \left(\frac{\gamma_{i+1}}{\gamma_i} \right). \tag{1.20}$$

Finally, premultiplying equation (1.14) by $\mathbf{p}_{i+1}^T \mathbf{G}$ and replacing i by $i+1$ gives

$$\mathbf{p}_{i-1}^T \mathbf{G}\mathbf{K}\mathbf{G}\mathbf{p}_i = \delta_i$$

so that equation (1.14) becomes

$$\mathbf{p}_{i+1} = \mathbf{K}\mathbf{G}\mathbf{p}_i - \mathbf{p}_i \left(\frac{\mathbf{p}_i^T \mathbf{G}\mathbf{K}\mathbf{G}\mathbf{p}_i}{\delta_i} \right) - \mathbf{p}_{i-1} \left(\frac{\delta_i}{\delta_{i-1}} \right). \tag{1.21}$$

The additional storage required for these two versions of the algorithm is no longer proportional to n^2 but only to n. If the matrix $\mathbf{G}$ is sparse, with at most m non-zero elements in any one row then the storage required for the matrix itself cannot exceed mn. This implies that much larger systems of equations may, in principle, be solved by these variations, but there is a penalty to be paid. Using the simplified form (1.14) means that the conjugacy between $\mathbf{p}_{i+1}$ and $\mathbf{p}_j$ is only implicit for $1 \leq j \leq i-2$ ($1 \leq j \leq i-1$ for equation (1.15)) and this opens the door to the accumulation of rounding error. The end result of this is that termination does not occur after n iterations, and convergence rather than termination becomes important. Thus with large n and in the presence of rounding error we are forced to regard these algorithms as being iterative rather than direct, this iterative nature being first noted by Reid [23].

Now we have already seen that if $\mathbf{G}$ is indefinite then the algorithm can crash when computing $\mathbf{x}_{i+1}$ due to a division by zero, and it can be shown [3] that if $\mathbf{K}$ is indefinite then the algorithm can stagnate. Thus convergence results may be obtained only if both $\mathbf{G}$ and $\mathbf{K}$ are definite, and it can be shown (see [5]) that in this case the rate of convergence is governed by the condition number of $\mathbf{G}^{\frac{1}{2}}\mathbf{K}\mathbf{G}^{\frac{1}{2}}$, the number of iterations required to effect a given reduction in the appropriate norm of $\mathbf{f}_i$ being roughly proportional to its square root.

At this point, apart from the memory-intensive algorithms Orthodir and GCR, we have not considered methods for solving $\mathbf{A}\mathbf{x} = \mathbf{b}$ where $\mathbf{A}$ is not symmetric. Neither have we considered block algorithms where the vectors $\mathbf{p}_i$ are replaced by matrices $\mathbf{P}_i$. Both these omissions will be rectified in the next section.

Other methods for solving equation (1.1) are based on "splitting". The idea behind this is the expression of $\mathbf{A}$ as $\mathbf{A} = \mathbf{B} + \mathbf{C}$, where $\mathbf{B}$ is arbitrary but is chosen to be readily invertible. Equation (1.1) may then be written

$$\mathbf{x} = \mathbf{B}^{-1}(\mathbf{b} - \mathbf{C}\mathbf{x}) \tag{1.22}$$

and this leads immediately to the iterative scheme

$$\mathbf{x}_{i+1} = \mathbf{B}^{-1}(\mathbf{b} - \mathbf{C}\mathbf{x}_i). \tag{1.23}$$

If the error $\mathbf{e}_i$ is defined by $\mathbf{e}_i = \mathbf{x}_i - \mathbf{x}^*$, where $\mathbf{x}^*$ is the solution of equation (1.1) then, since $\mathbf{x}^*$ also satisfies equation (1.22), it follows from this equation and the following one that

$$\mathbf{e}_{i+1} = -\mathbf{B}^{-1}\mathbf{C}\mathbf{e}_i \tag{1.24}$$

so that the iteration (1.23) converges if and only if $\rho\left(\mathbf{B}^{-1}\mathbf{C}\right) < 1$. Thus a clever (or lucky) choice of $\mathbf{B}$ results in a rapidly-converging algorithm. This method has been studied by, among others, Galligani [13, 12] and Galligani and Buffoni [14]. Other ideas for solving equation (1.1) have been explored by Galligani and Ruggiero [15] and Galligani and Trigiante [17, 16] while if $\mathbf{A}$ is sparse some theoretical aspects of its solution are considered in [11].

2. Block forms. We sometimes wish to solve the equation

$$\mathbf{G}\mathbf{X} = \mathbf{H} \tag{2.1}$$

where $\mathbf{X}$, $\mathbf{H} \in \mathbb{R}^{n\times s}$ for some small positive integer s. The motivation for this may be the desire to solve $\mathbf{G}\mathbf{x}_i = \mathbf{h}_i$ for several given right-hand sides $\mathbf{h}_i$, the need to solve $\mathbf{A}\mathbf{x} = \mathbf{b}$ where $\mathbf{A} \neq \mathbf{A}^T$ (see below) or the wish to exploit parallelism when solving $\mathbf{G}\mathbf{x} = \mathbf{h}$. The extra theory involved in constructing the block methods amounts to little more than replacing $\mathbf{p}_i$ by $\mathbf{P}_i \in \mathbb{R}^{n\times s}$ and γ_i and δ_i by $\mathbf{C}_i$ and $\mathbf{D}_i$, where

$$\mathbf{C}_i = \mathbf{F}_i^T\mathbf{K}\mathbf{F}_i, \qquad \mathbf{D}_i = \mathbf{P}_i^T\mathbf{G}\mathbf{P}_i \tag{2.2}$$

and

$$\mathbf{F}_i = \mathbf{G}\mathbf{X}_i - \mathbf{H}. \tag{2.3}$$

Both $\mathbf{C}_i$ and $\mathbf{D}_i$ are assumed to be nonsingular. Equations (1.4), (1.5) and (1.21) then become

$$\mathbf{X}_{i+1} = \mathbf{X}_i - \mathbf{P}_i\mathbf{D}_i^{-1}\mathbf{P}_i^T\mathbf{F}_i \tag{2.4}$$

$$\mathbf{F}_{i+1} = \mathbf{F}_i - \mathbf{G}\mathbf{P}_i\mathbf{D}_i^{-1}\mathbf{P}_i^T\mathbf{F}_i \tag{2.5}$$

and

$$\mathbf{P}_{i+1} = \mathbf{K}\mathbf{G}\mathbf{P}_i - \mathbf{P}_i\mathbf{D}_i^{-1}\mathbf{P}_i^T\mathbf{G}\mathbf{K}\mathbf{G}\mathbf{P}_i - \mathbf{P}_{i-1}\mathbf{D}_{i-1}^{-1}\mathbf{D}_i \tag{2.6}$$

while equations (1.18) - (1.20) generalise to

$$\mathbf{X}_{i+1} = \mathbf{X}_i - \mathbf{P}_i\mathbf{D}_i^{-1}\mathbf{C}_i \tag{2.7}$$

$$\mathbf{F}_{i+1} = \mathbf{F}_i - \mathbf{G}\mathbf{P}_i\mathbf{D}_i^{-1}\mathbf{C}_i \tag{2.8}$$

and

$$\mathbf{P}_{i+1} = \mathbf{K}\mathbf{F}_{i+1} + \mathbf{P}_i\mathbf{C}_i^{-1}\mathbf{C}_{i+1}. \tag{2.9}$$

One reason for wanting to solve equation (2.1) is to be able to obtain the solution of $\mathbf{Ax} = \mathbf{b}$ by solving

$$\begin{bmatrix} \mathbf{O} & \mathbf{A}^T \\ \mathbf{A} & \mathbf{O} \end{bmatrix} \begin{bmatrix} \mathbf{x} & \mathbf{0} \\ \mathbf{0} & \mathbf{z} \end{bmatrix} = \begin{bmatrix} \mathbf{0} & \mathbf{c} \\ \mathbf{b} & \mathbf{0} \end{bmatrix} \tag{2.10}$$

or

$$\begin{bmatrix} \mathbf{A}^T\mathbf{A} & \mathbf{O} \\ \mathbf{O} & \mathbf{A}\mathbf{A}^T \end{bmatrix} \begin{bmatrix} \mathbf{x} & \mathbf{0} \\ \mathbf{0} & \mathbf{z} \end{bmatrix} = \begin{bmatrix} \mathbf{A}^T\mathbf{b} & \mathbf{0} \\ \mathbf{0} & \mathbf{Ac} \end{bmatrix}. \tag{2.11}$$

These values of $\mathbf{G}$ are chosen to be not only themselves symmetric but also to enable a workable algorithm to be constructed with $\mathbf{K}$ symmetric, thereby reaping the benefits of short recurrences. Various values of $\mathbf{G}$ and $\mathbf{K}$ are given in the following table:

Name	G	H	K
CG	$\mathbf{A}$	$\mathbf{b}$	$\mathbf{I}$
CR	$\mathbf{A}^2$	$\mathbf{Ab}$	$\mathbf{A}^{-1}$
CGNR	$\mathbf{A}^T\mathbf{A}$	$\mathbf{A}^T\mathbf{b}$	$\mathbf{I}$
BCG	$\begin{bmatrix} \mathbf{O} & \mathbf{A}^T \\ \mathbf{A} & \mathbf{O} \end{bmatrix}$	$\begin{bmatrix} \mathbf{0} & \mathbf{c} \\ \mathbf{b} & \mathbf{0} \end{bmatrix}$	$\begin{bmatrix} \mathbf{O} & \mathbf{I} \\ \mathbf{I} & \mathbf{O} \end{bmatrix}$
HG	$\begin{bmatrix} \mathbf{O} & \mathbf{A}^T \\ \mathbf{A} & \mathbf{O} \end{bmatrix}$	$\begin{bmatrix} \mathbf{0} & \mathbf{c} \\ \mathbf{b} & \mathbf{0} \end{bmatrix}$	$\begin{bmatrix} \mathbf{I} & \mathbf{O} \\ \mathbf{O} & \mathbf{I} \end{bmatrix}$
BCR	$\begin{bmatrix} \mathbf{A}^T\mathbf{A} & \mathbf{O} \\ \mathbf{O} & \mathbf{A}\mathbf{A}^T \end{bmatrix}$	$\begin{bmatrix} \mathbf{Ab} & \mathbf{0} \\ \mathbf{0} & \mathbf{A}^T\mathbf{c} \end{bmatrix}$	$\begin{bmatrix} \mathbf{O} & \mathbf{A}^{-1} \\ \mathbf{A}^{-T} & \mathbf{O} \end{bmatrix}$

These relate to the un-preconditioned algorithms and will need to be modified to take preconditioning into account.

For the conjugate gradient and conjugate residual methods (CG and CR), $\mathbf{A}$ is symmetric but for the conjugate gradient normal residual, biconjugate gradient, Hegedüs Galerkin and biconjugate residual methods (CGNR, BCG, HG and BCR), $\mathbf{A}$ is only required to be nonsingular.

The actual performance of these methods depends critically on the product $\mathbf{KG}$ although only when both of these factors are positive definite is a simple theoretical analysis of convergence possible. For CG and CR, $\mathbf{KG} = \mathbf{A}$, for CGNR $\mathbf{KG} = \mathbf{A}^T\mathbf{A}$ while for BCG, HG and BCR

$$(\mathbf{KG})_{\text{BCG}} = \begin{bmatrix} \mathbf{A} & \mathbf{O} \\ \mathbf{O} & \mathbf{A}^T \end{bmatrix} \quad \text{and} \quad (\mathbf{KG})_{\text{HG, BCR}} = \begin{bmatrix} \mathbf{O} & \mathbf{A}^T \\ \mathbf{A} & \mathbf{O} \end{bmatrix}. \tag{2.12}$$

Thus for CG, CR, CGNR, HG and BCR, **KG** is symmetric but for BCG, since $\mathbf{A} \neq \mathbf{A}^T$, it is not. Intuitively one would expect a symmetric value of **KG** to give superior performance but this does not seem to be the case. For the majority of problems tested by the authors [5] the BCG algorithm outperforms the Hegedüs ones (HG and BCR) and for this no explanation is offered. Since the performance of the algorithms is so dependent on **KG** the choice of **K** is crucial. In the above table the quoted values of **K** give the unpreconditioned versions, but much work has been done on the problem of devising an easily-computed **K** that then gives rapid convergence. The optimal **K** is $\mathbf{G}^{-1}$ and many of the most successful methods of computing **K** are based on approximate LU decompositions of **G**. However none of the existing methods is so good that the search for better ones has been abandoned.

3. Transpose-free methods. If we wish to solve equation (1.1), where $\mathbf{A} \neq \mathbf{A}^T$, by using methods based on equations (2.10) or (2.11) then $\mathbf{P}_i$ has the form

$$\mathbf{P}_i = \begin{bmatrix} \mathbf{u}_i & \mathbf{0} \\ \mathbf{0} & \mathbf{v}_i \end{bmatrix}$$

and we obtain, on substituting the appropriate values of **G** and **K** in equations (2.7) - (2.9), individual (vector) equations for $\mathbf{u}_{i+1}$, $\mathbf{v}_{i+1}$, $\mathbf{r}_{i+1}$, $\mathbf{s}_{i+1}$, $\mathbf{x}_{i+1}$ and (if required) $\mathbf{z}_{i+1}$. For example, for BCG we get

$$\mathbf{u}_{i+1} = \mathbf{r}_{i+1} + \mathbf{u}_i \left(\frac{\mathbf{s}_{i+1}^T \mathbf{r}_{i+1}}{\mathbf{s}_i^T \mathbf{r}_i} \right) \tag{3.1}$$

$$\mathbf{r}_{i+1} = \mathbf{r}_i - \mathbf{A}\mathbf{u}_i \left(\frac{\mathbf{s}_i^T \mathbf{r}_i}{\mathbf{v}_i^T \mathbf{A}\mathbf{u}_i} \right) \tag{3.2}$$

$$\mathbf{x}_{i+1} = \mathbf{x}_i - \mathbf{u}_i \left(\frac{\mathbf{s}_i^T \mathbf{r}_i}{\mathbf{v}_i^T \mathbf{A}\mathbf{u}_i} \right) \tag{3.3}$$

with similar equations for $\mathbf{v}_{i+1}$ and $\mathbf{s}_{i+1}$. Examination of these equations shows that when computing the sequences $\{\mathbf{u}_i\}$, $\{\mathbf{r}_i\}$ and $\{\mathbf{x}_i\}$ the corresponding vectors $\{\mathbf{v}_i\}$ and $\{\mathbf{s}_i\}$ are used only in the calculation of the scalars $\mathbf{s}_i^T\mathbf{r}_i$ and $\mathbf{v}_i^T\mathbf{A}\mathbf{u}_i$ (note: this is *not* the case for the Hegedüs algorithms and is one of their weaknesses). Thus if another way of computing these scalars could be found we could avoid the calculation of $\mathbf{v}_i$, etc., and with it the tiresome calculation of the vectors $\mathbf{A}^T\mathbf{v}_i$, tiresome simply because it requires a multiplication by the transpose of a matrix (see [25]).

To see how this might be achieved, we return to equation (1.5), the block form of which is

$$\mathbf{F}_{i+1} = \left(\mathbf{I} - \mathbf{G}\mathbf{P}_i\mathbf{D}_i^{-1}\mathbf{P}_i^T\right)\mathbf{F}_i \tag{3.4}$$

so that from the second equation (2.2) the block form of equation (1.6) is trivially satisfied. Immediately we have

$$\mathbf{F}_{i+1} = \prod_{j=1}^{i} \left(\mathbf{I} - \mathbf{G}\mathbf{P}_j\mathbf{D}_j^{-1}\mathbf{P}_j^T\right)\mathbf{F}_1$$

or, from the block form of equation (1.10) and since $\mathbf{P}_j^T\mathbf{G}\mathbf{P}_k = \mathbf{O}$, $1 \le j < k \le i$,

$$\mathbf{F}_{i+1} = \mathbf{Q}_i\mathbf{F}_1. \tag{3.5}$$

Substituting the values appropriate to BCG in equation (2.3) gives, from equation (2.10),

$$\mathbf{F}_i = \begin{bmatrix} \mathbf{0} & \mathbf{s}_i \\ \mathbf{r}_i & \mathbf{0} \end{bmatrix} \tag{3.6}$$

where

$$\mathbf{r}_i = \mathbf{A}\mathbf{x}_i - \mathbf{b} \quad \text{and} \quad \mathbf{s}_i = \mathbf{A}^T\mathbf{z}_i - \mathbf{c}$$

and substitution of the appropriate value of $\mathbf{G}$ in the block form of equation (1.10) gives, after some manipulation,

$$\mathbf{Q}_i = \begin{bmatrix} \widehat{\mathbf{Q}}_i & \mathbf{O} \\ \mathbf{O} & \widetilde{\mathbf{Q}}_i \end{bmatrix} \tag{3.7}$$

where

$$\widetilde{\mathbf{Q}}_i = \mathbf{I} - \sum_{j=1}^{i} \frac{\mathbf{A}\mathbf{u}_j\mathbf{v}_j^T}{\mathbf{v}_j^T\mathbf{A}\mathbf{u}_j} \tag{3.8}$$

and

$$\widehat{\mathbf{Q}}_i = \mathbf{I} - \sum_{j=1}^{i} \frac{\mathbf{A}^T\mathbf{v}_j\mathbf{u}_j^T}{\mathbf{v}_j^T\mathbf{A}\mathbf{u}_j}. \tag{3.9}$$

Thus, from equations (3.5) - (3.7), $\mathbf{r}_{i+1} = \widetilde{\mathbf{Q}}_i\mathbf{r}_1$ and $\mathbf{s}_{i+1} = \widehat{\mathbf{Q}}_i\mathbf{s}_1$ so that $\mathbf{s}_{i+1}^T\mathbf{r}_{i+1} = \mathbf{s}_1^T\widehat{\mathbf{r}}_{i+1}$ where

$$\widehat{\mathbf{r}}_{i+1} = \widehat{\mathbf{Q}}_i^T\widetilde{\mathbf{Q}}_i\mathbf{r}_1. \tag{3.10}$$

Now the matrices $\widehat{\mathbf{Q}}_i^T$ and $\widetilde{\mathbf{Q}}_i$ only involve the matrix $\mathbf{A}$ (and not $\mathbf{A}^T$) and it was shown by Sonneveld [25] in a landmark paper that the sequence $\{\widehat{\mathbf{r}}_i\}$ could be computed recursively using only matrix-vector multiplications of the type $\mathbf{A}\mathbf{w}$ for some vector $\mathbf{w}$. The other sequences, $\{\widehat{\mathbf{u}}_i\}$ and $\{\widehat{\mathbf{x}}_i\}$ are computed at the same time, where $\widehat{\mathbf{r}}_i = \mathbf{A}\widehat{\mathbf{x}}_i - \mathbf{b}$. Moreover it is found that, in many cases, $\|\widehat{\mathbf{r}}_i\| \ll \|\mathbf{r}_i\|$ where $\mathbf{r}_i$ is the residual computed by BCG. Sonneveld's algorithm is called CGS (conjugate gradient squared) although it is, as was pointed out in [8], really the "square" of BCG. For further discussion see the original paper or, for greater detail, [5].

Subsequent to the publication of the CGS method, several variants have been proposed, in particular the Bi-CGSTAB method of Van der Vorst [26] in which a different sequence of vectors $\{\mathbf{s}_i\}$ is used in an attempt to suppress the natural *joie de vivre* of the original. However all these methods, including CGS, stand or fall with BCG and the basic BCG is probably the least stable conjugate gradient method after CGS.

4. Finale. In this brief account we have endeavoured to present some of the main ideas informing the CG methods. If $\mathbf{A}$ is symmetric and positive definite then equation (1.1) may be solved by the basic CG method with a positive definite preconditioner, and provided this preconditioner is well-chosen the algorithm is fast and stable. If $\mathbf{A}$ is indefinite or not symmetric it is not normally possible for both $\mathbf{G}$ and $\mathbf{K}$ to be positive definite and any attempt to make them so, as by applying CG to the normal equations or by using Craig's method [7] where $\mathbf{G} = \mathbf{I}$ and $\mathbf{K} = \mathbf{A}^T\mathbf{A}$, generally results in very slow convergence. In fact for some difficult problems, for these two methods, the presence of rounding error can destroy convergence completely and only a marginal reduction in the residual norm can be achieved.

For problems where $\mathbf{A}$ is not symmetric the methods currently in favour are GMRES, restarted every so many iterations to prevent "esaurimento" of memory, and BiCGSTAB. For both these methods $\mathbf{KG} = \mathbf{A}$, which seems in most cases to be a better bet than $\mathbf{KG} = \left(\mathbf{A}^T\mathbf{A}\right)^{\frac{1}{2}}$ which is essentially the value for the Hegedüs algorithms. This seems to imply that, since the spectral radius of $\mathbf{A}$ cannot exceed that of $\left(\mathbf{A}^T\mathbf{A}\right)^{\frac{1}{2}}$ and is sometimes much smaller, the convergence of the conjugate gradient methods depends more on the eigenvalues of $\mathbf{A}$ than its singular values. If this is correct the performance of the Hegedüs methods should approach that of GMRES and BiCGSTAB for $\mathbf{A}$ nearly symmetric, but the former methods are so little known that to date few comparisons are available. One paper that did tackle this problem was that of Broyden and Boschetti [4] and it is hoped to report further work in [5], but there is to date very little convergence theory when $\mathbf{A}$ is not symmetric (but see [2, 6, 18, 19, 27] for GMRES). A comprehensive theoretical analysis of BCG, CGS and the Hegedüs methods must be one of the goals for the many CG specialists in the short and medium term.

Acknowledgements. One author (CGB) thanks the Consiglio Nazionale di Ricerca, Italy (CNR), for partial financial support under Il Progetto Finalizzato Trasporti 2, Contract No. 93.01861.PF74.

REFERENCES

[1] J. Abaffy and E. Spedicato. *ABS Projection Algorithms.* Ellis Horwood, Chichester, 1989.

[2] M. Arioli, V. Ptak, and Z. Strakoš. *Krylov sequences of maximal length and convergence of GMRES.* BIT, 38:1–9, 1998.

[3] C. G. Broyden. *A breakdown of the block-CG method.* Optimization Methods and Software, 7:41–55, 1996.

[4] C. G. Broyden and M. A. Boschetti. *A comparison of three basic conjugate direction methods.* Numerical Linear Algebra with Applications, 3(6):473–489, 1996.

[5] C. G. Broyden and M. T. Vespucci. *Krylov Methods for Solving Linear Algebraic Equations.* In preparation.

[6] S. L. Campbell, I. C. F. Ipsen, C. T. Kelley, and C. D. Meyer. *GMRES and the minimal polynomial.* BIT, 36(4):664–675, 1996.

[7] E. J. Craig. *The n-step iteration procedure.* J. Math. Phys., 34:64–73, 1955.

[8] H. A. V. der Vorst and G. L. G. Sleipen. *Iterative Bi-CG type methods and implementational aspects.* In G. W. Althaus and E. Spedicato, editors, Algorithms for Large Scale Linear Algebraic Systems, pages 217–253. NATO Advanced Study Institute, Kluwer Academic Publishers, 1996.

[9] S. C. Eisenstat, H. C. Elman, and M. H. Schultz. *Variational iterative methods for nonsymmetric systems of linear equations.* SIAM J. Numer. Anal., 20:345–357, 1983.

[10] L. Fox, H. D. Huskey, and J. H. Wilkinson. *Notes on the solution of algebraic linear simultaneous equations.* Quart. J. Mech. Appl. Math., 1:149–173, 1948.

[11] I. Galligani. *Un teorema sulle matrici sparse stabili.* Pubbl. Ser III Ist. Appl. Calcolo, 140, 1978.

[12] I. Galligani. *A splitting method for large sparse linear systems on a multivector computer.* In Computer Algorithms for Solving Linear Algebraic Equations: The State of the Art, pages 207–215. NATO Advanced Study Institute, 1991.

[13] I. Galligani. *Splitting methods for the solution of systems of linear equations with singular matrices.* Rend. Mat. Appl., VII Ser. 14(2):341–353, 1994.

[14] I. Galligani and G. Buffoni. *Splitting methods for solving singular linear systems of a special class.* Rend. Mat. Appl., VII Ser. 16(3):375–386, 1996.

[15] I. Galligani and V. Ruggiero. *The arithmetic mean method for solving essentially positive systems on a vector computer.* Int. J. Comput. Math., 32(1/2):113–121, 1990.

[16] I. Galligani and D. Trigiante. *Numerical methods for solving large algebraic systems.* Pubbl. Ser III Ist. Appl. Calcolo, 98, 1974.

[17] I. Galligani and D. Trigiante. *Integration techniques for solving algebraic systems.* In Proceedings of 1975 Rome Conference on Math. Aspects Finite Elem. Meth., pages 113–121, 1977. Lecture Notes in Mathematics 606.

[18] A. Greenbaum, V. Ptak, and Z. Strakoš. *Any nonincreasing convergence curve is possible for GMRES.* SIAM J. Matrix Anal. Appl., 17(3):465–469, July 1996.

[19] A. Greenbaum and Z. Strakoš. *Matrices that generate the same residual subspaces.* In G. Golub, editor, Recent Advances in Iterative Methods, pages 95–118, New York, N.Y., 1994. IMA, Springer.

[20] M. R. Hestenes and E. Stiefel. *Methods of conjugate gradients for solving linear systems.* J. Res. Nat. Bureau of Standards, 49:409–436, 1952.

[21] C. Lanczos. *Solution of systems of linear equations by minimized iterations.* J. Res. Nat. Bureau of Standards, 49:33–53, 1952.

[22] J. Morris. *An escalator process for the solution of linear simultaneous equations.* Philos. Mag., VII Ser. 37:106–120, 1946.

[23] J. K. Reid. *On the method of conjugate gradients for the solution of large sparse linear equations.* In J. K. Reid, editor, Large Sparse Sets of Linear Equations, pages 231–254. Academic Press, New York, 1971.

[24] Y. Saad and M. H. Schultz. *GMRES: A generalized minimal residual algorithm for solving nonsymmetric linear systems.* SIAM J. Sci. Stat. Comput., 7(3):856–869, July 1986.

[25] P. Sonneveld. *CGS, a fast Lanczos-type solver for nonsymmetric linear systems.* SIAM J. Sci. Statist. Comput., 10:36–52, 1989.

[26] H. A. van der Vorst. *Bi-CGSTAB: A fast and smoothly-convergent variant of Bi-CG for the solution of nonsymmetric linear systems.* SIAM J. Sci. Stat. Comput., 13:631–644, 1992.

[27] H. A. van der Vorst and C. Vuik. *The superlinear convergence behaviour of GMRES.* J. Computational and Applied Maths., 48:327–341, 1993.

[28] P. K. W. Vinsome. *Orthomin, an iterative method for solving sparse sets of linear equations.* In Proc. Fourth Symposium on Reservoir Simulation, Society of Petroleum, Engineers of AIME, pages 149–159, 1976.

[29] D. M. Young and K. C. Jea. *Generalized conjugate gradient acceleration of nonsymmetrizable iterative methods.* Linear Algebra Applics., 34:159–194, 1980.

BLOCK IMPLICIT METHODS FOR ODES.*

LUIGI BRUGNANO† AND DONATO TRIGIANTE‡

Abstract. In this paper we study some important features of block implicit methods for ODEs. By using a unified framework, properties such as stability, blending of methods, A-convergence and parallelism are characterized. Major attention has been devoted to the implementation properties in view of their use in solving large-size problems.

Key words. Numerical Methods for ODEs, Stiff Problems, Implicit Methods, Iterative Solution of Linear Systems, Parallel Computing.

AMS subject classifications. 65L06, 65L05, 65L20, 65H10, 65F10.

1. Introduction. The numerical approximation of ODEs is still a very active field of research, as shown by the very rich amount of significant contributions in the last forty years. The demand for new methods well-suited for particular classes of applicative problems is relevant, as well as the need for the efficient implementation of the methods on modern computers, including parallel computers.

Many of the obtained results have been collected in several books, among which we quote [2, 7, 10, 12, 15, 17, 23]. Across the years, the required properties for the numerical methods have had an interesting evolution. Until the fifties, major attention was devoted to the order of accuracy. Later on, stability and convergence became more important. In the last years, properties of the methods such as the definition of efficient splittings, degree of parallelism, etc., more relevant for the implementation purposes, have become focal. This because such features are essential when solving large-size problems. In this paper we shall deal with such topics, namely the definition of methods that, in addition to classical requirements, do have favorable properties from the point of view of the implementation. In particular, we focus our attention on *r-block implicit methods* for approximating the solution of the problem

$$y' = f(t,y), \quad t \in (t_0, T], \qquad y(t_0) = y_0 \in \mathbb{R}^m, \tag{1}$$

over the discrete set of points $t_i = t_0 + ih$, $i = 1, 2, \ldots, N$, where $h = (T - t_0)/N$ is the stepsize. An r-block method is a method which provides the approximations at the first r mesh-points as the solution of the following discrete problem,

$$F(\mathbf{y}) \equiv A \otimes I_m \mathbf{y} - hB \otimes I_m \mathbf{f} + \mathbf{a} \otimes y_0 - h\mathbf{b} \otimes f_0 = \mathbf{0}, \tag{2}$$

where

$$\mathbf{y} = \begin{pmatrix} y_1 \\ \vdots \\ y_r \end{pmatrix}, \qquad \mathbf{f} = \begin{pmatrix} f_1 \\ \vdots \\ f_r \end{pmatrix},$$

*Work supported by GNIM-INDAM and MURST.

†Dipartimento di Matematica "U. Dini", Viale Morgagni 67/A, 50134 Firenze (Italy), E-mail: na.brugnano@na-net.ornl.gov

‡Dipartimento di Energetica "S. Stecco", Via C. Lombroso 6/17, 50134 Firenze (Italy), E-mail: na.dtrigiante@na-net.ornl.gov

y_i is the approximation to $y(t_i)$, and $f_i = f(t_i, y_i)$. Finally, A and B are $r \times r$ matrices, $\mathbf{a}$, $\mathbf{b}$ are vectors in $\mathbb{R}^r$, and in general, for any integer j, I_j will denote the identity matrix of size j. Moreover, in the following we shall also consider the augmented matrices

$$\hat{A} = [\mathbf{a}\ A], \qquad \hat{B} = [\mathbf{b}\ B]. \tag{3}$$

Methods in the form (2) are a generalization of those proposed originally in [3, 4, 27], where the normalization $A = I_r$ is considered. With such a normalization, they are also related to Runge-Kutta methods (see, for example, [5, 8, 9, 10, 16, 17]). We prefer, however, to consider the more general formulation (2) for two main reasons:

- in such formulation they are also related to block Boundary Value Methods (see [7]),
- it presents some advantages in discussing the implementation issues.

The latter point will be discussed in Sections 5 and 6. For such purpose, it will be hereafter assumed that both the two matrices A and B are nonsingular. Before that, in Sections 2 and 3 we collect some results, some of them known, concerning the construction of the methods. Additional considerations are reported in Section 4. Most results in the two sections dedicated to the implementation of the methods are new. Finally, in Section 7 we summarize the main facts and provide some final remark.

2. Construction of block implicit methods. An implicit r-block method is characterized by the fact that the rows of the augmented matrices (3) are defined through the characteristic polynomials of a suitable set of (distinct) r-step LMFs:

$$\hat{A} = \begin{pmatrix} \alpha_0^{(1)} & \dots & \alpha_r^{(1)} \\ \vdots & & \vdots \\ \alpha_0^{(r)} & \dots & \alpha_r^{(r)} \end{pmatrix}, \qquad \hat{B} = \begin{pmatrix} \beta_0^{(1)} & \dots & \beta_r^{(1)} \\ \vdots & & \vdots \\ \beta_0^{(r)} & \dots & \beta_r^{(r)} \end{pmatrix}. \tag{4}$$

On the coefficients of such matrices we shall impose order conditions. Let us consider at first the case where all such methods have (at least) order $p \geq 1$. The following result then holds true.

THEOREM 1. *Let the matrices (3) satisfy the following set of equations,*

$$\hat{A}\hat{\mathbf{q}}_i = i\hat{B}\hat{\mathbf{q}}_{i-1}, \qquad i = 0, \dots, p, \tag{5}$$

where

$$\hat{\mathbf{q}}_{-1} = \mathbf{0}, \qquad \hat{\mathbf{q}}_i = \begin{pmatrix} 0^i \\ 1^i \\ \vdots \\ r^i \end{pmatrix} \equiv \begin{pmatrix} 0^i \\ \mathbf{q}_i \end{pmatrix}, \quad i = 0, 1, \dots. \tag{6}$$

Then the LMFs defining the block method have a truncation error which is at least $O(h^{p+1})$.

Proof. The equations (5) are nothing but the usual order p conditions for LMFs, simultaneously imposed for all the r LMFs corresponding to (4). □

From the above result, when all methods in (4) are consistent ($p \geq 1$), one obtains relations between the first columns of the augmented matrices $\hat{A}$ and $\hat{B}$, and the square matrices A and B, respectively. In fact, from (5) we obtain, for $i = 0, 1$, the following conditions:

$$\mathbf{a} = -A\mathbf{e}, \qquad \mathbf{b} = A\mathbf{q}_1 - B\mathbf{e}, \tag{7}$$

where $\mathbf{e} \equiv \mathbf{q}_0$ denotes the vector with all unit entries (see (6)). Having spent the consistency conditions to relate the vectors $\mathbf{a}$ and $\mathbf{b}$ to the matrices A and B, we can use the subsequent order conditions to define the entries of such matrices. In particular, it is an easy matter to verify the following result.

COROLLARY 1. *Let the matrices defined in (3) satisfy (7) and the following set of equations,*

$$A\mathbf{q}_i = iB\mathbf{q}_{i-1}, \qquad i = 2, \ldots, p. \tag{8}$$

Then the LMFs defining the block method have a truncation error which is at least $O(h^{p+1})$.

Note that if (8) holds also true for $i = 1$, from (7) it follows that $\mathbf{b} = \mathbf{0}$. Such condition would simplify the derivation of the methods, moreover it implies the L-stability of A-stable methods. Nevertheless, we prefer not to impose it directly, in order to obtain a more general framework.

Let now define the following matrices:

$$D_j = \mathrm{diag}\begin{pmatrix} 1 & 2 & \ldots & j \end{pmatrix}, \quad Q_j = \begin{pmatrix} \mathbf{q}_1 & \ldots & \mathbf{q}_j \end{pmatrix}, \qquad j = 1, 2, \ldots.$$

As a consequence, the order conditions (8) can be rewritten as

$$AD_rQ_{p-1} = BQ_{p-1}\left(I_{p-1} + D_{p-1}\right). \tag{9}$$

The following result then holds true (see also [6]).

THEOREM 2. *If $p = r + 1$, the matrix $A^{-1}B$ is uniquely determined.*

Proof. In fact, when $p = r + 1$, the matrix Q_r in (9) is a nonsingular Vandermonde matrix. Consequently, one obtains that

$$A^{-1}B = D_rQ_r(I_r + D_r)^{-1}Q_r^{-1}, \tag{10}$$

whose right-hand side only depends on r. □

We observe that the matrix $A^{-1}B$ in (10) is similar to a Frobenius-type matrix. In fact, we have that

$$A^{-1}B = Q_r\left(Q_r^{-1}D_rQ_r(I_r + D_r)^{-1}\right)Q_r^{-1} \tag{11}$$

and

$$Q_r^{-1} D_r Q_r = \begin{pmatrix} & & & -\alpha_0 \\ 1 & & & -\alpha_1 \\ & \ddots & & \vdots \\ & & 1 & -\alpha_{r-1} \end{pmatrix}, \tag{12}$$

where

$$\sum_{i=0}^{r} \alpha_i x^i \equiv (x-1)(x-2)\cdots(x-r)$$

is the r-th order Wilkinson's polynomial. Consequently,

$$Q_r^{-1} D_r Q_r (I_r + D_r)^{-1} =$$

$$\begin{pmatrix} & & & -\frac{\alpha_0}{r+1} \\ \frac{1}{2} & & & -\frac{\alpha_1}{r+1} \\ & \ddots & & \vdots \\ & & \frac{1}{r} & -\frac{\alpha_{r-1}}{r+1} \end{pmatrix} = G^{-1} \begin{pmatrix} & & & -\frac{1!\alpha_0}{(r+1)!} \\ 1 & & & -\frac{2!\alpha_1}{(r+1)!} \\ & \ddots & & \vdots \\ & & 1 & -\frac{r!\alpha_{r-1}}{(r+1)!} \end{pmatrix} G, \tag{13}$$

where

$$G = \text{diag}(\; 1! \quad 2! \quad \ldots \quad r! \;), \tag{14}$$

From (2) and (7), it follows that methods having the same matrix $A^{-1}B$ provide the same discrete solution. For this reason, we shall call as *equivalent* methods sharing the same matrix

$$C = A^{-1}B. \tag{15}$$

In what follows we shall exploit the possibility of writing a specific method by using different equivalent forms. Moreover, it is important to point out that equivalent methods do have the same order and stability properties. For an r-block method, it is customary to define such properties by taking into account only the last entry of the solution vector $\mathbf{y}$, because it represents the starting point for subsequent applications of the same method. In particular, concerning the stability, if we consider the usual test equation

$$y' = \lambda y, \qquad y(t_0) = \eta \neq 0, \tag{16}$$

the application of the block method (2) generates the discrete problem

$$(A - qB)\mathbf{y} = (q\mathbf{b} - \mathbf{a})\eta, \qquad q = h\lambda. \tag{17}$$

From such equation one obtains that the usual property of A-stability is equivalent to require, for all $q \in \mathbb{C}^-$,

$$|y_r(q)| \equiv \left|\mathbf{e}_r^T(A - qB)^{-1}(q\mathbf{b} - \mathbf{a})\eta\right| < |\eta|,$$

where $\mathbf{e}_r$ is the last unit vector in $\mathbb{R}^r$, i.e.

(18) $$Re(q) < 0 \quad \Rightarrow \quad g(q) \equiv \left|\mathbf{e}_r^T(A - qB)^{-1}(q\mathbf{b} - \mathbf{a})\right| < 1.$$

A necessary requirement for this purpose, is to have problem (17) well-posed for all such q. For this reason, we give the following definition:

DEFINITION 1. A block method is said to be *pre-stable* if the spectrum of the corresponding matrix pencil $A - \mu B$ is contained in $\mathbb{C}^+$.

This fact implies that the result of Theorem 2 is useful only to define pre-stable methods up to $r = 8$; as matter of fact, by direct inspection one verifies that the matrix on the right-hand side of (10) has eigenvalues with negative real part, when $r \geq 9$. Consequently, the corresponding method cannot be pre-stable: in fact, the spectrum of the pencil $(A - \mu B)$ coincides with that of C^{-1} (see (15)), since both the two matrices A and B are assumed to be nonsingular.

In order to obtain alternative criteria for choosing the matrix C, we shall relax the order conditions for the LMFs on each row of the block method. In particular, it will be convenient to impose only an order r condition: i.e. (see (9))

(19) $$AD_rQ_{r-1} = BQ_{r-1}\left(I_{r-1} + D_{r-1}\right).$$

It remains one more condition to be imposed. It will be used to fix the spectrum of the matrix C. Such a condition will be written as

(20) $$AQ_r\hat{\mathbf{d}} = BQ_r\mathbf{e}_r,$$

where $\hat{\mathbf{d}}$ is a suitably chosen vector. The two conditions (19)-(20) can be also written as (see (15))

$$Q_r \begin{pmatrix} \mathbf{0}^T \\ (I_{r-1} + D_{r-1})^{-1} \end{pmatrix} = CQ_r \begin{pmatrix} I_{r-1} \\ \mathbf{0}^T \end{pmatrix},$$

and

$$Q_r\hat{\mathbf{d}} = CQ_r\mathbf{e}_r,$$

respectively. Altogether, we then obtain:

$$Q_r\left(\begin{pmatrix} \mathbf{0}^T \\ (I_{r-1} + D_{r-1})^{-1} \end{pmatrix} \middle| \, \hat{\mathbf{d}} \right) = CQ_r\left(\begin{pmatrix} I_{r-1} \\ \mathbf{0}^T \end{pmatrix} \middle| \, \mathbf{e}_r \right) = CQ_r.$$

Consequently, considering that Q_r is a nonsingular Vandermonde matrix, we have that (see (14))

(21)
$$C = Q_r \left(\left(\begin{array}{c} \mathbf{0}^T \\ (I_{r-1} + D_{r-1})^{-1} \end{array} \right) \Big| \, \hat{\mathbf{d}} \right) Q_r^{-1} = Q_r G^{-1} F G Q_r^{-1}$$

where

(22)
$$F = \left(\left(\begin{array}{c} \mathbf{0}^T \\ I_{r-1} \end{array} \right) \Big| -\mathbf{d} \right), \qquad \mathbf{d} \equiv \begin{pmatrix} d_0 \\ \vdots \\ d_{r-1} \end{pmatrix} = -\frac{1}{r!} G \hat{\mathbf{d}}.$$

One then concludes that the characteristic polynomial of C is given by

(23)
$$d(z) = \sum_{i=0}^{r} d_i z^i, \qquad d_r = 1.$$

Moreover, from (22) it follows that the vector $\hat{\mathbf{d}}$ in (20) and (21) is given by

$$\hat{\mathbf{d}} = -r! G^{-1} \mathbf{d} \equiv -r! \begin{pmatrix} \frac{d_0}{1!} \\ \vdots \\ \frac{d_{r-1}}{r!} \end{pmatrix}.$$

By the way, we observe that the choice (see (10)–(13))

$$\hat{\mathbf{d}} = \frac{-1}{r+1} \begin{pmatrix} \alpha_0 \\ \vdots \\ \alpha_{r-1} \end{pmatrix}$$

corresponds to the method obtained by imposing the order $r+1$ conditions on each row.

Let us now study the problem of appropriately choosing the characteristic polynomial $d(z)$ in (23). We surely will choose it in order to have all the roots contained in $\mathbb{C}^+$, so that the method is pre-stable. This is not enough, however, to define a "good" method. An additional requirement may be obtained by considering that, in order the Absolute stability region of the method (see (18)),

$$\mathcal{D} = \{ q \in \mathbb{C} \; : \; g(q) < 1 \},$$

to be unbounded, it would be preferable to have the point at ∞ inside $\mathcal{D}$. From (18), such a requirement is easily proved to be equivalent to

(24)
$$g(\infty) \equiv \left| \mathbf{e}_r^T B^{-1} \mathbf{b} \right| < 1.$$

The next result relates $g(\infty)$ to the characteristic polynomial (23).

THEOREM 3.

(25)
$$g(\infty) = \left| \frac{1}{d_0} \sum_{i=0}^{r} \frac{r^i}{i!} d_i \right| .$$

Proof. From (7), (15), (14), (21) and (22), we obtain that

$$\begin{aligned} B^{-1}\mathbf{b} &= B^{-1}AA^{-1}\mathbf{b} = C^{-1}(\mathbf{q}_1 - C\mathbf{e}) = C^{-1}D_r\mathbf{e} - \mathbf{e} = Q_rG^{-1}F^{-1}GQ_r^{-1}D_r\mathbf{e} - \mathbf{e} \\ &= Q_rG^{-1}F^{-1}G \begin{pmatrix} \mathbf{q}_0 & \dots & \mathbf{q}_{r-1} \end{pmatrix}^{-1} \mathbf{e} - \mathbf{e} = Q_rG^{-1}F^{-1}G\mathbf{e}_1 - \mathbf{e} \\ &= Q_rG^{-1}F^{-1}\mathbf{e}_1 - \mathbf{e} = -Q_rG^{-1} \begin{pmatrix} \frac{d_1}{d_0} \\ \vdots \\ \frac{d_r}{d_0} \end{pmatrix} - \mathbf{e} = -Q_r \begin{pmatrix} \frac{d_1}{1!d_0} \\ \vdots \\ \frac{d_r}{r!d_0} \end{pmatrix} - \mathbf{e}. \end{aligned}$$

Consequently, one obtains

$$\begin{aligned} g(\infty) &= \left|\mathbf{e}_r^T B^{-1}\mathbf{b}\right| = \left| \begin{pmatrix} r^1 & \dots & r^r \end{pmatrix} \begin{pmatrix} \frac{d_1}{1!d_0} \\ \vdots \\ \frac{d_r}{r!d_0} \end{pmatrix} + 1 \right| = \left| 1 + \frac{1}{d_0} \sum_{i=1}^{r} \frac{r^i}{i!} d_i \right| \\ &= \left| \frac{1}{d_0} \sum_{i=0}^{r} \frac{r^i}{i!} d_i \right| . \end{aligned}$$

□

The above result then provides an additional condition for the characteristic polynomial (23) of the matrix C, namely (see (24)-(25))

(26)
$$\left| \frac{1}{d_0} \sum_{i=0}^{r} \frac{r^i}{i!} d_i \right| < 1.$$

A further requirement for choosing the polynomial $d(z)$ follows by observing that, from (7) and (15)–(17), one obtains

(27)
$$y_r(q) = \frac{\det(M(q))}{\det(I_r - qC)} \eta \approx e^{rq}\eta,$$

where $M(q)$ is the matrix obtained from the matrix $I_r - qC$, whose last column has been substituted by $\mathbf{e} + q(\mathbf{q}_1 - C\mathbf{e})$. From the above equation, one obtains

(28)
$$e^{rz} \approx \frac{\det(M(z))}{\det(I_r - zC)} = \frac{\varphi(z)}{z^r d(z^{-1})} \equiv \frac{\varphi(z)}{\mu(z)},$$

where $\varphi(z) = \det(-M(z))$ is a polynomial of maximum degree r, and

(29) $$\mu(z) = \sum_{i=0}^{r} d_i z^{r-i}, \qquad d_r = 1,$$

is a polynomial of exact degree r, since we assume $(I_r - qC)$ to be nonsingular. The above arguments can then be summarized by saying that the characteristic polynomial of the matrix C (see (23)) coincides with the inverse of the polynomial at the denominator of a rational approximation to the exponential. By considering that, from (19), the above approximation to e^{rz} is at least $O(z^{r+1})$, the following result easily follows.

THEOREM 4. *Let the denominator $\mu(z)$ of the rational approximation (28) be given (i.e., the characteristic polynomial (23) is given). Then, the polynomial $\varphi(z)$ is uniquely determined.*

We observe that, from (27)-(29), one obtains that (24) is always satisfied, when $\deg(\varphi) < \deg(\mu) \equiv r$: in such a case, in fact, one obtains $g(\infty) = 0$ and A-stability implies L-stability. A complementary result is the following (see also [27]).

THEOREM 5. *If $\mu(z) \equiv \varphi(-z)$ and all the roots of $\mu(z)$ have positive real part, then the corresponding block method is perfectly A-stable and $g(\infty) = 1$.*

In other words, the above result states that, under the above condition of symmetry for $\mu(z)$ and $\varphi(z)$, pre-stability and A-stability are equivalent. Moreover, this result is complementary to the condition (26).

All the above facts then imply that a partial list of good criteria for choosing the characteristic polynomial $d(z)$ in (23) are the following:

- it must have all the roots contained in $\mathbb{C}^+$ (pre-stability);
- $z^r d(z^{-1})$ must be the denominator of a (at least) $O(z^{r+1})$ rational approximation to e^{rz};
- the polynomial should satisfy (26) (point at ∞ inside the absolute stability region $\mathcal{D}$). Alternatively, the conditions of Theorem 5 should be satisfied.

In view of the above criteria, and in particular of the second one, in the next section we shall review some rational approximations to the exponential. Before that, we conclude this section by considering the problem of determining the order of a block method. For this purpose, let us denote by

$$\hat{\mathbf{y}} = \begin{pmatrix} y(t_1) \\ \vdots \\ y(t_r) \end{pmatrix}, \qquad \hat{\mathbf{f}} = \begin{pmatrix} f(t_1, y(t_1)) \\ \vdots \\ f(t_r, y(t_r)) \end{pmatrix},$$

where $y(t)$ is the solution of problem (1). From (2) one then obtains that

$$A \otimes I_m \hat{\mathbf{y}} - hB \otimes I_m \hat{\mathbf{f}} + \mathbf{a} \otimes y_0 - h\mathbf{b} \otimes f_0 = \boldsymbol{\tau},$$

where $\boldsymbol{\tau}$ is the vector with the truncation errors of the method. By assuming that $y(t)$ is suitably smooth, the entries of the latter vector are given by

$$\tau_i = \sum_{j>r} \frac{y^{(j)}(t_n)}{j!} h^j \left(\sum_{k=0}^{r} k^{j-1} (k\alpha_k^{(i)} - j\beta_k^{(i)}) \right)$$

(30) $$\equiv \sum_{j>r} y^{(j)}(t_n)h^j v_{ji}, \qquad i = 1, \ldots, r,$$

because of the conditions (19). Consequently, we obtain that

$$A \otimes I_m(\hat{\mathbf{y}} - \mathbf{y}) - hB \otimes I_m(\hat{\mathbf{f}} - \mathbf{f}) = \boldsymbol{\tau}.$$

By introducing the vector $\hat{\mathbf{e}} = \hat{\mathbf{y}} - \mathbf{y}$, one then concludes that it satisfies the equation

(31) $$(A \otimes I_m - hB \otimes I_m \hat{\mathbf{J}})\, \hat{\mathbf{e}} = \boldsymbol{\tau},$$

where

$$\hat{\mathbf{J}} = \begin{pmatrix} \hat{J}_1 & & \\ & \ddots & \\ & & \hat{J}_r \end{pmatrix},$$

(32)

$$\hat{J}_i = \int_0^1 J(t_i, sy(t_i) + (1-s)y_i)ds \equiv J_0 + O(h),$$

$J(t,y) = \frac{\partial}{\partial y} f(t,y)$ and $J_0 = J(t_0, y_0)$. The order of the block method is then defined according to the next definition.

DEFINITION 2. The block method corresponding to (31) has order p provided that $\hat{e}_r = O(h^{p+1})$, where $\hat{e}_r$ is the last entry of $\hat{\mathbf{e}}$.

It is obvious that, from (30) and (31), we have that the order of the method is $p \geq r$. In general, the relations between the order conditions (30) and the global order of the method may be very entangled, as the Butcher theory for Runge-Kutta methods (see, for example, [12]) shows. Nevertheless, in case we look for values of p only slightly greater than r, the following result may be useful.

THEOREM 6. *Consider the following possible cases for the method corresponding to (30)-(31)*

0) $\mathbf{e}_r^T A^{-1}\mathbf{v}_{r+1} \neq 0$;

1) $\mathbf{e}_r^T A^{-1}\mathbf{v}_{r+1} = 0,$ *and* $\mathbf{e}_r^T A^{-1}\mathbf{v}_{r+2} \neq 0$ *or* $\mathbf{e}_r^T CA^{-1}\mathbf{v}_{r+1} \neq 0$;

2) $\mathbf{e}_r^T A^{-1}\mathbf{v}_{r+1} = 0, \quad \mathbf{e}_r^T A^{-1}\mathbf{v}_{r+2} = 0,$ *and* $\mathbf{e}_r^T CA^{-1}\mathbf{v}_{r+1} = 0,$

where $\mathbf{v}_j = (\, v_{j1} \ \ldots \ v_{jr} \,)^T$. *Then the global order of the method is exactly* $p = r + i$ *in cases* $i = 0, 1,$ *and* $p \geq r + 2$ *in case 2.*

Proof. From (15), (30) and (31)-(32), by posing

$$\hat{\mathbf{y}}^{(j)} = (\, y^{(j)}(t_1) \ \ldots \ y^{(j)}(t_r) \,)^T,$$

one obtains

$$\begin{aligned}\hat{\mathbf{e}} &= (I_r \otimes I_m - hC \otimes I_m \hat{\mathbf{J}})^{-1} A^{-1} \otimes I_m\, \boldsymbol{\tau} \\ &= h^{r+1}(A^{-1}\mathbf{v}_{r+1}) \otimes I_m\, \hat{\mathbf{y}}^{(r+1)} +\end{aligned}$$

$$
\begin{aligned}
&h^{r+2}\left((A^{-1}\mathbf{v}_{r+2})\otimes I_m\,\hat{\mathbf{y}}^{(r+2)} + C\otimes I_m\,\hat{\mathbf{J}}\,(A^{-1}\mathbf{v}_{r+1})\otimes I_m\,\hat{\mathbf{y}}^{(r+1)}\right) + O(h^{r+3})\\
&= h^{r+1}(A^{-1}\mathbf{v}_{r+1})\otimes I_m\,\hat{\mathbf{y}}^{(r+1)} +\\
&h^{r+2}\left((A^{-1}\mathbf{v}_{r+2})\otimes I_m\,\hat{\mathbf{y}}^{(r+2)} + (CA^{-1}\mathbf{v}_{r+1})\otimes J_0\,\hat{\mathbf{y}}^{(r+1)}\right) + O(h^{r+3}),
\end{aligned}
$$

from which, in view of Definition 2, the thesis easily follows. □

3. Approximating the exponential. In this section we review some classical rational approximation to the exponential. In particular, we look for polynomials described by (29), i.e. for characteristic polynomials (23), suitable to define "good" block methods, according to the criteria previously introduced.

3.1. Padé approximation. One of the most classical rational approximation to the exponential is the Padé (ν, r), where (see (28))

$$
\varphi(z) \equiv \varphi_{\nu,r}(rz), \qquad \mu(z) \equiv \mu_{\nu,r}(rz) \tag{33}
$$

are polynomials of degree ν and r, respectively, such that

$$
\varphi_{\nu,r}(z) = \mu_{\nu,r}(z)e^z + O(z^{\nu+r+1}).
$$

The expression of the two polynomial is well-known, and is given by

$$
\begin{aligned}
\varphi_{\nu,r}(z) &= \sum_{i=0}^{\nu} \frac{(\nu+r-i)!\nu!}{(\nu+r)!i!(\nu-i)!} z^i,\\
\mu_{\nu,r}(z) &= \sum_{i=0}^{r} (-1)^i \frac{(\nu+r-i)!r!}{(\nu+r)!i!(r-i)!} z^i.
\end{aligned} \tag{34}
$$

Such choice has been already considered in [27]. The following properties hold true for the polynomials $\varphi_{\nu,r}$ and $\mu_{\nu,r}$ (see [24] and the references therein).

THEOREM 7. *For all $\nu, r \geq 0$:*

1. $\mu_{\nu,r}(z) \equiv \varphi_{r,\nu}(-z)$;
2. *if $r \geq 1$, all the zeros of the polynomial $\mu_{\nu,r}(z)$ lie in the annulus*

$$
(r+\nu)\xi < |z| < r+\nu+4/3,
$$

where $\xi \approx 0.278465$ is the unique positive root of $x\,e^{x+1} = 1$.

Moreover, it is also known (*Ehle conjecture*, see also [26]) that, for $\nu \in \{r-2, r-1, r\}$, one obtains A-stable methods. Such methods are also L-stable when $\nu < r$, conversely, when $\nu = r$ one obtains $g(\infty) = 1$, since the result of Theorem 5 applies.

Concerning the order of convergence of the corresponding block methods, by applying the criteria in Theorem 6, it can be verified that it is given by

$$
(35) \qquad \begin{array}{ll} p = r+1, & \text{for } r \text{ odd}, \\ p \geq r+2, & \text{for } r \text{ even}, \end{array} \qquad r \geq 3, \qquad \nu = r-2, r-1, r.
$$

We shall consider again such a choice in the next sections.

3.2. Bernoulli numbers. A way of defining Bernoulli numbers is through the following formal power series:

$$
P(x) = \sum_{i=0}^{\infty} \frac{x^i}{(i+1)!},
$$

whose sum is $x^{-1}(e^x - 1)$. Its reciprocal, in fact, is given by

$$
P^{-1}(x) = \sum_{i=0}^{\infty} \frac{B_i}{i!} x^i,
$$

where $\{B_i\}$ is the sequence of Bernoulli numbers. Then, by means of simple calculations, one obtains that

$$
(36) \qquad e^{rq} = \frac{P^{-1}(-rq)}{P^{-1}(rq)} = \frac{\sum_{i=0}^{\infty} \frac{B_i}{i!}(-rq)^i}{\sum_{i=0}^{\infty} \frac{B_i}{i!}(rq)^i}.
$$

Consequently (see (28)-(29)), we choose the polynomial (23) as

$$
d(z) = \sum_{i=0}^{r} \frac{B_i}{i!}(rz)^{r-i},
$$

obtaining corresponding block methods. By considering that $B_{2i+1} = 0$, for $i \geq 1$, we can take into account only methods with blocks of even size. In such a case, by using the result of Theorem 6, it can be verified that the order of the corresponding methods is $p \geq r+2$, for all even values of r.

However, there is a severe drawback concerning such methods. In fact, for $r \geq 4$, the characteristic polynomials have roots with negative real parts, so that the corresponding block methods are not pre-stable. For this reason, it is not appropriate to use such an approximation if high order methods are needed, and we shall not consider it further.

3.3. RITZ fractions. In this case, the rational approximation to the exponential is obtained by using a truncated continuous fraction. To begin with, let us look for an approximation defined through a continuous fraction [18]

$$e^z = \cfrac{a_0}{b_0 + \cfrac{za_1}{b_1 + \cfrac{za_2}{b_2 + \dots}}}, \tag{37}$$

where the sequences $\{a_i\}$, and $\{b_i\}$ are suitably defined. In more detail, by defining the Moebius transformation

$$\Phi_i : u \longrightarrow \frac{a_i}{b_i + zu}, \qquad i \geq 0, \tag{38}$$

we obtain that the right-hand side in (37) can be written as

$$\Phi_0 \circ \Phi_1 \circ \Phi_2 \circ \cdots (0).$$

Since the transformation (38) involves a rotation (R), an inversion (I), a translation (T) and, moreover, it depends on a complex parameter z, the corresponding fraction (37) is called a *RITZ fraction*. The entries of the two sequences $\{a_i\}$ and $\{b_i\}$ are determined in order the ith truncated fraction,

$$\frac{p_i(z)}{q_i(z)} \equiv \Phi_0 \circ \Phi_1 \circ \ldots \Phi_{i-1}(0) = \cfrac{a_0}{b_0 + \cfrac{za_1}{b_1 + \cfrac{\ddots}{b_{i-2} + \cfrac{za_{i-1}}{b_{i-1}}}}}, \tag{39}$$

to be the best possible approximation to e^z. It can be shown that this lead to this (not unique) choice for the two sequences in (37):

$$a_i = (-1)^i, \qquad b_0 = 1, \quad b_{2i+1} = 2i + 1, \quad b_{2i+2} = 2, \qquad i = 0, 1, \ldots.$$

Since (38) are Moebius transformations, the corresponding matrix of the transformation,

$$M_i = \begin{pmatrix} 0 & a_i \\ 1 & b_i \end{pmatrix}, \qquad i \geq 0,$$

provides the following recursion for the polynomials (39):

$$\begin{aligned} p_{i+1}(z) &= za_i p_{i-1}(z) + b_i p_i(z), \\ q_{i+1}(z) &= za_i q_{i-1}(z) + b_i q_i(z), \qquad i \geq 1, \end{aligned}$$

with the initial conditions $p_0 = 0, q_0 = 1, p_1 = a_1, q_1 = b_1$. An easy induction argument shows that

$$\nu \equiv \deg(p_i) = \lfloor \frac{i-1}{2} \rfloor, \quad r \equiv \deg(q_i) = \lfloor \frac{i}{2} \rfloor, \quad q_i(0) = 1, \qquad i \geq 1. \tag{40}$$

However, in such a way we obtain again Padé approximations: in fact, one verifies that

$$\frac{p_i(z)}{q_i(z)} \equiv \frac{\varphi_{\nu,r}(z)}{\mu_{\nu,r}(z)}, \qquad i \geq 2,$$

where ν and r are defined according to (40) and the polynomials $\varphi_{\nu,r}(z)$ and $\mu_{\nu,r}(z)$ are respectively the numerator and denominator of the Padé approximation (see (34)). It is an easy matter to verify that, for $i \geq 2$, the couples (ν, r) assume the values $(r-1, r), (r, r)$, $r \geq 1$.

4. Conditioning and nonlinear iteration. From the results in the previous section, we can conclude that the choice of the Padé approximation for constructing A-stable (or L-stable) block methods of arbitrary high order is appropriate. We now shall study aspects related to the conditioning of the discrete problem, which make such a choice appealing from this point of view as well. Such aspects are related to the solution of the discrete problem (2), which is usually carried out by means of the Newton method (or its variants), tipically through the iteration

$$\begin{aligned} (A \otimes I_m - hB \otimes J_0)\, \delta^{(i)} &= F(\mathbf{y}^{(i)}), \\ \mathbf{y}^{(i+1)} &= \mathbf{y}^{(i)} - \delta^{(i)}, \qquad i = 0, 1, \ldots. \end{aligned} \tag{41}$$

We have, therefore, the problem of having the iteration (41) converging as fast as possible, as well as to accurately solve the linear system at each iteration. By means of a suitable modification of the Newton-Kantorovich theorem (see, for example, [22]) and of standard arguments of Numerical Linear Algebra, one obtains that both the two facts can be accomplished by requiring the condition number of the matrix $(A \otimes I_m - hB \otimes J_0)$, which we rewrite in the equivalent form (see (15))

$$(I_r \otimes I_m - hC \otimes J_0),$$

to be as small as possible. Obviously, when h is small there are no problems, since such matrix reduces to the identity matrix. Conversely, when h is large, the main contribute is due to $hC \otimes J_0$. By considering that

$$\kappa(hC \otimes J_0) = \kappa(C)\kappa(J_0),$$

(where obviously κ denotes the condition number of the specified matrix) we then conclude that it would be very appreciable to have the matrix C with a small condition number. We observe that the case where the stepsize h is large is a significant one, since it corresponds to the use of large stepsizes when, for example, in a stiff problem an asymptotically stable

TABLE 1
Condition numbers of the matrix C for the block methods derived from the Padé (r, r) and $(r-1, r)$.

r	1	2	3	4	5	6
(r, r)	1.00E0	6.88E0	1.49E1	3.10E1	5.62E1	1.15E2
$(r-1, r)$	1.00E0	4.19E0	8.15E0	1.67E1	3.03E1	6.08E1
r	7	8	9	10	11	12
(r, r)	2.19E2	5.61E2	1.28E3	4.23E3	1.14E4	4.08E4
$(r-1, r)$	1.17E2	2.95E2	6.86E2	2.26E3	6.16E3	2.23E4

stationary solution has been approached. In such a case, in fact, the solution is approximately constant and, consequently, any (appropriate) nonlinear iteration should converge rapidly.

Moreover, a necessary condition for having the matrix C well-conditioned is to have its eigenvalues bounded away from zero and infinity, as the size r increases. In this respect, the block methods derived by the Padé (ν, r) have this property. In fact, from (33)-(34) and the result of point 2 in Theorem 7, one obtains that the eigenvalues of C (which have positive real part) belong to the annulus

$$\frac{r+\nu}{r}\xi < |z| < \frac{r+\nu+4/3}{r}, \qquad \xi \approx 0.278465. \tag{42}$$

If, in addition, we consider that the best values for ν are $r-1$ and r, we conclude that all the eigenvalues of C have positive real part and are approximately contained in the annulus

$$2\xi < |z| < 2, \qquad r \gg 0. \tag{43}$$

In Table 1 we list the condition numbers of the matrix C corresponding to the block methods obtained from the Padé (r, r) and $(r-1, r)$, $r = 1, \ldots, 12$.

5. Implementation of block implicit methods: blended schemes. We now consider the problem of the efficient implementation of the block methods (2), namely we shall devise procedures for the solution of the corresponding discrete problem. This topic is focal for a numerical method to be competitive and has been, therefore, extensively studied for various classes of numerical methods (see, e.g., [1, 6, 11, 19, 20, 25]), even when implemented on different computer platforms. Concerning the the latter field, we mention Galligani's contributions [13, 14].

From the point of view of the implementation, the main problem is the efficient solution of the linear systems required by the iterative solution of the discrete problem. According to the general ways of solving linear systems, i.e. by using direct or iterative procedures, it is possible to classify the currently used approaches in two main categories:

- diagonalization (or block diagonalization) of the matrix C [11, 17];
- definition of a suitable splitting [1, 6, 19, 20, 21].

Concerning the first category, it would be preferable to have the matrix C with all real eigenvalues. This because in such a way it is possible to solve real linear systems of the same size of the continuous problem. Nevertheless, there is some drawback on this point, since the matrix which diagonalize the matrix C may be ill-conditioned. As matter of fact, from (21), by considering that (assuming for simplicity all distinct eigenvalues)

TABLE 2

Condition numbers for the matrix (44) in the case of the real eigenvalues (45) (first column) and for the block methods derived from the Padé $(r-1,r)$ (second column).

r	$\kappa(W)$	
2	4.66E0	3.03E0
3	6.65E1	1.01E1
4	3.71E2	3.47E1
5	1.34E4	1.37E2
6	1.24E5	6.16E2
7	3.49E6	3.18E3
8	5.29E7	1.82E4
9	9.58E8	1.08E5
10	2.26E10	6.51E5

$$F = V^{-1}\Lambda V,$$

where

$$\Lambda = \begin{pmatrix} \lambda_1 & & \\ & \ddots & \\ & & \lambda_r \end{pmatrix}, \qquad V = \begin{pmatrix} \lambda_1^0 & \dots & \lambda_1^{r-1} \\ \vdots & & \vdots \\ \lambda_r^0 & \dots & \lambda_r^{r-1} \end{pmatrix},$$

one obtains that the matrix

$$W = VGQ_r^{-1} \tag{44}$$

diagonalizes the matrix C. If we assume, for example, that the eigenvalues are equally spaced in the interval $[2\xi, 2]$ (see (42)-(43)), i.e.

$$\lambda_i = 2\xi + (i-1)\frac{2(1-\xi)}{r-1}, \qquad i = 1, \dots, r, \tag{45}$$

we obtain the condition numbers listed in the first column of Table 2. One may observe that the matrix W becomes rapidly ill-conditioned, as r grows. This feature is less evident when the eigenvalues are chosen more appropriately: for example, in the second column we list the condition numbers of the matrices W corresponding to the block methods derived from the Padé $(r-1,r)$, $r = 2, \dots, 10$.

Concerning the second approach, in general there is the problem of finding a splitting having satisfactory convergence properties.

The first strategy is used, for example, in the code RADAU5 [17], whereas the second strategy is used in the code GAM [21]. In this paper, after the introduction of the notion of "blended block method", we shall obtain a natural way to define appropriate splittings for the methods. Our approach, which is derived from the results in [6], is based on the definition of methods obtained as the combination of a couple of equivalent block methods. For this reason, such methods will be called *blended block methods*.

In order not to unnecessarily complicate the notation, we shall derive the new methods when the continuous problem to be solved is the usual test equation (16). Then, we look for methods generating a discrete problem in the following form,

$$M(q)\mathbf{y} \equiv (A(q) - qB(q))\mathbf{y} = (q\mathbf{b}(q) - \mathbf{a}(q))\eta \equiv \mathbf{g}, \tag{46}$$

where, being $\theta = \theta(q)$ a suitable "weight" matrix, $M(q)$ is the $r \times r$ matrix defined by

$$A(q) = \theta A_1 + (I_r - \theta)A_2, \qquad B(q) \;=\; \theta B_1 + (I_r - \theta)B_2, \tag{47}$$

and

$$\mathbf{a}(q) = \theta \mathbf{a}_1 + (I_r - \theta)\mathbf{a}_2, \qquad \mathbf{b}(q) \;=\; \theta \mathbf{b}_1 + (I_r - \theta)\mathbf{b}_2. \tag{48}$$

The couples of augmented matrices

$$\hat{A}_1 = [\mathbf{a}_1 \,|\, A_1], \quad \hat{B}_1 = [\mathbf{b}_1 \,|\, B_1], \qquad \hat{A}_2 = [\mathbf{a}_2 \,|\, A_2], \quad \hat{B}_2 = [\mathbf{b}_2 \,|\, B_2], \tag{49}$$

define two suitable equivalent block methods. In particular, we start considering the following choices:

$$A_1 = I_r, \quad B_1 = C, \quad A_2 = C^{-1}, \quad B_2 = I_r, \quad \theta(q) = (1-q)^{-1} I_r, \tag{50}$$

with the remaining vectors obtained through the consistency conditions (7). We shall choose the block methods in (49) so that the discrete problem (46) can be solved by using an iterative procedure,

$$N(q)\mathbf{y}^{(i+1)} = (N(q) - M(q))\mathbf{y}^{(i)} + \mathbf{g}, \qquad N(q) = A_1 - qB_2. \tag{51}$$

In view of the choice (50), we obtain

$$N(q)^{-1} = (1-q)^{-1} I_r \equiv \theta(q), \tag{52}$$

i.e. (51) defines a diagonal splitting. Moreover, the diagonal entries are equal, so that the resulting computational cost per iteration is relatively low.

The iteration (51) converges to the solution of (46) if and only if the spectral radius, say $\rho(q)$, of the iteration matrix

$$I_r - N(q)^{-1} M(q) \tag{53}$$

is smaller than 1. According to [19, 20], the *region of convergence* of the iteration (51) is given by

$$\Gamma = \{q \in \mathbb{C} : \rho(q) < 1\}.$$

Moreover, the iteration is said to be *A-convergent* if $\mathbb{C}^- \subseteq \Gamma$ ($A(\alpha)$-convergence is similarly defined). From the definition of $N(q)$, A-convergence is equivalent to require that

$$\rho^* \equiv \sup_{x>0} \rho(ix) < 1,$$

where, as usual, i is the imaginary unit. Another interesting property of the iteration (51) is that from the definition of $N(q)$ one obtains

$$\rho(q) \to 0, \quad \text{as} \quad q \to \infty.$$

Such property is very welcome, in order to have the iteration rapidly converging when the method is applied to stiff problems [19, 20].

A simple expression for the parameter ρ^* may be derived from (51)-(53). In fact, if λ is an eigenvalue of C, then

$$\zeta(\lambda) = \frac{q(\lambda - 1)^2}{\lambda(q-1)^2} \tag{54}$$

is an eigenvalue of the iteration matrix (53). Consequently, by assuming $q = ix$, we obtain that

$$\rho^* = \max_{x>0} \max_{\lambda \in \sigma(C)} g(x;\lambda), \tag{55}$$

where, by posing $\lambda = a + ib$,

$$g(x;\lambda) = \frac{x((1-a)^2 + b^2)}{\sqrt{a^2+b^2}(x^2+1)} \tag{56}$$

is the modulus of $\zeta(\lambda)$. By considering that $g(0;\lambda) = g(\infty;\lambda) = 0$, one concludes that if there is one stationary point, this must be a maximum. By differentiating $g(x;\lambda)$, we obtain

$$g'(x;\lambda) = \frac{(1-x^2)((1-a)^2+b^2)}{\sqrt{a^2+b^2}(x^2+1)},$$

from which we conclude that $x = 1$ is the point of maximum for $g(x;\lambda)$. As a consequence, we get

$$\rho^* = \max_{\lambda \in \sigma(C)} \frac{(1-a)^2 + b^2}{2\sqrt{a^2+b^2}}, \tag{57}$$

TABLE 3
Value of ρ^* for the blended block methods corresponding to the Padé (r,r) and $(r-1,r)$.

r	1	2	3	4	5	6	7	8	9	10	11	12
(r,r)	.250	.289	.419	.522	.600	.661	.710	.749	.783	.811	.835	.856
$(r-1,r)$	.0	.204	.386	.508	.595	.661	.712	.753	.787	.815	.840	.861

which is easy to compute, since we have the possibility of choosing the spectrum of C. In Table 3 we list the value of ρ^* for the block methods obtained from the Padé approximations (r,r) and $(r-1,r)$, $r=1,\ldots,12$. One then concludes that the iterations corresponding to all such methods are A-convergent. Moreover, it can be verified that this property holds true at least up to $r=24$, for both type of methods.

An improvement can be obtained by considering an alternative choice to (50), still providing us with a diagonal splitting, namely

$$A_1 = I_r, \quad B_1 = C, \quad A_2 = DC^{-1}, \quad B_2 = D, \quad \theta(q) = (I_r - qD)^{-1}, \tag{58}$$

where D is a diagonal matrix with positive real entries. The first possibility that we consider is the following:

$$D = \alpha I_r, \qquad \alpha > 0, \tag{59}$$

which reduces to the previous choice (50) in the case $\alpha = 1$. In such a case, (54), (55), (56) and (57) become, respectively,

$$\zeta(\lambda;\alpha) = \frac{q(\lambda-\alpha)^2}{\lambda(\alpha q-1)^2}, \ \rho^* = \max_{x>0}\max_{\lambda\in\sigma(C)} g(x;\lambda,\alpha), \ g(x;\lambda,\alpha) = \frac{x((\alpha-a)^2+b^2)}{\sqrt{a^2+b^2}(\alpha^2x^2+1)},$$

and, considering that $x=\alpha^{-1}$ is the point of maximum for $g(x;\lambda,\alpha)$,

$$\rho^*_\alpha = \max_{\lambda\in\sigma(C)} \frac{(\alpha-a)^2+b^2}{2\alpha\sqrt{a^2+b^2}}. \tag{60}$$

Consequently, the convergence properties of the iteration (51) now depend on the parameter $\alpha>0$, which can be chosen in order to minimize the corresponding value of ρ^*_α. In Tables 4 and 5 we list, respectively, the optimum parameters α_r, along with the corresponding parameter $\rho^*_{\alpha_r}$, for the block methods obtained from the Padé (r,r) and $(r-1,r)$, $r=1,\ldots,12$. One verifies that there has been an improvement over the previous values in Table 3. A different choice for the matrix D will be considered in the next section.

6. Implementation of block implicit methods: further analysis. We now consider a particular case of the above approach, namely

$$\lambda_i(C) = \alpha, \qquad i=1,\ldots,r, \tag{61}$$

where α is the same parameter in (59). From equation (60), we then obtain $\rho^*_\alpha = 0$. This conclusion may appear astonishing, since it implies that the iteration (51) converges in one

TABLE 4
Values of ρ^* for the modified blended block methods corresponding to the Padé (r, r).

r	1	2	3	4	5	6	7	8	9	10	11	12
α_r	.5	.5774	.5902	.5901	.5867	.5826	.5786	.5748	.5713	.5682	.5653	.5628
$\rho^*_{\alpha_r}$	.0	.134	.277	.379	.454	.511	.556	.592	.622	.647	.668	.687

TABLE 5
Values of ρ^* for the modified blended block methods corresponding to the Padé $(r-1, r)$.

r	1	2	3	4	5	6	7	8	9	10	11	12
α_r	1.	.8165	.7387	.6952	.6671	.6471	.6322	.6205	.6111	.6033	.5968	.5911
$\rho^*_{\alpha_r}$	.0	.184	.340	.442	.512	.564	.605	.637	.663	.685	.703	.720

iteration, when $q \in \mathbb{C}^-$ (actually, for all $q \neq \alpha^{-1}$). Nevertheless, the choice of a r-fold eigenvalue for the matrix C has a severe drawback, since the matrix C becomes very ill-conditioned, for increasing values of r. In fact, let us first consider the problem of choosing appropriately the parameter α. From (23) and (61), one obtains that

$$d(z) = \sum_{i=0}^{r} \binom{r}{i} (-\alpha)^{r-i} z^i,$$

i.e.

$$d_i = \binom{r}{i} (-\alpha)^{r-i}, \qquad i = 0, \ldots, r.$$

In order to obtain a method with $\infty \in \mathcal{D}$, we impose (see (24)-(25))

$$0 = \frac{1}{d_0} \sum_{i=0}^{r} \frac{r^i}{i!} d_i = \sum_{i=0}^{r} \frac{(-1)^i}{i!} \binom{r}{i} \left(\frac{r}{\alpha}\right)^i \equiv w(r\alpha^{-1}). \tag{62}$$

In such a case, in fact, A-stability implies L-stability. The polynomial $w(z)$ turns out to have all positive and real zeros. Among them, we obviously shall choose the one which generates the method with the best stability properties. Let it be $\bar{z}$. In correspondence of such root, we get the parameter

$$\bar{\alpha} = r\,\bar{z}^{-1}. \tag{63}$$

When more than one zero fulfills the above requirement, in view of the arguments in Section 4, we shall choose the one which minimizes the condition number of the matrix C. As an example, for $r = 2$ from (62)-(63) one obtains, after a few calculations, that the two values of α are

$$\alpha_1 = \frac{2}{2+\sqrt{2}}, \qquad \alpha_2 = 2 + \sqrt{2},$$

both generating A-stable methods. The parameter which minimizes the value of $\kappa(C)$ is α_1. From such value, one obtains an L-stable block method of order 2 having the following matrix

TABLE 6
Condition numbers for the matrix C obtained from (61)-(63).

r	$\bar{\alpha}$	$\kappa(C)$
2	.58578643762690497	3.08E0
3	1.3075995645253780	4.19E1
4	2.2912642499285432	2.15E3
5	1.3902692056822543	2.16E3
6	2.0048542024083011	6.92E4
7	2.7259875357196162	3.98E6
8	1.8749852768446345	3.86E6

$$A^{-1}\hat{B} = \begin{pmatrix} 4.9264068711928521\mathrm{E}-1 & 5.1471862576142957\mathrm{E}-1 & -7.3593128807148411\mathrm{E}-3 \\ 6.5685424949238014\mathrm{E}-1 & 6.8629150101523972\mathrm{E}-1 & 6.5685424949238014\mathrm{E}-1 \end{pmatrix},$$

for which (see (3) and (15)) $\kappa(C) \approx 3$. Nonetheless, for increasing values of r one realizes that the condition number of the matrix C grows very rapidly with r, as it is shown in Table 6. For each listed value of r, the corresponding block method has order r and is A-stable (or $A(\alpha)$-stable). Nevertheless, due to the growth of $\kappa(C)$, it is advisable to use such methods only for r very small.

REMARK 1. *We observe that the method corresponding, for example, to $r = 8$ in Table 6 is A-stable (see Figure 1). Nevertheless, the corresponding matrix C is ill-conditioned (see the third column in Table 6). The reason for this is the large value of* departure from normality *of the matrix, which can be measured by*

$$\Delta(C) = \|C\,C^T - C^T C\|_2.$$

In fact, for this method one obtains $\Delta(C) \approx 3.0\mathrm{E}9$, whereas the corresponding value for the method derived from the Padé (7,8) (see Table 1, $r = 8$) is $\Delta(C) \approx 66$.

Let us now consider, instead of (59), the more general choice

$$D = \mathrm{diag}(\,\gamma_1 \quad \dots \quad \gamma_r\,), \qquad \gamma_i > 0, \quad i = 1, \dots, r, \tag{64}$$

for the iteration (51)-(58). The diagonal entries of D will be chosen in order to reduce the value of the parameter ρ^*. Unfortunately, the spectral analysis is not as simple as in the case of the choice (59) and we must resort to a computational approach to obtain such entries. In particular, we have used the Matlab function `fmins` to minimize $\rho^* \equiv \rho^*(\gamma_1, \dots, \gamma_r)$, for the block methods obtained from the Padé (r, r) and $(r-1, r)$, $r = 1, \dots, 10$. The obtained diagonal entries are listed in Tables 7 and 8. The corresponding values of the parameter ρ^* are listed in Table 9: in all cases we obtain an improvement over the corresponding values listed in Tables 3 and 4-5. Moreover, for comparison we also plot in Figure 2 the function $\rho(ix)$ for $x > 0$, for the block methods obtained from the Padé $(7, 8)$ with the choices (50) and (58)-(59), corresponding respectively to $\alpha = 1$ and $\alpha = 0.6205$ (see Table 5), and to the choice (58)-(64), where the diagonal entries of D are those listed in Table 8, $r = 8$. From the plots, one may infer that the choice (58)-(64) is definitely an improvement over (50), since the curve in solid is always below the dashed one. Conversely, the dotted curve,

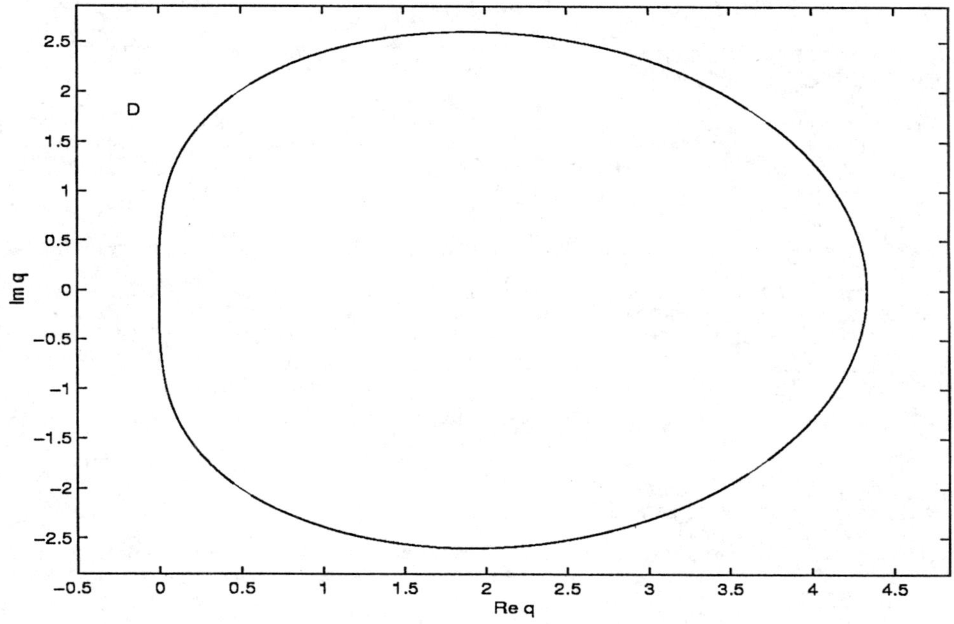

FIG. 1. Stability region $\mathcal{D}$ for the block method corresponding to $r = 8$ in Table 6.

corresponding to the choice (58)-(59) has a maximum greater than that of the curve in solid; nevertheless, for $x < 0.77$ it is under such curve. This means that, for the corresponding values of q, the iteration (58)-(59) is faster than (58)-(64), even though for larger values of x the roles are reversed.

Let us now end the section by studying the application of a blended block method to the more general problem (1). In such a case, the discrete problem corresponding to the application of the method (46), (47)-(48), (58) over the first r mesh points is given by

$$\tilde{A}\,\mathbf{y} - h\tilde{B}\,\mathbf{f} = h\tilde{\mathbf{b}}\,f(t_0, y_0) - \tilde{\mathbf{a}}\,y_0, \tag{65}$$

where, by setting $I = I_r \otimes I_m$,

$$\begin{aligned}
N &= I - hD \otimes J_0\\
\tilde{A} &= N^{-1}I + (I - N^{-1})\ \left((DC^{-1}) \otimes I_m\right),\\
\tilde{B} &= N^{-1}\ (C \otimes I_m) + (I - N^{-1})\ (D \otimes I_m),\\
\tilde{\mathbf{a}} &= N^{-1}\ (\mathbf{e} \otimes I_m) + (I - N^{-1})\ \left((DC^{-1}\mathbf{e}) \otimes I_m\right),\\
\tilde{\mathbf{b}} &= N^{-1}\ \left((\mathbf{q}_1 - C\mathbf{e}) \otimes I_m\right) + (I - N^{-1})\ \left((D(C^{-1}\mathbf{q}_1 - \mathbf{e})) \otimes I_m\right).
\end{aligned}$$

From the above expressions, it is evident that the application of the method requires the factorization of the matrix N, i.e. of the block

$$I_m - h\alpha J_0 \tag{66}$$

TABLE 7

Diagonal entries of the matrix D for the blended block methods derived from the Padé (r, r).

r	$\gamma_i, \quad i = 1, \dots, r$				
2	0.66275204106065	0.99880381565137			
3	0.79177839434600	0.66142441654254	1.09728375380713		
4	0.71248323670683	0.68572979240044	1.00835280450325	0.94336189309863	
5	0.73794709503502	0.69667939234304	0.94678950632936	0.96925988167893	1.08556894280691
6	0.78960666760857	0.77764317616699	1.10379439358622	1.12287112979631	1.04425410677717
	0.97747272393049				
7	0.90328518460926	0.93588189061015	1.27378358368408	1.06482337773942	0.95699568328484
	0.87962577636270	0.88647177187620			
8	0.90189079913764	0.96029017717008	1.20044261367058	1.06182151859105	0.98279767187713
	0.95130581186793	0.95477396330476	0.97709719034535		
9	0.78890956014085	0.98692274344950	1.16031736214545	1.06429394889840	0.97558449511996
	0.97973125397795	0.97616792047847	1.01092465767460	1.02624249573684	
10	0.98194772536412	0.91245934163159	1.04481326127609	1.00603860548456	1.00089107693710
	0.99217500074166	1.00032852544433	1.00860473013924	1.02306653120170	1.05905836178089

TABLE 8

Diagonal entries of the matrix D for the blended block methods derived from the Padé $(r-1, r)$.

r	$\gamma_i, \quad i = 1, \dots, r$				
2	0.96936165526039	1.50848352835328			
3	0.81868602814689	0.87495096575515	1.44998332454753		
4	0.35337640949037	0.80121932984835	1.27762338976231	1.66446327607499	
5	0.63946725983491	0.82968839663997	0.91909524862064	1.42126433313185	1.52443079963375
6	0.79474199557416	0.69885470794190	0.96328076538840	1.17906063317010	0.96006333245803
	1.34062294955658				
7	0.76154788274216	0.77052101658868	1.03651805522176	1.06062392757843	0.98001202471669
	1.07157541314092	1.33204542634207			
8	0.90280588616216	0.94667898522012	1.16837318551763	0.97151531383934	0.94628652026249
	0.92599286317791	1.01107530401679	1.15188588598478		
9	0.95021396531815	0.96824923001551	1.16031142832145	1.00905109392683	0.98542777497508
	0.95758015675129	0.97574516430175	0.97853576014775	0.97686463476444	
10	0.89639578999006	0.95322083486291	1.09782034111785	1.01308990374127	0.97688329481101
	0.97154048045308	0.96942938230334	0.99731505715738	0.97199798518272	1.08135584615257

TABLE 9

Values of ρ^* *for the blended block methods corresponding to the Padé* (r, r) *and* $(r-1, r)$ *with the diagonal scaling D.*

r	2	3	4	5	6	7	8	9	10
(r, r)	.055	.230	.273	.359	.414	.444	.479	.527	.622
$(r-1, r)$	.086	.280	.317	.450	.501	.536	.515	.541	.593

in case we use the method corresponding to (59); alternatively, when (64) is considered, we need to factorize the blocks

$$I_m - h\gamma_i J_0, \qquad i = 1, \dots, r. \tag{67}$$

It is customary to solve the equation (65) by using the modified Newton method, then solving linear systems with the matrix

$$M = \tilde{A} - h\tilde{B}\,(I_s \otimes J_0).$$

In place of such linear systems, we solve an *inner iteration* similar to (51), thus involving only linear systems with the matrix N, which has already been factored.

We then conclude that, leaving aside for simplicity function and Jacobian evaluations, the arithmetic complexity for solving (65) when ℓ Newton iterations are performed, each requiring μ inner iterations, amounts to

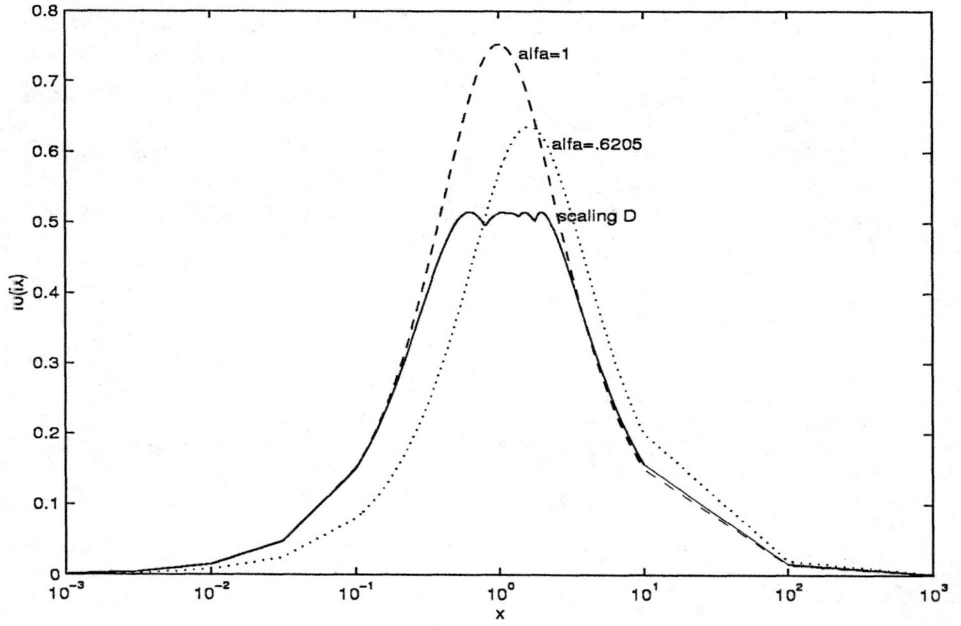

FIG. 2. Spectral radius of the iteration matrix for the blended block method derived from the Padé $(7,8)$, in its original formulation (dashed line), with the scaling $\alpha_8 I_8$ (dotted line) and with the scaling D (solid line).

$$\frac{2}{3}m^3 + O(\ell\mu r m^2) \qquad \text{floating operations,}$$

in the case of (59), or

$$\frac{2}{3}rm^3 + O(\ell\mu r m^2) \qquad \text{floating operations,}$$

in the case of (64). The leading term is obviously due to the factorizations of the matrices (66) and (67), respectively. In the latter case, however, by observing that the r factorizations are independent each other, it is possible to use r parallel processors to execute them concurrently. Consequently, the choice (64) will result in a method having a natural *parallelism across the method.*

7. Conclusions. In this paper we have reviewed the derivation of r-block methods for ODEs. The derivation of such methods has been done in a unified framework, which allows to discuss many important theoretical properties of the methods.

Among the possible choices, it has been confirmed that the methods derived from the Padé approximation to the exponential have good stability properties. In more detail, the methods derived from the Padé (r,r) are perfectly A-stable and those derived from the Padé $(r-1,r)$ are L-stable, for all $r \geq 1$. Moreover, from (35) one obtains that even values of r are preferable, because they provide higher order methods than the subsequent odd values.

The implementation of the block methods as blended block methods has also been introduced. Such an implementation exploits the possibility of writing a given block method in different equivalent forms. The blended block methods here studied are characterized by a diagonal splitting, which is A-convergent for all values of r of practical interest. Improvements of the basic splitting have been also considered, which can be tailored for an efficient implementation either on sequential or parallel computers. In both cases, we obtain methods with a (possibly parallel) complexity whose leading term is $2m^3/3$ flops per integration step (apart function and Jacobian evaluations), when the size m of the continuous problem is large.

REFERENCES

[1] P. Amodio, L. Brugnano. A Note on the Efficient Implementation of Implicit Methods for ODEs, *Jour. Comput. Appl. Math.* **87** (1997) 1–9.

[2] U. M. Ascher, L. R. Petzold. *Computer Methods for Ordinary Differential Equations and Differential-Algebraic Equations*, SIAM, Philadelphia, 1998.

[3] O. Axelsson. Global Integration of Differential Equations Through Lobatto Quadrature, *BIT* **4** (1964) 69–86.

[4] O. Axelsson. A Class of A-stable Methods, *BIT* **9** (1969) 185–199.

[5] C. Bendsten. On Implicit Runge-Kutta Methods with High Stage Order, *BIT* **37** (1997) 212–217.

[6] L. Brugnano. Blended Block BVMs (B_3VMs): a Family of Economical Implicit Methods for ODEs, *Jour. Comput. Appl. Math.* (1999) in press.

[7] L. Brugnano, D. Trigiante. *Solving Differential Problems by Multistep Initial and Boundary Value Methods*, Gordon and Breach, Amsterdam, 1998.

[8] K. Burrage. A Special Family of Runge-Kutta Methods for Solving Stiff Differential Equations, *BIT* **18** (1978) 22–41.

[9] K. Burrage. High Order Algebraically Stable Runge-Kutta Methods, *BIT* **18** (1978) 373–383.

[10] K. Burrage. *Parallel and Sequential Methods for Ordinary Differential Equations*, Clarendon Press, Oxford, 1995.

[11] J. C. Butcher. On the Implementation of Implicit Runge-Kutta Methods, *BIT* **6** (1976) 237–240.

[12] J. C. Butcher. *The Numerical Analysis of Ordinary Differential Equations: Runge-Kutta Methods and General Linear Methods*, John Wiley, Chichester, 1987.

[13] I. Galligani. Splitting Methods for Solving Large Systems of Linear Ordinary Differential Equations on a Vector Computer, *WSSIAA* **2** (1993) 165–176.

[14] I. Galligani, V. Ruggiero. Solving Large Systems of Linear Ordinary Differential Equations on a Vector Computer. *Parallel Comput.* **9** (1989) 359–365.

[15] E. Hairer, S. P. Norsett, G. Wanner. *Solving Ordinary Differential Equations I, 2nd ed.*, Springer Series in Computational Mathematics, vol. 8, Springer-Verlag, Berlin, 1993.

[16] E. Hairer, G. Wanner. Algebraically Stable and Implementable Runge-Kutta Methods of High Order, *SIAM J. Numer. Anal.* **18** (1981) 1098–1108.

[17] E. Hairer, G. Wanner. *Solving Ordinary Differential Equations II, Stiff and Differential-Algebraic Problems*, Springer-Verlag, Berlin, 1991.

[18] P. Henrici. *Applied and Computational Complex Analysis, Vol. 2*, John Wiley & Sons, New York, 1977.

[19] P. J. van der Houwen, J. J. B. de Swart. Triangularly Implicit Iteration Methods for ODE-IVP Solvers, *SIAM J. Sci. Comput.* **18** (1997) 41–55.

[20] P. J. van der Houwen, J. J. B. de Swart. Parallel Linear System Solvers for Runge-Kutta Methods, *Adv. Comput. Math.* **7**,1-2 (1997) 157–181.

[21] F. Iavernaro, F. Mazzia. Solving Ordinary Differential Equations by Generalized Adams Methods: Properties and Implementation Techniques. *Appl. Numer. Math.* **28** (1998) 107–126.

[22] V. Lakshmikantham, D. Trigiante. *Theory of Difference Equations: Numerical Methods and Applications*, Series "Mathematics is Science and Engineering", vol. 181, Academic Press, San Diego, 1988.

[23] J. D. Lambert. *Numerical methods for Ordinary Differential Systems*, John Wiley & Sons, New York, 1991.

[24] E. B. Saff, R. S. Varga. On Zeros and Poles of Padé Approximants to e^z. III, *Numer. Math.* **30** (1978) 241–266.

[25] R. D. Skeel, A. K. Kong. Blended Linear Multistep Methods, *ACM TOMS* **3** (1977) 326–345.

[26] G. Wanner, E. Hairer, S.P. Nørsett. Order Stars and Stability Theorems, *BIT* **18** (1978) 475–489.
[27] H. A. Watts, L. F. Shampine. *A*-stable Block One-step Methods, *BIT* **12** (1972) 252–266.

THE TRIMMED NURBS AGE *

G. CASCIOLA † AND S. MORIGI ‡

Abstract. In this work we report our experience in the geometric modeling field, using trimmed NURBS surfaces and we describe how these surfaces can be used both for solid and surface modeling purposes. Our experience is based on the design and implementation of *xcmodel*, an interactive graphics system merging new and well-known methods for modeling.

Key words. trimmed NURBS, rendering, surface modeling, solid modeling.

AMS subject classifications. 65D17, 65Y25, 65Y20

1. Introduction. In geometric modeling there are two fundamental representation schemes which have been successful and have been incorporated in the most popular CAD/CAM software packages: solid modeling and surface modeling. A solid modeler represents an object unambiguously by describing its surface boundary and by topologically orienting it, so that we can tell, at each surface point, on which side the solid interior lies. Whereas, a surface modeler gives only a geometrical description of the object boundary without any topological information.

The Non-Uniform Rational B-Splines (NURBS) is a well-established tool for geometric design. NURBS have become the de facto industry standards for the representation, design, and data exchange of geometric information. NURBS have been added to several international standards, and many packages include NURBS as the primitive for designing simple and free form curves and surfaces. However, the most useful NURBS paradigm is limited by the requirement that the surfaces are defined over rectangular regions and this leads to topological rectangular patches. A generalization for an arbitrary topology can be obtained by collapsing some of the control mesh edges, but this creates surfaces with ambiguous surface normal and degenerated parametrization.

The introduction of a 'trimmed surface' data type in the description of free form objects or parts of solids has provided greater power and flexibility to these representational schemes. A trimmed surface is an ordinary tensor product surface that has a restricted parameter domain, thus overcoming the limit of tensor product surfaces defined over rectangular regions, and allowing for arbitrary domains.

This work is not intended as an exhaustive survey on trimmed NURBS surfaces, but as a description of our experience of geometric modeling using trimmed NURBS surfaces. Our experimental testbed is *xcmodel*, a system developed at the University of Bologna [9] that is based on trimmed NURBS surfaces, which tests new modeling ideas and where techniques can easily be experimented and evaluated. This system is the result of our experience of over 10 years in NURBS research and development and it combines new and well-tried approaches. In fact, our work methods are well expressed by the following sentence from Voelker and Requicha [44]:

"It is important to do both theoretical research and experimental system building. They are synergistic, and the exclusive pursuit of either can lead to sterile theory or quirky, opaque systems."

---

*This work was supported by MURST, Cofin97, Numerical Analysis: Methods and Mathematical Software.

†Dept. of Mathematics, University of Bologna, Italy; E-mail: `casciola@dm.unibo.it`.

‡Dept. of Mathematics, University of Bologna, Italy; E-mail: `morigi@dm.unibo.it`.

Everyone agrees with this idea, although, in the academic world, it is hardly ever rewarded.

The paper is structured as follows: in Section 2 the basic definitions of trimmed NURBS surface are given. In Section 3 we address the problem of representing the trimmed regions, while Section 4 deals with several approaches used for rendering trimmed surfaces. In Section 5 and 6 we describe how these surfaces can be used for solid and surface modeling purposes.

2. Trimmed NURBS surfaces. A trimmed NURBS surface can be defined by a tensor product NURBS surface and a set of trimming curves in the parametric space of the surface.

A NURBS surface of order (m, n) defined in the parametric domain $U \times V = [0, 1] \times [0, 1]$, can be represented as:

$$\mathbf{r}(u, v) = \frac{\sum_{i=1}^{n+H} \sum_{j=1}^{m+K} w_{ij} \mathbf{P}_{ij} N_{i,n}(u) N_{j,m}(v)}{\sum_{i=1}^{n+H} \sum_{j=1}^{m+K} w_{ij} N_{i,n}(u) N_{j,m}(v)} \tag{2.1}$$

where w_{ij} are the weights, $\mathbf{P}_{ij}$ are the control points, and $N_{i,n}(u), N_{j,m}(v)$ are the order n and m B-splines, respectively, defined over the knot vectors

$$U = (\underbrace{0, \cdots, 0}_{n}, u_{n+1}, \cdots, u_{n+H}, \underbrace{1, \cdots, 1}_{n})$$

$$V = (\underbrace{0, \cdots, 0}_{m}, v_{m+1}, \cdots, v_{m+K}, \underbrace{1, \cdots, 1}_{m}).$$

Associated with a NURBS surface is a set of planar, closed, non-intersecting curves that can be conveniently represented as NURBS curves defined in the parameter domain $[0, 1]$ by

$$\mathbf{c}_k(t) = (x_k(t), y_k(t)) = \frac{\sum_{i=1}^{p_k+L_k} w_i \mathbf{P}_i N_{i,p_k}(t)}{\sum_{i=1}^{p_k+L_k} w_i N_{i,p_k}(t)}, \quad k = 1, \cdots, M, \tag{2.2}$$

with the knot vectors

$$T_k = (\underbrace{0, \cdots, 0}_{p_k}, t_{p_k+1}, \cdots, t_{p_k+L_k}, \underbrace{1, \cdots, 1}_{p_k}).$$

A trimmed NURBS surface is given by the restriction of $\mathbf{r}(u, v)$ to a subdomain $D \subset U \times V$, of the parametric space, named *trimmed region*. This domain D is defined as the set of regions on $U \times V$ whose boundaries are specified by the trimming curves and a given criteria. This criteria allows us to identify the part of the surface that remains when discarding all the holes defined by the trimming curves. The trimmed surface boundaries are obtained by mapping the 2D trimming curves onto the surface, that is,

$$\mathbf{r}(x_k(t), y_k(t)), \quad k = 1, \cdots, M.$$

Trimmed NURBS surfaces have been adopted by the CAD/CAM industry, and included in graphics standards. As such, trimmed NURBS surfaces are provided as modeling primitives in several geometric modeling software systems, and the rendering of trimmed NURBS surfaces is supported by international standards, such as PHIGS+, as well as graphics programming interfaces, such as Iris-GL and OpenGL (Silicon Graphics, Inc.), Starbase (Hewlett-Packard Corp.), and Renderman (Pixar).

Trimmed surfaces are essential for the modeling of non-regular boundary objects, generated by trimming away part of the rectangular patch. Trimming patches also play a fundamental role in the boundary description of solid models where the trimmed surfaces can give a complete representation of the boundary of a sculptured solid primitive by means of the union of surfaces restricted to suitable domains. They are the result of Boolean operations on solid objects bounded by NURBS surfaces [17].

3. Representing of trimmed NURBS surfaces. We focus now on the problem of representing the trimmed region D, providing three different representations that define the same domain D.

Representation 1: handedness rule of trimming

The curves $\mathbf{c}_k(t)$ are considered all properly oriented and joined to form a $N < M$ numbers of outer or inner loops. Outer loops are oriented counter-clockwise, whereas inner loops are oriented clockwise. The domain of the trimmed surface is defined as the common region within the outer boundary (corresponding to the outer loops) and outside the inner boundaries (corresponding to inner loops), including boundary curves [32]. Figure 3.1a presents an example of trimmed region obtained by this rule; the area that is part of the trimmed region is shaded.

Representation 2: winding rule of trimming

The curves $\mathbf{c}_k(t)$ are considered all properly joined to form $N < M$ closed loops (see Figure 3.1b). The set of contours divides the supporting surface into an inner or retained region and an outer or discarded region based on the odd-winding rule. According to this rule, the region of the surface that is enclosed by an even number of loops is trimmed out [29].

Representation 3: 2D CSG tree

The trimmed domain is represented by the 2D equivalent of a CSG (Constructive Solid Geometry [5]) tree, which, like its 3D counterpart, is a collection of half-spaces and Boolean operation symbols [17, 5]. Each leaf of the CSG tree structure is an embedded non-intersecting closed curve. Curves at the same level are disjoint to each other.

The trimmed region that is kept can be determined by classifying a single point. Conventions can be assumed to consider alternate levels of the tree to be outer and inner loops, starting from an inner/outer top level. In Figure 3.1c the 2D CSG tree representing the trimmed region illustrated in Figure 3.1a is shown with the top level set to be outer.

The trimmed region looks like a set of islands and lakes, where the islands represent part of the trimmed region, while the lakes are the holes in it. The algorithm used to build a 2D CSG tree $\mathcal{T}$ for a given trimmed region forms a union of islands (R_i), and subtracts from each island all of its lakes (S_{ij}):

$$\mathcal{T} = \bigcup_{i=1}^{I} (R_i \bigcap (\bigcup_{j=1}^{L_i} S_{ij})),$$

where I is the number of islands and L_i is the number of the lakes S_{ij} children of island i.

Another approach for representing a trimmed surface is based on the idea of decomposing the domain D of the parameter space using a set of planar subregions, whose union defines the entire domain D (see Figure 3.1d). This approach does not follow an intuitive idea of trimming out the part of the entire patch bounded by the trimming curves, that is preserved somehow in the previous definitions, but it allows us to manage a set of surfaces, defined on an irregular domain, that do not contain any trimming curves. This can represent an advantage during the evaluation/rendering phase, as well as in the mesh generation phase, because it allows for a representation of a trimmed tensor product patch as a collection of untrimmed patches. In [22] the authors consider this approach by defining a trimmed surface as the union of a set of planar, ruled (four-sided) subregions, addressing the problem of data exchange between systems. More generally, this approach can be applied using S-patches [28, 45]. S-patches are rational generalizations of Bezier-surfaces that admit any number of boundary curves.

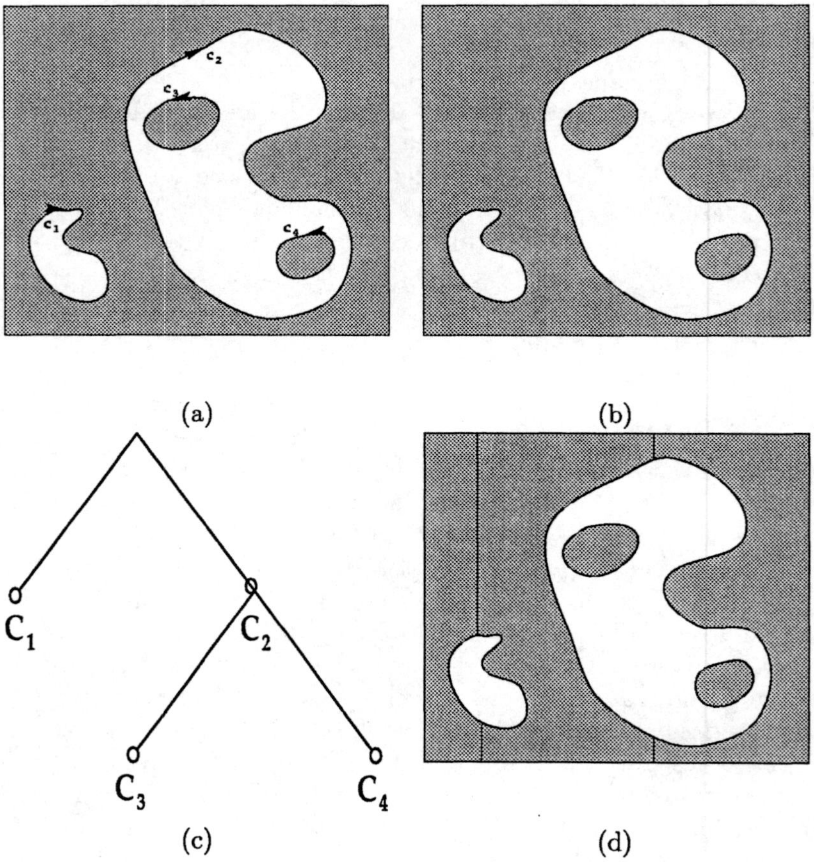

FIG. 3.1. *Example of trimmed region obtained by: (a) definition 1, (b) definition 2, (c) definition 3, and by untrimmed regions approach (d).*

4. Rendering of trimmed NURBS surfaces. This section deals with the problem of visualizing the trimmed NURBS surfaces. As we know, the visualization algorithms differ in rendering quality with a computational cost that increases as the quality improves. A modeling system needs both medium/low-level quality algorithms, allowing for real-time visualization (such as depth cueing, hidden-line or z-buffer), and high quality algorithms for a realistic final rendering (such as ray-tracing).

Medium/low-level quality algorithms use a piecewise planar approximation (tessellation or triangulation) of the trimmed surface within a given tolerance. After the triangulation phase a surface can be rendered in real time using the triangle rendering capabilities common in current graphics systems. Several algorithms have been developed in the past to obtain a triangulation of a trimmed NURBS surface [26, 32, 35, 40, 29]. These methods are parametrization dependent and use the same tolerance to polygonize the trimming curves and to triangulate the surface. Only Piegl and Tiller in [33] present an algorithm, which is not parametrization dependent, that solves the problem with different tolerances for the trimming curves and the surface. The triangulation algorithm implemented in *xcmodel* was inspired by the latter proposal and consists in the following steps:

- *grid surface*; a rectangular grid in the entire domain is created adaptively. By splitting each rectangular into two triangles, we obtain a piecewise linear approximation of the surface. The goal is to approximate curved regions using more triangles than for the flat regions, while ensuring that the approximation satisfies the requirements of a given tolerance [1];
- *polygonize trimming curves*; consists in obtaining a piecewise linear approximation of the trimming curves within a given tolerance.
- *merge trimming polygon and rectangular grid*; given a rectangle in the grid of the parameter domain, along with its status and a set of trimming polygons, the goal is to compute a set of closed polygons that bound the trimmed domain lying inside the given rectangle.
- *triangulating domain polygons*; this problem has been solved in a simple and heuristic way. Our approach distinguishes two cases: the points on the rectangular grid may or may not be consecutive polygon vertices. In the first case, the triangles open fanwise from one of the grid points; in the second case, where two grid points are opposite, the triangles form a seam.

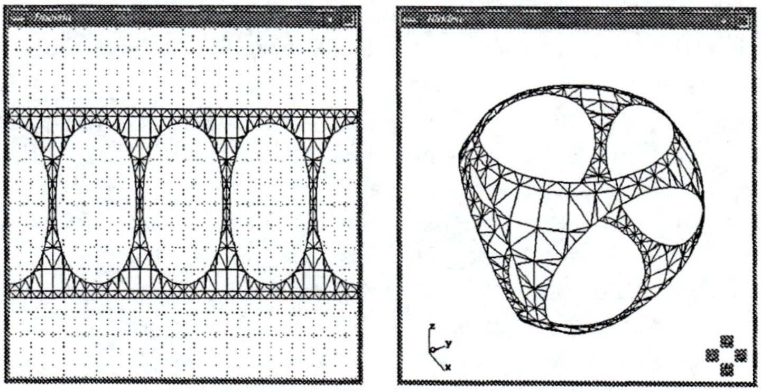

FIG. 4.1. *Rendering of trimmed NURBS sphere with a hidden-line algorithm; a triangulation of the trimmed domain (left), the rendered trimmed surface (right).*

Figure 4.1 illustrates an example of a trimmed NURBS sphere rendered by a hidden-line algorithm obtained using *xcmodel*.

High quality rendering algorithms use an exact representation. The history, theory and capabilities of ray-tracing algorithms are well known and documented [20]. The basic problems in ray-tracing trimmed NURBS surfaces are the following:

FIG. 4.2. *Rendering of trimmed NURBS sphere with a ray-tracing algorithm.*

- ray/patch intersection
- determining if the intersection points belong to a trimmed patch.

If the trimming is caused by a Boolean operation involving solid geometric models, the latter step can be performed using conventional CSG methods [37]. Our *xcmodel* system manages boundary representations for solids by using trimmed surfaces and therefore the second step determines wether an intersection point lies inside or outside a trimmed region using a 2D PMC (Point Membership Classification) algorithm (see section 5).

Solutions to the ray/patch intersection problem can be categorized as being based on subdivision, numerical or hybrid techniques. Subdivision approaches are described by Whitted [46], Rogers [36] and Woodward [47]. These algorithms are based on the convex-hull properties of NURBS surfaces. If the ray does not intersect the convex-hull of the control points, it does not intersect the patch. By recursively subdividing the patch and checking convex-hulls, the intersection points can be computed at a linear convergent rate amounting to a binary search. Whitted's algorithm operates in three dimensions, whereas Rogers and Woodward map the problem in two dimensions.

Numerical solutions of the ray/patch intersection problem include those developed by Toth [43], Sweeny and Bartels [42] and Joy and Bhetanabhotla [24]. Toth's algorithm is based on interval Newton iterations. It works robustly on any parametric surface for which bounds onto the surface and its first derivative can be obtained. Sweeny and Bartels proposed refining the control mesh using the Oslo algorithm [15] until the mesh closely approximates the surface. The ray intersection is then computed by intersecting the control mesh with the ray, and using that intersection point as a starting point for Newton iteration. Joy and Bhetanabhotla's algorithm uses quasi Newton optimization to capture the point/s on the patch nearest the ray, including the intersection points.

By hybrid solutions, we refer to Bezier clipping [31]. This algorithm uses the convex-hull property in a more powerful manner, by determining parameter ranges which guarantee that they do not include points of intersection. Bezier clipping has the flavor of a geometrically based interval Newton method, and thus may be categorized as partly a subdivision based algorithm and partly a numerical method.

From a numerical point of view all these methods find the ray/patch intersections by solving a non-linear system of equations. The above methods are global and when several solutions occur, they split the problem. The only one which uses a local convergence method after applying an initial global localization step is Toth's algorithm. However, the localization step is very expensive from a computational point of view, as well as the fact that, in the case of near solutions in the parametric domain this localization step can fail, requiring the application of a binary subdivision method of the function.

In our system we implemented first the Toth's proposal, then Bezier clipping and recently we added a new algorithm following one of our ideas. This new approach was called "Toth speed". It is a combination of the Toth and Bezier clipping algorithms that outperforms their performances [30, 41]. The idea is to reduce the application of the interval Newton iterations of the Toth method, trying to find the solution using a simple Newton, in the knowledge that, if a solution exists, it is unique. If this fails, an interval Newton iteration is applied. Moreover, when the Toth algorithm is forced to use binary subdivision, our method uses Bezier clipping.

All these algorithms are included in the system and can be compared on complex ray traced scenes. Note that a ray-tracing algorithm exploits a certain number of optimization techniques in order to speed up performance, such as subdividing a 3D scene in voxels, subdividing the NURBS surface in rational Bezier patches, further subdividing each rational Bezier patch until a given flat tolerance is reached [19]. These optimizations can improve a ray/patch intersection in different ways.

The optimizations mentioned in the case of trimmed surfaces, introduce a new problem: the subdivision of a patch implies the redefinition of the trimmed domain. The solution to this problem may be more or less complex, depending on the type of representation used for the trimmed region.

Figure 4.2 shows an example of a trimmed NURBS sphere rendered by a ray-tracing algorithm obtained by *xcmodel* .

5. Solid modeling. Following the B-rep (Boundary representation) of a solid, we can define as 'primitive solid' any solid that has, as its boundary, a single surface that is simply closed in order to separate the space into two parts, one of which will be enclosed. According to convention, the boundary surface must be parametrized in such a way that the surface normal vector indicates the area outside the solid. For example, spheres, cylinders, etc. are included in this definition, as well as closed sculptured solids. Primitive solids can also be combined by Boolean operations, such as intersection, union and difference, in order to create a complex solid. The difficulties posed by Boolean combinations of free form objects have been overcome through the use of trimmed surfaces.

The boundary of a solid object S is defined as follows

$$bS = \bigcup_{i=1}^{k} \mathbf{r}_i(D_i) \qquad (5.1)$$

where D_i is the trimmed region associated with the surface $\mathbf{r}_i$. A primitive solid is then defined by $k = 1$ and $D_1 = U \times V$. The boundary of the solid S is a closed surface.

Given two solids A and B, respectively defined by their boundary surfaces, the Boolean operation problem consists in determining the boundaries of the solids $A \bigcup B$, $A \bigcap B$, $A - B$ starting from bA and bB. We will refer to this process as *the set operation algorithm.*

The boundary of the resulting solid is given by the equations known as the boundary formula [34]:

$$\begin{aligned} b(A \bigcup B) &= (bA \bigcap cB) \bigcup (bB \bigcap cA) \\ b(A \bigcap B) &= (bA \bigcap iB) \bigcup (bB \bigcap iA) \\ b(A - B) &= (bA \bigcap cB) \bigcup (bB \bigcap iA) \end{aligned} \qquad (5.2)$$

where iX and cX represent, respectively, the interior and the complement of the solid X.

The trimmed patches of the resulting solid are the same as those of bA and bB, except for the portion of the patches that have been trimmed away (see Figure 5.1).

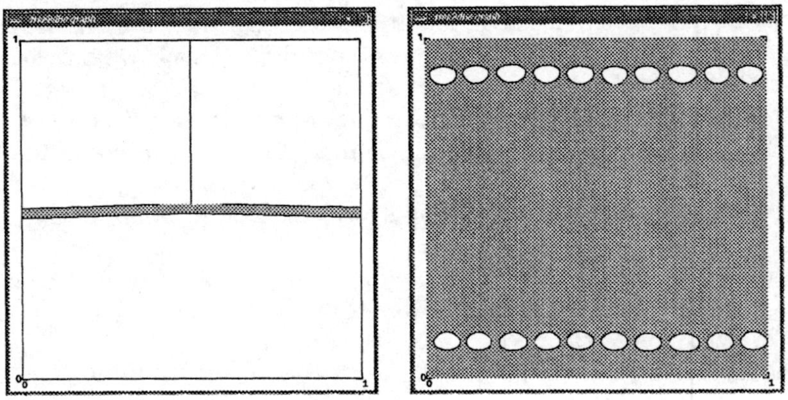

FIG. 5.1. *Solid modeling; trimmed domains for the solid finger disk in Fig.5.2; the cylinder part of the solid (left), the disk part of the solid (right).*

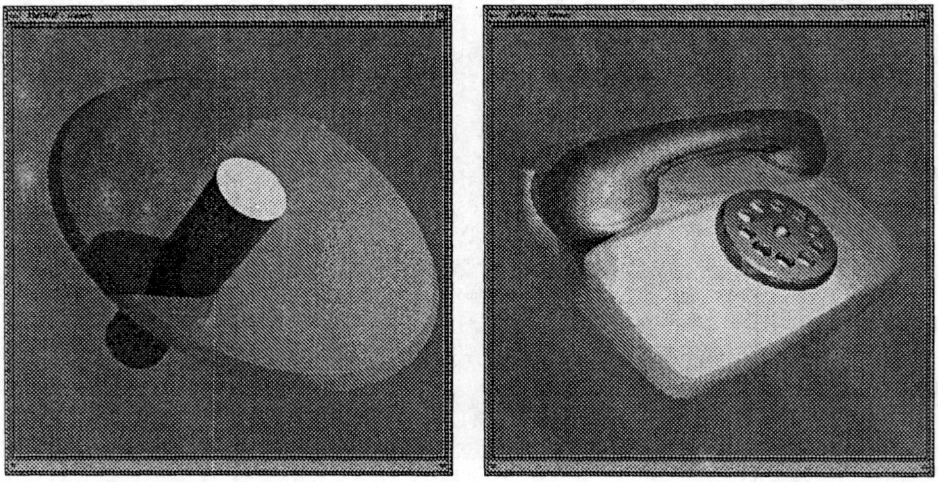

FIG. 5.2. *Solid modeling; primitive solids to compose the finger disk by difference (left), the resulting scene (right).*

In [5] and [6], a set operation algorithm is given, that operates on solids modelled with trimmed patches. In [17], the trimmed patch is represented by a dual representation, i.e. the trimmed patch is given both by an implicit surface and a parametric patch, and a basic theory is presented for a set operation algorithm, which operates on the dual representation. In [7] the proposal in [5] is taken up, adapted for NURBS surfaces and implemented in the *xcmodel* system. A complete description and implementation of a set operation algorithm is technically complex. The basic idea in [7] consists in avoiding the intersections between trimmed patches by considering the intersections between the patches over the whole domain. After this, the trimmed regions (2D CSG tree) resulting from the intersection between the trimmed regions of the solids and the trimmed regions obtained by the patch/patch intersection operations, are determined. In the following, we will give a brief description of untrimmed surface/surface intersection (SSI), 2D and 3D PMC and 2D curve/curve intersection (CCI), which represent the basic steps involved in this process.

5.1. SSI. The methods available for intersecting patches can be divided into two categories: curve following and subdivision. Firstly, *the curve following method* finds some points of intersection. Then, the points on the intersection curve are detected using a numerical method [1, 2]. This method is especially useful for processing singular cases of intersection [3]. The *subdivision method* divides the problem into smaller problems by approximating the patch into simpler linear or quadratic sub-patches [23, 6]. The patches are intersected, resulting in curves that are then refined by another numerical method. A modified version proposed in [1] has been implemented in the *xcmodel* system [8]. Here, the balance between robustness and efficiency, a problem for SSI algorithms, is solved by a "little trick". The four main steps are the following:

- *Adaptive mesh generation*; this consists in the approximation of each of the surfaces using an adaptive grid of isoparametric curves, approximated by piecewise linear curves within a given tolerance. From this grid we easily obtain a surface triangulation;
- *Initial intersection point generation*; the intersections between the grid segments of a surface and the triangles of the other surface are determined in order to obtain at least one starting point for each intersection curve;
- *Following an intersection curve*; with an initial point on the intersection curve, we move along the curve by steps. To do this, first of all we find an estimate for the next point on the intersection curve, and then we evaluate an exact intersection within a given tolerance;
- *Sorting*; finally, the curve intersection segments found in the preceding step are linked together in order to obtain complete intersection curves.

5.2. 2D and 3D PMC. Point Membership Classification is a function that takes, as its input, a point P and a closed set S, and which returns one of three possible outcomes: P is inside S, P is on the boundary of S, or P is outside S. The set operation algorithm makes extensive use of PMC in both two and three dimensions. During the construction of the 2D CSG tree, a single 3D point is classified with respect to the opposing solid. The result will be used for further 2D classifications in order to build a collection of half-spaces which, ultimately, become the leaves of the tree. A single concept is used for both 2D and 3D classification: a ray is extended from the point to be classified, and the number of intersections of the ray with the boundary of the set determines the membership status of the point. An even number of intersections means that the point is outside the set, while an odd number implies that the point is inside. While the concept is the same, its implementation is very different for two or three dimensions. In the 2D case, the ray is intersected with polygons or curved boundaries. In the case of curved boundaries all the roots in the parametric interval [0,1] of the scalar NURBS function resulting from the ray/curve intersection must be found. To solve this problem the methods in [21, 27, 4, 31] are explored and compared [41]. In the 3D case, the boundary consists of trimmed patches, and a ray/patch method is required (see section 4).

5.3. CCI. This problem can either be tackled geometrically, that is, using subdivision techniques for the curves, that exploit the convex-hull property of the NURBS curves [27, 25, 38], or numerically. The latter is done either by finding all the zeros of the function resulting from the system, when a curve has been implicitized [39] (this is profitably applicable only for low degree curves), or by finding the zeros of the vector function in two variables (2D surface), obtained from the difference between the curves [16]. The problem of finding the zeros of a 2D surface can be solved using the same methods as those used to determine the ray/patch intersections, (see section 4).

Figure 5.2 shows an example of solid modelling generated by the *xcmodel* system.

6. Surface modeling. The refinement technique for spline surfaces is a well known tool used to represent a surface defined by a finer control mesh with the feature that each single control point will have a more local effect on the surface. This allows us to model small patches and to modify their shape without affecting the rest. This idea led to the definition of Hierarchical Spline Surfaces (HSS) [18], proposed in 1988 by Forsey and Bartels, enabling a surface to be refined and details added, using a hierarchy, whose levels correspond to the varying levels of refinement of the surface patches. This technique enables us to restrict the influence of refinement to the relevant parts of the surface.

Our paper illustrates how the *xcmodel* system uses trimmed NURBS surfaces to represent HSS and enables modeling to be carried out with these surfaces. For this purpose *rectangular trimmed NURBS surfaces* are defined. They consists in a particular class of trimmed surface, whose trimmed away regions are defined as rectangular in the parametric domain.

We can represent an HSS as a NURBS surface $\mathbf{r}$ and a set of regions D_i as follows:

$$\bigcup_{i=1}^{n} \mathbf{r}(D_i)$$

where each region D_i represents the ith level of detail of an HSS and is defined by

$$D_i = A_{i-1} - A_i \quad \text{with} \quad A_0 = U \times V, \quad A_i = \bigcup_{j=1}^{m_i} \hat{A}_{ij}$$

and where $\hat{A}_{ij} = [u_{a_j}, u_{b_j}] \times [v_{c_j}, v_{d_j}]$ represents the jth rectangular trimmed region at level i, and is a subset of A_{i-1}. Note that $\bigcup_{i=1}^{n} D_i = U \times V$.

The modeling phase, by adding or editing details, consists in first selecting the rectangular patch on the surface to be modified, then determining the corresponding rectangular subdomain and trimming it so that the domain becomes restricted to the chosen patch, which now acts as an autonomous and untrimmed surface. The appropriate refinement is carried out and the control points that can be moved without altering their order of continuity with the original surface are determined. In the initial HSS work [18], the control points that cannot be altered in order to avoid losing the original continuity are already clearly specified. These form a frame $m-1$ wide and $n-1$ high in the control mesh, where m and n are the NURBS surface orders. Any change in position and/or weight of all or only some of the control points within this frame results in a local modification of the patch, while maintaining the original continuity with the rest of the surface.

In Figure 6.1 we show an example of HSS modeling using trimmed NURBS.

Furthermore, in order to design multi-sided surfaces, it makes perfect sense to consider a modeling procedure that trims one surface by means of another surface, even if no set operation is indicated. To do this, the *xcmodel* system allows the user to model the 2D trimming curves and define a 2D CSG tree for a given surface.

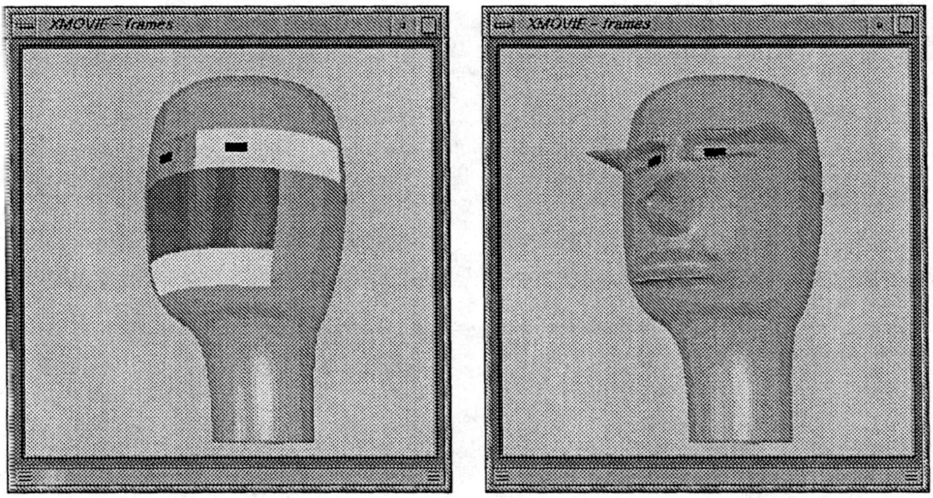

FIG. 6.1. *HSS modeling using trimmed NURBS; on the left three levels of rectangular patches are shown, on the right the modeled surface is given.*

7. Conclusions and new trends. In this paper we have discussed the usefulness of the trimmed NURBS paradigm for modeling.

Trimmed NURBS present at least two difficulties:

- trimming is expensive and prone to numerical errors;
- it becomes difficult to maintain smoothness at the seams of the patchwork, when the model is animated.

Recently, subdivision surfaces have been considered as a way of overcoming NURBS topological limitations without mesh degeneracy. They do not require trimming, and model smoothness is automatically guaranteed, even in animated models. The idea is to refine an irregular/regular mesh, by creating a new mesh that approximates the old one. By repeating this process, a smooth surface is formed as the limit of the process itself. Subdivision algorithms are quite simple, and are generally able to produce quite complex objects, although the basic theoretical modeling background has yet to be explored, and effective, intuitive user interface, designed to exploit the potential of the representation to the full, requires further development. Although trends in geometric modeling community research are moving towards an exhaustive exploration of the subdivision paradigm, at the moment, trimming modeling still remains the most complete and powerful means for free form object modeling.

REFERENCES

[1] R.E.Barnhill, G.Farin, M.Jordan, B.R.Piper, *Surface Surface Intersection*, Computer Aided Geometric Design, 4 (1987), pp. 3–16.

[2] R.E.Barnhill and S.N.Kersey, *A marching method for parametric surface/surface intersection*, Computer Aided Geometric Design, 7 (1990), pp. 257–280.

[3] C.L.Bajaj, J.E.Hoffman, R.E.Lynch, *Tracking surface intersection*, Computer Aided Geometric Design, 5 (1988), pp. 285–307.

[4] R.Brent, *Algorithm for minimization without derivatives*, Prentice-Hall, 1973.

[5] M.S. Casale, *Free-form solid modeling with trimmed surface patches*, IEEE Computer Graphics & Applications, Jan. (1987), pp. 33–43.

[6] M.S. Casale and J.E.Bobrow, *A set operation algorithm for sculptured solids modeled with trimmed patches*, Computer Aided Geometric Design, 6 (1989), pp. 235–247.

[7] G.Casciola and B.Quaquarelli, *Primitive solide, operazioni Booleane e NURBS*, Proceedings of ICOGRAPHICS'90, Milano, (1990).

[8] G.Casciola and S.Morigi, *Il problema SSI nella modellazione solida con superfici NURBS*, Atti dell'Accademia delle Scienze dell'Istituto di Bologna, serie V, n.6, (1995).

[9] G.Casciola and S.Morigi, *xcmodel: an aCADemic system*, Proceedings of Cofin.97, Ferrara, January 19-21, (2000).

[10] G.Casciola, *xcmodel: a system to model and render NURBS curves and surfaces* User's Guide (1999), http://www.dm.unibo.it/~casciola/html/xcmodel.html

[11] G.Casciola, *xccurv: the 2D modeller* User's Guide (1999), http://www.dm.unibo.it/~casciola /html/xcmodel.html

[12] G.Casciola, *xcsurf: the 3D modeller* User's Guide (1999), http://www.dm.unibo.it/~casciola /html/xcmodel.html

[13] G.Casciola, *xcbool: the object composer* User's Guide (1999), http://www.dm.unibo.it/~casciola /html/xcmodel.html

[14] G.Casciola, *xcrayt: the scene descriptor* User's Guide (1999), http://www.dm.unibo.it/~casciola /html/xcmodel.html

[15] E.Cohen, T.Lyche, R.F.Riesenfeld, *Discrete B-splines and subdivision techniques in computer-aided geometric design and computer graphics*, Computer Graphics and Image Processing, 14 (1980), pp. 87–111.

[16] T.Dokken, *Finding intersections of B-splines represented geometries using recursive subdivision techniques*, Computer Aided Geometric Design, 2 (1985), pp. 189–195.

[17] R.T.Farouki, *Trimmed-surface algorithms for the evaluation and interrogation of solid boundary representations*, IBM J.RES.DEVELOP. , 31 (1987), pp. 314–333.

[18] D.R.Forsey and R.H.Bartels, *Hierarchical B-spline refinement*, Proceeding SIGGRAPH'88, (1988). In Computer Graphics, 22 (1988), pp 205–212.

[19] A.Fujimoto, T. Tanaka, and K. Iwata, *ARTS: accelerated ray-tracing system*, IEEE Comp. Graphics and Appl., Apr. (1986), pp. 16–26.

[20] A.Glassner ed., *Introduction to ray tracing*, Academic Press, 1989.

[21] A.Grandine, *Computing zeroes of spline functions*, Computer-Aided Geometric Design, 6 (1989), pp 129–136.

[22] B.Hamann, P.-Y. Tsai, *A tessellation algorithm for the representation of trimmed NURBS surfaces with arbitrary trimming curves*, Computer-Aided Design, 28 (1996), pp. 461–472.

[23] E.G.Houghton, R.F.Emnett, J.D.Factor, C.L.Sabharwal, *Implementation of a divide-and-conquer method for intersection of parametric surfaces*, in R.E.Barnhill, W.Boehem, eds. Surfaces in Computer Aided Geometric Design (1984), North-Holland, Amsterdam.

[24] K.Joy, M.Bhetanabhotla, *Ray tracing parametric patches using numerical techniques and ray coherence*, Proceeding of SIGGRAPH'86, (1986). In Computer Graphics, 20 (1986), pp.279–285.

[25] P.A.Koparkar and S.P.Mudur, *A new call of algorithms for the processing of parametric curves*, Computer-Aided Design, 15 (1983), pp 35–46.

[26] S.Kumar and D.Manocha, *Efficient rendering of trimmed NURBS surfaces*, Computer-Aided Design, 27 (1995), pp. 509–521.

[27] J.M.Lane, R.F.Riesenfeld, *A theoretical development for the computer generation and display of piecewise polynomial surfaces*, IEEE Transaction on PAMI, 2 (1980), pp.35–46.

[28] C.T.Loop and T.D.DeRose, *A multisided generalization of Bezier surfaces*, ACM Transaction on Graphics, 8 (1989), pp.204–234.

[29] W.L.Luken, *Tessellation of trimmed NURBS surfaces*, Computer Aided Geometric Design, 13 (1996), pp 163–177.

[30] M.Marini, D.Grilli, *Un algoritmo di ray-tracing per superfici spline razionali trimmate*, Master Thesis in Compuer Science, University of Bologna, (1995).

[31] T.Nishita, T.W.Sederberg and M. Kakimoto, *Ray tracing trimmed rational surface patches*, ACM Computer Graphics, 24 (1990), pp. 337–345.

[32] L.A.Piegl and A.M.Richard, *Tessellating trimmed NURBS surfaces*, Computer-Aided Design, 27 (1995), pp. 16–26.

[33] L.A.Piegl and W. Tiller, *Geometry-based triangulation of trimmed NURBS surfaces*, Computer-Aided Design, 30 (1998), pp. 11–18.

[34] A.A.G.Requicha, *Mathematical models of rigid solids*, Tech. Memo. 28, Production Automation Project, University of Rochester, (1977).

[35] A.Rockwood, K.Heaton and T.Davis, *Real time rendering of trimmed surfaces*, Proceeding of SIG-

GRAPH'89, (1989). In Computer Graphics, 23 (1989), pp. 107–116.

[36] D.Rogers, *Procedural elements for computer graphics*, McGraw-Hill, 1985.

[37] S.Roth, *Ray casting for solid modeling*, Computer Graphics and Image Processing, 18 (1982), pp. 109–144.

[38] T.W.Sederberg and T.Nishita, *Curve intersection using Bezier-clipping*, Computer-Aided Design, 24 (1990), pp. 538–549.

[39] T.W.Sederberg and Parry, *Comparison on three curve intersection algorithm*, Computer-Aided Design, 18 (1986), pp 58–63.

[40] X.Sheng and B.E.Hirsch, *Triangulation of trimmed surfaces in parametric space*, Computer-Aided Design, 24 (1992), pp. 437–444.

[41] S.Spagna, *Analisi, sviluppo e applicazione di metodi di intersezione per spline razionali*, Master Thesis in Compuer Science, University of Bologna, (1998).

[42] M.Sweeney and R.Bartels, *Ray tracing free-form B-spline surfaces*, IEEE Computer Graphics & Applications, 6 (1986) pp.41–49.

[43] D.Toth, *On ray tracing parametric surfaces*, Proceeding of SIGGRAPH'85, (1985). In Computer Graphics, 19 (1985) pp.171–179.

[44] H.B.Voelcker, A.G.Requicha, *Research in solid modelling at the University of Rochester: 1972-87*, in Fundamental developments of Computer-Aided Geometric Modeling, L.A.Piegl ed., London, Academic Press 1993.

[45] J.Warren, *Creating multisided rational Bezier surfaces using base points*, ACM Transaction on Graphics, 11 (1992) pp.127–139.

[46] T.Whitted, *An improved illumination model for shaded display*, Communication ACM, 23 (1980), pp. 96–102.

[47] C.Woodward, *Ray tracing parametric surfaces by subdivision in viewing plane*, in W.Strasser and H.-P.Seidel, eds. Theory and Practice of Geometric Modeling, Springer Verlag (1989), pp.273–290.

MULTISCALE HYPERBOLIC EQUATIONS: NUMERICAL APPROXIMATION IN THE DIFFUSIVE REGIME *

V. COMINCIOLI†, G. NALDI‡, T. SCAPOLLA†, AND G. TOSCANI†

Abstract. Various problems occurring in applied sciences involve different scales in space or time. In many physical applications, the scaling parameter may differ in several order of magnitude from the rarefied regime to the hydrodynamic (diffusive) regime within the same problem. In these cases, it is desirable to develop numerical methods that work uniformly with respect to the scaling parameter. In the present paper we introduce and discuss recent results in this field. In particular, we illustrate the basic numerical techniques by means of the analysis of models of the kinetic theory of rarefied gases.

1. Introduction. A noticeable amount of problems occurring throughout engineering and applied sciences involve multiscales in space or time. Classical examples are the transport equations, such as the neutron transport and the radiative transfer, gas dynamics far from thermo or chemical equilibrium, fluid flows at different Reynolds numbers, etc. In many physical applications, the scaling parameter, i.e., the mean free path, may differ in several order of magnitude from the rarefied regime to the hydrodynamic (diffusive) regime within the same problem.

Usually, the first attempt to understand these multiscale phenomena relies in its asymptotic approximation, which leads (at least formally) to a substantially simpler macroscopic equation. When successful, the asymptotic technique provides a deep insight on the solution in certain asymptotic regimes.

A different attempt consists in the computation of a multiscale physical problem, with certain numerical resolution of small scales. In these cases, it is desirable to develop numerical methods that work uniformly with respect to the scaling parameter. In many applications, however, the small scale is not of primary physical interest, and it is satisfactory to capture the main physical properties at the macroscopic level. Therefore it is interesting to design underresolved numerical methods which describe accurately the desired macroscopic physics without resolving the small scales.

In last years, many authors started to develop systematically numerical techniques for solving kinetic equations in the diffusive regimes. The description of these tecniques will be the subject of this note. Although the approach is applicable to many physical problems of greater complexity, we will illustrate these basic techniques by means of a few simple models, of interest for the kinetic theory of rarefied gases.

In general, two types of scalings for hyperbolic transport equations are considered:

$$\partial_t U + \partial_x F(u) = \frac{1}{\epsilon} Q(U), \tag{1.1}$$

or

$$\epsilon\partial_t U + \partial_x F(u) = \frac{1}{\epsilon} Q(U). \tag{1.2}$$

*This work was supported by Progetto Nazionale di Ricerca MURST "Analisi numerica: metodi e software matematico" and by the European TMR Project Asymptotic Methods in Kinetic Theory, contract ERBFRMXCT 970157

†Dipartimento di Matematica, Università di Pavia, Pavia Italy, ({comi, scapolla, toscani} @dimat.unipv.it).

‡Dipartimento di Matematica e Applicazioni, Università di Milano-Bicocca, Milano Italy, (naldi@matapp.unimib.it).

These equations describe the evolution of a distribution function $U = U(x, v, t)$, where x, v, t are respectively the space, velocity and time variables. $Q(U)$ describes the variation of U due to interactions.

The first scaling is the usual Euler scaling, while the second one is the Navier-Stokes or diffusion scaling. In the present paper we will deal with equation (1.2). To be precise, we want to introduce underresolved discretizations of (1.2), i.e. discretizations such that $\Delta t, \Delta x/c \gg \epsilon$, where c is a typical characteristic speed. With this approach, we are mainly interested in capturing the macroscopic behaviour determined by the diffusion equation. Such a solution will be referred to as the *underresolved* solution. If $\epsilon \ll 1$, it is clear that one can solve directly the macroscopic equation. However, many problems require the simultaneous solution of equations like (1.1) or (1.2), where ϵ is small and large in different regions of the domain.

To handle these problems one can use a decomposition method, and separate the domain where ϵ is large from that where ϵ is small. In these regions, one just uses the macroscopic equation. This procedure requires to write consistent boundary conditions at the interfaces, which is practically impossible in nonlinear situations.

From a different viewpoint one can look for a scheme that works uniformly with respect to ϵ, which allows the use of underresolved discretizations. Even this second approach contains numerical challenges. As we shall see in the next sections, one of the key ingredients to overcome difficulties due to the loss of uniform consistency, is to consider only *asymptotic-preserving* (AP) schemes, namely schemes which, as $\epsilon \to 0$, become a consistent and stable discretization of the macroscopic equation. The definition of AP–schemes is quite recent, and is due to Jin, [22]. Schemes with the AP-property have been developed to solve PDE's of interest in different areas, and exist (without this classification) since many years. Among others, stationary problems for transport equations in diffusive regimes [18, 19, 27, 28, 29], hyperbolic systems with stiff relaxations [7, 17, 20, 37, 40], and kinetic modeling of semiconductor devices [16] have been studied in this way.

Let us now discuss in some details the main features of the method. A well-known nonlinear prototype is given by the Carleman model [9] in the diffusive scaling

$$
\begin{aligned}
\epsilon\partial_t u + \partial_x u &= \frac{1}{\epsilon}\,(v^2 - u^2),\\
\epsilon\partial_t v - \partial_x v &= \frac{1}{\epsilon}\,(u^2 - v^2),
\end{aligned}
\tag{1.3}
$$

where $(x, t) \in \mathbb{R} \times \mathbb{R}^+$, with initial data

$$
u(x, 0) = u_0(x) \geq 0 \quad v(x, 0) = v_0(x) \geq 0\,. \tag{1.4}
$$

Carleman equations model the evolution of a fictitious gas composed by two kinds of particles that move parallel to the x-axis with constant and equal speeds, either in the positive x-direction with a density u, or in the negative x-direction with a density v. Hence, the diffusive scaling leads to the hydrodynamics generated by this simple gas. Following the pionering paper by McKean [33], in recent years a rigorous justification of diffusive relaxation limits for Carleman-like kinetic models has been object of the studies of several authors [14, 31, 34, 42].

Comparing with the Euler scaling (1.1), this scaling is a long time scaling. To derive the asymptotic equation as $\epsilon \to 0$, we pass to the macroscopic variables, the mass ρ and the flux j defined by

$$\rho = u + v, \quad j = \frac{u - v}{\epsilon}. \tag{1.5}$$

We obtain the following hyperbolic system

$$\begin{aligned} &\partial_t \rho + \partial_x j = 0, \\ &\partial_t j + \frac{1}{\epsilon^2}\partial_x \rho = -\frac{2}{\epsilon^2}\rho j. \end{aligned} \tag{1.6}$$

When $\epsilon \ll 1$ system (1.6) is said to be *stiff*. In the small relaxation limit, $\epsilon \to 0^+$, the relaxation system (1.6) formally approximates to leading order

$$j = -\frac{1}{2}\partial_x \log \rho, \tag{1.7}$$

$$\partial_t \rho = \frac{1}{2}\partial_{xx} \log \rho. \tag{1.8}$$

As usual, the state satisfying (1.7) is called the *local equilibrium* while (1.8) corresponds to the equation of continuum mechanics generated by the kinetic model. Since the characteristic speeds of system (1.6) are inversely proportional to ϵ the *subcharacteristic condition* by Liu [10, 32] is automatically satisfied in the small relaxation limit. This can be seen applying the Chapman-Enskog expansion in (1.6). The key idea to deal with the problem (1.6) has been proposed in [23]. We rewrite (1.6) as a usual combination of a non-stiff advection part and a stiff relaxation part in the form

$$\begin{aligned} &\partial_t \rho + \partial_x j = 0, \\ &\partial_t j + a\partial_x \rho = -\frac{2}{\epsilon^2}\rho\left(j + \frac{1 - a\epsilon^2}{2}\partial_x \log \rho\right), \end{aligned} \tag{1.9}$$

where $a \leq 1/\epsilon^2$ is a positive constant. The new formulation (1.9) allows to construct relaxation schemes following some of the standard numerical techniques introduced for hyperbolic systems with stiff source terms [21, 26]. The first step is to apply the splitting method, considering separately the transport

$$\begin{aligned} &\partial_t \rho + \partial_x j = 0, \\ &\partial_t j + a\partial_x \rho = 0, \end{aligned} \tag{1.10}$$

and the relaxation

$$\begin{aligned} &\partial_t \rho = 0, \\ &\partial_t j = -\frac{2}{\epsilon^2}\left(j + \frac{1 - a\epsilon^2}{2}\partial_x \log \rho\right). \end{aligned} \tag{1.11}$$

For example, the standard explicit shock capturing methods can be applied to solve the transport step (1.10), while an implicit scheme can be used to solve the relaxation, with central differencing used to handle the derivative in the second equation of (1.11). Such a method has many interesting properties. In the small relaxation limit, we have a parabolic CFL condition $\Delta t \approx (\Delta x)^2$. In addition, in other regimes, the schemes maintain all the nice properties of those constructed earlier, such as the ability to capture shocks and other types of discontinuities with high resolutions, and avoiding solving nonlinear algebraic systems. Furthermore they are asymptotically preserving (AP). As $\epsilon \to 0$ the relaxation step (1.11) forces $j = -\frac{1}{2}\partial_x \log \rho$. Substituting this into (1.10) we obtain the desired diffusion limit. Finally, the relaxation step can be implemented explicitly, since ρ remains unchanged during the time step.

Moreover, we outline an important consequence of our procedure. Looking for an asymptotically preserving underresolved schemes that work well for $\epsilon \ll 1$ we obtain a way to solve numerically (1.3). The main advantage of numerically solving the relaxation system (1.6) instead of the original nonlinear diffusion equation (1.3) lies in the localized lower order term. This permits to reduce a nonlinear second order system to a semi-linear first order system and to avoid the use of Riemann solvers.

The rest of the paper is organized as follows. In Section 2 we review the diffusion limits and the relaxation approximation for various problems. In particular, the heat equation, the porous media equation, Burgers equation with viscosity and in general linear and nonlinear reaction-diffusion equations can be obtained by this limit procedure. In the same section, we look at the radiative transfer problem and its diffusion approximation. For many of these multiscale problems, we outline their numerical approximation. Finally, Section 3 presents several numerical experiments for nonlinear problems.

2. Multiscale problems in the diffusive regime. In this section we review different multiscale hyperbolic systems together with their Navier-Stokes macroscopic equations. The common feature of these systems is that they are represented at a mesoscopic level by discrete velocity models of the Boltzmann equation. For most of these models we illustrate the main tools used to overcome the numerical difficulties.

2.1. Nonlinear diffusion equations. A discrete-velocity model of the Boltzmann equation, in different regimes and in different multiscale limits was introduced and investigated in [11]. More precisely, in [11] was considered a modified Broadwell model with both non trivial fluid dynamical and hydrodynamical limits. In the fluid dynamical scaling, the macroscopic equations are the isothermal Euler system, while in the diffusive scaling the macroscopic density satisfies a nonlinear diffusion equation similar to (1.8). For numerical experiments a splitting method coupled with the Wild's sum approach [13] for the computation of the stiff collision term was used.

In [12] the equation of diffusion in a porous medium was obtained in a suitable scaling limit by a two velocity model of the Boltzmann equation (Carleman's model). A satisfactory analysis of the convergence was performed in [31] and, for the fast diffusion case, in [45]. A first attempt to construct a numerical approximation of this kind for multiscale hyperbolic systems in the small relaxation time, can be found in [37].

In these papers, the kinetic model is described by the equations

$$\begin{aligned} \partial_t u + c\partial_x u &= K(u+v, x)(v-u), \\ \partial_t v - c\partial_x v &= K(u+v, x)(u-v), \end{aligned} \tag{2.1}$$

where c is the modulus of the constant speed of the particles, and K is a non-negative rate function. It is possible to consider (2.1) as a generalized Carleman's model (1.3)(as discussed in the Introduction, Carleman's equations correspond to the choice $K(u+v,x) = u+v$. In the diffusive scaling, if we consider for simplicity $c = 1$, we have

$$\begin{aligned} \partial_t u + \frac{1}{\epsilon}\partial_x u &= \frac{1}{\epsilon^2}K(u+v,x)(v-u), \\ \partial_t v - \frac{1}{\epsilon}\partial_x v &= \frac{1}{\epsilon^2}K(u+v,x)(u-v). \end{aligned} \tag{2.2}$$

Moving to the macroscopic variables, the mass density $\rho = u+v$ and the flux $j = (u-v)/\epsilon$, system (2.2) can be rewritten as

$$\begin{aligned} &\partial_t \rho + \partial_x j = 0\,, \\ &\partial_t j + \frac{1}{\epsilon^2}\partial_x \rho = -\frac{2}{\epsilon^2}K(\rho,x)j\,. \end{aligned} \tag{2.3}$$

In the zero relaxation limit, as $\epsilon \to 0^+$, system (2.3) can be approximated to a leading order by the following equilibrium and, respectively, continuum equation

$$\begin{aligned} &j = -\frac{1}{2K(\rho,x)}\partial_x \rho\,, \\ &\partial_t \rho - \partial_x\left[\frac{1}{2K(\rho,x)}\partial_x \rho\right] = 0. \end{aligned} \tag{2.4}$$

In [31] the collisional rate $K = \rho^\alpha$, $\alpha < 0$, has been considered. In this case, the limit equation for the density is the well-known porous media equation

$$\partial_t \rho = \frac{1}{2(1+|\alpha|)}\partial_{xx}\rho^{1+|\alpha|}. \tag{2.5}$$

A more general collisional term, including linear and nonlinear collisions, has been considered in [23]

$$K(u,v,x) = k(u,v)\,(-\alpha_\epsilon u + \beta_\epsilon v + \gamma_\epsilon uv)\,. \tag{2.6}$$

In (2.6) $k(u,v)$, α_ϵ, β_ϵ, γ_ϵ are suitable nonnegative functions describing the various rates of collisions. With these equations it is possible to reach at the limit a large set of linear/nonlinear advection diffusion equations. The various cases are summarized in Table 2.1.

The main advantage of numerically solving relaxation system (2.3) over the original nonlinear convection–diffusion lies in the localized lower order term. This permits to reduce a nonlinear second order system to a semi-linear first order and to avoid the use of Riemann solvers. However, as observed recently [23, 24, 37, 38], the numerical solution to (2.3) is challenging due to the stiffness of the problem for both the convection and the relaxation terms. If a semi-implicit scheme (convection terms explicit and relaxation terms implicit) is used one needs a CFL condition of type $\Delta t < \epsilon\Delta x$, which is very expensive when ϵ is small. If a fully implicit scheme is used, due to the nonlinear nature of the numerical flux for the convection term, a major ingredient for all modern high order shock capturing schemes for hyperbolic conservation laws, one needs an iterative method to solve the resulting nonlinear

name	values of rate constants	equation for density ρ
Classical heat equation	$k(u,v) = k_0,\ \alpha_\epsilon = \beta\epsilon = 1,\ \gamma_\epsilon = 0.$	$\partial_t \rho = \frac{1}{2k_0}\partial_{xx}\rho$
Advection-diffusion equation	$k(u,v) = k_0,\ \alpha_\epsilon = 1 + A\epsilon, \beta\epsilon = 1 - A\epsilon,\ \gamma_\epsilon = 0.$	$\partial_t \rho + A\partial_x \rho = \frac{1}{2k_0}\partial_{xx}\rho$
The porous media equation	$k(u,v) = (u+v)^m,\ \alpha_\epsilon = \beta\epsilon = 1,\ \gamma_\epsilon = 0.$	$\partial_t \rho = \frac{1}{2(1-m)}\partial_{xx}\rho^{1-m}$
The Burgers equation	$k(u,v) = k_0,\ \alpha_\epsilon = \beta\epsilon = 1,\ \gamma_\epsilon = \epsilon C.$	$\partial_t \rho + \frac{C}{4}\partial_x \rho^2 = \frac{1}{2k_0}\partial_{xx}\rho$

TABLE 2.1
List of some possible diffusive limit.

systems, which significantly affects the algorithm.

Moreover in [37] it is pointed out that, on the contrary to the Euler scaling where only one stiff term is present in the collisional operator, the common splitting of the operator into the transport and relaxation parts does not provide the correct diffusive behavior as $\epsilon \to 0^+$. In details, the usual splitting approach applied to system (2.3) consists in solving first the homogeneous (transport) problem

$$\begin{aligned} &\partial_t \rho + \partial_x j = 0, \\ &\partial_t j + \frac{1}{\epsilon^2}\partial_x \rho = 0, \end{aligned} \tag{2.7}$$

and then the following system of ordinary differential equations (relaxation)

$$\begin{aligned} &\partial_t \rho = 0, \\ &\partial_t j = -\frac{2}{\epsilon^2} K(\rho, x) j. \end{aligned} \tag{2.8}$$

Hence, in the limit $\epsilon \to 0^+$ the equilibrium equations become

$$j = 0, \quad \rho = \rho_0, \tag{2.9}$$

which are inconsistent with the limit state expressed by the second equation (2.4). On the other hand the natural splitting of problem (2.3) consists in the separation of the non-stiff part

$$\begin{aligned} &\partial_t \rho + \partial_x j = 0, \\ &\partial_t j = 0, \end{aligned} \tag{2.10}$$

from the stiff relaxation part

$$\begin{aligned} &\partial_t \rho = 0, \\ &\epsilon^2 \partial_t j = -2K(\rho, x) j + \partial_x \rho. \end{aligned} \tag{2.11}$$

This kind of scale splitting leads to the right asymptotic behavior but not provide any simplification to the numerical approach.

A characteristic based method for hyperbolic system with diffusive scaling is discussed in [43] by using the so called hyperbolic heat equation as a test problem. In particular in [43] it is shown that this method performs better than the usual splitting scheme and yields accurate solutions even in the case of values of the parameter ϵ smaller than the time discretization step. Unlikely the geometry of the characteristic mesh does not allow to obtain in the diffusive scaling a stability constraint independent of the parameter ϵ.

As observed in [20], the long time behavior of equations (1.2) is governed by a reduced system which is parabolic in nature. Then it follows that the standard high resolution methods for hyperbolic consevation laws fail to capture this parabolic behavior unless the small relaxation scale is resolved by a fine spatial mesh. In the linear case (that is by taking in (2.6) $k(u,v) = k_0$, $\alpha_\epsilon = \beta\epsilon = 1$, $\gamma_\epsilon = 0$), for example, the classical first order upwind method on the characteristic variables $\rho \pm \epsilon j$ leads to a truncation error with the term $(\Delta x/\epsilon)\partial_{xx}\rho$, where Δx is the size of the space mesh. When $\Delta x \gg \epsilon$ the numerical dissipation dominates the physical dissipation and the numerical solution will dissipate faster than the physical one. A similar behaviour remains also for second order upwind schemes. In order to avoid these phenomena in [20] a modified high order Godunov scheme with a correct asymptotic behavior was used. This idea is difficult to generalize for systems like (2.3) because suitable numerical fluxes have to be computed by using asymptotics which are dependent of the problem.

More recently, a fully discrete high order scheme for problem (2.3) has been studied in [37]. In particular, the paper introduces a underresolved numerical scheme that works in the stiff regime. The idea is to couple a modified upwind selection with a fully implicit centered based method. In the limit $\epsilon \to 0^+$, and for the linear problem, it is shown that the scheme is equivalent to a second order discretization of the equilibrium diffusion equation under a stability constraint independent of ϵ. Some numerical examples on one dimensional systems support this analysis.

This short review clarifies the main difficulties in working with diffusive multiscale problems. To overcome most of these difficulties, the key idea was introduced in [23]. The starting point is to reformulate (2.3) as a usual combination of a non-stiff advection part and a stiff relaxation part in the form

$$\begin{aligned} &\partial_t\rho + \partial_x j = 0, \\ &\partial_t j + \phi\partial_x\rho = -\frac{1}{\epsilon^2}F_\epsilon(\rho,\partial_x\rho,j,\epsilon), \end{aligned} \tag{2.12}$$

where $F_\epsilon(\rho,\partial_x\rho,j,\epsilon) = 2K(\rho,j)j + (1-\epsilon^2\phi)\partial_x\rho$, and $\phi = \phi(\epsilon)$ is such that $0 \le \phi \le 1/\epsilon^2$. This restriction on ϕ guarantees the positivity of $\phi(\epsilon)$ and $(1-\epsilon^2\phi(\epsilon))$ so that the problem remains well-posed uniformly in ϵ. The simplest choice of ϕ is

$$\phi(\epsilon) = \min\left\{1, \frac{1}{\epsilon}\right\}, \tag{2.13}$$

but other and smoother functions ϕ could be taken. As $\epsilon \to 0$; the second equation in (2.12) forces the flux to get the correct diffusion limit. This is of course mantained numerically by discretizing the system in an appropriate way. A natural splitting on (2.12) consists of a combination of the relaxation step

$$\begin{aligned} \partial_t \rho &= 0, \\ \partial_t j &= -\frac{1}{\epsilon^2} F_\epsilon(\rho, \partial_x \rho, j, \epsilon). \end{aligned} \tag{2.14}$$

with the transport step

$$\begin{aligned} \partial_t \rho + \partial_x j &= 0, \\ \partial_t j + \phi \partial_x \rho &= 0. \end{aligned} \tag{2.15}$$

Step (2.15) can be solved using an explicit scheme, whereas for step (2.14) we have to use an implicit scheme. For example, by using the backward Euler method for the relaxation step (2.14) we have the following semidiscrete scheme

$$\begin{aligned} \frac{\rho^{n+1} - \rho^n}{\Delta t} &= 0, \\ \frac{j^{n+1} - j^n}{\Delta t} &= -\frac{1}{\epsilon^2} F_\epsilon(\rho^{n+1}, \partial_x \rho^{n+1}, j^{n+1}, \epsilon). \end{aligned} \tag{2.16}$$

As usual, Δt is the time step. We consider for semplicity a uniform discretization, and with the index n we indicate the $n-$th time level corresponding to $t_n = n\Delta t$. Although we have an implicit relaxation term, (2.16) can be implemented explicitly, since ρ is preserved in the first equation in (2.16), i.e., $\rho^{n+1} = \rho^n$. Moreover the function F_ϵ is linear with respect to the variable j. Thus our implicit scheme, yielding an ϵ-independent numerical stability, can be implemented explicitly. The spatial derivative that appears in the function F_ϵ in (2.16) can be approximated using central differences in the diffusive limit, for small values of ϵ. Moreover, it is possible to show that this scheme is AP. In fact, as $\epsilon \to 0$, we get the correct diffusion equation. Finally, the new formulation (2.12) gives the way to construct high order relaxation schemes following the numerical techniques introduced for hyperbolic systems with stiff source terms [21, 26].

To obtain a fully discrete scheme we consider here an uniform discretization both in space and in time. Let $x_{i+1/2}$, $i = \dots, -1, 0, 1, \dots$ with space step $\Delta x = x_{i+1/2} - x_{i-1/2}$, t_n, $n = 0, 1, 2, \dots$ with $\Delta t = t_{n+1} - t_n$. As usual $U^n_{i+1/2}$ is the nodal value of $U(x,t)$ at $x = x_{i+1/2}$, $t = t_n$, and U^n_i is the cell average value on the $i-$th cell at time t_n defined by

$$U^n_i = \frac{1}{\Delta x} \int_{x_{i-1/2}}^{x_{i+1/2}} U(x, t_n) dx.$$

Then the first order scheme for the relaxation step reads

$$\begin{aligned} \frac{\rho_i^{n+1/2} - \rho_i^n}{\Delta t} &= 0, \\ \frac{j_i^{n+1/2} - j_i^n}{\Delta t} &= -\frac{1}{\epsilon^2} F_\epsilon^{n+1/2}, \end{aligned} \tag{2.17}$$

where $F_\epsilon^{n+1/2} = F_\epsilon(\rho_i^{n+1/2}, (\rho_{i+1/2}^{C,n+1/2} - \rho_{i-1/2}^{C,n+1/2})/\Delta x, j_i^{n+1/2}, \epsilon)$. The convective step takes the form

$$\begin{aligned} \frac{\rho_i^{n+1} - \rho_i^{n+1/2}}{\Delta t} + \frac{j_{i+1/2}^{U,n+1/2} - j_{i-1/2}^{U,n+1/2}}{\Delta x} &= 0, \\ \frac{j_i^{n+1} - j_i^{n+1/2}}{\Delta t} + \phi \frac{\rho_{i+1/2}^{U,n+1/2} - \rho_{i-1/2}^{U,n+1/2}}{\Delta x} &= 0. \end{aligned} \tag{2.18}$$

In (2.17), to agree with the nature of the limit system, we select the central difference operator for the numerical fluxes $\rho_{i\pm1/2}^{C,n+1/2}$. Moreover we remark that if $\rho_{i\pm1/2}^{C,n+1/2}$ is linear and F_ϵ is linear in the variable $j_i^{n+1/2}$, the scheme becomes explicit.

Since (2.18) is an hyperbolic system, we consider instead some high order upwind selection for $\rho_{i\pm1/2}^{U,n+1/2}, j_{i\pm1/2}^{U,n+1/2}$. In the limit $\epsilon \to 0^+$, from (2.17) and (2.18) it is possible to obtain the numerical asymptotic limit which agrees with the continuous one.

Let

$$\partial_t U + A(U) = B(U), \tag{2.19}$$

with U the numerical solution, $A(U)$ the convective term and $B(U)$ the relaxed term in the diffusive regime. Starting from (2.17)(2.18) it is possible [26, 7, 21] to build up a second order scheme by considering the following Runge-Kutta type method

$$\begin{aligned} U^{(1)} &= U^n - \zeta_1 \Delta t A(U^n), \\ U^* &= U^{(1)} - \eta_1 \Delta t B(U^*) - \eta_2 \Delta t B(U^n), \\ U^{(2)} &= U^* - \zeta_2 \Delta t A(U^*), \\ U^{**} &= \beta U^n + \gamma U^{(2)}, \\ U^{n+1} &= U^{**} - \eta_3 \Delta t B(U^{n+1}), \end{aligned} \tag{2.20}$$

with $\zeta_1 = 3/2$, $\eta_1 = 1/4$, $\eta_2 = 5/4$, $\zeta_2 = 3/4$, $\beta = 5/9$, $\gamma = 4/9$ and $\eta_3 = 1/3$.

2.1.1. Numerical examples. We present here numerical examples of some simple prototype of hyperbolic systems with diffusive relaxation. In the first test we compare the classical upwind approach (UW-scheme) coupled with the splitting without the function ϕ but with the classical CFL stability condition, $\Delta t = \epsilon \Delta x/2$, and the first order relaxation scheme (DR-scheme) with splitting function ϕ as in (2.13). We compute the different schemes in the stiff regime for $\epsilon = 10^{-8}$ for the linear heat equation as limit state, by solving a Riemann problem with the initial data

$$\begin{cases} \rho_L = 2.0, \quad j_L = 0, & 0 < x < 0.5, \\ \rho_R = 1.0, \quad j_R = 0, & 0.5 < x < 1, \end{cases} \tag{2.21}$$

on the interval $[0, 1]$. We take 100 spatial cells and $\Delta t = 0.005$. We compare the schemes with the 'exact' solution obtained with a fine spatial grid of 5000 cells. The solution output at $t = 0.03$, depicted in Figure 2.1b, shows that, although the relaxation time is underresolved, the DR-scheme captures the correct parabolic behavior given by the heat equation. Conversely, in Figure 2.1a we see that at $t = 0.02$ the upwind scheme decays at unphysical rate.

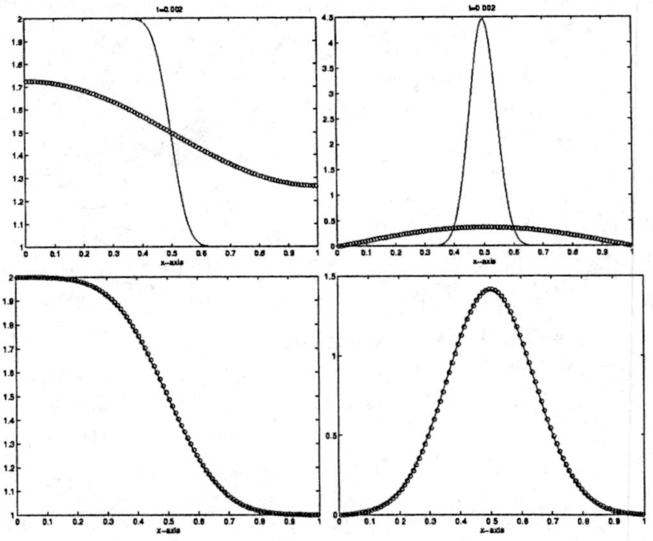

FIG. 2.1. *Numerical solutions of equations (2.3) for* $K(\rho) = 1$, $\epsilon = 10^{-8}$, $\Delta x = 0.01$ *with initial data (2.21). From top to bottom* ρ *(left) and* j *(right): (a) UW-scheme at time* $t = 0.02$ *and* $\Delta t = \epsilon\Delta x/2$, *(b) DR-scheme at* $t = 0.03$ *with* $\Delta t = \Delta x^2/4$.

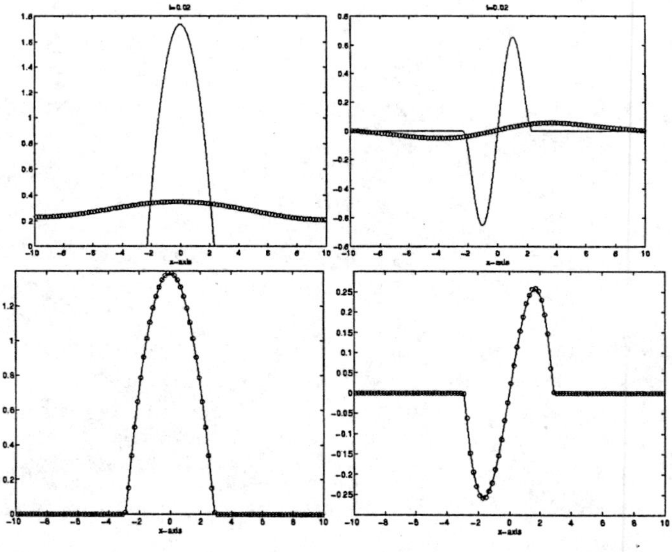

FIG. 2.2. *Numerical solutions of equations (2.3) for* $K(\rho) = 1/\rho$, $\epsilon = 10^{-8}$, $\Delta x = 0.01$ *with initial data (2.23) at* $t = 0$. *From top to bottom* ρ *(left) and* j *(right): (a) UW scheme at time* $t = 0.02$ *and* $\Delta t = \epsilon\Delta x/2$, *(b) DR scheme at* $t = 0.03$ *with* $\Delta t = \Delta x^2/4$.

Next we take a nonlinear case by choosing $K(\rho) = 1/\rho$. The limiting equation for the density ρ is the porous media equation [4]

$$\partial_t \rho = \partial_{xx} \rho^2. \tag{2.22}$$

We test the schemes by comparing the numerical solution with the solution of the porous medium equation given analytically by Barenblatt [4]

$$\rho(x,t) = \frac{1}{r(t)}\left(1 - \left(\frac{x}{r(t)}\right)^2\right) \quad \text{if } |x| \le r(t), \quad \rho(x,t) = 0 \text{ if } |x| > r(t), \tag{2.23}$$

where $r(t) = (12(t+1))^{1/3}$, $t \ge 0$ and $x \in [-10, 10]$. The numerical solutions are plotted in Figure 2.2 for $\epsilon = 10^{-8}$, with 100 spatial cells. The DR-scheme describes accurately the dynamics of the two viscous shock fronts moving to the left and to the right respectively, while one can clearly see the effect of the numerical viscosity in the UW-scheme.

2.2. Reaction-diffusion equations. In this section we consider an hyperbolic relaxation system in the diffusive scaling with a macroscopic limit represented by a nonlinear convection-reaction-diffusion equation [44, 46]. Using the relaxation scheme idea it is possible to develop stable numerical schemes for this equation. In particular, we can construct a relaxing hyperbolic system which has a precise kinetic interpretation as a generalized Carleman model with source terms [9]. The simplest prototype is given by the one-dimensional equation

$$\partial_t u + \partial_x f(u) = \partial_{xx} p(u) + s(u), \tag{2.24}$$

where $(x,t) \in \mathbb{R} \times \mathbb{R}^+$, p, f and s are given smooth functions such that $p(0) \ge 0$ and $p'(u) > 0$, with initial data

$$u(x,0) = u_0(x). \tag{2.25}$$

By introducing a new variable v, one can couple v and u through the following linear hyperbolic system

$$\begin{aligned} &\partial_t u + \partial_x v = s(u), \\ &\partial_t v + \frac{1}{\epsilon^2}\partial_x u = -\frac{1}{\epsilon^2} k(u)(v - f(u)), \end{aligned} \tag{2.26}$$

with the additional initial condition $v(x,0) = v_0(x)$. Here ϵ is a small positive parameter called the *relaxation time* and $k(u) = (p'(u))^{-1}$. For non negative initial data in (2.25), when the initial condition for v is choosed to satisfy

$$|v_0(x)| \le \frac{u_0(x)}{\epsilon}, \tag{2.27}$$

and the convection term in (2.24) is such that $|f(u)| \le u/\epsilon$, $u \ge 0$, system (2.26) is nothing than the system of macroscopic equations of a generalized Carleman model with a source term. In fact if $U = (u + \epsilon v)/2$ and $V = (u - \epsilon v)/2$, U and V satisfy

$$\begin{aligned} &\partial_t U + \frac{1}{\epsilon}\partial_x U = -\frac{1}{2\epsilon^2} k(U+V)\,(U - V - \epsilon f(U+V)) + s(U+V), \\ &\partial_t V - \frac{1}{\epsilon}\partial_x V = \frac{1}{2\epsilon^2} k(U+V)\,(U - V - \epsilon f(U+V)) + s(U+V). \end{aligned} \tag{2.28}$$

Even in this case, the functions U and V represent the densities of two species of particles that move parallel to the x-axis with a collision rate characterized by $k = k(U+V)$ and that are in local equilibrium when $U = (u + \epsilon f(u))/2$ and $V = (u - \epsilon f(u))/2$. The special initial condition (2.27) guarantees the positivity of the initial data for the kinetic system (2.28). In the small relaxation limit, $\epsilon \to 0^+$, the relaxation system (2.26) formally approximates to leading order

$$v = f(u) - \partial_x p(u), \tag{2.29}$$

$$\partial_t u + \partial_x f(u) = \partial_{xx} p(u) + s(u). \tag{2.30}$$

2.2.1. Relaxation approximation to special systems. Consider now a system of convection-reaction-diffusion equations in one space variable in the form

$$\partial_t \mathbf{u} + \partial_x F(\mathbf{u}) = \partial_{xx} P(\mathbf{u}) + S(\mathbf{u}), \quad x \in \mathbb{R}, \quad t \in \mathbb{R}^+, \quad \mathbf{u} \in \mathbb{R}^n. \tag{2.31}$$

In (2.31), $F(\mathbf{u})$, $P(\mathbf{u})$ and $S(\mathbf{u})$ are sufficiently smooth vector-valued functions of $\mathbb{R}^n$. We will also assume $P(0) \geq 0$ and that the Jacobian matrix $P'(\mathbf{u})$ of the diffusion function P is a positive and diagonal matrix. In this situation, the relaxation approximation to (2.31) reads

$$\begin{aligned} &\partial_t \mathbf{u} + \partial_x \mathbf{v} = S(\mathbf{u}), \\ &\partial_t \mathbf{v} + \frac{1}{\epsilon^2}\partial_x \mathbf{u} = -\frac{1}{\epsilon^2} K(\mathbf{u})(\mathbf{v} - F(\mathbf{u})), \end{aligned} \tag{2.32}$$

where $\mathbf{v} \in \mathbb{R}^n$ and $K(\mathbf{u}) = (P'(\mathbf{u}))^{-1}$. The corresponding kinetic system can be derived by introducing the vector variables $\mathbf{U} = (\mathbf{u} + \epsilon\mathbf{v})/2$ and $\mathbf{V} = (\mathbf{u} - \epsilon\mathbf{v})/2$

$$\begin{aligned} &\partial_t \mathbf{U} + \frac{1}{\epsilon}\partial_x \mathbf{U} = -\frac{1}{2\epsilon^2} K(\mathbf{U}+\mathbf{V})\,(\mathbf{U} - \mathbf{V} - \epsilon F(\mathbf{U}+\mathbf{V})) + S(\mathbf{U}+\mathbf{V}), \\ &\partial_t \mathbf{V} - \frac{1}{\epsilon}\partial_x \mathbf{V} = \frac{1}{2\epsilon^2} K(\mathbf{U}+\mathbf{V})\,(\mathbf{U} - \mathbf{V} - \epsilon F(\mathbf{U}+\mathbf{V})) + S(\mathbf{U}+\mathbf{V}). \end{aligned} \tag{2.33}$$

Now, for $i = 1, \ldots, n$, the functions U_i and V_i represents the densities of $2n$ species of particles that move parallel to the x-axis with different collision rates characterized by $K_i = K_i(U_i + V_i)$ which are in local equilibrium when $U_i = (\mathbf{u}_i + \epsilon F_i(\mathbf{u}))/2$ and $V_i = (\mathbf{u}_i - \epsilon F_i(\mathbf{u}))/2$. A special initial condition for $\mathbf{v}$ of the form (2.27) can be chosen in order to guarantee the positivity of the initial data for the kinetic system (2.33). Similarly if $|F_i(\mathbf{u})| \leq \mathbf{u}_i/\epsilon$, $\mathbf{u}_i \geq 0$, the positivity of the local equilibrium is also guaranteed.

2.2.2. General nonlinear advection-reaction-diffusion system. We have seen before a general rule to construct a kinetic relaxation approximation to a system of convection reaction-diffusion equations when the Jacobian matrix of the diffusion function P is positive and diagonal. This approach can not be used for more general diffusion functions P. For this reason we consider a different relaxation approximation to a more general convection-reaction-diffusion systems. The idea [39] is to introduce a third variable w and to consider the relaxing system

$$\partial_t u + \partial_x v = s(u),$$
$$\partial_t v + \frac{1}{\epsilon^2}\partial_x w = -\frac{1}{\epsilon^2}(v - f(u)), \tag{2.34}$$
$$\partial_t w + \partial_x v = -\frac{1}{\epsilon^2}(w - p(u)).$$

Now, (2.34) is a first order semi-linear system and the advection operator is linear. This system has the same advantages of the previous ones, with one more rate equation.

By introducing the variables $U = (w + \epsilon v)/2$, $V = (w - \epsilon v)/2$ and $W = (u - w)/2$ we can now rewrite system (2.34) in diagonal form

$$\partial_t U + \frac{1}{\epsilon}\partial_x U = -\frac{1}{2\epsilon^2}\left[2U - p(u) - \epsilon f(u)\right],$$
$$\partial_t V - \frac{1}{\epsilon}\partial_x V = -\frac{1}{2\epsilon^2}\left[2V - p(u) + \epsilon f(u)\right], \tag{2.35}$$
$$\partial_t W = \frac{1}{2\epsilon^2}\left[U + V - p(u)\right] + s(u),$$

where $u = U + V + 2W$. System (2.35) presents several analogies with the one-dimensional Broadwell model [6] of the Boltzmann equation. From a kinetic point of view it describes the evolution of a fictitious a gas of three species of particles moving parallel to the x-axis with velocities 0 and $\pm 1/\epsilon$ which are in local equilibrium when $U = (p(u) + \epsilon f(u))/2$, $V = (p(u) - \epsilon f(u))/2$ and $W = (u - p(u))/2$. The positivity of the initial data requires $|v_0| \leq w_0/\epsilon \leq u_0/\epsilon$, $u_0 \geq 0$, whereas positivity of the local equilibrium is guaranteed if the convection term and the diffusion term satisfy $|f(u)| \leq p(u)/\epsilon \leq u/\epsilon$, $u \geq 0$.

In the small relaxation limit, $\epsilon \to 0^+$, it is clear that the relaxation system (2.34) formally approximates to leading order (2.29) and (2.30) with $w = p(u)$. The generalization to systems of the form (2.31) when the Jacobian matrix of the diffusion function P has not a diagonal form follows straightforwardly by considering the relaxing system

$$\partial_t \mathbf{u} + \partial_x \mathbf{v} = S(\mathbf{u}),$$
$$\partial_t \mathbf{v} + \frac{1}{\epsilon^2}\partial_x \mathbf{w} = -\frac{1}{\epsilon^2}(\mathbf{v} - F(\mathbf{u})), \tag{2.36}$$
$$\partial_t \mathbf{w} + \partial_x \mathbf{v} = -\frac{1}{\epsilon^2}(\mathbf{w} - P(\mathbf{u})),$$

where $\mathbf{u}, \mathbf{v}, \mathbf{w} \in \mathbb{R}^n$. The corresponding kinetic system can be obtained introducing the vector variables $\mathbf{U} = (\mathbf{w} + \epsilon\mathbf{v})/2$, $\mathbf{V} = (\mathbf{w} - \epsilon\mathbf{v})/2$ and $\mathbf{W} = (\mathbf{u} - \mathbf{w})/2$

$$\partial_t \mathbf{U} + \frac{1}{\epsilon}\partial_x \mathbf{U} = -\frac{1}{2\epsilon^2}\left[2\mathbf{U} - P(\mathbf{u}) - \epsilon F(\mathbf{u})\right],$$

$$\partial_t \mathbf{V} - \frac{1}{\epsilon}\partial_x \mathbf{V} = -\frac{1}{2\epsilon^2}\left[2\mathbf{V} - P(\mathbf{u}) + \epsilon F(\mathbf{u})\right], \tag{2.37}$$

$$\partial_t \mathbf{W} = \frac{1}{2\epsilon^2}\left[\mathbf{U} + \mathbf{V} - P(\mathbf{u})\right] + S(\mathbf{u}),$$

where $\mathbf{u} = \mathbf{U} + \mathbf{V} + 2\mathbf{W}$. The kinetic interpretation of system (2.37) as evolution system for the densities of $2n$ species of particles follows directly from the previous remarks. In order to use our framework for a construction of a suitable numerical schemes for the system (2.36) we consider the new formulation as a diffusive relaxation system

$$\partial_t \mathbf{u} + \partial_x \mathbf{v} = S(\mathbf{u}),$$

$$\partial_t \mathbf{v} + \Phi\partial_x \mathbf{w} = -\frac{1}{\epsilon^2}(\mathbf{v} - F_\epsilon(\mathbf{u}, \partial_x \mathbf{w})), \tag{2.38}$$

$$\partial_t \mathbf{w} + \partial_x \mathbf{v} = -\frac{1}{\epsilon^2}(\mathbf{w} - P(\mathbf{u})),$$

where $\mathbf{u}, \mathbf{v}, \mathbf{w} \in \mathbb{R}^n$, $\Phi = diag\{\phi_1, \phi_2, \dots, \phi_n\}$, $0 < \phi_i \le 1/\epsilon^2$, $\forall\, i$, and

$$F_\epsilon(\mathbf{u}, \partial_x \mathbf{u}) = F(\mathbf{u}) + (I - \epsilon^2\Phi)\partial_x \mathbf{w}. \tag{2.39}$$

As was done normally for a kinetic equation or a hyperbolic system with relaxation, (2.38) can be splitted into a relaxation part and a convective part. The main reason for introducing this splitting is given by the fact that, when $\epsilon \to 0$, the relaxation step always projects the solution to the correct local equilibrium

$$\mathbf{w} = P(\mathbf{u}), \qquad \mathbf{v} = F(\mathbf{u}) - \partial_x P(\mathbf{u}), \tag{2.40}$$

which is a sufficient condition to guarantee that the splitting preserves the asymptotic limit from the relaxation system to the convection-reaction-diffusion equations. As done above an implicit time discretization may be used for the relaxation term. For the numerical treatment of the source terms $S_i(\mathbf{u})$, $i = 1, \dots, n$ we can use explicit schemes if $S_i(\mathbf{u})$ is non stiff, or implicit schemes if $S_i(\mathbf{u})$ is stiff. In this second situation we will assume that the corresponding implicit scheme can be solved explicitly. Hence, although we have an implicit scheme on the stiff terms, solving a nonlinear algebraic system can be completely avoided. For the spatial derivatives in the third equation of (2.38) to avoid oscillations one can make use of a *MinMod* limiter [30], although in most situations the simple central differencing yields second order non oscillatory solutions [23, 24, 37, 38]. In the convective step, high order upwind spatial discretizations [30] can be applied without difficulties to the characteristic variables

$$\mathbf{v} \pm \Phi^{1/2}\mathbf{w}, \tag{2.41}$$

on the diagonal form of (2.38). Finally we remark that, as a consequence of the splitting procedure on the reformulated problem, when $P_i(\mathbf{u})$ is close to zero for some i to ensure the dissipative nature of the modified system it is necessary that

$$\Phi - F'(\mathbf{u})^2 \ge 0, \tag{2.42}$$

where $F'(\mathbf{u})$ is the Jacobian matrix of the flux function $F(\mathbf{u})$. Second order accuracy in time can be again achieved using suitable Runge-Kutta methods as in (2.20).

2.3. Nonlinear diffusion and viscous conservation laws. If we consider different scaling factors for time and space, it is possible to deal with the following diffusive relaxation system

$$\begin{cases} \partial_t \rho + \partial_x j = 0, \\ \partial_t j + \dfrac{1}{\epsilon^2}\partial_x p(\rho) = -\dfrac{1}{\epsilon^{2+\alpha}}\left(j - f(\rho)\right), \end{cases} \tag{2.43}$$

taking $\alpha = 0$ in the small relaxation limit we get

$$\begin{cases} j = f(\rho) - \partial_x p(\rho), \\ \partial_t \rho + \partial_x f(\rho) = \partial_{xx} p(\rho). \end{cases} \tag{2.44}$$

Whereas when $\alpha > 0$, the limit behavior has a different behavior and its solutions is governed by

$$\begin{cases} j = f(\rho), \\ \partial_t \rho + \partial_x f(\rho) = 0. \end{cases} \tag{2.45}$$

In particular, for small values of ϵ, using the Chapmann-Enskog expansion one may get the following nonlinear parabolic equation

$$\begin{cases} j = f(\rho) - \epsilon^\alpha \partial_x p(\rho), \\ \partial_t \rho + \partial_x f(\rho) = \epsilon^\alpha \partial_{xx} p(\rho). \end{cases} \tag{2.46}$$

In this situation, although we have a relaxation process from a hyperbolic system to the reduced hyperbolic system, the solution satisfies a parabolic system. Moreover we can link the physical viscosity ϵ^α in (2.46) to the parameter ϵ. The splitting scheme for the reformulated system, by using the function ϕ, can still apply for the numerical discretization of (2.46). The additional difficulty of the system (2.46) is that now the convective part is nonlinear. Then in the homogeneous transport step we have to use a solver for nonlinear conservation laws. In order to avoid Riemann solvers and to follow the phylosophy described here we can use in our numerical computations the TVD relaxed schemes developed in [26].

2.4. Models with a continuous set of velocities: the radiative transfer. In this chapter, we show how to extend the method to kinetic equations with a continuous set of velocities. The prototype of this class of problems is the one-dimensional transport equation

$$\epsilon \partial_t f + v \partial_x f = \frac{1}{\epsilon}\left(\frac{1}{2}\int_{-1}^{1} f\,dv' - f\right), \qquad x_L < x < x_R, \quad -1 \le v \le 1, \tag{2.47}$$

with boundary conditions

$$\begin{aligned} f(t, x_L, v) &= F_L(v), \qquad \text{for} \quad v > 0, \\ f(t, x_R, -v) &= F_R(v), \qquad \text{for} \quad v > 0. \end{aligned} \tag{2.48}$$

Transport equations have been studied in the diffusive regime from many authors [1, 18, 19, 27, 28, 29, 35]. Many of these papers make use of the even and odd parities. If we split this equation as two equations, each for $v > 0$:

$$\begin{aligned} \epsilon\partial_t f(v) + v\partial_x f(v) &= \frac{1}{\epsilon}\left(\frac{1}{2}\int_{-1}^{1} f dv' - f(v)\right), \\ \epsilon\partial_t f(-v) - v\partial_x f(-v) &= \frac{1}{\epsilon}\left(\frac{1}{2}\int_{-1}^{1} f dv' - f(-v)\right). \end{aligned} \tag{2.49}$$

In this case the even and odd parities are

$$\begin{aligned} r(t,x,v) &= \frac{1}{2}\left(f(t,x,v) + f(t,x,-v)\right), \\ j(t,x,v) &= \frac{1}{2\epsilon}\left(f(t,x,v) - f(t,x,-v)\right). \end{aligned} \tag{2.50}$$

Adding and subtracting the two equations in (2.49) lead to

$$\begin{aligned} \partial_t r + v\partial_x j &= \frac{1}{\epsilon^2}(\rho - r), \\ \partial_t j + \frac{v}{\epsilon^2}\partial_x r &= -\frac{1}{\epsilon^2} j, \end{aligned} \tag{2.51}$$

where

$$\rho(t,x) = \int_0^1 r dv. \tag{2.52}$$

This system looks similar to the hyperbolic heat equation introduced in Subsection 2.1, except for the right hand side of the first equation which is not critical in the numerical implementation. Now, we follow the method in [24], and as stated in (2.12), we rewrite (2.51) into the following diffusive relaxation system

$$\begin{aligned} \partial_t r + v\partial_x j &= \frac{1}{\epsilon^2}(\rho - r), \\ \partial_t j + \phi v\partial_x r &= -\frac{1}{\epsilon^2}\left(j + (1-\epsilon^2\phi)v\partial_x r\right). \end{aligned} \tag{2.53}$$

where $\phi = \phi(\epsilon)$ has the same property as described in the previous section. As $\epsilon \to 0$, system (2.53) gives

$$r = \rho, \qquad j = -v\partial_x r. \tag{2.54}$$

Applying this in the first equation of (2.53) and integrating over v we get the diffusion equation

$$\partial_t \rho = \frac{1}{3}\partial_{xx}\rho. \tag{2.55}$$

Then (2.53) guarantees the correct diffusion limit. As above natural splitting on (2.53) consists of combining the relaxation step

$$\begin{aligned} \partial_t r &= \frac{1}{\epsilon^2}(\rho - r), \\ \partial_t j &= -\frac{1}{\epsilon^2}\left(j + (1-\epsilon^2\phi)v\partial_x r\right), \end{aligned} \tag{2.56}$$

with the transport step

$$\begin{aligned} &\partial_t r + v\partial_x j = 0, \\ &\partial_t j + \phi v \partial_x r = 0. \end{aligned} \tag{2.57}$$

Now the backward Euler method gives

$$\begin{aligned} \frac{r^{n+1} - r^n}{\Delta t} &= -\frac{1}{\epsilon^2}(r^{n+1} - \rho^{n+1}), \\ \frac{j^{n+1} - j^n}{\Delta t} &= -\frac{1}{\epsilon^2}(j^{n+1} + (1 - \epsilon^2\phi)v\partial_x r^{n+1}). \end{aligned} \tag{2.58}$$

The variable ρ is preserved in (2.58), i.e. $\rho^{n+1} = \rho^n$ if one integrate the first equation of (2.58) over v, thus it can be replaced by

$$\frac{r^* - r^n}{\Delta t} = -\frac{1}{\epsilon^2}(r^* - \rho^n). \tag{2.59}$$

Now we can solve (2.59) for r^* explicitly, and then get j^* from (2.58) also explicitly provided the spatial derivative on the right hand side is discretized using linear schemes. Thus our implicit scheme can be still implemented as an explicit method. A similar procedure applies to problems in higher dimensions, as well as to more general transport equations in presence of sources. Recently, the same methodology has been applied for semiconductor devices [25].

3. Numerical applications. We apply the second order scheme for hyperbolic system in the diffusive regime to different physical problems. In all tests the spatial derivatives in the relaxation step have been approximated by using central differences. In the diffusive stiff regime the CFL condition is of parabolic nature, that is $\Delta t \approx (\Delta x)^2$. The 'exact' solution as usual is obtained by a much finer grid.

3.1. Sedimentation problem. We consider a model for the dynamics of a compressible sediment (see for example [2]). Here $\rho(x,t)$ is the volume fraction of solid material in a suspending fluid, the dimensionless continuity equation which governs the behaviour of particles is given by

$$\partial_t \rho + \partial_x \left[(1-\rho)^a \left(\rho - \frac{1}{Pe} \frac{b\rho_M}{(\rho_M - \rho)^2} \partial_x \rho \right) \right] = 0,$$

where the x axis has the origin at the top of the liquid ($x = 0$), while the constant Pe is the ratio of the gravitational potential over the osmotic pressure. In our experiments we take $a = 6.55$, $\rho_M = 0.64$, $b = 1.85$, and a constant initial data $\rho(x,0) = 0.1$. The equation is solved in the interval $x \in [0,1]$ with zero flux at the boundary. For small values of Pe, u varies smoothly from top to bottom and remains below ρ_M at the bottom, see Figure 3.1. As Pe increases the interface contact with the supernatant fluid becomes sharper and a region with a constant volume fraction of solid material now appears, see Figure 3.1.

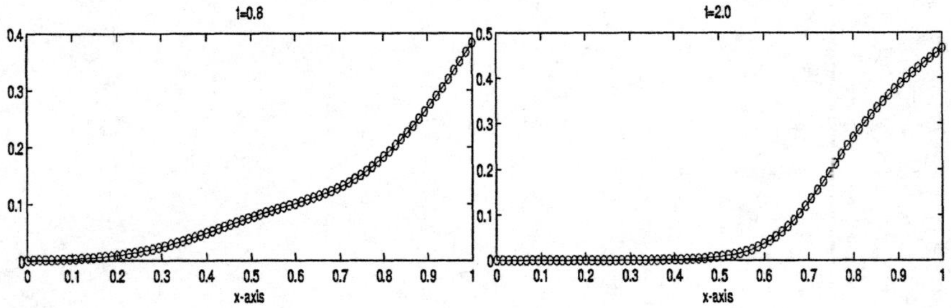

FIG. 3.1. *Numerical solution for the sedimentation problem, $Pe = 50$.*

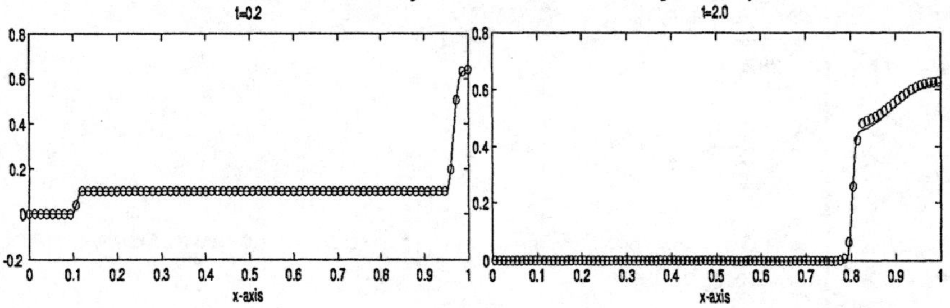

FIG. 3.2. *Numerical solution for the sedimentation problem, $Pe = 50000$.*

3.2. Fisher's equation. We consider the approximation of Fisher's equation modeling the dynamics of a population $u(x,t)$ with a linear diffusion effect and a nonlinear law of reproduction

$$\partial_t u = \partial_{xx} u + u(1-u), \qquad x \in \mathbb{R},\ t > 0.$$

Many numerical approaches have been proposed including pseudo-spectral methods, Galerkin finite element methods, and finite difference methods. Recently, Carey and Shen [8] developed a least-squares finite element approximation with a particular interest in the numerical approximation of progressive wave solution.

The test problem we consider refers to the initial data [8]

$$u(x,0) = \begin{cases} e^{10(x+1)} & x \le -1, \\ 1 & -1 \le x \le 1, \\ e^{-10(x-1)} & x \ge 1. \end{cases}$$

The short time behaviour of the numerical solution, with $\Delta x = 0.25$ is shown in Figure 3.3 for final time $T = 5$. Initially the solution near the step corner decays and then recovers. Then the front evolves smoothly and propagates, as indicated in the second part of Figure 3.3.

We consider also a generalization of the Fisher's equation in the case of a nonlinear convection phenomena. More precisely we present a numerical test for the following porous-Fisher's equation [41, 47, 44],

$$\partial_t u = \partial_x \left(u^\alpha \partial_x u\right) + u(1-u^\alpha), \tag{3.1}$$

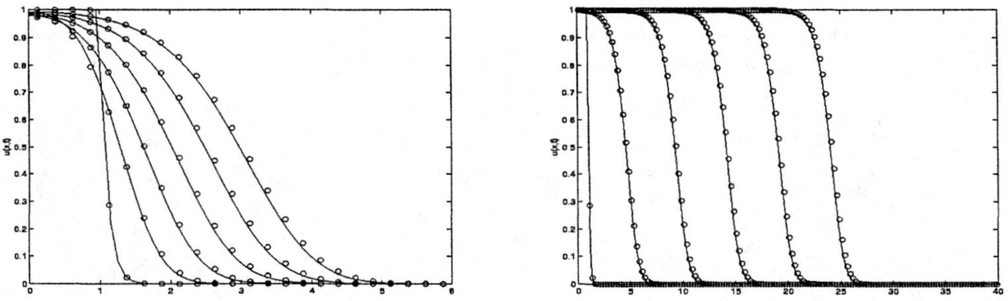

FIG. 3.3. *Short and long time behaviour of Fisher's equation, $T = 5$ (left) and $T = 40$ (right).*

on $-\infty < x < \infty$, where α is a positive real parameter.
Such equation can be written in the form (2.24) by taking the following data:

$$f(u) = 0,\ p(u) = \frac{1}{\alpha+1}u^{\alpha+1},\ s(u) = u(1-u^{\alpha}).$$

The equations of the porous-Fisher type arise, for example, in some model of combustion phenomena and in population dynamics when the interaction between individuals and the behavior of the population flow depend on the density function $u(x,t)$ [15, 36].

Newman [41] has shown that (3.1) has the unique traveling wave solution

$$u(x,t) = \left(\left(1 - exp\left[\frac{\alpha z}{\sqrt{(\alpha+1)}}\right]\right)\right)^{+}, \quad z = x - ct, \tag{3.2}$$

where $(x)^+ \equiv max(x,0)$ and the velocity c is defined as

$$c = \frac{1}{\sqrt{(\alpha+1)}}.$$

Using suitable translation and reflection [47] of the wave (3.2) it is possible to write a family of solutions of (3.1). For example it is possible to describe a traveling wave $u_0(x,t)$ moving to the right, starting from some point x_0, and a traveling wave $u_1(x,t)$ starting from the point x_1 and moving to the left. Using these two wave solutions, we can produce merging populations. While the two populations, represented by $u_0(x,t)$, and $u_1(x,t)$, remain separated there will be no interaction and the whole population density, at each point, is merely the maximum between them. At time $\hat{t} = (x_1 - x_0)/2c$ and position $\hat{x} = (x_0 + x_1)/2$, the two populations first meet and begin their interaction. An analysis of such behavior was done by Witelski [47] by using perturbation theory and the method of matched asymptotic expansions.

The approximation of the macroscopic system corresponding to the porous-Fisher's as the limit state, was performed by using the relaxation scheme described in this paper. If we chose the parameter $\alpha = 2$ in the original equation (3.1) then the velocity c of the traveling waves, the critical time $\hat{t}$ and the critical position $\hat{x}$ are, respectively, $c = 1/\sqrt{3}$, $\hat{t} = 15\sqrt{3}$ and $\hat{x} = 5$. The system is solved in the interval $x \in [-20, 30]$ with suitable fixed values of

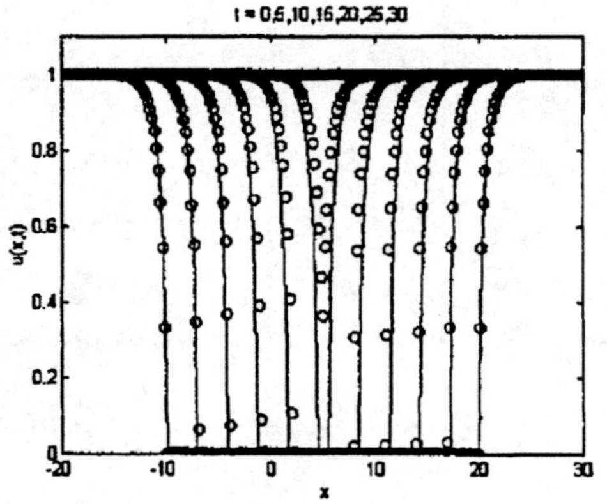

FIG. 3.4. *Numerical simulation of the evolution of two traveling waves solutions of the porous-Fisher's equation for different times before the interaction.*

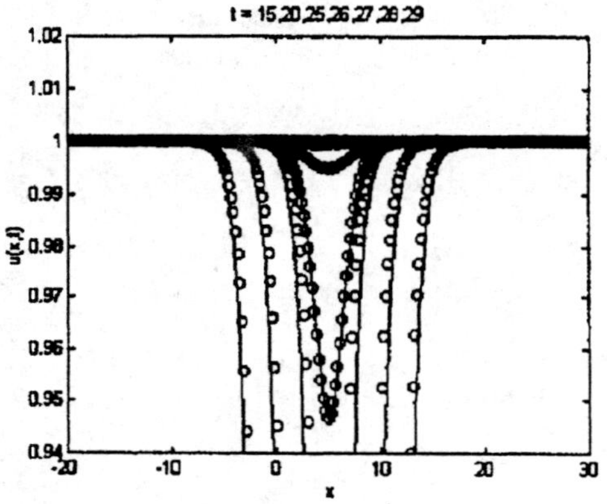

FIG. 3.5. *Numerical simulation of the evolution of two traveling waves solutions of the porous-Fisher's equation for different times after the critical time $\hat{t}$ when the two waves first meet.*

density $u(x,t)$ at the boundary which agree with the values of the interacting waves. In Figure 3.4 we show the short time behaviour of the numerical solution, with no interaction of the waves, while the long time behavior of the numerical solution is plotted in Figure 3.5. In the computation we used a uniform space-time mesh where $\Delta x = 0.2$, $\Delta t = 0.02$ and $\epsilon = 10^{-8}$. As true solution, we considered the asymptotic solution computed by Witelski in [47].

REFERENCES

[1] M.L. Adams, *Even-Parity Finite-Element Transport Methods in the Diffusion Limit*, Progress in Nuclear Energy **25**, 1991, pp. 159–198.

[2] F.M. Auzerais, R. Jackson and W.B. Russel, *The resolution of shocks and the effects of compressible*

sediments in transient settling, J. Fluid Mech. **195**, 1988, pp. 437–462.

[3] C. Bardos, F. Golse and D. Levermore,*Fluid Dynamic Limits of discrete velocity Kinetic equations* in Advances in Kinetic Theory and Continuum Mechanics, R. Gatignol and Soubbaramayer Eds., Springer-Verlag Heidelberg, pp 57–72 (1991).

[4] G.I. Barenblatt, *On some unsteady motion of a liquid or a gas in porous media*, Prikl. Mat. Mekh., **16** (1952) 67–79.

[5] J.T. Beale, *Large-Time Behavior of the Broadwell Model of a Discrete Velocity Gas*, Comm. Math. Phys. **102**, pp 217–235 (1985).

[6] J.E.Broadwell, *Shock structure in a simple discrete velocity gas*, Phys. Fluids **7**, **8** (1964), 1243–1247.

[7] R. Caflish, S. Jin and G. Russo *Uniformly accurate schemes for hyperbolic systems with relaxation*, SIAM J. Num. Anal. **34** (1997) 246–281.

[8] G.F. Carey and Y. Shen, *Least-squares finite element method approximation of Fisher's reaction-diffusion equation*, Num. Methods for Part. Diff. Equations, **11**, (1995), 175–186.

[9] T. Carleman, *Problémes mathèmatiques dans la thèorie cinètique des gas*, Publ. Scient. Inst. Mittag-Leffler (1957).

[10] G.Q. Chen, C.D. Levermore and T.P. Liu, *Hyperbolic conservation laws with stiff relaxation terms and entropy*, Comm. Pure and App. Math., XLVII, (1994), 787–83.

[11] V. Comincioli, G. Naldi, and G. Toscani, *Nonlinear diffusion and fluid dynamical limit from discrete velocity models*, Commun. Appl. Nonlinear Anal. **2**, No. 4, 1995, pp. 1–29.

[12] V. Comincioli, G. Naldi, and G. Toscani, *Diffusive limit of two velocity models: the porous medium equation*, Transp. Theory Stat. Phys. **26**, No. 1-2, 1997, pp. 49–63.

[13] E. Gabetta, L. Pareschi, and G. Toscani, *Wild's sums and numerical approximation of nonlinear kinetic equation*, Transp. Theory Stat. Phys. **25**, No.3-5, 1996, pp. 515–530.

[14] E. Gabetta and B. Perthame, *Scaling limits of the Ruijgrook-Wu model of the Boltzmann equation* Proceedings of the International Conference on Nonlinear Equations and Applications, Bangalore 19-23 August 1996. Springer-Verlag.

[15] W.S. Gurney and R.M. Nisbet,*The regulation of inhomogeneous populations*, J. Theor. Biol., **52**, 441–457 (1975).

[16] Klar, A. *A numerical method for nonstationary transport equations in diffusive regimes*. Proceedings of the Fifteenth International Conference on Transport Theory, Part II (Gteborg, 1997). Transp. The. Stat. Phys. **27** (1998), 653–666.

[17] S. Jin *Runge-Kutta methods for hyperbolic conservation laws with stiff relaxation terms*, J. Comp. Phys. **122**, (1995) 51–67.

[18] S. Jin and C.D. Levermore, *The discrete-ordinate method in diffusive regimes*, Transp. The. Stat. Phys. **20** (1991) 413-439.

[19] S. Jin and C.D. Levermore, *Fully discrete numerical transfer in diffusive regimes*, J. Comp. Phys. **126** (1996) 449–467.

[20] S. Jin and C.D. Levermore, *Numerical schemes for hyperbolic conservation laws with stiff relaxation terms*, Transp. The. Stat. Phys. **22** (1993) 739–791.

[21] S. Jin, *Runge-Kutta methods for hyperbolic conservation laws with stiff relaxation terms*, J. Comput. Phys. **122**, (1995), 51-67.

[22] S. Jin, *Asymptotic and numerical approximations to multiscale hyperbolic and transport equations*, Preprint (1997).

[23] S. Jin, L. Pareschi and G. Toscani, *Diffusive relaxation schemes for multiscale discrete-velocity kinetic equations*, SIAM J. Num. Anal. **35** (1998), no. 6, 2405–2439.

[24] S. Jin, L. Pareschi and G. Toscani, *Uniformly accurate diffusive relaxation schemes for multiscale transport equations*, SIAM J. Num. Anal., to appear.

[25] S. Jin and L. Pareschi, *Discretizations of the Semiconductor Boltzmann Equation by Diffusive Relaxation Schemes*, preprint (1999).

[26] S. Jin and Z. Xin, *The relaxation schemes for systems of conservation laws in arbitrary space dimension*, Comm. Pure and Appl. Math. **48**, (1995), 235–276.

[27] E.W. Larsen *The asymptotic diffusion limit of discretized transport problems*, Nucl. Sci. Eng. **112**, (1992) 336-346.

[28] E.W. Larsen, J.E. Morel and W.F. Miller *Asymptotic solutions of numerical transport problems in optically thick, diffusive regimes*, J. Comp. Phys. **69**, (1987) 283–324.

[29] E.W. Larsen and J.E. Morel *Asymptotic solutions of numerical transport problems in optically thick, diffusive regimes II*, J. Comp. Phys. **83**, (1989) 212–236.

[30] R.J. LeVeque, *Numerical methods for conservation laws*, Birkhäuser Verlag, (1992), Basel.

[31] P. L. Lions and G. Toscani, *Diffusive limit for finite velocity Boltzmann kinetic models*, Rev. Mat. Iberoam. **13**, No.3, 1997, pp. 473–513.

[32] T.P. Liu, *Hyperbolic conservation laws with relaxation*, Comm. Math. Phys., **108**, (1987).
[33] H.P. McKean, *The central limit theorem for Carleman's equation*, Jsrael J. Math. **21** (1975), 54–92.
[34] P. Marcati and B. Rubino, *Hyperbolic to parabolic relaxation theory for quasi-linear first order systems*, preprint.
[35] W.F. Miller Jr., *An Analysis of the Finite-Differenced, Even-Parity Discrete-Ordinate Equations in Slab Geometry*, Nucl. Sci. & Eng. **108** (1990), 247–266.
[36] J.D. Murray, *Lecture on Nonlinear Differential Equation Models in Biology,* Oxford University Press, (1977), Oxford.
[37] G. Naldi and L. Pareschi, *Numerical schemes for kinetic equations in diffusive regimes*, Appl. Math. Letters **11** (1998) 29–35.
[38] G. Naldi and L. Pareschi, *Numerical schemes for hyperbolic systems of conservation laws with stiff diffusive relaxation*, SIAM J. Num. Anal., to appear.
[39] G. Naldi, L. Pareschi and G. Toscani *Convergence of kinetic relaxation approximation to nonlinear parabolic problems* preprint (1999).
[40] G. Naldi, L. Pareschi and G. Toscani *Hyperbolic relaxation approximation to nonlinear parabolic problems*, Inter. Series of Numer. Math. **130**, Birkhäuser Verlag, Basel (1999) 747–7456.
[41] W.I. Newman, *Some exact solutions to a nonlinear diffusion problem in population genetics and combustion*, J. Theor. Biol., **85**, 325–334, (1980).
[42] A. Pulvirenti and G. Toscani *Fast diffusion as a limit of a two-velocity kinetic model.* Proceedings of the VIII International Conference on Waves and Stability in Continuous Media, Part II (Palermo, 1995). Rend. Circ. Mat. Palermo (2) Suppl. **45**, (1996), 521–528.
[43] P.L. Roe and M. Arora, *Characteristic-based schemes for dispersive waves I. The method of characteristics for smooth solutions*, Numerical Meth. for PDE's, **9**, 1993, pp.459–505.
[44] J. Smoller, *Shock waves and reaction-diffusion equations*, Springer-Verlag, (1983), New-York.
[45] G. Toscani and A. Pulvirenti, *Fast diffusion as a limit of a two-velocity kinetic model*, Circolo Matematico di Palermo, Suppl. Rend. Circ. Mat. Palermo, II. Ser. 45, 1996, pp. 521–528.
[46] G.B. Whitham, *Linear and nonlinear waves*, Wiley New York (1974).
[47] T.P. Witelski, *Merging traveling waves for the Porous-Fisher's equation*, Appl. Math, Lett., Vol. **8**, No. 4, 57–62, (1995).

INTERPROXIMATING CURVES BY MEANS OF MIXED SPLINES

C. CONTI[‡], R. MORANDI[‡], AND A. SESTINI[‡]

Abstract. The interproximation problem for curves consists in determining a curve interpolating some data points and passing through some specified regions. In this work mixed splines, that is smoothing splines with additional interpolation constraints, are used to obtain interproximating $2D$ or $3D$ curves. The solution is obtained by solving a linear system whose structure can be improved by using B-splines basis.

Key words. Interproximation, Mixed splines, B-splines

AMS subject classifications. 65D07, 65D10

1. Introduction. It is well known that interpolation and/or approximation methods are used to model the shape of an object. In this field, a problem commonly encountered is the uncertainty of the data points, e.g. deriving from measurement errors. This is especially true, for example, in the reconstruction of digitized images or in CAD applications. Consequently, the data to be fitted appear as exact points and/or regions specifying the possible ranges of the data to be approximated. Furthermore, with regard to the curve reconstruction, the specification of additional regions besides the interpolation points can be also used to control the shape of the interpolating curve. These considerations motivate the study of the interproximation problem, i.e. the problem of interpolating some data $\{D_i\}_{i=1}^{n+m}$, where each D_i can be either a point or a region in $\mathbb{R}^2$ or in $\mathbb{R}^3$. More precisely, the curve interpolates D_i if D_i is a point, otherwise the curve passes through the region specified by D_i. This problem was first studied in [8] and in [9], then in [4], [5], [10], [11].

In this work, we discuss the application of mixed splines to solve the interproximation problem in the case of rectangular and circular regions even though it could be easily extended to any convex region. The presented approach has the advantage of dealing with a linear system, making the method particularly attractive compared with the previously quoted references, where the curves are obtained via non linear techniques. In addition, it has a natural extension to a multivariate setting even for the case of scattered data [6].

The idea of the mixed splines is to combine given interpolation constraints with a natural measure of smoothness and a standard measure of goodness-of-fit [2]. For example, a cubic mixed spline, can be obtained as solution of the constrained variational problem

$$\min_{f(x_i)=z_i,\ i=1,\ldots,n} \{\alpha \int_{\mathbb{R}} (f'')^2 + \sum_{j=1}^{m} (f(\xi_j) - \zeta_j)^2\},$$

where $\{x_i, z_i\}_{i=1}^{n}$ and $\{\xi_j, \zeta_j\}_{j=1}^{m}$ with $x_i, z_i, \xi_j, \zeta_j \in \mathbb{R}$ are the data and $\alpha \geq 0$ is a smoothing parameter.

In particular, in this paper we prove the existence of mixed splines with positive smoothing parameter α solutions of the stated interproximation problem. In addition, special attention is given to the choice of suitable basis functions, i.e. B-spline functions.

The paper is organized as follows. In Section 2, the mixed splines are introduced in the space $D^{-2}L^2(\mathbb{R})$. In Section 3, their application to the interproximation problem for the functional case is investigated. In Section 4, mixed splines are used to solve the interproximation problem for the parametric case. Finally, in Section 5, some graphs are shown to illustrate the method for the functional and the parametric cases in $\mathbb{R}^2$ and $\mathbb{R}^3$.

[‡]Dipartimento di Energetica, Università di Firenze, via C. Lombroso 6/17, 50134 Firenze, Italia

2. Definition and characterization of mixed splines. In this section, we recall the definition of the space $D^{-2}L^2(\mathbb{R})$, then we state a characterization theorem for a mixed spline in that space. We refer to [2] for all the proofs of the presented results. Let $D^{-2}L^2(\mathbb{R})$ be the space of all the functions with second derivatives in $L^2(\mathbb{R})$. This is a semi-Hilbert space if equipped with the scalar semi-product

$$(f,g) = \int_{\mathbb{R}} f''(y)g''(y)dy, \ \forall f,g \in D^{-2}L^2(\mathbb{R}), \tag{2.1}$$

inducing the semi-norm $|f| = (f,f)^{1/2}$. Following [1], it can be proved that any reproducing kernel of $D^{-2}L^2(\mathbb{R})$ is in the form

$$H(x,y) = v(x-y) \tag{2.2}$$

where v satisfies $v^{(iv)} = \delta$ and δ is the Dirac distribution, e.g. $v(x) = \frac{|x|^3}{12}$, or $v(x) = \frac{(x)_+^3}{6}$.

THEOREM 2.1. *Let $k_1,\dots,k_n$, $l_1,\dots,l_m$ be $n+m$ linear independent functionals from $D^{-2}L^2(\mathbb{R})$ to $\mathbb{R}$ and z_i, $i=1,\dots,n$, ζ_j, $j=1,\dots,m$ be given real data. Let $H(x,y)$ be a reproducing kernel of $D^{-2}L^2(\mathbb{R})$ and p_1,p_2 be a basis for $\mathbb{P}_1$, the linear polynomial space. Then, the variational problem*

$$\sigma_\alpha = arg \min_{k_i(u)=z_i,\ i=1,\dots,n} \{\alpha|u|^2 + \sum_{j=1}^{m}(l_j(u)-\zeta_j)^2\}, \tag{2.3}$$

has a unique solution $\sigma_\alpha \in D^{-2}L^2(\mathbb{R})$, called mixed spline, which has the following form

$$\sigma_\alpha(x) = \sum_{i=1}^{n}\lambda_i^\alpha k_i(H(x,\bullet)) + \sum_{j=1}^{m}\mu_j^\alpha l_j(H(x,\bullet)) + \sum_{r=1}^{2}\nu_r^\alpha p_r(x). \tag{2.4}$$

If $\boldsymbol{\lambda}^\alpha = (\lambda_1^\alpha,\dots,\lambda_n^\alpha)^T$, $\boldsymbol{\mu}^\alpha = (\mu_1^\alpha,\dots,\mu_m^\alpha)^T$ and $\boldsymbol{\nu}^\alpha = (\nu_1^\alpha,\nu_2^\alpha)^T$, the vector

$$\eta^\alpha = ((\boldsymbol{\mu}^\alpha)^T, (\boldsymbol{\lambda}^\alpha)^T, (\boldsymbol{\nu}^\alpha)^T)^T \in \mathbb{R}^{n+m+2}$$

is the solution of the linear system

$$(G+\alpha I')\eta^\alpha = \mathbf{e} \tag{2.5}$$

where G and I' are the block matrices of order $n+m+2$

$$G = \begin{pmatrix} G_{11} & G_{12} & C_1 \\ G_{21} & G_{22} & C_2 \\ C_1^T & C_2^T & 0 \end{pmatrix}, \quad I' = \begin{pmatrix} 0 & 0 & 0 \\ 0 & I & 0 \\ 0^T & 0^T & 0 \end{pmatrix} \tag{2.6}$$

with I identity matrix of order m and $\mathbf{e}$ the vector

$$\mathbf{e} = (z_1,\dots,z_n,\zeta_1,\dots,\zeta_m,0,0)^T.$$

The blocks G_{ij}, C_i, $i,j=1,2$ are obtained by applying the linear operators to the reproducing kernel and to the basis p_1, p_2 of $\mathbb{P}_1$, that is

$$\begin{aligned}
(G_{11})_{ij} &= k_ik_j(H(\bullet,\bullet)), & i&=1,\dots,n, & j&=1,\dots,n,\\
(G_{12})_{ij} &= k_il_j(H(\bullet,\bullet)), & i&=1,\dots,n, & j&=1,\dots,m,\\
(G_{21})_{ji} &= l_jk_i(H(\bullet,\bullet)), & i&=1,\dots,n, & j&=1,\dots,m,\\
(G_{22})_{ij} &= l_il_j(H(\bullet,\bullet)), & i&=1,\dots,m, & j&=1,\dots,m,\\
(C_1)_{ir} &= k_i(p_r), & i&=1,\dots,n, & r&=1,2,\\
(C_2)_{jr} &= l_j(p_r), & j&=1,\dots,m, & r&=1,2.
\end{aligned}$$

Furthermore, the mixed spline σ_α satisfies the following conditions

$$\left\{\begin{array}{lll} k_i(\sigma_\alpha) & = z_i, & i = 1, \dots, n, \\ l_j(\sigma_\alpha) & = \zeta_j - \alpha\mu_j^\alpha, & j = 1, \dots, m. \end{array}\right. \tag{2.7}$$

3. Mixed splines interproximating functional data. In this section, first we prove the existence of a positive smoothing parameter α which guarantees the solvability of the interproximation problem in the case of functional data via the solution of the linear system (2.5). Then, we introduce a basis transformation to improve the structure of (2.5).

Given $\{x_i, z_i\}_{i=1}^n$, $x_i, z_i \in \mathbb{R}$, a set of points to be interpolated and $\{\xi_j, I_j\}_{j=1}^m$, $\xi_j \in \mathbb{R}$, $I_j = [a_j, b_j]$ and a_j, $b_j \in \mathbb{R}$, a set of regions to be approximated, a function σ_α is a solution of the interproximation problem for the given data if it satisfies

$$\left\{\begin{array}{ll} \sigma_\alpha(x_i) = z_i, & i = 1, \dots, n, \\ \sigma_\alpha(\xi_j) \in I_j, & j = 1, \dots, m. \end{array}\right. \tag{3.1}$$

To this purpose, we define the $n+m$ linear independent functionals $k_1, \dots, k_n$, $l_1, \dots, l_m$ from $D^{-2}L^2(\mathbb{R})$ to $\mathbb{R}$ as

$$k_i(f) := f(x_i),\ i = 1, \dots, n \quad l_j(f) := f(\xi_j),\ j = 1, \dots, m.$$

With the previous choice, the mixed spline σ_α satisfies

$$\left\{\begin{array}{ll} \sigma_\alpha(x_i) = z_i, & i = 1, \dots, n, \\ \sigma_\alpha(\xi_j) = \zeta_j - \alpha\mu_j^\alpha, & j = 1, \dots, m. \end{array}\right. \tag{3.2}$$

Hence the choice $\zeta_j := \frac{a_j+b_j}{2}$, $j = 1, \dots, m$ ensures the existence of a mixed spline solving (3.1), at least for $\alpha = 0$. In that case, we are, actually, dealing with the natural cubic spline interpolating all the data $\{x_i, z_i\}_{i=1}^n$, $\{\xi_j, \zeta_j\}_{j=1}^m$. Since a non zero smoothing parameter is preferable, Theorem 2.1 is used with $\alpha > 0$ to obtain an interproximating mixed spline expressed as

$$\sigma_\alpha(x) = \sum_{i=1}^n \lambda_i^\alpha H(x, x_i) + \sum_{j=1}^m \mu_j^\alpha H(x, \xi_j) + \sum_{r=1}^2 \nu_r^\alpha x^{r-1}. \tag{3.3}$$

The vector $\eta^\alpha = ((\boldsymbol{\lambda}^\alpha)^T, (\boldsymbol{\mu}^\alpha)^T, (\boldsymbol{\nu}^\alpha)^T)^T \in \mathbb{R}^{n+m+2}$ is the solution of the linear system

$$(G + \alpha I')\eta^\alpha = \mathbf{e} \tag{3.4}$$

where the blocks of the matrix G defined in (2.6) are given by

$$\begin{array}{llll} (G_{11})_{ij} & = H(x_i, x_j), & i = 1, \dots, n, & j = 1, \dots, n, \\ (G_{12})_{ij} & = H(x_i, \xi_j), & i = 1, \dots, n, & j = 1, \dots, m, \\ (G_{21})_{ji} & = H(x_i, \xi_j), & i = 1, \dots, n, & j = 1, \dots, m, \\ (G_{22})_{ij} & = H(\xi_i, \xi_j), & i = 1, \dots, m, & j = 1, \dots, m, \\ (C_1)_{ir} & = x_i^{r-1}, & i = 1, \dots, n, & r = 1, 2, \\ (C_2)_{jr} & = \xi_j^{r-1}, & j = 1, \dots, m, & r = 1, 2, \end{array}$$

and $H(x, y) = v(x - y) = \frac{|x-y|^3}{12}$.

We can now state the following theorem where, for the sake of simplicity, the symbol $\|\cdot\|$ will denote $\|\cdot\|_\infty$.

THEOREM 3.1. *Let* $\alpha \leq \frac{l}{\|\eta^0\|+l\|G^{-1}\|}$, *where* $l = \min_{j=1,\ldots,m}\{\frac{(b_j-a_j)}{2}\}$ *and* $\eta^0 := G^{-1}\mathbf{e}$. *Then, the interproximation problem (3.1) is solvable.*

Proof Any mixed spline σ_α, defined in (3.3), solution of the interproximation problem (3.1) has to verify $a_j \leq \sigma_\alpha(\xi_j) \leq b_j$, $j = 1, \ldots, m$. So, using $\sigma_\alpha(\xi_j) = \zeta_j - \alpha\mu_j^\alpha$, $j = 1, \ldots, m$, we get the following conditions

$$|\mu_j^\alpha| \leq \frac{(b_j - a_j)}{2\alpha}, \; j = 1, \ldots, m.$$

Thus, a sufficient condition for the previous inequality is obviously given by

$$\|\eta^\alpha\|_\infty \leq \frac{l}{\alpha}. \tag{3.5}$$

Regarding the matrix $\alpha I'$ as a perturbation of the matrix G, it can be easily shown [3] that, if $\alpha \leq \frac{1}{\|G^{-1}\|}$ then

$$\|\eta^\alpha - \eta^0\| \leq \frac{\|G^{-1}\| \cdot (\alpha\|\eta^0\|)}{(1 - \alpha\|G^{-1}\|)}. \tag{3.6}$$

Thus, we arrive at

$$\|\eta^\alpha\| \leq \frac{\|\eta^0\|}{(1 - \alpha\|G^{-1}\|)}. \tag{3.7}$$

To satisfy (3.5), we can write

$$\frac{\|\eta^0\|}{(1 - \alpha\|G^{-1}\|)} \leq \frac{l}{\alpha} \tag{3.8}$$

which is equivalent to the condition

$$\alpha \leq \frac{l}{(\|\eta^0\| + l\|G^{-1}\|)}. \qquad \square \tag{3.9}$$

Since, from a computational point of view, the solution of (3.4) is not recommended, we are now looking for a more suitable basis, such as the B-spline basis. Thus, we define a knot vector $\mathbf{t} \in \mathbb{R}^{n+m}$ by ordering the vector $(x_1, \ldots, x_n, \xi_1, \ldots, \xi_m)^T$ in increasing order and the integer vectors, $\phi \in \mathbb{N}^n$ and $\psi \in \mathbb{N}^m$, so that

$$t_{\phi(i)} = x_i, \; i = 1, \ldots, n, \quad t_{\psi(j)} = \xi_j, \; j = 1, \ldots, m.$$

Let us now define the $(n+m+2) \times (n+m+2)$ permutation matrix P as

$$P \begin{bmatrix} v(x - x_1) \\ \vdots \\ v(x - x_n) \\ v(x - \xi_1) \\ \vdots \\ v(x - \xi_m) \\ 1 \\ x \end{bmatrix} = \begin{bmatrix} v(x - t_1) \\ \vdots \\ \vdots \\ \vdots \\ v(x - t_{n+m}) \\ 1 \\ x \end{bmatrix},$$

and the $(n+m+2)\times(n+m+2)$ matrix T such that

$$T\begin{bmatrix} v(x-t_1) \\ \vdots \\ v(x-t_{n+m}) \\ 1 \\ x \end{bmatrix} = \begin{bmatrix} B_0(x) \\ \vdots \\ \vdots \\ B_{n+m+1}(x) \end{bmatrix}.$$

The functions $B_i(x)$, $i=0,\ldots,n+m+1$ are the usual cubic B-splines with knots t_{i-2}, t_{i-1}, t_i, t_{i+1}, t_{i+2} written as

$$B_i(x) = \tfrac{(t_{i+2}-t_{i-2})}{4}\Big(\ \theta^i_{i-2}v(x-t_{i-2}) + \theta^i_{i-1}v(x-t_{i-1}) + \theta^i_i v(x-t_i) + \theta^i_{i+1}v(x-t_{i+1}) + \theta^i_{i+2}v(x-t_{i+2})\ \Big) \tag{3.10}$$

and θ^i_{i+j}, $j=-2,\ldots,2$ are the coefficients of the fourth order divided difference. The auxiliary knots t_i, $i=-2,-1,0$ and t_{n+m+i}, $i=1,2,3$ are defined in a suitable way depending on dealing with open or closed curves.
The matrix T has the following block structure

$$T = \begin{bmatrix} T_{11} \\ T_{21} \\ T_{31} \end{bmatrix}, \tag{3.11}$$

where T_{11} and T_{31} are $3\times(n+m+2)$ matrix with non zero entries given by the coefficients of the basis $B_0(x)$, $B_1(x)$, $B_2(x)$ and $B_{n+m-1}(x)$, $B_{n+m}(x)$, $B_{n+m+1}(x)$ respectively, when expressed as linear combination of 1, $x, v(x-t_i)$, $i=1,\ldots,n+m$. T_{21} is a $(n+m-4)\times(n+m+2)$ matrix with non zero entries given by

$$(T_{21})_{ij} = \theta^{i+2}_j, \quad j=i,\ldots,i+4, \quad i=1,\ldots,n+m-4.$$

Setting $\tilde{T} := T\,P$ and the vector $\tilde{\eta}^\alpha := \tilde{T}^{-1}\eta^\alpha$, the system (3.4) becomes

$$(G\tilde{T} + \alpha I'\tilde{T})\tilde{\eta}^\alpha = \mathbf{e}. \tag{3.12}$$

The mixed spline σ_α is then

$$\sigma_\alpha(x) = \sum_{i=0}^{n+m+1} \tilde{\eta}^\alpha_i B_i(x). \tag{3.13}$$

4. Mixed spline interproximating non-functional data. When dealing with parametric curves, the interproximation problem can be formulated as follows. Given n points $\{\mathbf{P}_i\}_{i=1}^n$, $\mathbf{P}_i \in \mathbb{R}^d$, $d=2,3$, and m regions $\{D_j\}_{j=1}^m$, $D_j \in \mathbb{R}^d$, $d=2,3$, a curve $\sigma_\alpha : [0,1] \to \mathbb{R}^d$, $d=2,3$ is a solution of the interproximation problem, if it satisfies the following conditions

$$\begin{aligned} \exists s_i &\in [0,1] : \ \sigma_\alpha(s_i) = \mathbf{P}_i, \quad i=1,\ldots,n \\ \exists \tau_j &\in [0,1] : \ \sigma_\alpha(\tau_j) \in D_j, \quad j=1,\ldots,m. \end{aligned} \tag{4.1}$$

To obtain such a σ_α, the technique described in the previous section is applied to each component of the curve. For the sake of simplicity, the approach is presented only for the two dimensional case, as trivial modifications are necessary for the three dimensional one. In addition, the same smoothing parameter is considered for each component of the curve. Here, rectangular and circular regions $D_j,\ j = 1, \dots, m$ are analyzed, even though any kind of convex region could be considered.

THEOREM 4.1. *Let* $D_j = [a_j^x, b_j^x] \times [a_j^y, b_j^y],\ j = 1, \dots, m$. *Let* $\alpha \le \min\{\frac{l^x}{\|\eta_x^0\| + l^x \|G^{-1}\|}, \frac{l^y}{\|\eta_y^0\| + l^y \|G^{-1}\|}\}$, *where* η_x^0, η_y^0 *are analogous to* $\eta^0 = G^{-1}\mathbf{e}$, *for the first and the second components of the curve, respectively and* $l^x := \min_{j=1\dots,m}\{\frac{(b_j^x - a_j^x)}{2}\}$, $l^y := \min_{j=1\dots,m}\{\frac{(b_j^y - a_j^y)}{2}\}$. *Then, the interproximation problem (4.1) is solvable.*

Proof From (3.2) we get

$$|(\mu_x^\alpha)_j| \le \frac{(b_j^x - a_j^x)}{2\alpha}, \quad |(\mu_y^\alpha)_j| \le \frac{(b_j^y - a_j^y)}{2\alpha},\ j = 1, \dots, m, \tag{4.2}$$

where μ_x^α and μ_y^α are obtained by working with the first and second components of the curve, respectively. With the same arguments given in Theorem 3.1, we obtain a solution of the problem (4.1) if the following conditions are verified

$$\begin{cases} \alpha \le \frac{l^x}{(\|\eta_x^0\| + l^x \|G^{-1}\|)} \\ \alpha \le \frac{l^y}{(\|\eta_y^0\| + l^y \|G^{-1}\|)}. \end{cases} \tag{4.3}$$

Thus, the theorem is proved. □

In the case of circular regions D_j with center $\mathbf{c}_j$ and radius r_j, the mixed spline has to satisfy

$$\min_{\tau \in [0,1]} \{\sigma_\alpha(\tau) - \mathbf{c}_j\} \le r_j,\ \ j = 1, \dots, m. \tag{4.4}$$

After a little algebra, the following sufficient conditions can be obtained

$$\alpha^2 (\mu_x^\alpha)_j^2 + \alpha^2 (\mu_y^\alpha)_j^2 \le r_j^2,\ j = 1, \dots, m. \tag{4.5}$$

A theorem analogous to the previous one can be stated taking in to account $\|\cdot\|_2$ instead of the infinity norm $\|\cdot\|$.

5. Numerical results. In this section some graphs are given to illustrate the method for both functional and parametric data, for $2D$ and $3D$ curves. From all the presented examples, it can be seen how the curve obtained choosing $\alpha = 0$ (i.e. the natural cubic interpolating spline) presents bumps and wiggles which reduce in the interproximating curve with a smoothing parameter $\alpha > 0$.

In all the figures, the results obtained for $\alpha = 0$ (*dotted line*) and $\alpha > 0$ (*continuous line*) are displayed stressing the advantage of using a non zero smoothing parameter. The symbol '$*$' denotes an interpolation point, while the symbol '$+$' denotes a barycenter point.

The first example, given in Figure 1, shows mixed splines interproximating functional data. The smoothing parameter α is chosen equal to 0.1 and the approximation regions are represented by vertical segments. The second example relates to parametric data in $\mathbb{R}^2$ with $\alpha = 0.0004$ for both the components of the curve. In this case, the approximation regions are circular. The last example deals with a $3D$ curve where the interpolation points are obtained by evaluating a helix at not uniformly distributed knots in the parameter space $[0, 1]$. The approximation regions are here parallelepipeds. The smoothing parameter is $\alpha = 1$ for all the components of the curve. The projection of the interproximating spline onto the xy-plane is also given. It should be noted that, in the parametric case, the parameter set is defined by using the chord length parametrization with uniform and periodic auxiliary knots in the case of open and closed curves respectively.

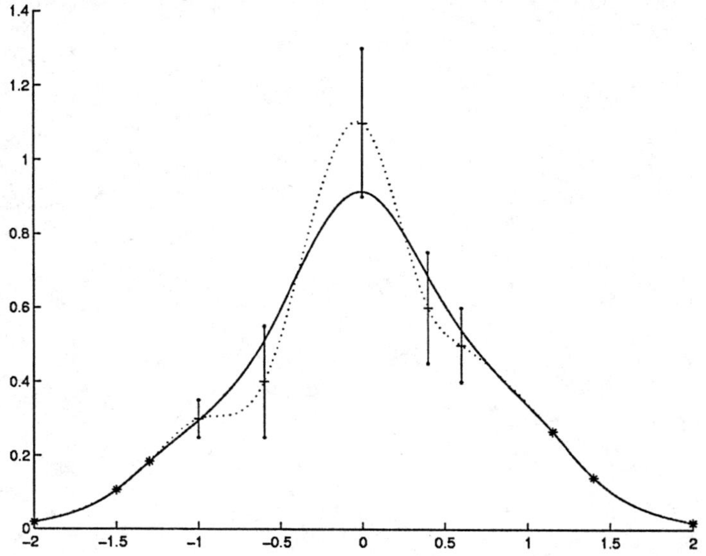

Fig. 1. *Example 1: functional case.*

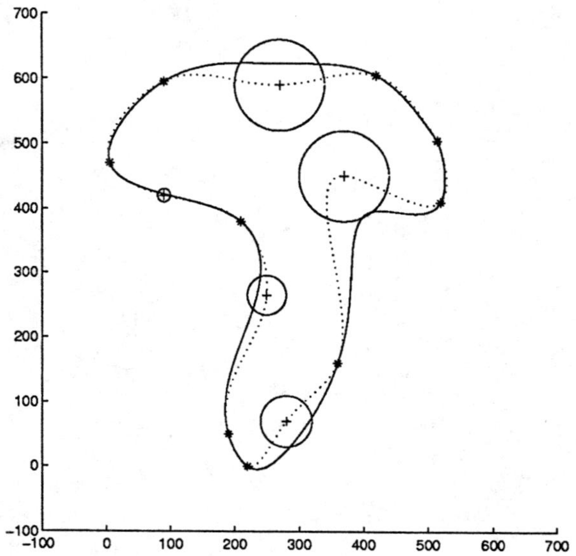

Fig. 2. *Example 2: parametric case in* $\mathbb{R}^2$.

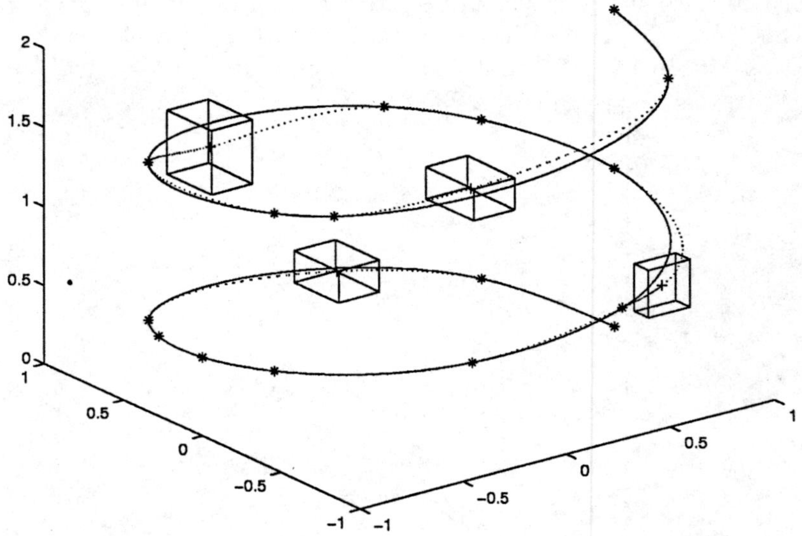

Fig. 3. *Example 3: parametric case in* $\mathbb{R}^3$.

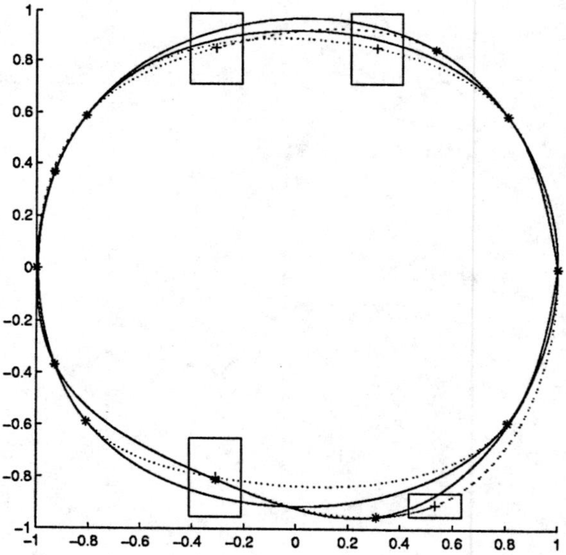

Fig. 4. *Example 3: parametric case in* $\mathbb{R}^3$*, projection onto the xy-plane.*

REFERENCES

[1] N. Aronszajn, *Theory of Reproducing Kernels*, Trans. Amer. Math. Soc. 68 (1968), pp. 337–404.

[2] A. Yu. Bezhaev, and V. A. Vasilenko, *Variational Splines Theory*, Bulletin of the Novosibirsk Computing Center, Series: Numerical Analysis, Special Issue, 3, NCC Publisher, Novosibirsk, 1993.

[3] D. Bini, M. Capovani, and O. Menchi, *Metodi Numerici per l'Algebra Lineare*, Zanichelli, Bologna, 1988.

[4] F. Cheng and B. A Barsky, *Interproximation: Interpolation and Approximation Using Cubic Spline Curves*, CAD, 23, 10 (1991), pp. 700–706.

[5] F. Cheng and B. A. Barsky, *Interproximation Using Cubic B-spline Curves*, in Modeling in Computer Graphics, B. Falcidieno, T. L. Kunii eds, Springer-Verlag, Berlin, 1993, pp. 359–374.

[6] C. Conti and R. Morandi, *The Bidimensional Interproximation Problem and Mixed Splines*, submitted.

[7] C. de Boor, *A Practical Guide to Splines*, Springer Verlag, New York, (1978).

[8] P. J. Laurent, *Construction of Spline Functions in a Convex Set*, in Approximation with Special Emphasis on Spline Functions, I. J. Schoenberg, ed., Academic Press, London, 1969, pp. 415–445.

[9] K. Ritter, 1969 *Generalized Spline Interpolants and Nonlinear Programming*, in Approximation with Special Emphasis on Spline Functions, I. J. Schoenberg ed., Academic Press, London, 1969, pp. 75–118.

[10] M. L. Sampoli, *Fair Interproximation of closed curves*, Technical Report, Firenze, (1997).

[11] X. Wang, F. Cheng, and B. A. Barsky, *Energy and B-spline interproximation*, CAD, 29, 7, (1997), pp. 485–496.

ON THE CONSTRUCTION OF BIVARIATE QUADRATIC C^1 SPLINES*

CATTERINA DAGNINO† AND PAOLA LAMBERTI‡

Abstract. In this paper local bivariate C^1 quasi-interpolating splines with a four directional mesh are considered and a computational procedure for their construction is proposed. Numerical and graphical examples are reported.

Key words. bivariate splines, algorithms.

AMS subject classifications. 65D07, 65-04.

1. Introduction. In several mathematical models, where functions of two variables are required, the class of piecewise polynomials, usually called splines, is very useful and it has been widely studied in the literature ([2, 9, 11, 12, 16] and references therein).

Let $S_2^1(\Delta_{mn}^{(2)})$ be the bivariate quadratic C^1 spline space defined on a four directional mesh $\Delta_{mn}^{(2)}$ and spanned by a sequence of B-splines that are the scaled translates of a finitely supported C^1 piecewise quadratic polynomial basic spline function, so called ZP-element [14, 19]. Recently quasi-interpolating (q-i) splines [9] of $S_2^1(\Delta_{mn}^{(2)})$ on uniform type-2 triangulations have been considered and studied in the literature ([3, 5, 6, 7, 8, 9] et al.).

This paper deals with a class of local bivariate C^1 q-i splines $\mathcal{A}f$ of $S_2^1(\Delta_{mn}^{(2)})$, defined on uniform type-2 triangulations and generated by a sequence of B-splines, that have interesting applications in science and technology [18]. First we introduce the spline operators $\mathcal{A}$ and we present some properties of theirs. Then we propose an algorithm, developed in Matlab environment, for their generation and we give numerical and graphical test results.

2. Local bivariate C^1 q-i spline operators. Let $S = \{(x, y) : 0 \le x, y \le 1\}$ and let $\Delta_{mn}^{(2)}$ be a uniform grid partition of S with type-2 triangulation (Fig.2.1), also called four-directional mesh, defined by the restriction on S of the $\mathbf{R}^2$ partition consisting of lines

$$mx - i = 0, \qquad ny - i = 0, \qquad ny - mx - i = 0, \qquad ny + mx - i = 0$$

$i = \ldots, -1, 0, 1, \ldots$, where m, n are given integers.

We assume $m = \alpha n$, where α is a rational positive constant.

Let $S_2^1(\Delta_{mn}^{(2)})$ be the space of bivariate C^1 spline functions that in every cell of $\Delta_{mn}^{(2)}$ are polynomials of total degree 2.

*Work sponsored by C.N.R. and M.U.R.S.T. of Italy.

†Dipartimento di Matematica, Università di Torino, via C.Alberto, 10 – 10123 Torino, Italy (dagnino@dm.unito.it)

‡Dipartimento di Matematica, Università di Torino, via C.Alberto, 10 – 10123 Torino, Italy (lamberti@dm.unito.it)

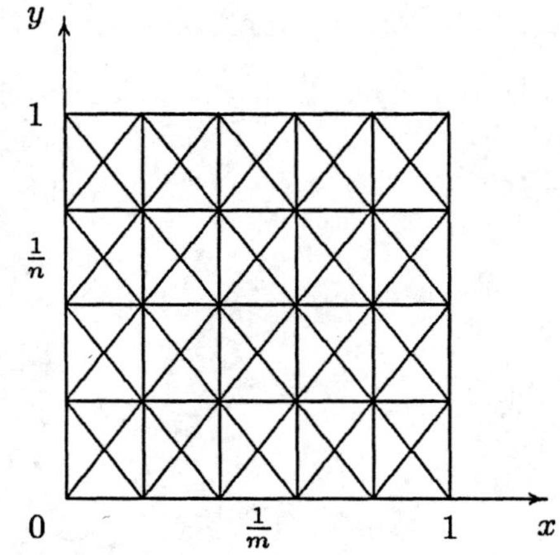

Fig. 2.1 A uniform type-2 triangulation of S.

By using the B-spline $B(x, y)$, also known as ZP-element [9, 14, 19], Chui and Wang [5] showed that the set

$$\left\{B_{ij}(x,y) = B\Big(mx - i + \frac{1}{2}, ny - j + \frac{1}{2}\Big),\ i = 0,\dots,m+1,\ j = 0,\dots,n+1\right\} \tag{2.1}$$

is a subset of $S_2^1(\Delta_{mn}^{(2)})$ and it spans all splines of $S_2^1(\Delta_{mn}^{(2)})$.

We recall that $B(x, y)$ is supported on the octagon Q with centre at $(0,0)$ and vertices at $(3/2, 1/2), (1/2, 3/2), (-1/2, 3/2), (-3/2, 1/2), (-3/2, -1/2), (-1/2, -3/2), (1/2, -3/2), (3/2, -1/2)$ It is strictly positive inside its support that is partitioned into 25 cells Δ_k labelled by k as in Fig.2.2.

On every cell Δ_k the function $B(x, y)$ is a polynomial $p_k(x, y)$ of total degree 2. In particular it results that [5]

$$\begin{aligned}
&p_1(x,y) = \frac{1}{2} - \frac{1}{2}x^2 - \frac{1}{2}y^2, \quad p_2(x,y) = \frac{5}{8} - \frac{1}{2}x - \frac{1}{2}y^2,\\
&p_6(x,y) = \Big(\frac{7}{8} - x + \frac{1}{4}x^2\Big) + \Big(\frac{1}{2} - \frac{1}{2}x\Big)y - \frac{1}{4}y^2,\\
&p_7(x,y) = \frac{9}{8} - \frac{3}{2}x + \frac{1}{2}x^2,\\
&p_9(x,y) = \Big(1 - x + \frac{1}{4}x^2\Big) + \Big(-1 + \frac{1}{2}x\Big)y + \frac{1}{4}y^2,
\end{aligned} \tag{2.2}$$

and the other p_k's are obtained from the above ones by simmetry. In particular:

$$\begin{array}{lll}
p_3(x,y) = p_2(y,x); & p_4(x,y) = p_2(-x,y); & p_5(x,y) = p_2(-y,x);\\
p_8(x,y) = p_6(x,-y); & p_{11}(x,y) = p_6(y,-x); & p_{13}(x,y) = p_6(y,x);\\
p_{16}(x,y) = p_6(-x,-y); & p_{18}(x,y) = p_6(-x,y); & p_{21}(x,y) = p_6(-y,x);\\
p_{23}(x,y) = p_6(-y,-x); & p_{12}(x,y) = p_7(y,x); & p_{17}(x,y) = p_7(-x,y);
\end{array}$$

$p_{22}(x,y) = p_7(-y,x)$; $p_{10}(x,y) = p_9(y,x)$; $p_{14}(x,y) = p_9(y,-x)$;
$p_{15}(x,y) = p_9(-x,y)$; $p_{19}(x,y) = p_9(-x,-y)$; $p_{20}(x,y) = p_9(-y,-x)$;
$p_{24}(x,y) = p_9(-y,x)$; $p_{25}(x,y) = p_9(x,-y)$.

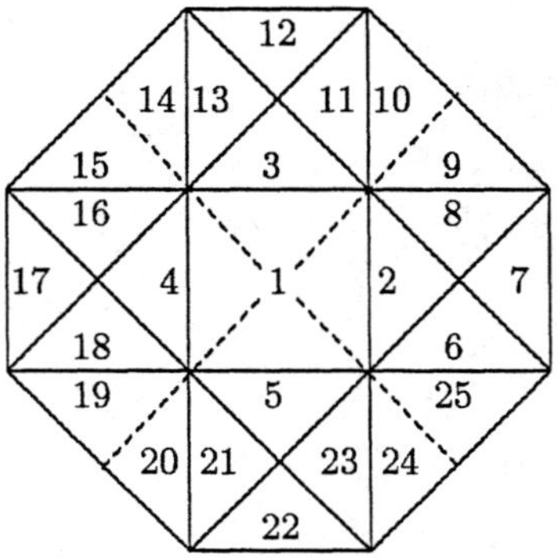

Fig. 2.2 The support Q of $B(x,y)$.

We recall that B is in $C^1(\mathbf{R}^2)$.

Therefore the B-spline $B_{ij}(x,y)$ has got a local support $Q_{ij} = \bigcup_{k=1}^{25} \Delta_k^{(ij)}$ that has centre at $(x_i, y_j) = \left(\frac{2i-1}{2m}, \frac{2j-1}{2n}\right)$, where $\Delta_k^{(ij)}$ is the k-th cell of Q_{ij}.

Let Ω be an open set containing S. Given $f \in C(\Omega)$, we consider an approximation $\mathcal{A}f$ to f defined by

$$\mathcal{A}f(x,y) = \sum_{ij} \Lambda_{ij} f B_{ij}(x,y) \tag{2.3}$$

where $\{\Lambda_{ij}\}_{\substack{i=0,\dots,m+1 \\ j=0,\dots,n+1}}$ is a set of linear functionals $\Lambda_{ij} : C(\Omega) \to \mathbf{R}$ of the following form

$$\Lambda_{ij} f = \sum_{\mu=1}^{p} w_\mu f(x_\mu^{(i)}, y_\mu^{(j)}), \tag{2.4}$$

involving only a finite fixed number $p \geq 1$ of triangular mesh-points $(x_\mu^{(i)}, y_\mu^{(j)})$ either in the support or close to the support Q_{ij} and of real non-zero weights w_μ. Moreover the Λ_{ij}'s are assumed to be such that $\mathcal{A}f = f$ for all $f \in \mathbf{P}_l$, $1 \leq l \leq 3$ with $\mathbf{P}_l$ the class of all polynomials in two variables of total degree less than l.

Therefore $\mathcal{A}$ defines a linear q-i operator mapping $C(\Omega)$ in $S_2^1(\Delta_{mn}^{(2)})$. Then it is a local approximation scheme, because the value of $\mathcal{A}f(x,y)$ depends only on the values of f in a neighborhood of (x,y). Indeed if we assume $(t,\bar{t}) \in S$ and if T is the triangular cell of $\Delta_{mn}^{(2)}$ containing $(t,\bar{t})$, since just seven B-splines B_{ij} are non zero on T, then it results:

$$\mathcal{A}f(t,\bar{t}) = \sum_{\substack{ij\ / \\ T \cap Q_{ij} \neq \emptyset}} \Lambda_{ij} f B_{ij}(t,\bar{t}). \tag{2.5}$$

Two interesting choices of functionals Λ_{ij} are the following [3, 5, 6, 7, 18].

i) If in (2.4) we assume $p = 1$, $w_1 = 1$ and $(x_1^{(i)}, y_1^{(j)}) = (x_i, y_j)$, (Fig.2.3*a*), then we obtain the following q-i spline for which $l = 2$:

(2.6)
$$V_{mn}f(x,y) = \sum_{i=0}^{m+1} \sum_{j=0}^{n+1} f(x_i, y_j) B_{ij}(x,y).$$

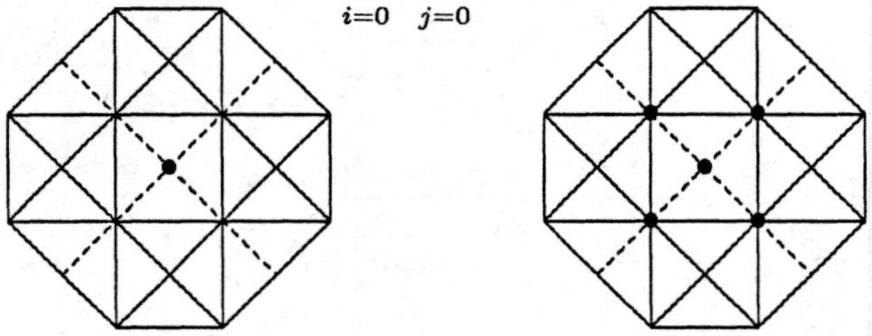

Fig. 2.3 Points pattern on Q_{ij} : (*a*) for V_{mn}, (*b*) for W_{mn}.

Some convergence properties and shape-preserving results for sequences of splines (2.6) are given in [5] and [3], respectively.

ii) If in (2.4) we let $p = 5$, $w_1 = 2$, $w_k = -\frac{1}{4}$, $k = 2, \ldots, 5$ and

$$(x_1^{(i)}, y_1^{(j)}) = \left(\frac{2i-1}{2m}, \frac{2j-1}{2n}\right), \; (x_2^{(i)}, y_2^{(j)}) = \left(\frac{i}{m}, \frac{j}{n}\right), \; (x_3^{(i)}, y_3^{(j)}) = \left(\frac{i-1}{m}, \frac{j}{n}\right),$$

$$(x_4^{(i)}, y_4^{(j)}) = \left(\frac{i-1}{m}, \frac{j-1}{n}\right), \; (x_5^{(i)}, y_5^{(j)}) = \left(\frac{i}{m}, \frac{j-1}{n}\right) \quad \text{(Fig.2.3}b\text{)},$$

then we obtain the following q-i spline for which $l = 3$:

(2.7)
$$W_{mn}f(x,y) = \sum_{i=0}^{m+1} \sum_{j=0}^{n+1} \left\{ 2f\left(\frac{2i-1}{2m}, \frac{2j-1}{2n}\right) - \frac{1}{4}\left[f\left(\frac{i}{m}, \frac{j}{n}\right) + f\left(\frac{i-1}{m}, \frac{j}{n}\right) \right.\right.$$
$$\left.\left. + f\left(\frac{i-1}{m}, \frac{j-1}{n}\right) + f\left(\frac{i}{m}, \frac{j-1}{n}\right) \right] \right\} B_{ij}(x,y).$$

Some convergence results for sequences of splines (2.7) are obtained in [5].

We recall that some other choices of functionals of kind (2.4) are proposed in [9].

For the spline operators of the form (2.3) interesting results on the approximation power can be shown. Indeed we let [10]:

(2.8)
$$E_{r\bar{r}s}(t,\bar{t}) = \begin{cases} D^{r\bar{r}}(f - Af)(t,\bar{t}), & 0 \le r + \bar{r} < s \\ D^{r\bar{r}}Af(t,\bar{t}), & s \le r + \bar{r} < 3 \end{cases}$$

where $(t,\bar{t}) \in S$, $D^{r\bar{r}} = \frac{\partial^{r+\bar{r}}}{\partial x^r \partial y^{\bar{r}}}$ and the parameter s is introduced since the approximation Af could be more differentiable than f;

$$Q^* = \bigcup_{ij \,/\, T \cap Q_{ij} \neq \emptyset} Q_{ij} \, ,$$

i.e. Q^* is the union of the supports Q_{ij} containing the triangular cell T, with $(t,\bar{t}) \in T$;

$$\omega(D^{s-1}f;\delta_{mn};Q^*) = \max_{0\leq\nu\leq s-1} \omega(D^{\nu,s-\nu-1}f;\delta_{mn};Q^*)$$

where $\delta_{mn} = \max\left\{\frac{1}{m},\frac{1}{n}\right\}$ and for a function ψ in $C(\Im)$, with $\Im$ a compact set in $\mathbf{R}^2$, the modulus of continuity of ψ on $\Im$ is defined by

$$\omega(\psi;\delta;\Im) = \max\{|\psi(x,y)-\psi(u,v)| : (x,y),(u,v)\in\Im, \| (x,y)-(u,v) \|\leq\delta\}$$

with $\| (x,y) \| = (x^2+y^2)^{1/2}$. Moreover let $\| \cdot \|_S$ be as usual the supremum norm over S. Then we can prove the following local estimate for (2.8) [7].

THEOREM 2.1. *Let* $1 \leq s \leq l \leq 3$. *If* $f \in C^{s-1}(Q^*)$, *then for* $0 \leq r+\bar{r} < 3$,

$$\| E_{r\bar{r}s} \|_{\dot{T}} \leq C_{\alpha s r\bar{r}}\ \delta_{mn}^{s-r-\bar{r}-1}\ \omega(D^{s-1}f;\delta_{mn};Q^*)$$

where $C_{\alpha s r\bar{r}} = \max\{1,\alpha^r\}\cdot\max\{1,\alpha^{-\bar{r}}\}\cdot C^*$, *with* C^* *a constant independent on* m *and* n *and* $\dot{T} = T$ *if* $r+\bar{r} = 0,1$, *while* $\dot{T} = \text{int}\, T$ *if* $r+\bar{r}=2$.

If, for $s = l$, $f \in C^s(Q^*)$, *then for* $0 \leq r+\bar{r} < 3$,

$$\| E_{r\bar{r}s} \|_{\dot{T}} \leq C_{\alpha s r\bar{r}}\ \delta_{mn}^{s-r-\bar{r}}\ \| D^s f \|_{Q^*}$$

where $\| D^s f \| = \max_{0\leq\nu\leq s} \| D^{\nu,s-\nu} f \|$.

The above local estimates lead immediately to the following global results.

COROLLARY 2.2. *Let* $1 \leq s \leq l \leq 3$. *If* $f \in C^{s-1}(K)$, *where* K *is a compact set, closure of* Ω, *then for* $0 \leq r+\bar{r} < 2$,

$$\| E_{r\bar{r}s} \|_S \leq C_{\alpha s r\bar{r}}\ \delta_{mn}^{s-r-\bar{r}-1}\ \omega(D^{s-1}f;\delta_{mn};K)$$

with $C_{\alpha s r\bar{r}}$ *defined as in Theorem 2.1.*

If, for $s = l$, $f \in C^s(K)$, *then for* $0 \leq r+\bar{r} < 2$,

$$\| E_{r\bar{r}s} \|_S \leq C_{\alpha s r\bar{r}}\ \delta_{mn}^{s-r-\bar{r}}\ \| D^s f \|_K .$$

We remark that Theorem 2.1 provides local convergence results of $D^{r\bar{r}}\mathcal{A}f$ to $D^{r\bar{r}}f$ for $f \in C^{s-1}(Q^*)$, $1 \leq s \leq l \leq 3$, $0 \leq r+\bar{r} < 3$ and $r+\bar{r} < s$. If $s = l$ and $f \in C^l(Q^*)$, then it results $\| D^{r\bar{r}}(f-\mathcal{A}f) \|_{\dot{T}} = O(\delta_{mn}^{l-r-\bar{r}})$.

Moreover from Corollary 2.2 we can deduce that for each $f \in C^{s-1}(K)$, $1 \leq s \leq l \leq 3$, the q-i spline $\mathcal{A}f \in S_2^1(\Delta_{mn}^{(2)})$ satisfies $\| D^{r\bar{r}}(f-\mathcal{A}f) \|_S = o(\delta_{mn}^{s-r-\bar{r}-1})$ for $0 \leq r+\bar{r} < 2$ and $r+\bar{r} < s$. If $s = l$ and $f \in C^l(K)$, then we have $\| D^{r\bar{r}}(f-\mathcal{A}f) \|_S = O(\delta_{mn}^{l-r-\bar{r}})$.

Finally local and global upper bounds for $\| D^{r\bar{r}}\mathcal{A}f \|$ can be given with $s \leq r+\bar{r} < 3$ by Theorem 2.1 and with $s \leq r+\bar{r} < 2$ by Corollary 2.2, respectively.

Thanks to the above general results we can add some new convergence properties [7] to the ones shown in [5] for sequences of splines (2.6) and (2.7). In particular it results that (2.6) and (2.7) approximate f and its partial derivatives to nearly optimal and optimal order, respectively.

3. Computational procedure. In this section we present an algorithm to generate bivariate splines of kind (2.3). The software, developed in the convenient interactive environment that MATLAB 5 provides, is modularized in a fashion similar to the structure of the Spline Toolbox function M-files [17]. Our main user-callable function M-files are as follows [8]:

- `amn`, which returns the value of a bivariate spline $\mathcal{A}f$ at an arbitrary grid of evaluation points in a domain $D = [a, b] \times [c, d]$, according to (2.5).
- `bij`, which returns the value of the B-spline B_{ij} at an arbitrary point in $\mathbf{R}^2$, according to (2.1).
- `bspl`, which returns the value of the ZP-element B at an arbitrary point in $\mathbf{R}^2$, according to (2.2).
- `lv` and `lw`, which return the matrix $\Lambda_{ij}f$, defined by (2.4), for the splines (2.6) and (2.7), respectively. The above matrix is an input argument for the function `amn`. For other choices of functionals of kind (2.4) the corresponding M-files generating the matrix $\Lambda_{ij}f$ have to be provided by the user.

The procedure `amn` calls `bij` which in turn calls `bspl`.

Other minor M-files are utilities that perform the visualization of $\mathcal{A}f$, $\{B_{ij}\}$ and the construction and the visualization of the supports $\{Q_{ij}\}$ [8].

All computations are carried out on a Pentium PC.

Now we test our procedure on twelve sets of data, taken from the following functions defined in the unit square and usually considered in approximation algorithms [1, 13, 15]:

- $f_1(x,y) = (r - .6)_+^4$
- $f_2(x,y) = e^{-1/(r-.6)_+^2}$
- $f_3(x,y) = .75e^{\frac{-(9x-2)^2+(9y-2)^2}{4}} + .75e^{-\frac{(9x+1)^2}{49}-\frac{(9y+1)}{10}}$
  $+.5e^{\frac{-(9x-7)^2+(9y-3)^2}{4}} - .2e^{-(9x-4)^2-(9y-7)^2}$
- $f_4(x,y) = \frac{1}{9}(\tanh(9y - 9x) + 1)$
- $f_5(x,y) = \frac{1.25+\cos(5.4y)}{6(1+(3x-1)^2)}$
- $f_6(x,y) = \frac{1}{3}e^{-\frac{81}{16}((x-\frac{1}{2})^2+(y-\frac{1}{2})^2)}$
- $f_7(x,y) = \frac{1}{3}e^{-\frac{81}{4}((x-\frac{1}{2})^2+(y-\frac{1}{2})^2)}$
- $f_8(x,y) = \frac{1}{9}\sqrt{64 - 81((x-\frac{1}{2})^2 + (y-\frac{1}{2})^2)} - \frac{1}{2}$
- $f_9(x,y) = \left(1 + 2e^{-3(10r-6.7)}\right)^{-\frac{1}{2}}$
- $f_{10}(x,y) = e^{-\frac{(5-10x)^2}{2}} + .75e^{-\frac{(5-10y)^2}{2}} + .75e^{-\frac{(5-10x)^2}{2}}e^{-\frac{(5-10y)^2}{2}}$
- $f_{11}(x,y) = 2\cos(10x)\sin(10y) + \sin(10xy)$
- $f_{12}(x,y) = e^{-.04\sqrt{(80x-40)^2+(90y-45)^2}}\cos(.15\sqrt{(80x-40)^2+(90y-45)^2})$

where $r = \sqrt{x^2 + y^2}$.

The functions $f_3 \div f_8$ are the well-known Franke's test functions.

The last three test functions are chosen for their multiple features and abrupt transitions. Thus, while all the test functions are smooth (except for a first derivative discontinuity at the point $(0.5, 0.5)$ in the case of function f_{12}), the last three are quite challenging.

We present numerical and graphical results obtained assuming either $\mathcal{A} = V_{mn}$ or $\mathcal{A} = W_{mn}$.

TABLE 3.1
Maximum absolute error for $f_1(x,y) - f_{12}(x,y)$.

$m=n$	$f_1(x,y)$ EV_{mn}	$f_1(x,y)$ EW_{mn}	$f_2(x,y)$ EV_{mn}	$f_2(x,y)$ EW_{mn}	$f_3(x,y)$ EV_{mn}	$f_3(x,y)$ EW_{mn}	$f_4(x,y)$ EV_{mn}	$f_4(x,y)$ EW_{mn}
2	3.07(−1)	4.76(−2)	7.01(−2)	3.52(−2)	3.61(−1)	5.73(−1)	6.52(−2)	5.63(−2)
4	7.47(−2)	3.63(−3)	2.25(−2)	6.62(−3)	3.21(−1)	2.14(−1)	4.03(−2)	3.18(−2)
8	1.85(−2)	3.37(−4)	6.20(−3)	6.97(−4)	9.96(−2)	1.91(−2)	1.90(−2)	9.83(−3)
16	4.63(−3)	2.27(−5)	1.60(−3)	5.61(−5)	2.76(−2)	2.23(−3)	5.66(−3)	8.28(−4)
32	1.15(−3)	2.39(−6)	4.03(−4)	4.60(−6)	7.10(−3)	2.42(−4)	1.51(−3)	6.95(−5)
64	2.89(−4)	3.07(−7)	1.00(−4)	4.58(−7)	1.79(−3)	2.83(−5)	3.82(−4)	3.86(−6)
$m=n$	$f_5(x,y)$ EV_{mn}	$f_5(x,y)$ EW_{mn}	$f_6(x,y)$ EV_{mn}	$f_6(x,y)$ EW_{mn}	$f_7(x,y)$ EV_{mn}	$f_7(x,y)$ EW_{mn}	$f_8(x,y)$ EV_{mn}	$f_8(x,y)$ EW_{mn}
2	1.78(−1)	1.21(−1)	1.56(−1)	1.16(−1)	3.06(−1)	3.64(−1)	1.95(−1)	1.46(−1)
4	6.96(−2)	2.12(−2)	4.87(−2)	1.32(−2)	1.56(−1)	1.16(−1)	8.23(−2)	8.15(−2)
8	1.99(−2)	2.79(−3)	1.29(−2)	9.83(−4)	4.87(−2)	1.32(−2)	1.42(−2)	6.70(−3)
16	5.41(−3)	2.32(−4)	3.27(−3)	6.42(−5)	1.29(−2)	9.83(−4)	3.41(−3)	1.66(−4)
32	1.35(−3)	1.22(−5)	8.22(−4)	5.96(−6)	3.27(−3)	6.42(−5)	8.46(−4)	9.27(−6)
64	3.39(−4)	1.43(−6)	2.05(−4)	5.52(−7)	8.22(−4)	4.05(−6)	2.11(−4)	5.64(−7)
$m=n$	$f_9(x,y)$ EV_{mn}	$f_9(x,y)$ EW_{mn}	$f_{10}(x,y)$ EV_{mn}	$f_{10}(x,y)$ EW_{mn}	$f_{11}(x,y)$ EV_{mn}	$f_{11}(x,y)$ EW_{mn}	$f_{12}(x,y)$ EV_{mn}	$f_{12}(x,y)$ EW_{mn}
4	2.08(−1)	2.34(−1)	1.54(0)	1.70(0)	1.89(0)	1.55(0)	1.34(0)	1.71(0)
8	7.87(−2)	4.72(−2)	5.53(−1)	2.80(−1)	6.87(−1)	2.24(−1)	6.83(−1)	5.23(−1)
16	2.46(−2)	6.58(−3)	1.53(−1)	2.39(−2)	1.95(−1))	1.71(−2)	2.73(−1)	1.57(−1)
32	6.71(−3)	6.57(−4)	3.93(−2)	1.62(−3)	5.00(−2)	1.16(−3)	1.09(−1)	6.43(−2)
64	1.72(−3)	6.98(−5)	9.89(−3)	1.04(−4)	1.25(−2)	9.24(−5)	4.65(−2)	3.04(−2)

The maximum absolute errors, computed using an 11×11 uniform rectangular grid of evaluation points in the unit square $[0,1] \times [0,1]$, are reported in Table 3.1 for increasing values of m and n. The measure of the error is denoted by EV_{mn} and EW_{mn} for the operators V_{mn} and W_{mn}, respectively.

We remark that the obtained results are consistent with our expectation on the approximation order given by Corollary 2.2, in most cases.

We also note that the maximum norm of $f_1 \div f_{12}$ is 0.439, 0.221, 1.220, 0.222, 0.375, 0.333, 0.333, 0.389, 1, 2.500, 2.996, 0.995, respectively.

Finally in Fig. 3.1−3.6, some bivariate spline functions for the tested cases are displayed.

4. Final remarks. In this paper we have considered local q-i splines defined in $S_2^1(\Delta_{mn}^{(2)})$ and generated by a sequence of B-splines. Moreover we have proposed a Matlab computational procedure for their construction. The generalization of our above results to local bivariate quadratic C^1 q-i splines on non-uniform type-2 triangulations [4] is in progress.

REFERENCES

[1] R. K. BEATSON AND Z. ZIEGLER, *Monotonicity preserving surface interpolation*, SIAM J. Numer. Anal., 22 (1985), pp. 401–411.

[2] C. K. CHUI, *Multivariate splines*, CBMS-NSF Regional Conference Series in Applied Mathematics, SIAM, Philadelphia, PA, 1988.

[3] C. K. CHUI, H. DIAMOND AND L. RAPHAEL, *Shape-preserving quasi-interpolation and interpolation by box spline surfaces*, J. Comp. Appl. Math., 25 (1989), pp. 169–198.

[4] C. K. CHUI AND R.-H. WANG, *Concerning C^1 B-splines on triangulations of non-uniform rectangular partition*, Approx. Th. Appl., 1 (1984), pp. 11–18.

[5] C. K. CHUI AND R.-H. WANG, *On a bivariate B-spline basis*, Scientia Sinica, XXVII (1984), pp. 1129–1142.

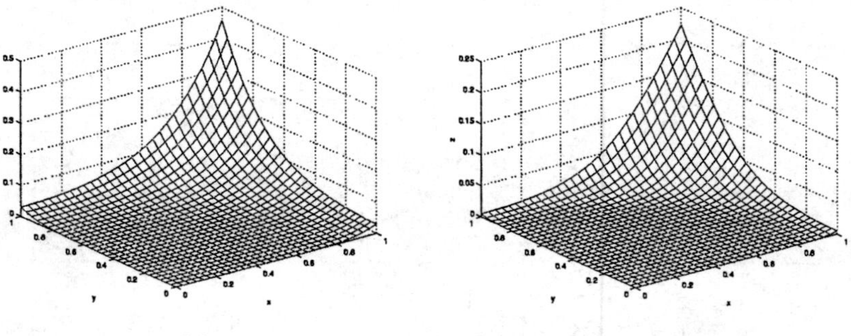

(a) (b)

FIG. 3.1. (a) $V_{88}f_1$ (b) $V_{88}f_2$

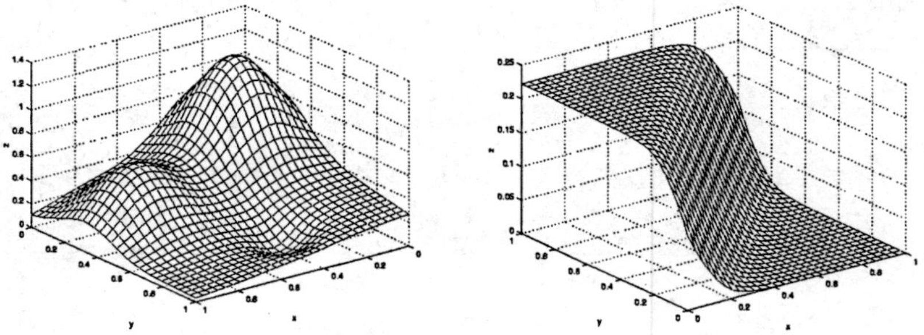

(a) (b)

FIG. 3.2. (a) $W_{88}f_3$ (b) $V_{88}f_4$

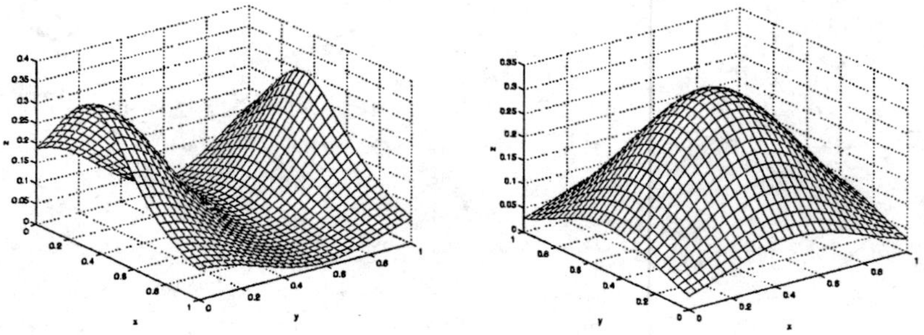

(a) (b)

FIG. 3.3. (a) $V_{88}f_5$ (b) $V_{88}f_6$

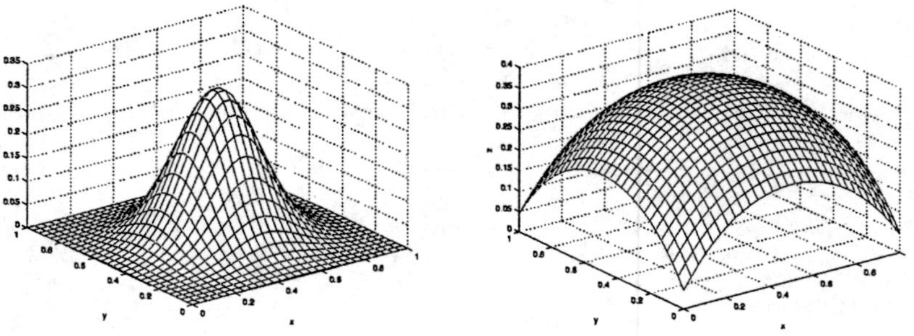

(a) (b)

FIG. 3.4. (a) $V_{16,16}f_7$ (b) $W_{88}f_8$

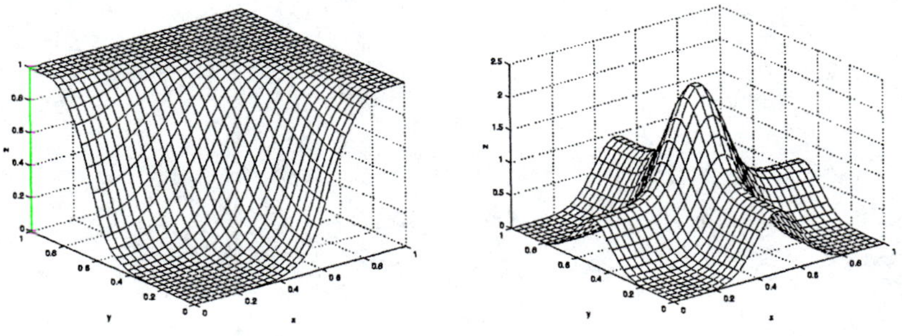

(*a*) (*b*)

FIG. 3.5. (*a*) $V_{88}f_9$ (*b*) $V_{16,16}f_{10}$

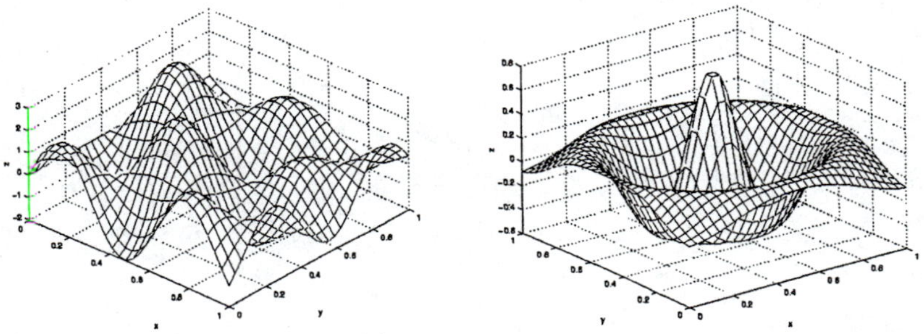

(*a*) (*b*)

FIG. 3.6. (*a*) $W_{16,16}f_{11}$ (*b*) $W_{16,16}f_{12}$

[6] C. DAGNINO AND P. LAMBERTI, *Numerical integration of 2-D integrals based on local bivariate C^1 quasi-interpolating splines*, Adv. Comp. Math., 8 (1998), pp. 19–31.

[7] C. DAGNINO AND P. LAMBERTI, *On the approximation power of bivariate quadratic C^1 splines*, submitted (Technical Report n.38/1999).

[8] C. DAGNINO AND P. LAMBERTI, *SPLISURF: un software per la rappresentazione di spline bivariate su triangolazioni di tipo 2*, Int. Rep. (1999).

[9] C. DE BOOR, K. HÖLLIG AND S. RIEMENSCHNEIDER, *Box splines*, Springer-Verlag, New York, 1993.

[10] T. LYCHE AND L.L. SCHUMAKER, *Local spline approximation methods*, J. Appr. Th., 15 (1975), pp. 294–325.

[11] G. MICULA AND S. MICULA, *Handbook of Splines*, Kluwer Academic Publishers, 1999.

[12] G. NÜRNBERGER, *Approximation by spline functions*, Springer-Verlag, 1989.

[13] G. NÜRNBERGER AND TH. RIESSINGER, *Bivariate spline interpolation at grid points*, Numer. Math., 71 (1995), pp. 91–119.

[14] M. J. D. POWELL, *Piecewise quadratic surface fitting for contour plotting, Software for Numerical Mathematics*, D.J.Evans ed., Academic Press, London, 1974.

[15] R. J. RENKA AND R. BROWN, *Algorithm 792: Accuracy tests of ACM algorithms for interpolation of scattered data in the plane*, ACM Trans. Math. Soft., 25 (1999), pp. 78–94.

[16] L. L. SCHUMAKER, *Spline functions: basic theory*, Krieger Publishing Company, Malabar Florida, 1993.

[17] *Spline Toolbox User's Guide, For Use with Matlab, Version 2*, The MathWorks Inc., 1999.

[18] R.-H. WANG, *Multivariate spline and its applications in science and technology*, Rend. Sem. Mat. Fis. Milano, 63 (1993), pp. 213–229 (1995).

[19] P. B. ZWART, *Multivariate splines with non-degenerate partitions*, SIAM J. Numer. Anal., 10 (1973), pp. 665–673.

ABOUT 3-D SPECTRAL ELEMENT COMPUTATIONS*

DANIELE FUNARO†

Abstract. A project for the numerical approximation by spectral elements of elliptic boundary value problems, defined in three dimensional geometries, is proposed. The domain is assumed to be decomposed in subdomains which are isomorphic to cubes. Polynomials of degree five in each variable are collocated inside each subdomain. Special grids are used to treat advection-diffusion equations. The general structure of the computational code and the first achievements are presented.

Key words. Spectral Elements, Partial Differential Equations, Domain Decomposition.

AMS subject classifications. 65N35, 65N55.

1. Introduction. This is a preliminary version of a more extended report concerning the use of the spectral element method in the numerical approximation of elliptic type linear boundary value problems, defined in non-trivial geometries of the three dimensional space. The aim is to give an idea of the level of complexity needed in 3-D computations, and describe the experience gained in these last few years in the development of a code, to implement numerically some typical differential problems. Without paying too much attention to the details, we shall introduce the main patterns and discuss the major achievements. Other numerical codes, based on the spectral element method, are already available. Concerning the solution techniques, our approach is quite different and could provide a reference point for reorganizing or improving existing software.

2. How to handle the complexity of the domain. Our first concern is to design the computational domain Ω. This will be the union of a certain number of *bricks* B_k, chosen according to the following assumptions. Each brick is obtained by smoothly deforming a reference cube. Thus, each brick will basically consists of an interior part, 6 faces, 12 edges and 8 vertices. We agree that the faces do not contain their boundary, and the edges do not contain their endpoints. The bricks will be not overlapping, that is their interior parts are all disjoint. The bricks will be connected in conforming way, which means that if two of them have some boundary parts in common, only these three cases may occur: they share a whole face, they share a single entire edge, they share only a vertex. An example of geometry is given in figure 2.1. The pieces in common, not belonging to the boundary of Ω, will be generically called *interfaces*.

The solution technique we shall implement to discretize a partial differential equation defined in Ω, uses an iterative procedure based on the domain decomposition individuated by the bricks. Each iteration will involve the parallel solution of subproblems, each one defined in his brick. This will be followed by an exchange of information between the bricks, in preparation for the next iteration. It is very important, at this point, to set up a file from where we can clearly read what kind of relations there are between the subdomains. For any brick, the file gives us the number of bricks next to it, the kind of matching existing with the neighboring bricks, specifying if there are parts belonging to the boundary of Ω.

---

*This work has been supported by Consiglio Nazionale delle Ricerche, Rome. Grants: 96.00173.CT01, 97.00886.CT01.

†Dipartimento di Matematica, Università di Modena, Via Campi 213/B, 41100 Modena (Italy), funaro@unimo.it

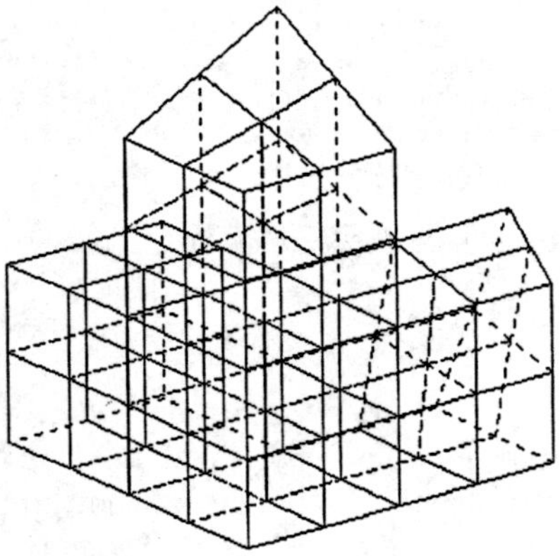

FIG. 2.1. *Example of computational domain decomposed into 32 bricks.*

The construction of the file is accomplished through the following steps:

1. the numbering of the bricks;
2. the numbering of the various parts belonging to each brick;
3. the determination of the interfaces and the kind of connections existing between the bricks;
4. the determination of the boundary of Ω.

The procedure may be rather complicated if the bricks, whose coordinates are supposed to be given, are quite deformed or they have a non standard orientation with respect to the coordinate axes.

3. The collocation grid. The next step is to relate each brick B_k with the reference cube $C = [-1, 1]^3$. To this end, we can define the vector map $\theta_k : B_k \to C$, whose components are first degree polynomials in the variables x, y and z, bringing in a specified order the vertices of B_k into the vertices of C. The map θ_k is uniquely determined by 24 parameters. As we are in the domain C we implement the collocation method at the zeros of

$$(1-x^2)(1-y^2)(1-z^2)P_n'(x)P_n'(y)P_n'(z)$$

where P_n is the Legendre polynomial of degree n. From now on, we shall take $n = 5$, although most of what we are going to say holds for any $n \geq 2$. We think that spectral elements of degree $n = 5$ have a sufficient number of degrees of freedom to allow accurate local approximations. Low polynomial degrees also allow us to handle more bricks in order to better follow the geometry of Ω.

We recall that the zeros of P_5 and P_5' are respectively:

$$-\sqrt{\tfrac{35+2\sqrt{70}}{63}},\ -\sqrt{\tfrac{35-2\sqrt{70}}{63}},\ 0,\ \sqrt{\tfrac{35-2\sqrt{70}}{63}},\ \sqrt{\tfrac{35+2\sqrt{70}}{63}},$$

$$\text{and}\ -\sqrt{\tfrac{21+2\sqrt{63}}{63}},\ -\sqrt{\tfrac{21-2\sqrt{63}}{63}},\ \sqrt{\tfrac{21-2\sqrt{63}}{63}},\ \sqrt{\tfrac{21+2\sqrt{63}}{63}}.$$

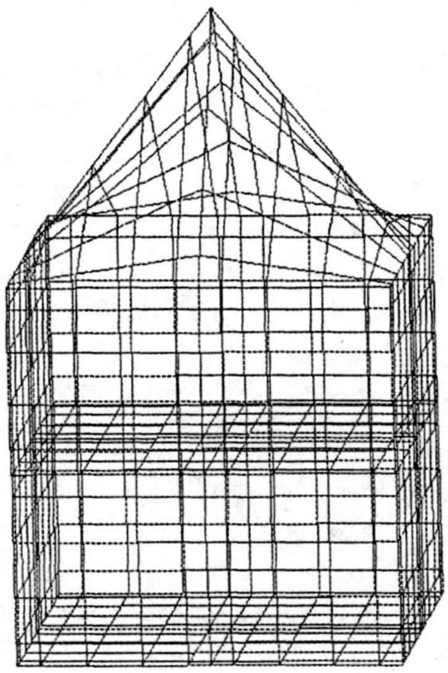

FIG. 3.1. *Grid points in a domain decomposed into 8 bricks.*

The 216 grid points in C (152 on the boundary and 64 inside) are mapped by θ_k^{-1} into the corresponding points in B_k. An example is given by figure 3.1. As we shall see in the coming sections, the spectral element method enforces, together with the continuity of the global approximating function, the continuity of the normal derivatives across the interfaces. It is convenient to compute in advance all the 152 outward normal vectors associated with the boundary of B_k, even if some of them may not be used since they are referred to grid points belonging to the boundary of Ω. The normal vector to a edge of B_k is obtained by averaging the two normal vectors related to the faces generating that edge. The normal vector to a vertex of B_k is obtained by averaging the three normal vectors related to the faces generating that vertex.

For advection-diffusion problems, the collocation of the equation, as well as the continuity of normal derivatives, will be imposed at certain upwind nodes, to be determined according to the coefficients of the differential operator. We refer to section 7 for such developments (see [1] for the treatment of the 2-D case). It is usually convenient to carry out the computation of the upwind grid before starting the iterative algorithm to determine the approximated solution.

4. The collocation method in the reference cube. Suppose first that we have to approximate the solution u of a linear boundary value problem $Lu = f$ defined in the reference cube C, where f is given and L is an elliptic second-order operator. Some boundary conditions are prescribed. In the collocation method, the approximated solution $\tilde{u}$ turns out to be a polynomial of degree $n = 5$ in each variable x, y and z. Such a polynomial is required to satisfy the differential equation at the 64 grid points inside C, and a set of suitable conditions at the 152 nodes belonging to the boundary of C. To construct the linear system associated with this discretization, it is necessary to write down the Lagrange polynomial basis $\phi_j(x,y,z)$, $j = 1, ..., 216$, with respect to all the nodes, and consider the expansion $\tilde{u} = \sum_{j=1}^{216} \tilde{u}_j \phi_j$, where $\tilde{u}_j$ is the value of $\tilde{u}$ at the jth node. We will mainly solve problems with Dirichlet boundary conditions. Hence, 152 values $\tilde{u}_j$ out of 216 are known and coincide with the boundary assignments.

The set of indexes for which the corresponding nodes are inside C is denoted by Θ_I. The set of indexes for which the corresponding nodes belong to the boundary of C is denoted by Θ_B. At any of the 64 nodes inside C we write $L\tilde{u} = f$, which yields:

$$\sum_{j\in\Theta_I} \tilde{u}_j\,(L\phi_j) = f - \sum_{j\in\Theta_B} \tilde{u}_j\,(L\phi_j)$$

which is a linear equation in the unknowns $\tilde{u}_j$, $j \in \Theta_I$. These equations are 64, so that we obtain a 64×64 linear system. It is an easy task to obtain an explicit expression of the polynomials ϕ_j and successively evaluate $L\phi_j$, with the aim of writing the coefficients of the matrix.

5. The spectral element method. The goal is now to solve the problem $Lu = f$ in the domain Ω which is the union of many bricks. Dirichlet conditions are assumed on the boundary of Ω, although different kind of conditions, for instance of Neumann type, can be treated with a little extra care. For any k, a grid of 216 nodes is defined in B_k as the image of the map θ_k^{-1}. Note that, when two bricks have a face in common, they also share the same set of 16 grid points. If, instead, they have an edge in common, they share the same set of 4 grid points.

We denote by u_k the restriction of u to B_k, so that one has $Lu_k = f$ in B_k, which yields: $(Lu_k)(\theta_k^{-1}) = f(\theta_k^{-1})$ in C. This last equation is equivalent to an equation of the type:

$$\mathcal{L}_k v_k \;=\; g_k \qquad \text{in } C \qquad \forall k$$

where $v_k = u_k(\theta_k^{-1})$, $g_k = f(\theta_k^{-1})$ and $\mathcal{L}_k$ is an elliptic, variable-coefficients, second-order differential operator. The coefficients of $\mathcal{L}_k$ depend on the partial derivatives of θ_k. They can be easily computed, although their expression in terms of the coefficients of L is rather long.

We can now introduce the function $\tilde{u}$ approximating the unknown u in Ω. This will be a continuous function in Ω satisfying the boundary conditions at the nodes belonging to the boundary of Ω. Moreover, $\tilde{u}$ is such that, for any k, $\tilde{v}_k = \tilde{u}(\theta_k^{-1})$ is a polynomial of degree 5, satisfying the equation $\mathcal{L}_k\tilde{v}_k = g_k$ at the nodes inside C. We have a last request, that is, for each interface node, the sum of the outward normals derivatives of $\tilde{u}$, with respect to the bricks sharing that node, must be zero. All these conditions are sufficient and not redundant to individuate a unique approximating function $\tilde{u}$. The determination of $\tilde{u}$ at the nodes is subjected to the resolution of a linear system, whose dimension coincides with the number of grid nodes strictly inside Ω (interfaces included) and considered with multiplicity one.

We do not explicitly write the matrix of such a linear system, which is solved by an iterative block relaxation procedure of Dirichlet-Neumann type (see [2], section 1.3), in the way we are going to describe. We initialize the iterative algorithm by prescribing some arbitrary Dirichlet data at the interface points. Then, for any k, we invert the equation $\mathcal{L}_k\tilde{v}_k = g_k$ in C, by solving a 64×64 linear system whose matrix has been previously factorized. The successive step is to compute the normal derivatives of $\tilde{u} = \tilde{v}_k(\theta_k)$ to the brick B_k, making use of spectral differentiation and using the Jacobian of θ_k. Up to this point the computation can be fully parallelized.

Now, we need an exchange of information between the bricks in order to sum up the outward normal derivatives. At this stage it is very important to access the file that takes into account the connections between the subdomains, as briefly mentioned in section 2. If the sum of normal derivatives is nearly zero then we stop the algorithm, otherwise we accordingly update the Dirichlet values at the interfaces in order to proceed with another iteration. The way the Dirichlet values are modified depends on the acceleration procedure chosen. We found very competitive the use of the Chebyshev acceleration method (see [3]), although other techniques can also perform very well. This method does not require, contrary to the conjugate gradient method, the symmetry of the matrix, a condition which is not guaranteed.

Suitable preconditioners have been introduced with the aim of speeding up the convergence. We skip the description, which is a little technical, in order not to complicate this exposition. Other tricks, such as the one of initializing the iterative procedure with the discrete solution obtained with lower degree polynomials ($n = 3$), have been successfully experimented.

6. Visualization. The graphical representation of 3-D output data deserves a lot of attention, since it usually requires a large amount of memory, expecially when one needs to interpolate the polynomial solution in each brick on a finer grid, in order to appreciate the accuracy of the spectral approximation. Animations also require a lot of efforts, both in computational time and memory.

In this section we will show the results of some numerical experiments. The figures are obtained with the *AVS/Express* software (see [4]) using an unstructured type data base (*Unstructured Cell Data*). The *network* that we mostly used is given in figure 6.1.

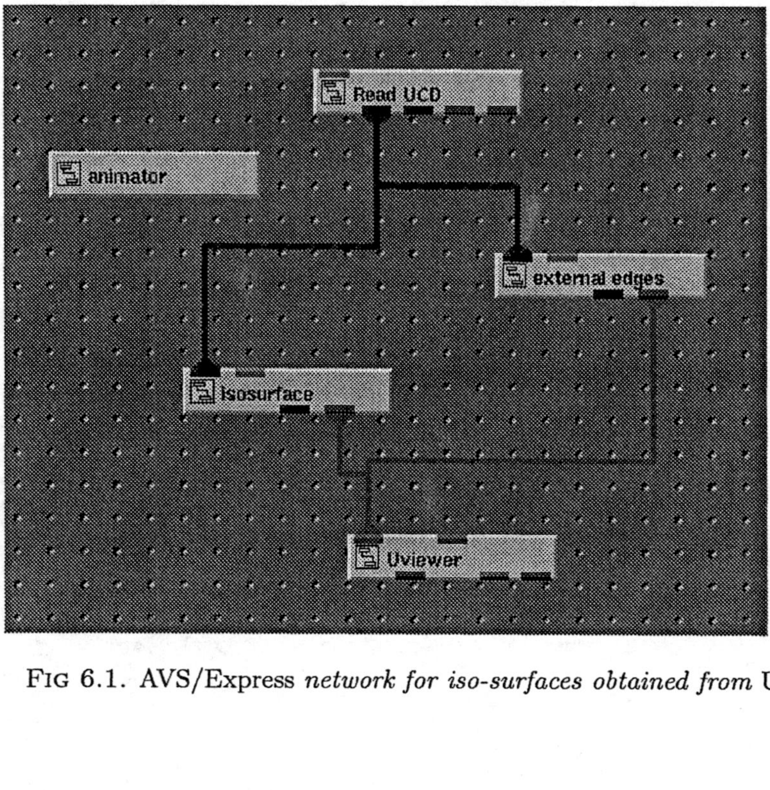

FIG 6.1. AVS/Express *network for iso-surfaces obtained from* UCD.

In figures 6.2 and 6.3 we show the iso-surfaces of some approximated solutions related to the problem $-\Delta u = 1$, with homogeneous Dirichlet boundary conditions, defined on two different domains Ω. In all the experiments the length of the edge of any cube is equal to the unity. In figure 6.4 we show the change of shape of the iso-surface corresponding to the first plot of figure 6.3, during the iteration process initialized with zero Dirichlet conditions at the interfaces. A few iterations are sufficient to ensure a good matching of the various pieces.

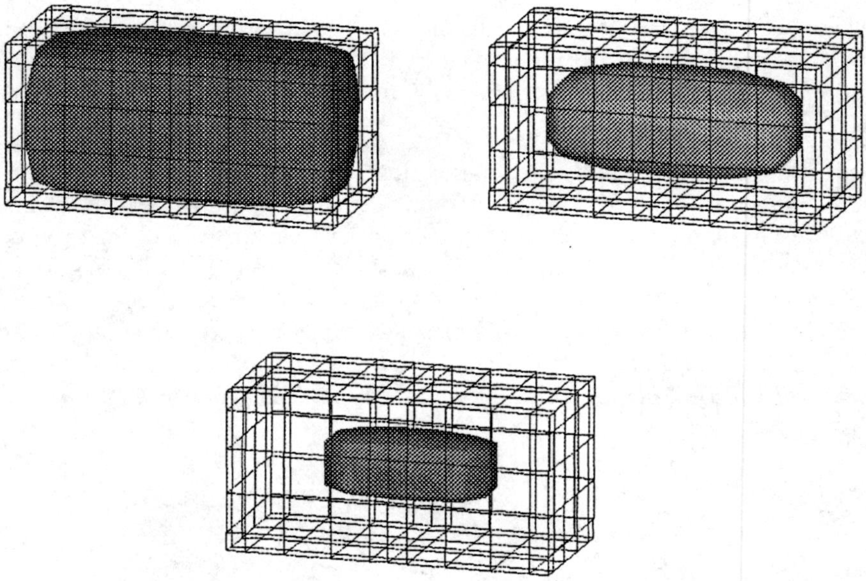

FIG. 6.2. *Iso-surfaces of the approximated solution to the equation* $-\Delta u = 1$.

7. Transport-dominated problems. Extra care has been devoted to the numerical approximation of linear transport-dominated advection-diffusion problems in the form $-\epsilon\Delta u + \vec{\beta} \cdot \vec{\nabla} u = f$, where $\epsilon > 0$ is small compared to the given vector $\vec{\beta}$. A stabilization of the discrete solution with respect to the parameter ϵ is obtained after the introduction of a certain amount of artificial viscosity, accomplished through the collocation at special upwind points defined in C. For any k, the collocation grid depends on the coefficients of the operator $\mathcal{L}_k$. Moreover, special grids at the interfaces, where the continuity of the normal derivatives is imposed, are also introduced. These boundary points, when mapped in B_k by θ_k^{-1}, are situated upwind with respect to the orthogonal projection of the transport vector $\vec{\beta}$ to the boundary of B_k.

This method extends to 3-D the one described in [1], section 4.6, in the case of domains in 2-D. The generalization is conceptually straightforward, but it presents technical difficulties, due to the numerous cases to be handled in defining the new grid inside C and in any of the 6 faces and 12 edges. Since here the aim is to give the main ideas, we will not proceed any further in the description. We just show in figure 7.1 what is the situation in the case of two domains for $\epsilon = .05$, where $\vec{\beta}$ has norm equal to 1 and it is orthogonal to the interface (first plot), or it has not orthogonal component to the interface (second plot).

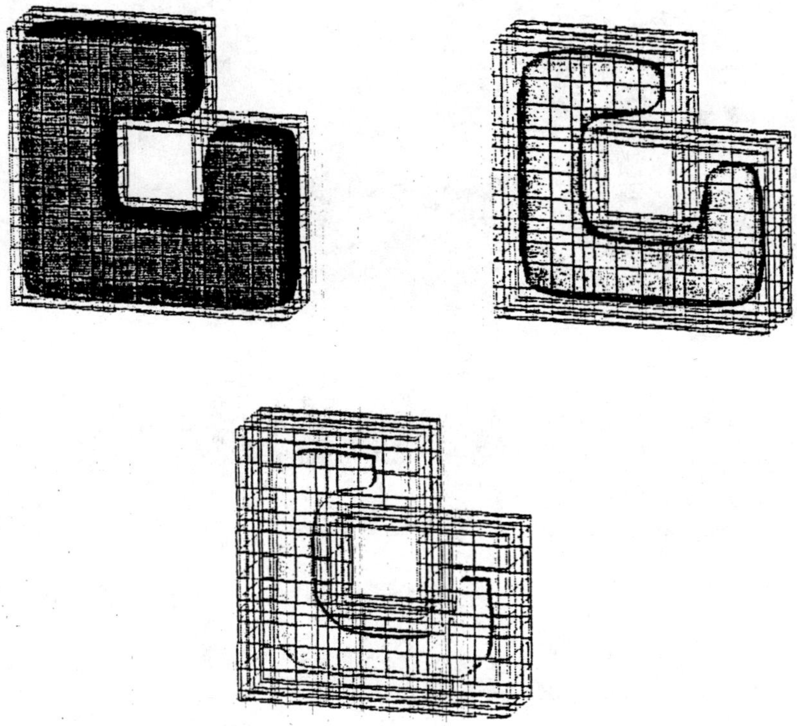

FIG. 6.3. *Iso-surfaces of the approximated solution to the equation* $-\Delta u = 1$.

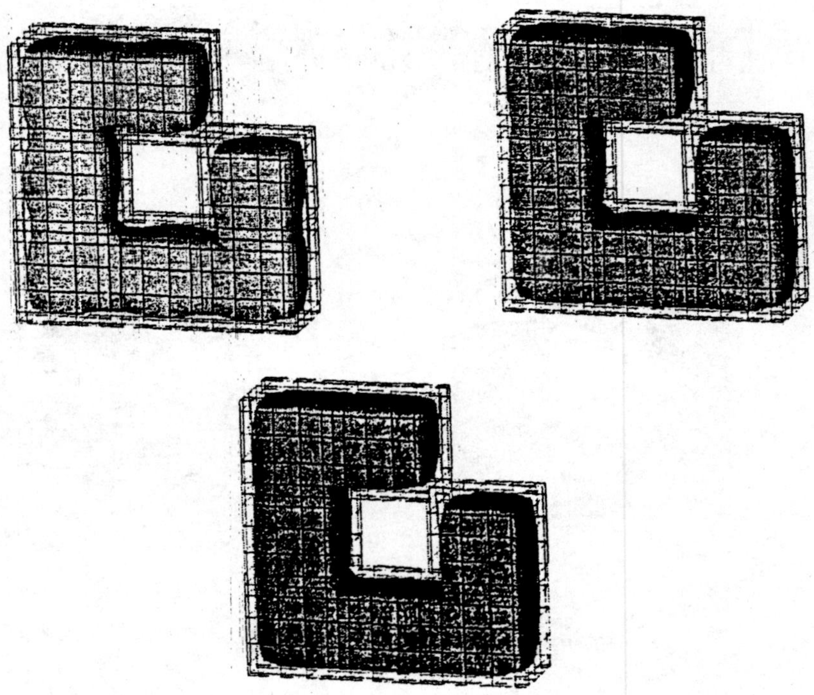

FIG. 6.4. *The making of an iso-surface during the first three iterations.*

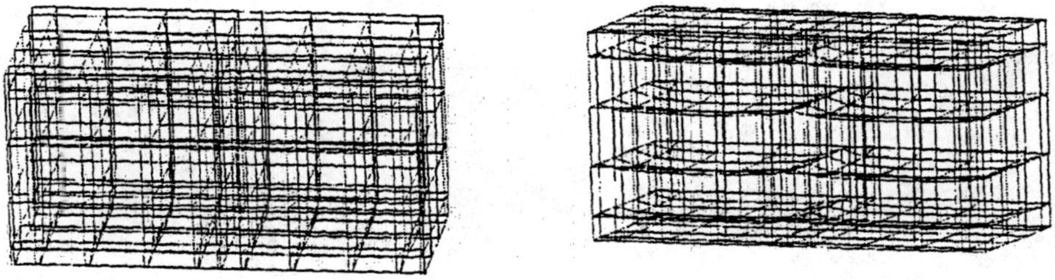

FIG. 7.1. *Upwind grids. In the first picture the flux is orthogonal to the interface. In the second one, the flux belongs to the plane of the interface.*

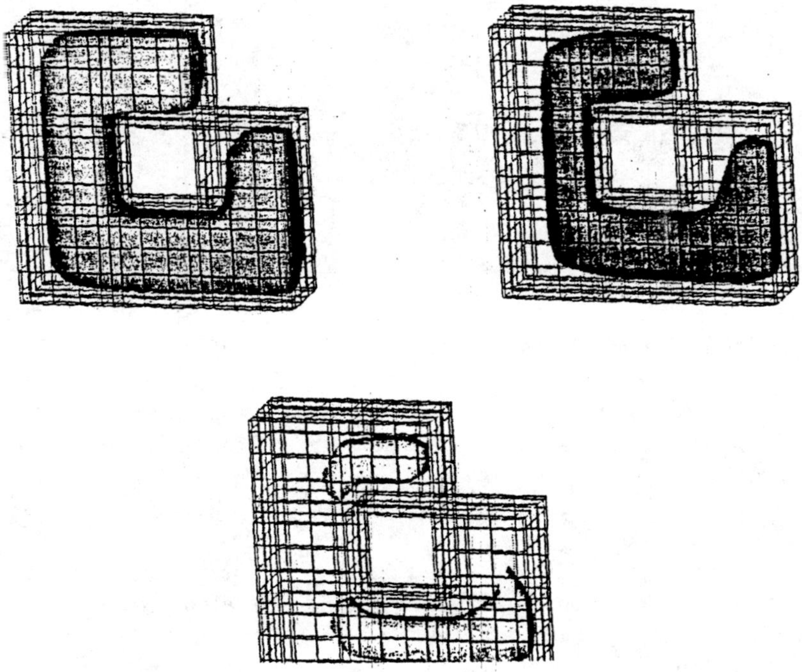

FIG. 7.2. *Iso-surfaces of the approximated solution to the equation* $-\frac{1}{10}\Delta u + u_x = 1$

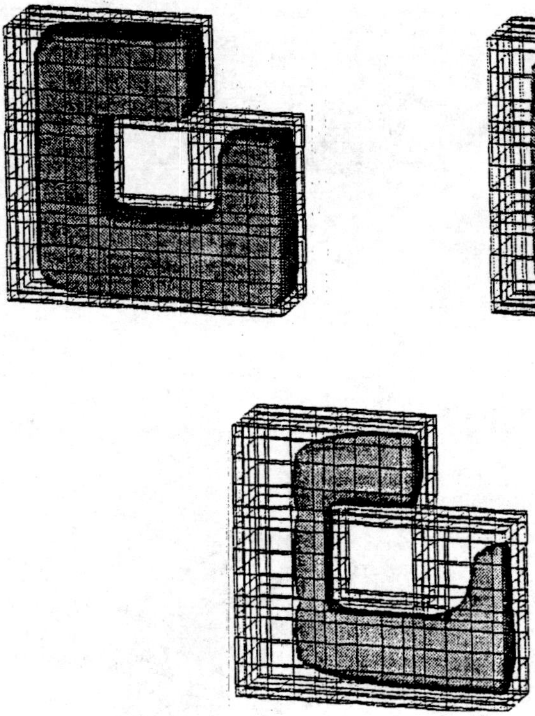

FIG. 7.3. *Iso-surfaces of the approximated solution to the equation* $-\frac{1}{20}\Delta u + u_x = 1$.

Finally, in figures 7.2 and 7.3, we show the iso-surfaces obtained in a more complicated domain for $\vec{\beta} = (1, 0, 0)$ with $\epsilon = .1$ and $\epsilon = .05$ respectively. In the first case, the degrees of freedom in Ω (544 nodes) are sufficient to resolve the boundary layer. In the second case, the degrees of freedom are not sufficient, the solution is affected by some artificial viscosity and presents little deformations at the interfaces. On the other hand, without the help of the upwind grid, for $\epsilon = .05$, high oscillations would pollute the approximated solution. In addition, the iterative procedure for the resolution of the linear system would turn out to be quite slow (in similar cases could not converge at all). Other comparisons between the stabilized and the non-stabilized approach, made in [1] for the 2-D case, also apply to 3-D computations.

REFERENCES

[1] FUNARO D., *Spectral Elements for Transport-Dominated Equations*, LNCSE n.1, Springer, Heidelberg (1997).

[2] Quarteroni A. & Valli A., *Domain Decomposition Methods for partial Differential Equations*, Clarendon Press, Oxford (1999).

[3] Manteuffel T. A., The Chebyshev iteration for nonsymmetric linear systems, Numer. Math., n. 28, pp. 307-327 (1997).

[4] *Using AVS/Express*, Advanced Visual System Inc., Release 4.0, Waltham MA (1998).

ON THE QUALITATIVE BEHAVIOUR OF NUMERICAL METHODS FOR DELAY DIFFERENTIAL EQUATIONS OF NEUTRAL TYPE. A CASE STUDY: Θ-METHODS.

NICOLA GUGLIELMI *

Abstract. This paper provides a geometric characterization of the asymptotic stability regions of Θ-methods for neutral delay differential equations (NDDEs), with respect to the test equation $y'(t) = a\,y(t) + b\,y(t-1) + c\,y'(t-1)$ with $a, b, c \in \mathbb{R}$ (we also shortly consider the case when $a, b, c \in \mathbb{C}$).

We are mainly interested in methods that produce unconditionally stable numerical solutions for all a, b and c, for which the exact solution tends to zero. To this aim we direct our attention to a novel stability analysis, recently introduced and investigated for the non-neutral case (see [6, 7, 8, 10]), which addresses the dependence of the stability on the delay. In particular, we direct our attention here to a subclass of *natural* Runge-Kutta methods for NDDEs, the Θ-methods.

Key words. neutral delay differential equations, NDDEs, asymptotic stability, Θ-methods, Runge-Kutta processes, $\tau(0)$-stability.

AMS subject classifications. 65L20.

1. Introduction. The two most important issues, relating to numerical formulæ for functional differential equations, address the accuracy and the stability of the formulæ. The title of this paper indicates that the emphasis is on the second aspect.

The main decisions to make when discussing the stability of numerical formulæ for delay differential equations are:

(1) to decide on a class of test equations;
(2) to decide on a definition of stability;
(3) to decide on a class of formulæ to analyze.

Stability is essentially concerned with the response of the solution of an equation to perturbations in the problem. Since the class of admissible perturbations has to be formulated, one can arrive at many different definitions of stability. The approach used when studying ordinary differential equations is, in the main, to consider perturbations in the initial value. For delay differential equations, the analogue is to consider certain types of perturbations in the initial function. In this case, for linear equations, the response to perturbations can be studied simply by asking how the solution behaves if the initial function has appropriate properties. Thus one arrives, as here, at concepts of stability and of stability regions that invoke the requirement that the solution is bounded (stability) or that it is bounded and tends to zero (asymptotic stability).

In choosing a class of formulæ to analyze, we have mainly directed our attention to the class of Θ-methods. The Runge-Kutta methods are flexible for coping with the possibility of discontinuous derivatives in the solutions, and the Θ-methods constitute a significant subclass.

In deciding upon the test equation

$$y'(t) = ay(t) + by(t-\tau) + cy'(t-\tau) \quad (t \geq 0) \quad \textit{with} \quad y(t) = g(t)\ t \in [-\tau, 0],$$

one is to some extent driven by mathematical necessity. This problem is the analogue of the classical test equations $y'(t) = ay(t)$ in ordinary differential equations and $y'(t) =$

*Dipartimento di Matematica Pura e Applicata, Università di L'Aquila, via Vetoio (Coppito), I-67010 L'Aquila, Italy, e_mail: **guglielm@univaq.it**

$ay(t)+by(t-\tau)$ in delay differential equations. If one seeks precise analytical statements, one is constrained to consider some form of idealised equation. As soon as one addresses more difficult classes of equations, is usually forced to make assumptions stronger than necessary and the results cannot be sharp anymore (see, e.g., [4]).

We remark that most of papers published on this subject assume conditions that are stronger than necessary for the asymptotic stability of the solution, which are derived by the requirement that stability holds for every delay $\tau > 0$ (for instance, $a + |b| < 0$ and $|c| < 1$ (see, e.g., [1])). Here, instead, we assume conditions which determine asymptotic stability only for an arbitrary but fixed delay.

The paper is organized as follows: In Sect. 2 we describe the stability region for the true solutions. In Sect. 3 we analyze the numerical root locus and relate it to the true stability boundary. In Sect. 4 we exploit a *geometric* stability condition to provide a suitable stability analysis for the case $\Theta > 1/2$.. In Sect. 5 we direct attention at the trapezoidal rule (unique scheme of order 2), for which we obtain a *global* stability result. Finally, in Sect. 6, we turn our attention to the complex coefficient test equation and conclude with some final remarks on the subject.

2. Background. When one has a parametrized test equation, as here (with parameters a, b, c, and possibly τ), one can address the question of a stability region. Note that by a change of scaling τ can be standardized to unity. A (strict-) stability region is a region of parameter space in which the solution is (respectively, asymptotically-) stable. For the analytical solution, transform theory reduces the analysis to a criterion on the location of the zeros of a parametrized quasi-polynomial (combinations of powers and exponentials). For the numerical solution, the analysis of the stability region reduces to a criterion on the location of the zeros of a (highly complicated) parametrized rational function that may be considered to approximate the quasi-polynomial. The most elegant way to describe a stability region is to give the equations of the curves that define its boundary. One is now involved with the root locus (see [5]) and its numerical analogue, the numerical root locus.

2.1. The asymptotic stability region for the true solutions. In order to simplify the notation, but without loosing generality, we can fix the delay equal to 1. Hence, in the sequel, we shall consider the test equation

$$(1)\qquad \begin{cases} y'(t) = ay(t) + by(t-1) + cy'(t-1), & t > 0, \\ y(t) = g(t), & -1 \le t \le 0. \end{cases}$$

with $a, b, c \in \mathbb{R}$ and $g(t)$ real-valued continuous and differentiable function.

By looking at solutions of the form $y(t) = e^{\lambda t}$ we are led to the quasipolynomial characteristic equation

$$(2)\qquad \lambda = a + be^{-\lambda} + c\lambda e^{-\lambda}.$$

An algebraic analysis (see, for example, [9]) shows that the set of terns (a, b, c) for which the solution $y(t)$ of (1) tends to 0 for $t \to \infty$, is given by

$$\Sigma_* = \{(a, b, c)\ ;\ \text{every root } \lambda \text{ of (2) satisfy (i) and (ii)}\},$$

being

(i) $\Re(\lambda) < 0$;
(ii) λ is uniformly bounded away from the imaginary axis.

We remark that a necessary condition to asymptotic stability (independently of $g(t)$) is $|c| < 1$. This is a direct consequence of Theorem 1.1. in [9, Chapter 3].

More in detail, it turns out that the set Σ_* is bounded above and below by the planes $c = 1$ and $c = -1$, respectively, and is bounded to the right by the plane $a + b = 0$ and by a transcendental surface,

$$\gamma_* = \left\{(a, b, c)\ ;\ c^2\phi^2 + b^2 = a^2 + \phi^2,\ \phi = \operatorname{arccot}\frac{ab - c\phi^2}{\phi(b + ac)},\quad \phi \in \mathbb{R}\right\}. \tag{3}$$

REMARK 2.1. *The analysis of (3) shows that all the points in the segment* $(a, b, c) = (1 - c, c - 1, c)$, $c \in (-1, 1)$ *are double points of the stability boundary* $\partial\Sigma_*$.

3. Natural R–K methods for NDDEs. When we apply a Runge-Kutta method with constant stepsize $h = 1/m$ (m positive integer) to the neutral delay differential equation

$$y'(t) = f\Big(t, y(t), y(t-1), y'(t-1)\Big),$$

the most natural choice consists in using the internal stage value $k_i^{(n-m)}$ as an approximation to $y'(t_n + c_i h - 1)$ and the value $g_i^{(n-m)} = y_{n-m} + h\sum_{j=1}^{s} a_{ij} k_j^{(n-m)}$ as an approximation to $y(t_n + c_i h - 1)$.

When applied to (1), this procedure gives rise to the method

$$\begin{aligned} k_i^{(n)} &= a\Big(y_n + h\sum_{j=1}^{s} a_{ij} k_j^{(n)}\Big) + b\Big(y_{n-m} + h\sum_{j=1}^{s} a_{ij} k_j^{(n-m)}\Big) + c\,k_i^{(n-m)} \\ y_{n+1} &= y_n + h\sum_{i=1}^{s} b_i k_i^{(n)}. \end{aligned} \tag{4}$$

where s is the number of stages and c_i, b_i, a_{ij} are the usual coefficients of the Runge-Kutta method.

For a stability analysis of the method (4) we obtain, in analogy to the non-neutral case (see [3]), that it is sufficient to look at numerical approximations of the form $y_n = \zeta^n y_0$, $k_i^{(n)} = \zeta^n k_i$.

In this way we are able to obtain the rational characteristic equation (see also Bellen *et al.* [1]),

$$\zeta = 1 + \frac{a + b\zeta^{-m}}{m}\mathbf{b}^T\left[\mathbf{I} - \frac{a + b\zeta^{-m}}{m}\mathbf{A} - c\zeta^{-m}\mathbf{I}\right]^{-1}\mathbf{u}, \tag{5}$$

where $\mathbf{b}^T$ and $\mathbf{A}$ are, respectively, the vector of weights and the the matrix of the Runge-Kutta method, $\mathbf{I}$ stands for the $s \times s$ identity matrix and $\mathbf{u} = [1, 1, \ldots, 1]^T$. With $\sigma = \dfrac{a + b\zeta^{-m}}{m}$, $\gamma = 1 - c\zeta^{-m}$ and $z = \sigma/\gamma$, we transform the characteristic equation (5) into the form

$$\zeta = R(z), \qquad z = \frac{1}{m}\left(\frac{a + b\zeta^{-m}}{1 - c\zeta^{-m}}\right), \tag{6}$$

$R(z)$ being the A-stability function of the method.

The numerical solution of (4) is therefore asymptotically stable if and only if $|\zeta| < 1$ whenever ζ satisfies (6). Then it is natural to set

$$\Sigma_m = \{(a, b, c)\ ;\ \text{all roots } \zeta \text{ of (6) satisfy } |\zeta| < 1\}. \tag{7}$$

The following definition extends that of $\tau(0)$-stability, recently given by the author [6] with reference to the non-neutral version of the considered test equation (that is (1) with $c = 0$).

DEFINITION 3.1.
The natural Runge-Kutta process (4) for NDDEs is $N\tau(0)$-stable *if*

$$\Sigma_* \subseteq \bigcap_{m=1}^{\infty} \Sigma_m.$$

In order to prove that $\Sigma_* \subset \Sigma_m$ we apply the so-called root locus technique (see, for instance [3]). Since z of (6) depends continuously on a, b and c, also $\zeta = R(z)$ depends continuously on a, b and c. Therefore, it is sufficient to prove that the values of (a, b, c) satisfying (6) with $|\zeta| = 1$ lie all outside the analytical stability region Σ_*. The set defined by such terns is the so-called *numerical* root locus.

It is hard to represent geometrically the set of all $(a, b, c) \in \mathbb{R} \times \mathbb{R} \times \mathbb{R}$ such that (1) is stable. For this reason, as we shall see in the next section, it is convenient to reduce the dimension and to get 2-dimensional parametrizations of the numerical root locus, choosing a, b, or c as parameters and developing the stability investigation in the (b, c)- either in the (a, c)- or in the (a, b)-plane respectively.

3.1. Stability with respect to c. We choose to fix $c \in \mathbb{R}$ and consider the set of all $(a, b) \in \mathbb{R} \times \mathbb{R}$ such that stability occurs. This choice makes more feasible the asymptotic stability analysis introduced in the previous section. We therefore define

$$\Sigma_{*,c} = \{(a, b)\ ;\ \text{all roots of (2) } (c \text{ being fixed) satisfy } \Re(\lambda) < 0\}. \tag{8}$$

This set $\Sigma_{*,c}$ is bounded to the right by the straight line $b = -a$ and by the transcendental curve

$$\gamma_{*,c} = \left\{(a_*(\varphi; c), b_*(\varphi; c); \quad \varphi \in (0, \pi)\right\}$$

where

$$\begin{cases} a_*(\varphi; c) = \dfrac{\varphi}{\sin\varphi}(\cos\varphi - c), \\ b_*(\varphi; c) = \dfrac{\varphi}{\sin\varphi}(c\cos\varphi - 1), \end{cases} \tag{9}$$

which is shown in Fig. 1 (left part), where we observe that $(a, b) = (1 - c, -1 + c)$ is a double point for the stability boundary.

We proceed in a similar way for investigating the numerical stability region. We set

$$\Sigma_{m,c} = \{(a, b)\ ;\ \text{all roots } \zeta \text{ of (6) } (c \text{ being fixed) satisfy } |\zeta| < 1\}. \tag{10}$$

In order to study the (a, b)-pairs determined by $|\zeta| = 1$ in (6), we insert $z = x + iy$ and $\zeta = e^{i\varphi}$ and we separate real and imaginary parts. This yields

$$mx(1 - c\cos m\varphi) - myc\sin m\varphi = a + b\cos m\varphi$$

$$my(1 - c\cos m\varphi) + mxc\sin m\varphi = -b\sin m\varphi.$$

According to Lemma 4.1. in [6], for $x = y = \varphi = 0$ we get the straight line

(11) $$a + b = 0,$$

as well as for the analytic stability region (see Fig. 1). If $\sin m\varphi = 0$ and $y = 0$ (this typically happens for $\varphi = s\pi$), we get another straight line

(12) $$a \pm b = mx(1 - c).$$

For $y \neq 0$, instead, we obtain the algebraic curve

$$\gamma_{m,c} = \Big\{(a_m(\varphi; c), b_m(\varphi; c); \quad \varphi \in (0, s\pi)\Big\},$$

with

(13) $$\begin{cases} a_m(\varphi; c) &= mx + \dfrac{my}{\sin m\varphi}(\cos m\varphi - c) \\ b_m(\varphi; c) &= -mxc + \dfrac{my}{\sin m\varphi}(c\cos m\varphi - 1), \end{cases}$$

which typically passes through infinity when $m\varphi$ is an integer multiple of π. Eq. (13) suggests to compare the values (a_m, b_m) to those of (2) corresponding to $\lambda = im\varphi$, which define the root locus for the true solution. They are given by (see (9))

$$a_*(m\varphi; c) = \frac{m\varphi}{\sin m\varphi}\,(\cos m\varphi - c), \qquad b_*(m\varphi; c) = -\frac{m\varphi}{\sin m\varphi}\,(c\,\cos m\varphi - 1).$$

Only for $0 < m\varphi < \pi$ the point (a_*, b_*) is on the boundary of $\Sigma_{*,c}$. For $m\varphi > \pi$ it lies on a curve that separates regions with different numbers of roots of (2) in the right half-plane.

The intensity of grey in Fig. 1 (left picture) indicates the increasing number of roots of (2) in the right half-plane as well as (in the right picture) it indicates the increasing number of roots of (6) outside the unit disk. The dotted curve in the same figure (right part) indicates the true stability boundary and the white regions represent the sets of unconditional asymptotic stability.

Comparing the obtained formulæ for (a_m, b_m) and (a_*, b_*) yields the useful representation

(14) $$a_m(\varphi; c) = \frac{y}{\varphi}\,a_*(m\varphi; c) + mx, \qquad b_m(\varphi; c) = \frac{y}{\varphi}\,b_*(m\varphi; c) - mxc.$$

3.2. The case of Θ-methods. We consider here the Θ-method. Usually assuming $\Theta \in [0, 1]$, this class of methods is conventionally represented by the Butcher tableau

$$\begin{array}{c|c} \mathbf{c} & \mathbf{A} \\ \hline & \mathbf{b}^T \end{array} \quad := \quad \begin{array}{c|cc} 0 & 0 & 0 \\ 1 & 1-\Theta & \Theta \\ \hline & 1-\Theta & \Theta \end{array}.$$

For this method we get

$$R(z) = \frac{1 + (1 - \Theta)z}{1 - \Theta z}.$$

In this paper we clearly assume A-stability of the methods; hence we take into consideration the case $\Theta \geq 1/2$. Whenever $\Theta > 1/2$, differently from the non-neutral DDE case, we are going to show that *numerical instabilities* can occur, in correspondence to certain values of the parameters, for which the true solutions are instead stable.

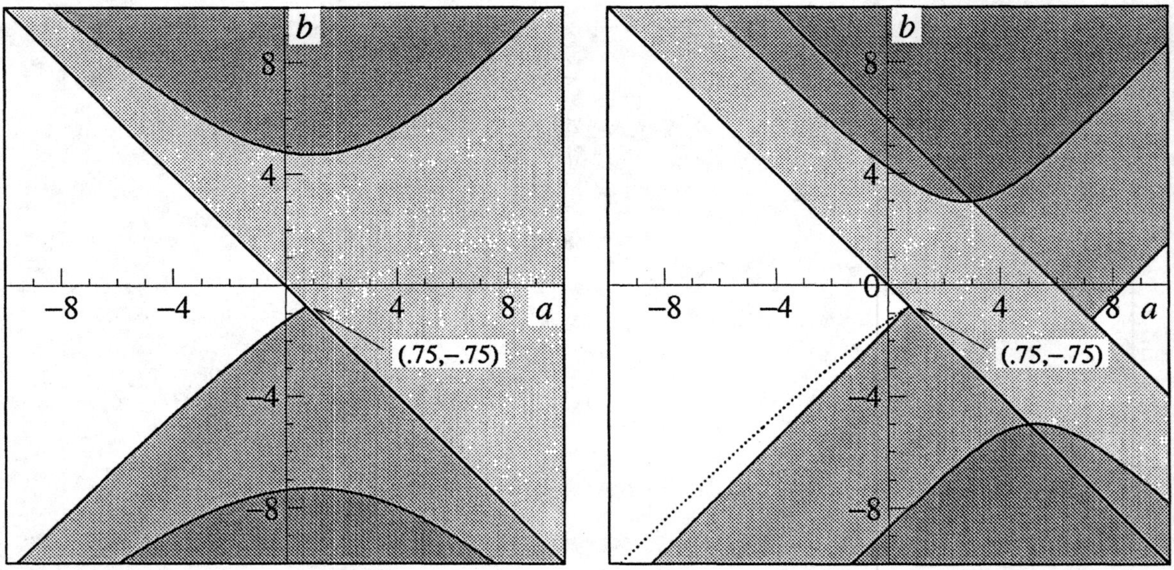

FIG. 1. Stability region and root locus for the analytical solution of (1) (left part) and example of stability region and numerical root locus ($c = 0.25, m = 4$) for the backward Euler method (right part).

4. Geometric investigation. We study now the condition $\Sigma_{*,c} \subset \Sigma_{m,c}$ close to the cusp point $(a, b) = (1 - c, c - 1)$, which corresponds to the case $z = 0$ and $\zeta = 1$ in (6). In the case of Θ-methods, by solving the equation $R(x + iy) = e^{i\varphi}$ with respect to φ, we get the following formulæ for x and y,

$$
(15) \qquad \begin{cases} x(\varphi) & = & \dfrac{(-1+2\,\Theta)\,(-1+\cos(\varphi))}{-1+2\,\Theta-2\,\Theta^2+2\,(-1+\Theta)\,\Theta\,\cos(\varphi)} \\ y(\varphi) & = & -\dfrac{\sin(\varphi)}{-1+2\,\Theta-2\,\Theta^2+2\,(-1+\Theta)\,\Theta\,\cos(\varphi)}. \end{cases}
$$

Partially extending a property shown in [6] for Runge-Kutta processes for DDEs, we provide the following useful geometric result.

THEOREM 4.1. *For any Θ-method and any positive integer m, the point $(a, b) = (1 - c, c - 1)$, $c \in \mathbb{R}$ is a double point for the corresponding numerical root locus.*

Proof. We fix m and analyze (6) with $(a, b) = (1-c, c-1)$. Straightforward calculations show that $\zeta = 1$ is a double zero of the *numerical* characteristic equation. In correspondence of it the line (11) and the curve $\gamma_{m,c}$ intersect (to check this, insert (15) into (13) and consider the limit as $\varphi \to 0$); hence the numerical root locus presents a double point.

With the exception of the case $\Theta = 1$ and $m = 1$, we get that $\gamma_{m,c}$ is a *true* curve, whose trace is different from (11). Hence, in correspondence to the considered point, the numerical root locus presents a cusp bifurcation (due to simmetry of (13), (15) w.r.t. φ) made of two different branches. In the special case $\Theta = 1$ and $m = 1$, where the curve $\gamma_{m,c}$ degenerates to a single point, we get instead that (11) and (12) intersect at $(1 - c, c - 1)$. ☐

The result is illustrated in the right picture in Fig. 1.

An important consequence of the previous theorem is that the curve $\gamma_{m,c}$ provides - in a neighbourhood of the double point $(a, b) = (1 - c, c - 1)$ - an algebraic approximation of the *true* stability boundary, $\gamma_{*,c}$. We can therefore compare the two curves and establish a necessary stability condition, by geometric considerations.

Whenever $m > 1$ in fact, the traces of both curves (over suitable ranges for the parameter φ) define two injective functions (with respect to a) on the left of the double point; hence we associate to the two curves $\gamma_{m,c}$ and $\gamma_{*,c}$, the two functions $\Upsilon_{m,c}(a)$ and $\Upsilon_{*,c}(a)$. By an exam of (9) and (13), we have that

(1) $\Upsilon_{m,c}(1-c) = \Upsilon_{*,c}(1-c) = c-1$;

(2) Both $\Upsilon_{m,c}(a)$ and $\Upsilon_{*,c}(a)$ are negatively unbounded as $a \to -\infty$, whenever $m > 1$.

Now look at the shape of the stability regions. By geometric considerations, based on the above properties (1) and (2), we have that a necessary stability condition is given by

$$\Upsilon_{m,c}(a) - \Upsilon_{*,c}(a) \leq 0, \tag{16}$$

in a left neighbourhood of $a = 1-c$, at least when $m > 1$ (see Fig. 1 and Fig. 2).

4.1. Methods of order $p = 1$: the case $\Theta > 1/2$. In general, in this case, the curves $\gamma_{m,c}$ and $\gamma_{*,c}$ are not tangent at $a = 1-c$. Hence we have the following necessary stability condition.

THEOREM 4.2. *Let*

$$\Delta_{m,c} = \left.\frac{d\left(\Upsilon_{m,c} - \Upsilon_{*,c}\right)}{da}\right|_{a=1-c}.$$

If the Θ-method ($\Theta > 1/2$) is $N\tau(0)$-stable then

$$\Delta_{m,c} > 0 \qquad \forall c \in (-1,1) \text{ and } m > 1. \tag{17}$$

Proof. This is a direct consequence of the stability condition (16) in a neighbourhood of the double point, which is obtained by considering first order Taylor expansions of $\Upsilon_{m,c}(a)$ and $\Upsilon_{*,c}(a)$ at $a = 1-c$. □

Routine calculations provide the following expression:

$$\Delta_{m,c} = \frac{-3\,(1+c)}{2+c} - r(m,c) \tag{18}$$

with

$$r(m,c) = \frac{3\,\left(m - c\,m + m^2 + c\,m^2 - 2\,m\,\Theta + 2\,c\,m\,\Theta\right)}{-1 + c - 3\,m - 2\,m^2 - c\,m^2 + 6\,\Theta - 6\,c\,\Theta + 6\,m\,\Theta - 6\,\Theta^2 + 6\,c\,\Theta^2}.$$

First set $m = 2$. In a right neighbourhood of $c = -1$, we get, by (18),

$$\Delta_{2,c} = \frac{1 - 2\,\Theta}{(-1+\Theta)^2} + \mathcal{O}(1+c).$$

This formula contradicts condition (17), when $\Theta \in (1/2, 1)$ and c is sufficiently close to -1.

Finally consider the case $\Theta = 1$, which corresponds to the backward Euler method. Expanding $\Delta_{m,c}$ at $m = +\infty$, we get

$$\Delta_{m,c} = \frac{3\,\left(1 + 4\,c + c^2\right)}{(2+c)^2\,m} + \mathcal{O}\left(\frac{1}{m}\right)^2,$$

which means *local* instability for an arbitrarily large number of choices of the integration stepsize, as soon as $c < -2 + \sqrt{3}$. In Fig. 2 (left part) we show the true stability region and the root locus curve for the problem with $c = -0.5$. In the right part, relevant to the backward Euler method, we can see $\Sigma_{4,-0.5}$ together with the numerical root locus curve $(a_4(\varphi; -0.5), b_4(\varphi; -0.5))$. The dotted curve in the same figure indicates the true stability boundary, which is shown to cross the numerical stability boundary on the left of $a = 0$.

In conclusion we have the following result.

THEOREM 4.3. *If* $\Theta \neq 1/2$, *the* Θ*-method is not* $N\tau(0)$*-stable.*

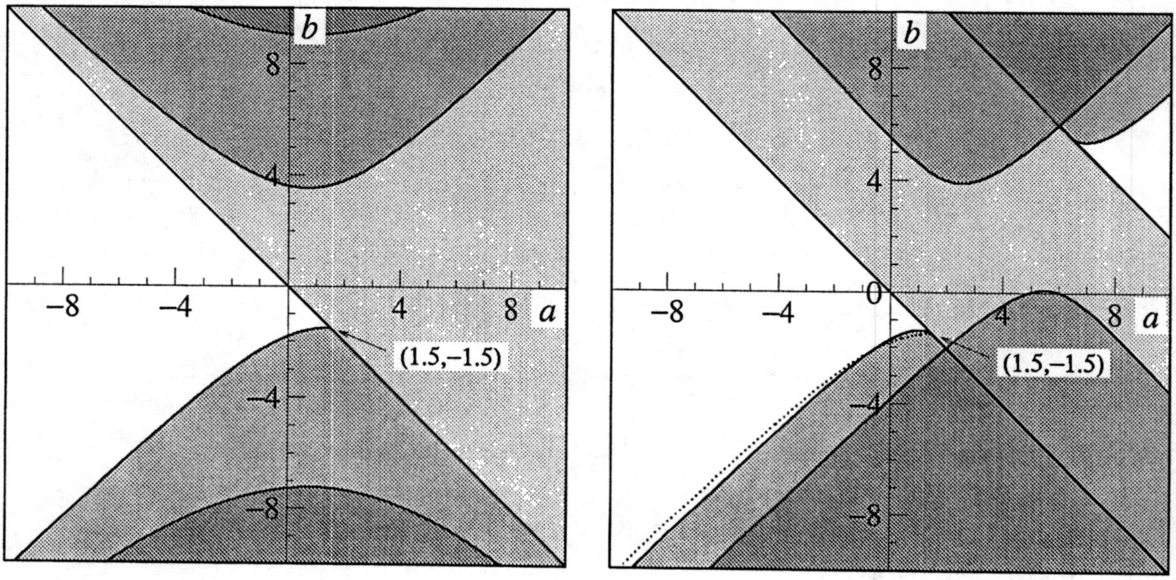

FIG. 2. Stability region and root locus for the analytical solution of (1) (left part) and example of stability region and numerical root locus ($c = -0.5, m = 4$) for the backward Euler method (right part). Instability occurs in a left neighbourhood of $(a, b) = (1.5, -1.5)$.

5. A global stability result for the trapezoidal rule. We investigate now the stability behaviour of the trapezoidal rule, that is of the unique symmetric method in the considered class. We begin with the following geometric property for the true stability region.

LEMMA 5.1. *Let* $c \in (-1, 1)$. *The set* $\Sigma_{*,c}$ *is star-shaped with respect to the origin.*

Proof. Under the assumptions, the set $\gamma_{*,c}$ is included in the cone $|a| + b < 0$; hence it is sufficient to show that the equation

$$a_*(\varphi; c) - \theta b_*(\varphi; c) = 0, \quad \theta \in \mathbb{R},$$

admits a unique solution $|\theta| < 1$ for any $\varphi \in (0, \pi)$. This yields the equation

$$\theta = \frac{c - \cos(\varphi)}{1 - c\cos(\varphi)}.$$

Since the function at the right hand side is bijective in the domain $\varphi \in (0, \pi)$ (for any fixed $c \in (-1, 1)$) and has modulus smaller than 1, the property is proved. □

Then we are in a position to state the following result.

THEOREM 5.2. *The trapezoidal rule is $N\tau(0)$-stable.*

Proof. For the rational function

$$R(z) = \frac{1 + z/2}{1 - z/2},$$

which corresponds to the trapezoidal rule or to the implicit midpoint rule, we consider, according to (15), the parametrization $x(\varphi) \equiv 0$ and $y(\varphi) = 2\tan(\varphi/2)$. In this case we have $y(\varphi) > \varphi$ for all $\varphi \in (0, \pi)$. Since $x(\varphi) \equiv 0$ and the set $\Sigma_{*,c}$ is star-shaped with respect to the origin, this relation together with (14) implies that (a_m, b_m) is always outside the set $\Sigma_{*,c}$. Consequently, we have $\Sigma_{*,c} \subset \Sigma_{m,c}$ for all $c \in (-1, 1)$ and $m \geq 1$. □

Hence, the unique $N\tau(0)$-stable Θ-method is the trapezoidal rule.

6. Final remarks. The conclusion of the paper is devoted to the complex coefficient case and to a general classification of stability properties of Θ-methods.

The first remark involves the essential instability of Θ-methods for NDDEs applied to the complex coefficient version of equation (1).

In fact, as a consequence of Theorem 4.3 and of Theorem 6.2. in [7], which asserts that the trapezoidal rule is not τ-stable, that is stable on (1) with $c = 0$ and $a, b \in \mathbb{C}$, we have the following negative result ($N\tau$-stability is an obvious extension of $N\tau(0)$-stability to the complex case).

THEOREM 6.1. *The Θ-methods are not $N\tau$-stable.*

On account of this, we conjecture that no natural R-K methods could be $N\tau$-stable. In particular, also the backward Euler method which has been recently proved to be τ-stable [10], has been proved here to be not $N\tau(0)$-stable.

We accomplish now a taxonomy of Θ-methods for delay equations, with relevance to the asymptotic stability for neutral (as well as non neutral) real and complex test equations of the form (1). Such classification is a consequence of the results obtained in the present paper together with those previously obtained in [6, 7, 8].

Table 1. A taxonomy for the asymptotic stability of Θ-methods

Method	Order	$a \in \mathbb{C}$, $b = 0$, $c = 0$	$a \in \mathbb{R}$, $b \in \mathbb{R}$, $c = 0$	$a \in \mathbb{C}$, $b \in \mathbb{C}$, $c = 0$	$a \in \mathbb{R}$, $b \in \mathbb{R}$, $c \in \mathbb{R}$	$a \in \mathbb{C}$, $b \in \mathbb{C}$, $c \in \mathbb{C}$
$\Theta \in (1/2, 1)$	1	S	S	?	N	N
$\Theta = 1/2$	2	S	S	N	S	N
$\Theta = 1$	1	S	S	S	N	N

In the above table, we have denoted stability with the acronym **S**, instability with the acronym **N** and open problems with the acronym **?**.

Although some issues remain to be investigated, the table gives an almost complete view on the qualitative behaviour of Θ-methods. For the complex coefficient case as well as for the case of linear systems of (neutral) delay differential equations, where standard Runge-Kutta processes can perform in an unstable way, we think that novel classes of methods need to be discovered in order to provide good stability properties (see, e.g., Bellen & Maset [2]).

REFERENCES

[1] A. Bellen, Z. Jackiewicz and M. Zennaro. Stability analysis of one-step methods for neutral delay differential equations, *Numer. Math*, 52 (1988), pp. 605–619.

[2] A. Bellen and S. Maset, A semigroup approach for the numerical solution of delay differential equations, *Numer. Math*, (1999), *in press.*

[3] C.T.H. Baker and C.A.H. Paul, Computing stability regions—Runge-Kutta methods for delay differential equations, *IMA J. Numer. Anal.*, 14 (1994), pp. 347–362.

[4] C.T.H. Baker, C.A.H. Paul and D.R. Willè, Issues in the numerical solution of evolutionary delay differential equations, *Adv. Comp. Math.*, 3 (1995), pp. 171–196.

[5] L.E. El'sgol'ts and S.B. Norkin, *Introduction to the theory and application of differential equations with deviating arguments*, Academic Press, New York, 1973.

[6] N. Guglielmi, On the asymptotic stability properties of Runge–Kutta methods for delay differential equations, *Numer. Math.*, 77 (1997), pp. 467–485.

[7] N. Guglielmi, Delay dependent stability regions of Θ-methods for delay differential equations, *IMA J. Numer. Anal.*, 18 (1998), pp. 399–418.

[8] N. Guglielmi and E. Hairer, Order stars and stability for delay differential equations, *Numer. Math.*, 83 (1999), pp. 371–383.

[9] Y. Kuang, *Delay Differential Equations with Application in Population Dynamics*, Academic Press, Boston, 1993.

[10] S. Maset, Stability of Runge-Kutta methods for linear delay differential equations, *Numer. Math*, (1999), *in press.*

POWERS OF GENERALIZED FIBONACCI MATRIX

M.X. HE* AND P.E. RICCI†

Abstract. The power of matrices plays important role in many areas of mathematics. The power of 2×2 Fibonacci matrix

$$Q_2 = \begin{bmatrix} 0 & 1 \\ 1 & 1 \end{bmatrix}$$

generates well-known Fibonacci numbers $F_n (n = 1, 2, 3, ...,)$. The classical Fibonacci polynomials $F_n(x)$ are generated by the following 2×2 matrix

$$Q_2(x) = \begin{bmatrix} 0 & 1 \\ 1 & x \end{bmatrix}.$$

In this paper we generalize the 2×2 matrix into the following $r \times r$ matrix

$$Q_r[u] := Q_r(u_1, u_2, ..., u_r) = \begin{bmatrix} 0 & 0 & 0 & \cdots & 0 & (-1)^{r-1}u_r \\ 1 & 0 & 0 & \cdots & 0 & (-1)^{r-2}u_{r-1} \\ 0 & 1 & 0 & \cdots & 0 & (-1)^{r-3}u_{r-2} \\ \vdots & \vdots & \vdots & & \vdots & \vdots \\ 0 & 0 & 0 & \cdots & 0 & -u_2 \\ 0 & 0 & 0 & \cdots & 1 & u_1 \end{bmatrix}.$$

This matrix establishes a connection with the generalized Lucas polynomials which leads us to determine the powers of the matrix $Q_r[u]$.

Key words. Powers of matrices, Fibonacci polynomials, generalized Lucas polynomials.

AMS subject classifications. 12E10, 30C15.

1. Introduction. Consider the following 2×2 Fibonacci matrix

$$Q_2 = \begin{bmatrix} 0 & 1 \\ 1 & 1 \end{bmatrix}. \tag{1}$$

It is well-known that this matrix generates Fibonacci numbers F_n $(n = 1, 2, 3, ...)$ by taking its powers,

$$Q_2^n = \begin{bmatrix} F_{n-1} & F_n \\ F_n & F_{n+1} \end{bmatrix}. \tag{2}$$

If we replace the matrix Q_2 by

$$Q_2(x) = \begin{bmatrix} 0 & 1 \\ 1 & x \end{bmatrix} \tag{3}$$

and take the power n of $Q_2(x)$, then similar to (1.2) we have

$$Q_2^n(x) = \begin{bmatrix} F_{n-1}(x) & F_n(x) \\ F_n(x) & F_{n+1}(x) \end{bmatrix}, \tag{4}$$

*Department of Mathematics, Nova Southeastern University, Ft. Lauderdale – FL 33314. E-mail: hem@polaris.nova.edu.

†Dipartimento di Matematica "Guido CASTLENUOVO", Università degli Studi di Roma "La Sapienza", Italy. E-mail: riccip@axrma.uniroma1.it.

where $F_n(x)$ is the classical Fibonacci polynomial. The matrix $Q_2(x)$ was generalized in [5] to a $r \times r$ matrix ($r = 2, 3, ...$),

$$Q_r(x) = \begin{bmatrix} 0 & 0 & 0 & \cdots & 0 & 1 \\ 1 & 0 & 0 & \cdots & 0 & x \\ 0 & 1 & 0 & \cdots & 0 & x^2 \\ \vdots & \vdots & \vdots & & \vdots & \vdots \\ 0 & 0 & 0 & \cdots & 0 & x^{r-2} \\ 0 & 0 & 0 & \cdots & 1 & x^{r-1} \end{bmatrix}. \tag{5}$$

Different type of polynomials corresponding to different matrix dimensions of $Q_r(x)$ are generated by matrix (1.5), such as Tribonacci polynomial $T_n(x)$ when $r = 3$, Quadranacci polynomial $Q_n(x)$ when $r = 4$, and R-Bonacci polynomial $R_n(x)$ in general. The powers of $Q_r(x)$ were studied in [5]. The matrix $Q_r(x)$ was used to study the zeros distribution of the polynomials $R_n(x)$ in [9]. Some dynamical properties of Fibonacci-type polynomials generated by taking powers of $Q_r(x)$ were investigated in [7, 8, 10].

In this paper we generalize matrix $Q_r(x)$ into the following $r \times r$ matrix with r parameters $u_1, u_2, \cdots, u_r$, denoted by $Q_r[u] := Q_r(u_1, u_2, ..., u_r)$,

$$Q_r[u] = \begin{bmatrix} 0 & 0 & 0 & \cdots & 0 & (-1)^{r-1}u_r \\ 1 & 0 & 0 & \cdots & 0 & (-1)^{r-2}u_{r-1} \\ 0 & 1 & 0 & \cdots & 0 & (-1)^{r-3}u_{r-2} \\ \vdots & \vdots & \vdots & & \vdots & \vdots \\ 0 & 0 & 0 & \cdots & 0 & -u_2 \\ 0 & 0 & 0 & \cdots & 1 & u_1 \end{bmatrix}. \tag{6}$$

As we shall see that this matrix $Q_r[u]$ establishes a connection with the generalized Lucas polynomials of second kind which can be used to determine the powers of the matrix $Q_r[u]$.

2. Generalized Lucas Polynomials of Second Kind. We first recall in this section some results on representation formulas for powers of a general matrix A, whose invariants are given by $u_1, u_2, \ldots, u_r$.

THEOREM 2.1. [1] *Let A be an $r \times r$ complex matrix and denote by*

$$P(\lambda) := \det(\lambda I - A) = \sum_{j=0}^{r} (-1)^j u_j \lambda^{r-j}$$

its characteristic polynomial (by convention $u_0 := 1$), then for the powers of A with integral exponents the following representation formula holds true:

$$A^n = F_{1,n-1}[u]A^{r-1} + F_{2,n-1}[u]A^{r-2} + \cdots + F_{r,n-1}[u]I. \tag{7}$$

The functions $F_{k,n}[u] := F_{k,n}(u_1, \ldots, u_r)$ which appear as coefficients in the preceding linear combination are defined by the recurrence relation

$$F_{k,n}[u] = u_1 F_{k,n-1}[u] - u_2 F_{k,n-2}[u] + \cdots + (-1)^{r-1} u_r F_{k,n-r}[u], \tag{8}$$

$(k = 1, \ldots, r;\ n \geq -1)$ and initial conditions:

$$F_{r-k+1,h-2}[u] = \delta_{k,h}, \quad (k, h = 1, \ldots, r).$$

Furthermore, if A is non-singular ($u_r \neq 0$), then formula (2.1) still holds for negative values of n, provided that we define the $F_{k,n}[u]$ functions for negative values of n as follows, for $k = 1, \ldots, r$, $n < -1$,

(9) $$F_{k,n}[u] = F_{r-k+1,-n+r-3}\left(\frac{u_{r-1}}{u_r}, \ldots, \frac{u_1}{u_r}, \frac{1}{u_r}\right).$$

The Generalized Lucas Polynomials of second kind are a standard mathematical tool for obtaining representation formulas for powers of square matrices (see [4]).

Consider the bilateral linear homogeneous recurrence relation with $r+1$ terms and constant complex coefficients u_k, $(k = 1, 2, \ldots, r)$, and suppose $u_r \neq 0$:

(10) $$X_n = u_1 X_{n-1} - u_2 X_{n-2} + \cdots + (-1)^{r-1} u_r X_{n-r}, \quad (n \in \mathcal{Z})$$

this is just the recurrence relation satisfied by the $F_{k,n}[u]$ functions. Furthermore, the $F_{k,n}[u]$ functions are determined by the initial conditions

$$(1, 0, 0, \ldots, 0),$$

$$(0, 1, 0, \ldots, 0),$$

$$\ldots\ldots\ldots\ldots$$

$$(0, 0, \ldots, 0, 1),$$

in fact,

$$F_{r-k+1,h-2}[u] = \delta_{k,h}, \quad (k, h = 1, \ldots, r).$$

Then they constitute a basis of the vectorial space $\mathcal{V}_r$ of all solutions of the recurrence relation (2.4). Usually a basis of $\mathcal{V}_r$ is obtained by considering the roots of the characteristic equation

(11) $$z^r - u_1 z^{r-1} + \cdots + (-1)^r u_r = 0.$$

In the simplest case, i.e. if equation (2.5) admits r distinct roots $z_1, \ldots, z_r$, then the general solution of the recurrence relation (2.4) is expressed by

(12) $$X_n = C_1 {z_1}^n + C_2 {z_2}^n + \cdots + C_r {z_r}^n, \quad (n \in \mathcal{Z}).$$

A little bit complicate formula holds when multiple roots appear. It was proved in [4] that the $F_{k,n}[u]$ function has the following representation formula, for $k = 1, 2, \cdots, r$, $n \in \mathcal{Z}$,

$$F_{k,n-1}[u] = \frac{\begin{vmatrix} 1 & 1 & \cdots & 1 \\ z_1 & z_2 & \cdots & z_r \\ \cdots & \cdots & \cdots & \cdots \\ {z_1}^n & {z_2}^n & \cdots & {z_r}^n \\ \cdots & \cdots & \cdots & \cdots \\ {z_1}^{r-1} & {z_2}^{r-1} & \cdots & {z_r}^{r-1} \end{vmatrix}}{V(z_1, z_2, \ldots, z_r)},$$

where, in the above determinant, the row $\{z_1{}^n, z_2{}^n, \ldots, z_r{}^n\}$ appears instead of the usual $(r-k+1)$-th one, and $V(z_1, z_2, \ldots, z_r)$, denotes the Vandermonde determinant of the numbers $z_1, z_2, \ldots, z_r$.

Another important property [4] is the $F_{k,n}[u]$ functions can be expressed by the relations

$$
(13) \qquad \begin{cases} F_{1,n}[u] = u_1 F_{1,n-1}[u] + F_{2,n-1}[u] \\ F_{2,n}[u] = -u_2 F_{1,n-1}[u] + F_{3,n-1}[u] \\ \cdots\cdots\cdots\cdots\cdots\cdots\cdots\cdots \\ F_{r-1,n}[u] = (-1)^{r-2} u_{r-1} F_{1,n-1}[u] + F_{r,n-1}[u] \\ F_{r,n}[u] = (-1)^{r-1} u_r F_{1,n-1}[u] \end{cases}
$$

by means of which all the $\{F_{k,n}[u]\}_{n\in \mathcal{Z}}$ can be computed in terms of the bilateral sequence $\{F_{1,n}[u]\}_{n\in \mathcal{Z}}$.

Then it is useful to introduce the following definition.

DEFINITION 2.2. *The bilateral sequence $\{F_{1,n}[u]\}_{n\in\mathcal{Z}}$, i.e. the solution of (2.4) corresponding to the initial conditions*

$$F_{1,-1}[u] = 0, \quad F_{1,0}[u] = 0, \quad F_{1,2}[u] = 0, \ldots, F_{1,r-2}[u] = 1$$

is called the "fundamental solution" of the recurrence relation (2.4) (see É. Lucas [3].)

In the following we will use the notation:

$$F_{1,n}[u] = F_{1,n}(u_1, \ldots, u_r) =: \Phi_n(u_1, \ldots, u_r) = \Phi_n, \quad (n \in \mathcal{Z}).$$

For $n \geq -1$, the $\Phi_n(u_1, \ldots, u_r)$ functions are a sequence of polynomials called in literature generalized Lucas polynomials of second kind.

Note that if $r = 2$, and $u_2 = 1$, by putting $u_1 = x$, we have:

$$\Phi_n(u_1, 1) = \Phi_n(x, 1) \equiv U_n\left(\frac{x}{2}\right), \quad (n \in N_0),$$

where $\{U_n(x)\}_{n\in N_0}$ are the classical Chebyshev polynomials of second kind. If $r \geq 3$, and putting $u_r = 1$, we obtain a sequence of polynomials in $r-1$ variables which appear as an extension of the classical Chebyshev polynomials of second kind:

$$\Phi_n(u_1, \ldots, u_{r-1}, 1) =: U_n(u_1, \ldots, u_{r-1}), \quad (n \in N_0).$$

The above mentioned results can be resumed as follows.

The generalized Lucas polynomials of second kind are defined as the solution of the bilateral linear homogeneous recurrence relation (2.4)

$$\Phi_n = u_1\Phi_{n-1} - u_2\Phi_{n-2} + \cdots + (-1)^r u_r \Phi_{n-r}, \quad (n \in \mathcal{Z})$$

corresponding to the initial conditions

$$\Phi_{-1} = 0, \Phi_0 = 0, \Phi_1 = 0, \ldots, \Phi_{r-2} = 1.$$

It is called the "fundamental solution" of the recurrence relation (2.4) since all solutions of (2.4) can be expressed in terms of this particular solution. The fundamental properties of $\Phi_n(u_1, \ldots, u_{r-1}, 1)$ can be found in [1, 2, 3]. The relations (2.1), (2.2), (2.3) and (2.7) are particularly useful for computing the powers of matrices.

3. Powers of Generalized Fibonacci Matrix. In this section we apply the results stated in Section 2 to determine the powers of $Q_r[u]$ using $F_{k,n}[u]$ sequence functions. We present our main results in two parts. We first show in details the powers of $Q_r[u]$ when $r = 3$. Then we extend our representations to the general case for all $r's$.

3.1. Powers of $Q_3[u]$. For the notation simplicity, we denote $F_{k,n}[u]$ by $F_{k,n}$, same as defined in (2.2). Let $r = 3$ in (1.6). Then

$$Q_3[u] := \begin{bmatrix} 0 & 0 & u_3 \\ 1 & 0 & -u_2 \\ 0 & 1 & u_1 \end{bmatrix}.$$

We now present a formula for computing the powers of $Q_3[u]$.

THEOREM 3.1. *For any integer k, integral powers of the matrix $Q_3[u]$ are given by the formula*

$$Q_3^k[u] = \begin{bmatrix} F_{3,k-1} & F_{3,k} & F_{3,k+1} \\ F_{2,k-1} & F_{2,k} & F_{2,k+1} \\ F_{1,k-1} & F_{1,k} & F_{1,k+1} \end{bmatrix}. \tag{14}$$

Proof. Let $r = 3$ in recurrence relation (2.4). We have

$$X_k = u_1 X_{k-1} - u_2 X_{k-2} + u_3 X_{k-3} \tag{15}$$

and the characteristic equation,

$$z^3 - u_1 z^2 + u_2 z - u_3 = 0. \tag{16}$$

The companion matrix of equation (3.3) is $Q_3[u]$, Then

$$\mathrm{Tr}(Q_3[u]) = u_1, \quad \Sigma_2(Q_3[u]) = u_2, \quad \det(Q_3[u]) = u_3,$$

where $\mathrm{Tr}(A)$ is the trace of matrix A, $\Sigma_2(A)$ denotes the sum of principal minors of order 2 of A, and $\det(A)$ is the determinant of matrix A. Let $\{J_k\}, \{K_l\}, \{L_m\}$ be sequences of functions satisfying the recurrence relation (3.2), then the following matrix product relation holds true:

$$\begin{bmatrix} J_k & J_{k+1} & J_{k+2} \\ K_l & K_{l+1} & K_{l+2} \\ L_m & L_{m+1} & L_{m+2} \end{bmatrix} Q_3[u] = \begin{bmatrix} J_{k+1} & J_{k+2} & J_{k+3} \\ K_{l+1} & K_{l+2} & K_{l+3} \\ L_{m+1} & L_{m+2} & L_{m+3} \end{bmatrix}. \tag{17}$$

This is by direct inspection taking into account of the hypothesis that functions $\{J_k\}, \{K_l\}$ $\{L_m\}$ must satisfy the recurrence (3.2).

Note that, by using the $F_{k,n}$ functions of section 2, the matrix $Q_3[u]$ can be written as

$$Q_3[u] = \begin{bmatrix} F_{3,0} & F_{3,1} & F_{3,2} \\ F_{2,0} & F_{2,1} & F_{2,2} \\ F_{1,0} & F_{1,1} & F_{1,2}. \end{bmatrix} \tag{18}$$

Since the $F_{k,n}$ functions satisfy the recurrence relation (3.2), the relation (3.4) can be applied to obtain (3.1), the powers of matrix $Q_3[u]$.□

We see from Theorem 3.1 that the powers of matrix $Q_3[u]$ are determined by the sequence functions $F_{k,n}[u]$. We next present a theorem concerning a nonlinear recurrence relation for $F_{k,n}[u]$.

THEOREM 3.2. *For every $s = 1, 2, 3$, and for every $t \geq l$ the following relations hold true*

$$F_{s,t}[u] = F_{1,l}[u]F_{s,t-l+1}[u] + F_{2,l}[u]F_{s,t-l}[u] + F_{3,l}[u]F_{s,t-l-1}[u]. \tag{19}$$

The same relations hold true when $t < l$ provided that $u_3 \neq 0$.

Proof. It is sufficient to compare the representation formula (3.1) with the general one (2.1) for matrices. Consequently,

$$Q_3^k[u] = \begin{bmatrix} F_{3,k-1} & F_{3,k} & F_{3,k+1} \\ F_{2,k-1} & F_{2,k} & F_{2,k+1} \\ F_{1,k-1} & F_{1,k} & F_{1,k+1} \end{bmatrix}$$

$$= F_{1,k-1}Q_3^2[u] + F_{2,k-1}Q_3[u] + F_{3,k-1}I$$

$$= F_{1,k-1}\begin{bmatrix} F_{3,1} & F_{3,2} & F_{3,3} \\ F_{2,1} & F_{2,2} & F_{2,3} \\ F_{1,1} & F_{1,2} & F_{1,3} \end{bmatrix} + F_{2,k-1}\begin{bmatrix} F_{3,0} & F_{3,1} & F_{3,2} \\ F_{2,0} & F_{2,1} & F_{2,2} \\ F_{1,0} & F_{1,1} & F_{1,2} \end{bmatrix} +$$

$$+ F_{3,k-1}\begin{bmatrix} F_{3,-1} & F_{3,0} & F_{3,1} \\ F_{2,-1} & F_{2,0} & F_{2,1} \\ F_{1,-1} & F_{1,0} & F_{1,1} \end{bmatrix}.$$

□

Finally we show how any particular solution of recurrence relation (3.2) with arbitary initial conditions can be obtained by means of powers of matrix $Q_3[u]$.

THEOREM 3.3. *If $\{G_k\}$ is a particular solution of the recurrence relation (3.2), corresponding to initial conditions*

$$\begin{aligned} G_{-1} &= \gamma = \gamma(u_1, u_2, u_3), \\ G_0 &= \beta = \beta(u_1, u_2, u_3), \\ G_1 &= \alpha = \alpha(u_1, u_2, u_3), \end{aligned}$$

then sequence $\{G_k\}$ is generated by powers of matrix $Q_3[u]$.

Proof. Since functions $F_{j,k}$ (j=1,2,3) constitute a basis of $\mathcal{V}_3$, then G_k can be written

$$G_k = \alpha F_{1,k} + \beta F_{2,k} + \gamma F_{3,k}, \quad (k \geq -1). \tag{20}$$

Using formula (3.1), we have

$$\begin{bmatrix} \alpha & 0 & 0 \\ 0 & \beta & 0 \\ 0 & 0 & \gamma \end{bmatrix} Q_3^k[u] = \begin{bmatrix} \gamma F_{3,k-1} & \gamma F_{3,k} & \gamma F_{3,k+1} \\ \beta F_{2,k-1} & \beta F_{2,k} & \beta F_{2,k+1} \\ \alpha F_{1,k-1} & \alpha F_{1,k} & \alpha F_{1,k+1} \end{bmatrix}. \tag{21}$$

Then summing columns of the above matrix the vector (G_{k-1}, G_k, G_{k-1}) appears, so that the sequence $\{G_k\}$ is obtained.

We remark that for $u_3 \neq 0$, the preceding formula (3.8) holds true even for other negative values of the index k, provided that the $F_{k,n}$ functions are defined by (2.3).□

3.2. Powers of $Q_r[u]$. All representation formulas presented in Section 3.1 for the 3×3 matrix $Q_3[u]$ can be extended to a general $r\times r$ matrix $Q_r[u]$. We now present our results in the following theorems with the key steps for the proofs because the similar proofs to the case $r=3$.

THEOREM 3.4. *Let $Q_r[u]$ be the matrix as in (1.5). Then*

$$Q_r^k[u]=\begin{bmatrix} F_{r,k-1} & F_{r,k} & \cdots & F_{r,k+r-2}\\ F_{r-1,k-1} & F_{r-1,k} & \cdots & F_{r-1,k+r-2}\\ \vdots & \vdots & \vdots & \vdots\\ F_{1,k-1} & F_{1,k} & \cdots & F_{1,k+r-2}\end{bmatrix}. \tag{22}$$

Proof. In this case the recurrence is (2.4), and the characteristic equation is (2.5). Then the companion matrix is

$$Q_r[u]=\begin{bmatrix} 0 & 0 & 0 & \cdots & 0 & (-1)^{r-1}u_r\\ 1 & 0 & 0 & \cdots & 0 & (-1)^{r-2}u_{r-1}\\ 0 & 1 & 0 & \cdots & 0 & (-1)^{r-3}u_{r-2}\\ \vdots & \vdots & \vdots & & \vdots & \vdots\\ 0 & 0 & 0 & \cdots & 0 & -u_2\\ 0 & 0 & 0 & \cdots & 1 & u_1\end{bmatrix}.$$

Note that

$$\mathrm{Tr}(Q_r[u])=u_1,\ldots,\Sigma_k(Q_r[u])=(-1)^{k-1}u_k,\ldots,\det(Q_r[u])=(-1)^{r-1}u_r,$$

where Σ_k denotes the sum of principal minors of order k.

If $\{J_k\},\{K_l\},\ldots,\{L_m\}$ are sequences of functions satisfying the recurrence relation (2.4), then (3.4) becomes:

$$\begin{bmatrix} J_k & J_{k+1} & \cdots & J_{k+r-1}\\ K_l & K_{l+1} & \cdots & K_{l+r-1}\\ \vdots & \vdots & \vdots & \vdots\\ L_m & L_{m+1} & \cdots & L_{m+r-1}\end{bmatrix} Q_r[u]=\begin{bmatrix} J_{k+1} & J_{k+2} & \cdots & J_{k+r}\\ K_{l+1} & K_{l+2} & \cdots & K_{l+r}\\ \vdots & \vdots & \vdots & \vdots\\ L_{m+1} & L_{m+2} & \cdots & L_{m+r}\end{bmatrix}. \tag{23}$$

Then we have (3.9).□

Comparing (3.9) with (2.1), a nonlinear recurrence relation which generalizes (3.6) is easily deduced.

THEOREM 3.5. *Let $F_{k,u}[u]$ be defined as in (2.3). For all $s=1,2,\ldots,r;t\geq l$, we have*

$$F_{s,t}[u]=F_{1,l}[u]F_{s,t-l+r-2}[u]+F_{2,l}[u]F_{s,t-l+r-3}[u]+\cdots+F_{r,l}[u]F_{s,t-l-1}[u].$$

These relations still hold for $t<l$ if $u_r\neq 0$. If we consider a particular solution $\{G_k\}$ of the recurrence relation (2.4), corresponding to initial conditions

$$\begin{aligned} G_{-1} &= \zeta=\zeta(u_1,u_2,\ldots,u_r),\\ \cdots & \quad \cdots\cdots\cdots\cdots\\ G_{r-3} &= \beta=\beta(u_1,u_2,\ldots,u_r),\\ G_{r-2} &= \alpha=\alpha(u_1,u_2,\ldots,u_r),\end{aligned}$$

then, since the $F_{k,n}[u]$ functions constitute a basis, G_k can be written:

$$G_k = \alpha F_{1,k} + \beta F_{2,k} + \cdots + \zeta F_{r,k}, \quad (k \geq -1).$$

For $u_r \neq 0$, the preceding formula holds true even for other negative values of the index k, provided that the $F_{k,n}[u]$ functions are defined by (2.3). Since

$$\begin{bmatrix} \zeta & 0 & \cdots & 0 \\ 0 & \beta & \cdots & 0 \\ \vdots & \vdots & \vdots & \vdots \\ 0 & 0 & \cdots & \alpha \end{bmatrix} Q_3^k[u] = \begin{bmatrix} \zeta F_{r,k-1} & \zeta F_{r,k} & \cdots & \zeta F_{r,k+r-2} \\ \beta F_{r-1,k-1} & \beta F_{r-1,k} & \cdots & \beta F_{r-1,k+r-2} \\ \vdots & \vdots & \vdots & \vdots \\ \alpha F_{1,k-1} & \alpha F_{1,k} & \cdots & \alpha F_{1,k+r-2} \end{bmatrix},$$

summing columns of the last matrix, the values of the sequence $\{G_k\}$, for every k, can be deduced.

Acknowledgement. This paper was concluded on a visit of the first author to the Dipartmento di Matematica Università degli Studi di Roma "La Sapienza", Italy, whose hospitality and partial support from the Italian National Research Council (G.N.I.M.) we wish to acknowledge.

REFERENCES

[1] M. Bruschi, P.E. Ricci, *An explicit formula for f(A) and the generating function of the generalized Lucas polynomials*, Siam J. Math. Anal. 13, 162-165, 1982.
[2] M. Bruschi, P.E. Ricci, *I polinomi di Lucas e di Tchebycheff in piú variabili*, Rend. Mat. VI, Vol. 13, 1980.
[3] É. Lucas, *Théorie des Nombres*, Gauthier Villars, Paris, 1891.
[4] I.V.V. Raghavacharyulu, A.R. Tekumalla, *Solution of the Difference Equations of Generalized Lucas Polynomials*, J. Math. Phys. 13, 321-324, 1972.
[5] M. Bicknell and V.E. Hoggatt, Jr., *Generalized Fibonacci Polynomials*, The Fibonacci Quarterly, 11.5, 457-468, 1973.
[6] M.X. He, P.E. Ricci,*Asymptotic Distribution of Zeros of Weighted Fibonacci Polynomials*, An International Journal in Complex Variables, Vol. 28, pp. 375-384, 1996.
[7] M.X. He, P.E. Ricci and D. Simon,*Dynamics of the Zeros of Fibonacci Polynomials*, Fibonacci Quarterly, 35.2, 160-168, 1997.
[8] M.X. He, P.E. Ricci and D. Simon,*Fibonacci Type Polynomial as a Trajectory of a Dynamical System*, submitted for publication.
[9] M.X. He, P.E. Ricci and D. Simon, *Numerical Results on the Zeros of Generalized Fibonacci Polynomials*, in Journal of Calcolo, A Quarterly on Numerical Analysis and Theory of Computation, Vol. 34, pp. 25-40, 1997.
[10] M.X. He, P.E. Ricci and D. Simon,*Entropy and Zeta Function of Fibonacci Matrix*, submitted for publication.

EXACT NUMERICAL METHODS AND THEIR APPLICATIONS TO BLENDING CURVES *

ZI-CAI LI †, CHIA-SHEN CHANG †, AND T. D. BUI ‡

Abstract. The cubic Hermite FEM is used to solve the 4th order ordinary differential equations (ODEs), to yield the superconvergence

$$\|\tilde{u}_h - u_I\|_2 = O(h^4),$$

where h is the maximal length of quasi-uniform subsections, and

$$\|\tilde{u}_h - u_I\|_2 = O(h^\ell), \qquad \ell > 4$$

in the case that no lower terms exist, where $\tilde{u}_h$ and u_I are the cubic Hermite FEM solutions and the cubic Hermite interpolant solutions of the true solutions, $\|\cdot\|_2$ is the Sobolev norm over the entire region, and ℓ is the order of accuracy in approximate integration rules used. When neither lower terms nor integration errors are involved, the equality $\tilde{u}_h = u_I$ is achieved, which implies the numerical solutions and their derivatives at the element nodes are exact! This is a remarkably distinctive from the traditional numerical solutions for ordinary and partial differential equations, where there always exist some errors at the element nodes (or grid nodes) between the numerical and true solutions. From this reason, the exact numerical methods are called in this paper.

The exact numerical methods are also applied to formulate the blending curves in 3D, which are described by an elastic beam satisfying ODEs with order 4, and satisfying the containment and tangent boundary requirements. Since the exact nodal solutions and the exact nodal derivatives can be obtained, much less computation is needed due to only a few elements used. Another advantage of the ODE description for blending curves is that the unique curves in 3D with global minimal energy can be achieved, contrasted to the traditional interpolation techniques in blending problems.

Key words. Ordinary differential equations, finite element method, blending curves.

AMS subject classifications. 65L60, 65S05

1. Introduction. High accuracy of numerical solutions of ordinary and partial differential equations is important to industrial problems. Of course, we may choose the high order FEMs, even the p-version FEMs, or the spectral method, etc. It is easier to employ the simple methods and to raise the numerical solutions obtained. Hence, many reports have appeared on superconvergence at specific points (see [26, 29, 35, 41]) and on the global region (see [25, 27, 28]). By Lin's techniques, under some reasonable limitations on elements, e.g., rectangular elements, the superconvergence on the entire region can be derived between the numerical solution $\tilde{u}_h$ and the interpolant solution u_I of u, and the global superconvergence can then be achieved by means of a posteriori interpolant on $\tilde{u}_h$.

For an elastic beam under an external load f, we report surprising results for the cubic Hermite FEM solutions:

$$\tilde{u}_h = u_I,$$

by following Lin's techniques, when no integration errors occur in the integral $\int_0^1 fv\,ds$, where $v(= v(s))$ is the functions of the cubic Hermite elements. This implies that at all the element nodes s_i, the cubic Hermite FEM solutions are exact:

$$u_h\Big|_{s_i} = u\Big|_{s_i}, \quad u_h'\Big|_{s_i} = u'\Big|_{s_i}, \quad \forall i.$$

---

* Partial results of this paper were presented at the SIAM Annual Meeting at Toronto, July 13-17, 1998

† Department of Applied Mathematics, National Sun Yat-sen University, Kaohsiung, Taiwan 80424

‡ Department of Computer Science, Concordia University, 1455 de Masionneune Blvd, West, Montreal, Quebec H3G 1M8, Canada

The solutions of an elastic beam are nontrivial under arbitrary f, where $\int_0^1 fv\,ds$ is integrable. In fact, the external load may be piecewise, for instance:

$$f = \begin{cases} f_1, & 0 \le s < s^*, \\ f_2, & s^* \le s < 1. \end{cases}$$

If the integration formulas, $\int f_1 s^k\,ds$, $\int f_2 s^k\,ds$, $0 \le k \le 3$, are also exact, or can be found in the integration tables of [17], the numerical solutions at $s = s_i$ are exact. Note that f can even be chosen as a delta function. In the problem of blending curves, the elastic beam is taken as the transferring curves, and a certain load f may be chosen to formulate the desired shape. The exact solutions obtained at $s = s_i$ are significant, because very high accuracy of the entire approximate solutions can be easily obtained by a posteriori interpolation on $\widetilde{u}_h$.

Note that the approaches yielding blending curves in this paper are different from the traditional, purely interpolating techniques or other techniques in [3, 9, 10, 13, 15, 16, 14, 18, 19, 21, 22, 23, 24, 29, 31, 32, 33, 34, 37, 38, 40]. We solicit the differential equations as in [5] but pursue the global minimal energy of blending curves. Moreover, by the techniques given in Section 2, the optimal blending curves are unique. This also solves a puzzle of multi-blendings, pointed out in the book of Foley, et al [16], p. 487. This paper also gives a framework of blending curves by ordinary differential equations (ODEs). Details of applications are given in [7].

To expose easily the accurate numerical methods, we describe them for the simple ODE boundary value problems in the next section. We then apply them to the problem of blending curves in Section 3, together with numerical examples.

2. Model Problems. For simplicity, let us begin with the solution $u = u(s)$ governed by the fourth order ODE in $s \in [0, 1]$:

(2.1) $$u'''' - k^2 u'' = f,$$

with the simple support boundary conditions:

(2.2) $$u(0) = u_0,\ u(1) = u_e,\ u'(0) = u_0',\ u'(1) = u_e'.$$

Eqs (2.1) and (2.2) are written in a weak form:

(2.3) $$a(u, v) = f(v), \quad v \in H_0^2[0, 1],$$

where

(2.4) $$a(u, v) = \int_0^1 \left(u''v'' + k^2 u'v'\right) ds,$$

(2.5) $$f(v) = \int_0^1 fv\,ds,$$

(2.6) $$H_0^2[0, 1] = \left\{v,\ v',\ v'' \in L^2[0, 1] \text{ and } v(0) = v(1) = v'(0) = v'(1) = 0\right\}.$$

Denote by V_h the space of the piecewise cubic Hermite interpolation functions on the subsections

(2.7) $$0 = s_0 < s_1 < \ldots < s_i < s_{i+1} < \ldots < s_N = 1$$

with $h_{i-1} = s_i - s_{i-1}$ satisfying $v(0) = v_0$ and $v(1) = v_1$, and by V_h^0 that satisfying

$$v(0) = v(1) = 0. \tag{2.8}$$

We assume that the subsections $[s_i, s_{i+1}]$ are quasi-uniform: $\frac{\max_i h_i}{\min_i h_i} \le C$. The cubic Hermite FEM is to seek the solution $u_h \in V_h$ such that

$$a(u_h, v) = f(v), \ \forall v \in V_h^0. \tag{2.9}$$

If choosing the integration rules for the integral $f(v)$ only, we obtain

$$a(\tilde{u}_h, v) = \widehat{f}(v), \ \forall \ v \in V_h^0. \tag{2.10}$$

Denote the norms

$$\|v\|_k = \left\{ \sum_{\ell=0}^{k} \int_0^1 \left(\frac{d^\ell v}{ds^\ell} \right)^2 ds \right\}^{1/2}, \quad |v|_k = \left\{ \int_0^1 \left(\frac{d^k v}{ds^k} \right)^2 ds \right\}^{1/2}.$$

Now, we prove the following lemma, by following the ideas in [27, 28].

LEMMA 2.1. *Denote by u, u_I and $\tilde{u}_h$ the true solution, the interpolation solution, and the FEM solution respectively, and $u \in H^4[0,1]$. Then errors between the interpolation solution and the FEM solution have the bound:*

$$\|\tilde{u}_h - u_I\|_2 \le Ck^2h^4|u|_4 + \sup_{w \in V_h^0} \frac{|f(w) - \widehat{f}(w)|}{\|w\|_2}. \tag{2.11}$$

Proof. The uniform V_h^0-elliptic inequality holds true (see [30]) :

$$C_0\|v\|_2^2 \le a(v, v), \quad \forall v \in V_h^0,$$

where C_0 is a bounded constant independent of v. From (2.3) and (2.10) we obtain

$$a(u - \tilde{u}_h, v) = f(v) - \widehat{f}(v), \ \ \forall v \in V_h^0.$$

Denote $w = \tilde{u}_h - u_I$, we have

$$\begin{aligned} C_0 \|\tilde{u}_h - u_I\|_2^2 &\le a(\tilde{u}_h - u_I, w) = a(u - u_I, w) - a(u - \tilde{u}_h, w) \\ &= a(u - u_I, w) - \left(f(w) - \widehat{f}(w) \right). \end{aligned} \tag{2.12}$$

By applying integration by parts, $w'''' = 0$, and

$$(u - u_I)'\Big|_{s_i} = (u - u_I)|_{s_i} = 0,$$

we obtain

$$a(u - u_I, w) = \int_0^1 (u - u_I)'' \, w'' \, ds + k^2 \int_0^1 (u - u_I)' \, w' \, ds$$

$$= \sum_{i=0}^{n-1} (u - u_I)' w'' \Big|_{s_i}^{s_{i+1}} - \int_0^1 (u - u_I)' w''' \, ds + k^2 \sum_{i=0}^{n-1} (u - u_I) w' \Big|_{s_i}^{s_{i+1}} -$$

$$-k^2\int_0^1 (u-u_I)w''ds$$

$$= -\int_0^1 (u-u_I)'\, w'''\, ds - k^2\int_0^1 (u-u_I)\, w''\, ds$$

$$= -\sum_{i=0}^{n-1}(u-u_I)w'''\Big|_{s_i}^{s_{i+1}} + \int_0^1 (u-u_I)w''''ds - k^2\int_0^1 (u-u_I)w''ds$$

$$= -k^2\int_0^1 (u-u_I)\, w''\, ds, \tag{2.13}$$

to lead to

$$|a(u-u_I,w)| \le k^2|u-u_I|_0\,|w|_2 \le C\,k^2h^4|u|_4|w|_2, \tag{2.14}$$

where $h=\max_i h_i$ and C is also a bounded constant independent of v.

It follows from (2.12) and (2.14) that

$$\|\tilde{u}_h-u_I\|_2 \le \frac{1}{C_0\|w\|_2}\left\{|a(u-u_I,w)|+|f(w)-\widehat{f}(w)|\right\}$$
$$\le C\left\{k^2h^4|u|_4+\frac{|f(w)-\widehat{f}(w)|}{\|w\|_2}\right\}.$$

This completes the proof of Lemma 2.1. ∎

Let n be even, when the solution $\tilde{u}_h$ is obtained, we may construct the a posteriori interpolant $\Pi_{2h}^5\tilde{u}_h$ such that ([8] and [28])

$$\left\|u-\Pi_{2h}^5 u_I\right\|_2 \le Ch^4|u|_6.$$

Here interpolant $\Pi_{2h}^5 u_I$ is the polynomials of order 5 on $[s_{2i}, s_{2i+2}]$ with the supports u_{2i}, u_{2i+1}, u_{2i+2}, u'_{2i}, u'_{2i+1} and u'_{2i+2}. Let us give an explicit formula for the a posteriori interpolation $\Pi_{2h}^5\,\tilde{u}_h$. Assume that the known solutions and their derivatives are:

$$\widehat{u_0}=u(0),\ \widehat{u_1}=u(1),\ \widehat{u_2}=u(2),$$
$$\widehat{u_0'}=u'(0), \widehat{u_1'}=u'(1), \widehat{u_2'}=u'(2).$$

Then the cubic Hermite interpolation of order 5 on $[0,2]$ can be obtained from [2], p.160:

$$P_5(s)=\widehat{u_0}\,T_0(s)+\widehat{u_1}\,T_1(s)+\widehat{u_2}\,T_2(s)+\widehat{u_0'}\,Q_0(s)+\widehat{u_1'}\,Q_1(s)+\widehat{u_2'}\,Q_2(s), \tag{2.15}$$

where

$$\begin{aligned} T_0(s) &= \frac{1}{4}(1+3s)(s-1)^2(s-2)^2, T_1(s)=s^2(s-2)^2,\\ T_2(s) &= \frac{1}{4}(7-3s)\,s^2\,(s-1)^2, Q_0(s)=\frac{1}{4}s\,(s-1)^2\,(s-2)^2,\\ Q_1(s) &= (s-1)\,s^2\,(s-2)^2, Q_2(s)=\frac{1}{4}(s-2)\,s^2\,(s-1)^2. \end{aligned} \tag{2.16}$$

The FEM solutions in $[s_i, s_{i+1}]$ with u_i and u_i' are expressed by

$$\begin{aligned} u_h(s) = u_i\,\phi_0\left(\frac{s-s_i}{h_i}\right) + u_{i+1}\,\phi_1\left(\frac{s-s_i}{h_i}\right) + \\ +u_i'\,h_i\,\psi_0\left(\frac{s-s_i}{h_i}\right) + u_{i+1}'\,h_i\,\psi_1\left(\frac{s-s_i}{h_i}\right), \end{aligned} \tag{2.17}$$

and

$$\begin{aligned} \phi_0(s) = 2s^3 - 3s^2 + 1, \phi_1(s) = -2s^3 + 3s^2, \\ \psi_0(s) = s^3 - 3s^2 + s, \quad \psi_1(s) = s^3 - s^2. \end{aligned} \tag{2.18}$$

Let the division number be even, i.e., N is even and $h_{2i} = h_{2i+1}$. Then we may establish the a posteriori interpolation on $[s_{2i}, s_{2i+2}]$ from [2],

$$\begin{aligned} \Pi_{2h}^5\,u_h(s) = u_{2i}\,T_0\left(\frac{s-s_{2i}}{h_{2i}}\right) + u_{2i+1}\,T_1\left(\frac{s-s_{2i}}{h_{2i}}\right) + u_{2i+2}\,T_2\left(\frac{s-s_{2i}}{h_{2i}}\right) \\ +u_{2i}'\,h_{2i}Q_0\left(\frac{s-s_{2i}}{h_{2i}}\right) + u_{2i+1}'\,h_{2i}Q_1\left(\frac{s-s_{2i}}{h_{2i}}\right) + u_{2i+2}'\,h_{2i}Q_2\left(\frac{s-s_{2i}}{h_{2i}}\right). \end{aligned} \tag{2.19}$$

Now we prove a lemma.

LEMMA 2.2. *Let u_h and $\Pi_{2h}^5\,u_h$ be given in (2.17) and (2.19). Then there exists a bounded constant C independent of h and u_h such that*

$$\left\|\Pi_{2h}^5 u_h\right\|_2 \le C\,\|u_h\|_2, \;\forall u_h \in V_h^0. \tag{2.20}$$

Proof: Denoting $v = \Pi_{2h}^5\,u_h$, we have

$$\|\Pi_{2h}^5\,u_h\|_2^2 = \|v\|_2^2 = \sum_{i=0}^{\frac{N}{2}-1} \|v\|_{2,[s_{2i},s_{2i+2}]}^2. \tag{2.21}$$

By using the linear transformation

$$t = \frac{s-s_{2i}}{h_{2i}}, \tag{2.22}$$

we obtain

$$\|v\|_{2,[s_{2i},s_{2i+2}]}^2 = \sum_{\ell=0}^{2} |v|_{\ell,[s_{2i},s_{2i+2}]}^2 = \sum_{\ell=0}^{2} h_{2i}^{1-2\ell}\,|\widehat{v}|_{\ell,[0,2]}^2, \tag{2.23}$$

where

$$\begin{aligned} \widehat{v_0} = \widehat{v}(0) = v(s_{2i}),\; \widehat{v_1} = \widehat{v}(1) = v(s_{2i+1}),\; \widehat{v_2} = \widehat{v}(2) = v(s_{2i+2}), \\ \widehat{v_0}' = \widehat{v}'(0) = v'(s_{2i}), \widehat{v_1}' = \widehat{v}'(1) = v'(s_{2i+1}), \widehat{v_2}' = \widehat{v}'(2) = v'(s_{2i+2}). \end{aligned}$$

Later, we shall prove the following bounds for $\ell = 0 \sim 2$

$$|\widehat{v}|_{\ell,[0,2]}^2 \le C\left(|\widehat{u_h}|_{\ell,[0,1]}^2 + |\widehat{u_h}|_{\ell,[1,2]}^2\right) = C|\widehat{u}_h|_{\ell,[0,2]}^2, \tag{2.24}$$

where $\widehat{u_h}$ is the cubic Hermite interpolation on $[0,1]$ and $[1,2]$,

$$\widehat{u_h} = \widehat{u_0}\,\phi_0(s) + \widehat{u_1}\,\phi_1(s) + \widehat{u_0}'\,\psi_0(s) + \widehat{u_1}'\,\psi_1(s) \quad \text{on } [0,1],$$
$$\widehat{u_h} = \widehat{u_1}\,\phi_0(s-1) + \widehat{u_2}\,\phi_1(s-1) + \widehat{u_1}'\,\psi_0(s-1) + \widehat{u_2}'\,\psi_1(s-1) \quad \text{on } [1,2].$$

By applying the inverse transformation of (2.22) we have for $\ell = 0 \sim 2$,

$$\begin{aligned} |\widehat{u_h}|^2_{\ell,[0,2]} &= |\widehat{u_h}|^2_{\ell,[0,1]} + |\widehat{u_h}|^2_{\ell,[1,2]} \\ &= h_{2i}^{2\ell-1}\left[|u_h|^2_{\ell,[s_{2i},s_{2i+1}]} + |u_h|^2_{\ell,[s_{2i+1},s_{2i+2}]}\right] = h_{2i}^{2\ell-1}|u_h|^2_{\ell,[s_{2i},s_{2i+2}]}. \end{aligned} \tag{2.25}$$

Hence we obtain from (2.20), (2.21), (2.24) and (2.25).

$$\begin{aligned} \|\Pi^5_{2h}u_h\|_2^2 &= \sum_{i=0}^{\frac{N}{2}-1}\sum_{\ell=0}^{2} h_{2i}^{1-2\ell}|\widehat{v}|^2_{\ell,[0,2]} \le C\sum_{i=0}^{\frac{N}{2}-1}\sum_{\ell=0}^{2} h_{2i}^{1-2\ell}|\widehat{u_h}|^2_{\ell,[0,2]} \\ &= C\sum_{i=0}^{\frac{N}{2}-1}\sum_{\ell=0}^{2}|u_h|^2_{\ell,[s_{2i},s_{2i+2}]} = C\,\|u_h\|_2^2. \end{aligned} \tag{2.26}$$

Finally, let us prove (2.24) only for the case $\ell = 0$ with the notation $|v|_{0,[0,2]} = \|v\|_{0,[0,2]}$, because the proof for $\ell > 0$ is similar.

$$\|\widehat{v}\|^2_{0,[0,2]} \le C\left(\|\widehat{u_h}\|^2_{0,[0,1]} + \|\widehat{u_h}\|^2_{0,[1,2]}\right). \tag{2.27}$$

Since the norm $\|\widehat{v}\|_{0,[0,2]}$ and the Euclidean norm

$$\|\widehat{v}\| = \left(\widehat{v_0}^2 + \widehat{v_1}^2 + \widehat{v_2}^2 + (\widehat{v_0}')^2 + (\widehat{v_1}')^2 + (\widehat{v_2}')^2\right)^{1/2} \tag{2.28}$$

are six-dimensional on $[0,2]$, they are equivalent to each other, i.e.,

$$C_0\|\widehat{v}\|^2 \le \|\widehat{v}\|^2_{0,[0,2]} \le C_1\|\widehat{v}\|^2, \tag{2.29}$$

where C_0 and C_1 are two bounded constants independent of h and v.

On the other hand

$$\|\widehat{u_h}\|^2_{0,[0,1]} = \vec{x}^T H\,\vec{x}, \tag{2.30}$$

where $\vec{x} = (\widehat{v_0},\ \widehat{v_1},\ \widehat{v_0}',\ \widehat{v_1}')^T$,

$$H = (H_{ij}),\quad H_{ij} = \int_0^1 \phi_i\,\phi_j\,ds,\quad 1 \le i,\ j \le 4. \tag{2.31}$$

In (2.31) the basic functions $\phi_0(s)$ and $\phi_1(s)$ are given in (2.18), and $\phi_3(s) = \psi_0(s)$ and $\phi_4(s) = \psi_1(s)$. The matrix H in (2.31) is positive definite and symmetric, whose upper triangular part is given explicitly by

$$\begin{bmatrix} \frac{13}{35} & \frac{9}{70} & \frac{11}{210} & -\frac{13}{420} \\ & \frac{13}{15} & \frac{13}{420} & -\frac{11}{210} \\ & & \frac{1}{105} & -\frac{1}{104} \\ & & & \frac{1}{105} \end{bmatrix}.$$

Then we have

$$\lambda_{\min}\|\vec{x}\|^2 \le \vec{x}^T H \vec{x}, \tag{2.32}$$

where $\lambda_{\min}(> 0)$ is the minimal eigenvalue of H, which is independent of v. Combining (2.30) and (2.32) gives

$$\|\widehat{u_h}\|^2_{0,[0,1]} \ge \lambda_{\min}[\widehat{v_0}^2 + \widehat{v_1}^2 + (\widehat{v_0}')^2 + (\widehat{v_1}')^2]. \tag{2.33}$$

Similarly we have

$$\|\widehat{u_h}\|^2_{0,[1,2]} \ge \lambda_{\min}[\widehat{v_1}^2 + \widehat{v_2}^2 + (\widehat{v_1}')^2 + (\widehat{v_2}')^2]. \tag{2.34}$$

From (2.33) and (2.34) it follows that

$$\|\widehat{v}\|^2 \le \frac{1}{\lambda_{\min}}\left(\|\widehat{u_h}\|^2_{0,[0,1]} + \|\widehat{u_h}\|^2_{0,[1,2]}\right). \tag{2.35}$$

The bound (2.27) is obtained from (2.29) and (2.35). This completes the proof of Lemma 2.2. ∎

From Lemmas 2.1 and 2.2 we obtain

$$\begin{aligned}
&\left\|u - \Pi^5_{2h}\tilde{u}_h\right\|_2 \le \left\|u - \Pi^5_{2h}u_I\right\|_2 + \left\|\Pi^5_{2h}(u_I - \tilde{u}_h)\right\|_2 \\
&\le Ch^4|u|_6 + C\|u_I - \tilde{u}_h\|_2 \\
&\le C\left\{h^4|u|_6 + k^2h^4|u|_4 + \sup_{w\in V_h^0}\frac{|f(w) - \widehat{f}(w)|}{\|w\|_2}\right\}.
\end{aligned}$$

We write this important result as a theorem on the global superconvergence rates.

THEOREM 2.3. *Let $u \in H^6[0,1]$, there exist the error bound of the cubic Hermite FEM solutions:*

$$\left\|u - \Pi^5_{2h}\tilde{u}_h\right\|_2 \le C\left\{h^4|u|_6 + k^2h^4|u|_4 + \sup_{w\in V_h^0}\frac{|f(w) - \widehat{f}(w)|}{\|w\|_2}\right\}. \tag{2.36}$$

Denoting by P_k the k-order polynomials, we provide more interesting results:

COROLLARY 2.4. *Let $u \in H^6[0,1]$ and choose the integration*

$$\int_0^1 v\,ds = \widehat{\int_0^1} v\,ds, \tag{2.37}$$

to be exact $\forall v \in P_4$. Then when $f \in H^5[0,1]$ the global superconvergence rates are obtained

$$\left\|u - \Pi^5_{2h}\tilde{u}_h\right\|_2 = O(h^4). \tag{2.38}$$

Proof. Since the integration

$$\int_0^1 fv\,ds = \widehat{\int_0^1} fv\,ds, \quad \forall fv \in P_4.$$

Then for $w \in V_h^0$ we have from Ciarlet [8] and $w'''' = 0$ that

$$\left|\left(\int_0^1 - \widehat{\int_0^1}\right) fw \, ds\right| \le Ch^5 \left\|(fw)'''''\right\|_0 \le Ch^5 \|f\|_5 \|w\|_3 \le Ch^4 \|f\|_5 \|w\|_2,$$

where we have used the bound $\|w\|_3 \le \frac{C}{h}\|w\|_2$, $\forall w \in V_h^0$ for quasi-uniform partition of $[s_i, s_{i+1}]$. The desired result (2.38) follows Theorem 2.3. ∎

Note that the global superconvergence rates $O(h^4)$ are two orders higher than $O(h^2)$, the latter is the optimal convergence rate (see Birkhoff and Lynch [4]).

Since the integrals

$$\int_0^1 u''v''ds, \quad \text{and} \quad \int_0^1 u'v'ds, \quad \forall u, \; v \in V_h^0$$

can be computed exactly by the integral rule (2.37), we may also rewrite the FEM (2.10) equivalently as follows

$$\widehat{\int_0^1} u''v''ds + k^2 \widehat{\int_0^1} u'v'ds = \widehat{\int_0^1} f \, vds. \tag{2.39}$$

Note that the integral, $\int_0^1 u''v'' \, ds$, can also be computed exactly by Simpson's rule in [2].

Next, we consider the simple case $k = 0$ to obtain the following surprising results.

COROLLARY 2.5. *Let $k = 0$, all conditions in Lemma 2.1 hold, and the integration (2.37) is exact for all $v \in P_l$, $l \ge 3$. Then*

$$\|\widehat{u}_h - u_I\|_2 \le C \sup_{w \in V_h^0} \frac{|f(w) - \widehat{f}(w)|}{\|w\|_2} \le Ch^l \|f\|_{l+1}, \tag{2.40}$$

where P_ℓ is the space of polynomials with order $\le \ell$. In particular, if $f \in P_{l-3}$, then

$$\tilde{u}_h = u_I. \tag{2.41}$$

Proof. Since

$$\left|\left(\int - \widehat{\int}\right) fw \, ds\right| \le Ch^{l+1} \|f\|_{l+1} \|w\|_3 \le Ch^l \|f\|_{l+1} \|w\|_2,$$

by noting $k = 0$ Eq. (2.40) follows from Lemma 2.1. If $f \in P_{\ell-3}$, then $fw \in P_\ell$ where $w \in V_h^0$. Hence

$$\widehat{\int_0^1} fw \, ds = \int_0^1 fw \, ds.$$

We conclude from (2.40) that $\|u_I - \tilde{u}_h\|_2 = 0$. By considering that the restriction of the function $u_I - \tilde{u}_h$ to each subinterval $[s_i, s_{i+1}]$ is a third degree polynomial, we then conclude that $u_I = \tilde{u}_h$. This completes the proof of Corollary 2.5. ∎

REMARK 2.1. *The result of Corollary 2.5 is surprising: when $k = 0$ the convergence rates depend only on the integration accuracy. Moreover, no integration errors occur, the nodal solutions including both the solutions u_i and their derivatives u_i' at all the nodes are exact! This is contrary to the approximate solutions by the traditional FEM or FDM, where there always exist some errors at the element nodes or the difference grid nodes (see [1, 2, 28, 36, 39]).*

Let us also consider the FEM (2.9), to obtain the following corollary, by following the arguments in Lemma 2.1.

COROLLARY 2.6. *Let $k = 0$, then the nodal solutions of the cubic Hermite FEM by (2.9) without approximate integration are exact, i.e.,*

$$u_h = u_I. \tag{2.42}$$

When f is chosen as some integrable function, the integrals, $\int f s^k ds$, can be found in the tables in [17] or evaluated exactly by some manipulation. The equality (2.42) holds true.

3. Blending Curves. In this section, we apply the exact numerical methods to important applications: the blending curve problems. In doing so, we first model the blending curves as an elastic beam in Section 3.1, and then provide the optimal convergence rates in Section 3.2 for comparison. In Section 3.3, the new exact methods proposed in Section 2 are adopted to blending curves. Moreover, numerical experiments are carried out to verify the theoretical analysis made.

3.1. Mathematical Modelling. Curves will be represented by the parametric forms $\{(x(s), y(s), z(s))\}$, here the parameter s can be interpreted as the arc length. Let the parameter s be normalized as $0 \le s \le 1$. Then the cases at $s = 0$ and $s = 1$ may denote the beginning point B and the ending point E, respectively. Therefore, the curve can also be denoted by the vector:

$$W(s) = (x(s), y(s), z(s))^T, \quad 0 \le s \le 1. \tag{3.1}$$

Formally, we intend to design a three dimensional curve passing through two ending points, $B = (x_0, y_0, z_0)$ and $E = (x_e, y_e, z_e)$, and satisfying the tangent requirements at the end-points. The tangent requirements are usually given as follows [7]:

$$\left.\frac{\partial y}{\partial x}\right|_B = \overline{y}_x(B), \quad \left.\frac{\partial z}{\partial x}\right|_B = \overline{z}_x(B); \quad \left.\frac{\partial y}{\partial x}\right|_E = \overline{y}_x(E), \quad \left.\frac{\partial z}{\partial x}\right|_E = \overline{z}_x(E). \tag{3.2}$$

Suppose that $x'(0) = \frac{dx(0)}{ds}$ and $x'(1) = \frac{dx(1)}{ds}$ are nonzero, and the values of $\overline{y}_x(B)$, $\overline{z}_x(B)$, $\overline{y}_x(E)$ and $\overline{z}_x(E)$ are known and given in Eqs. (3.2). The boundary conditions at B and E can be expressed by

$$W_0 = W(0) = (x_0, y_0, z_0)^T, \qquad W_1 = W(1) = (x_1, y_1, z_1)^T, \tag{3.3}$$

$$W_0' = \frac{d}{ds}W(0) = \alpha_0(1, \overline{y}_x(B), \overline{z}_x(B))^T, \; W_1' = \frac{d}{ds}W(1) = \alpha_1(1, \overline{y}_x(E), \overline{z}_x(E))^T, \tag{3.4}$$

where the constants α_0 and α_1 are arbitrary. If the curve has no singularities, the two constants α_0 and α_1 cannot be zero.

The problem of blending curves can also be formulated as follows. Find a function $W(s) = (x(s), y(s), z(s))^T \in (C^1[0,1])^3$ satisfying Eqs. (3.3) and (3.4). There arise two questions:

1. It can be seen from [16], p. 486, that the constants α_0 and α_1 affect greatly the shape of the curve. How can we choose the *optimal* values of α_0 and α_1 ?
2. If it is required that the curve be piecewise low degree polynomials, the cubic Hermite interpolation can be used, provide that the positions and directions of the curve on all support points $s = s_i$ in (2.7) are known. However, $W(s_i)$ and $W'(s_i)$ are usually not known in the design of the curves. How can we select them appropriately?

We try to provide answers to the above two questions by ordinary differential equations (ODEs) to model the functions $W(s)$ of the curve. From differential geometry point of view [11, 12], we have

$$\frac{1}{\rho^2} \approx \left(\frac{d^2x}{d\bar{s}^2}\right)^2 + \left(\frac{d^2y}{d\bar{s}^2}\right)^2 + \left(\frac{d^2z}{d\bar{s}^2}\right)^2,$$

where ρ is the curvature radius of the curve, and $\bar{s}$ is the arc length of the curve. Let c be the total arc length and $\bar{s} = cs$, then

$$\frac{1}{\rho^2} \approx \frac{1}{c^4}\left[\left(\frac{d^2x}{ds^2}\right)^2 + \left(\frac{d^2y}{ds^2}\right)^2 + \left(\frac{d^2z}{ds^2}\right)^2\right]. \tag{3.5}$$

If we require the curvature of the curve to be small, the solution $W(s)$ in (3.1) may be requested to satisfy a system of fourth order ODEs:

$$\frac{d^2}{ds^2}P\frac{d^2}{ds^2}W = F, \tag{3.6}$$

where

$$P = \begin{pmatrix} p_x(s) & 0 & 0 \\ 0 & p_y(s) & 0 \\ 0 & 0 & p_z(s) \end{pmatrix}, \quad F = \begin{pmatrix} f_x(s) \\ f_y(s) \\ f_z(s) \end{pmatrix}, \tag{3.7}$$

and $p_x(s)$, $p_y(s)$ and $p_z(s)$ are positive functions. The containment boundary conditions are given by

$$W(s)\Big|_{s=0} = W(0), \quad W(s)\Big|_{s=1} = W(1), \tag{3.8}$$

and the tangent requirement may be written as

$$y'(0) = a_1x'(0), \quad z'(0) = a_2x'(0), \quad y'(1) = a_3x'(1), \quad z'(1) = a_4x'(1), \tag{3.9}$$

where the values of $a_1 = \frac{dy}{dx}(0)$, $a_2 = \frac{dz}{dx}(0)$, $a_3 = \frac{dy}{dx}(1)$ and $a_4 = \frac{dz}{dx}(1)$ are known. In fact, Eqs. (3.9) are equivalent to Eq. (3.4), but without the unknown constants α_0 and α_1. If the unique solution W can be found, then $\frac{dW(0)}{ds}$ and $\frac{dW(1)}{ds}$ are known, and the values of α_0 and α_1 can then be determined uniquely.

Denote the space

$$H = \left\{W \middle| W \in (H^2[0,1])^3, \text{ and } W \text{ satisfying (3.8) and (3.9)}\right\}, \tag{3.10}$$

$$H_0 = \left\{W \middle| W \in (H^2[0,1])^3, \text{ and } W(0) = W(1) = \vec{0} \text{ and satisfying (3.9)}\right\}, \tag{3.11}$$

and the total energy

$$T_t(W) = \frac{1}{2}\int_0^1 \left(P\frac{d^2W}{ds^2}\right)\cdot\left(\frac{d^2W}{ds^2}\right) ds - \int_0^1 F\cdot W\, ds, \tag{3.12}$$

where $F \cdot W$ is the dot product of vectors F and W.

Denote the constraints (3.9) as the linear constraints among x_0', y_0', z_0' as well as x_n', y_n', z_n' by

$$L_0 = L_0(x'(0), y'(0), z'(0)) = 0; \quad L_1 = L_1(x'(1), y'(1), z'(1)) = 0. \tag{3.13}$$

The solution W can be represented in the following variational form: to find a solution W such that

$$T_t(W) = \min_{\substack{V\in H\\ L_0=0\\ L_1=0}} T_t(V). \tag{3.14}$$

It is proven in [7] that the optimal solution W in (3.14) exists uniquely.

3.2. The Cubic Hermite Elements and Optimal Convergence Rates. The true solution $W \in H$ can also be written as

$$A(W,U) = F(U), \quad U \in H_0, \tag{3.15}$$

where

$$A(W,U) = \int_0^1 P\frac{d^2}{ds^2}W \cdot \frac{d^2}{ds^2}U\, ds, \quad F(U) = \int_0^1 F\cdot U\, ds. \tag{3.16}$$

Choose the piecewise cubic Hermite functions

$$x_h(s) = \sum_{i=0}^{n-1} x_h^{(i)}(s), \tag{3.17}$$

where

$$\begin{aligned} x_h^{(i)}(s) = {} & x_i\phi_0\left(\frac{s-s_i}{h_i}\right) + x_{i+1}\phi_1\left(\frac{s-s_i}{h_i}\right) \\ & + h_{i+1}\left[x_i'\psi_0\left(\frac{s-s_i}{h_i}\right) + x_{i+1}'\psi_1\left(\frac{s-s_i}{h_i}\right)\right], \end{aligned} \tag{3.18}$$

and $x_h^{(i)}(s) = 0$, $s \notin (s_i, s_{i+1}]$. When $i = 0$, Eq. (3.18) holds for $[s_0, s_1]$. Denote by V_h the space of $W_h(s) = (x_h(s), y_h(s), z_h(s))^T$, where $x_h(s)$, $y_h(s)$ and $z_h(s)$ are given by (3.17) satisfying (3.8) and (3.9). Also denote by V_h^0 the space of $W_h(s)$ satisfying $W_h(0) = W_h(1) = 0$ and (3.9). Then the cubic Hermite FEM is to seek $W_h \in V_h$ such that

$$A(W_h, U_h) = F(U_h), \quad \forall U_h \in V_h^0. \tag{3.19}$$

In fact, the final solution $\widetilde{W}_h \in V_h$ of FEM involving approximate integrals can be represented as

$$\widetilde{A}(\widetilde{W}_h, U_h) = \widetilde{F}(U_h), \quad \forall U_h \in V_h^0, \tag{3.20}$$

where

$$\widetilde{A}_h(W,U) = \widetilde{\int_0^1} P\frac{d^2W}{ds^2}\cdot\frac{d^2V}{ds^2}\,ds,\quad \widetilde{F}(U) = \widetilde{\int_0^1} F\cdot U\,ds. \tag{3.21}$$

The following Sobolev norms are needed for error analysis:

$$\|W\|_p = \left(\int_0^1 \sum_{k=1}^{p}\left\|\frac{d^kW}{ds^k}\right\|^2 ds\right)^{1/2},\quad |W|_p = \left(\int_0^1 \left\|\frac{d^pW}{ds^p}\right\|^2 ds\right)^{1/2}, \tag{3.22}$$

where $\|W\|^2 = W\cdot W$. We have the following theorem from Strang and Fix [36].

LEMMA 3.1. *Assume that the following bounds are given:*

$$\widetilde{A}(W,U) \le C_1\|W\|_2\|U\|_2,\quad \forall W\in H,\ \forall\, U\in V_h^0, \tag{3.23}$$

$$C_0\|U\|_2^2 \le \widetilde{A}(W,U),\quad \forall U\in V_h^0, \tag{3.24}$$

where C_0 and C_1 are two bounded constants independent of h, W and U. Then there exists a bounded constant C independent of h, W and U such that

$$\|W-\widetilde{W}_h\|_2 \le C\left\{\inf_{U_h\in V_h}|W-U|_2 + \sup_{V\in V_h^0}\frac{|\widetilde{A}(W,V)-A(W,V)|}{\|V\|_2} + \sup_{V\in V_h^0}\frac{|\widetilde{F}(V)-F(V)|}{\|F\|_2}\right\}. \tag{3.25}$$

We may choose Simpson's rule: $\widetilde{\int_{s_i}^{s_{i+1}}} = \frac{h_i}{6}(g_i + 4g_{i+\frac12} + g_{i+1})$, where $g_i = g(s_i)$ and $g_{i+\frac12} = g((s_i+s_{i+1})/2)$. It is important to verify (3.24). In fact, we have (also see Marti [30])

$$\widetilde{A}(U,U) \ge C_1|U|_2^2. \tag{3.26}$$

Note that Friedrichs' inequality involves only the displacement boundary conditions, then

$$\|U\|_2^2 \le C_0\left(x^2(0)+x^2(1)+y^2(0)+y^2(1)+z^2(0)+z^2(1)+|U|_2^2\right) = C_0|U|_2^2,\quad U\in H_0. \tag{3.27}$$

Then, we have from (3.26) and (3.27)

$$\widetilde{A}(U,U) \ge \frac{C_1}{C_0}\|U\|_2^2. \tag{3.28}$$

This is Eq. (3.24). We below provide a lemma, a theorem and a corollary without proof since the proof may follow the standard analysis in [8].

LEMMA 3.2. *Let $W\in(H^4[0,1])^3$, $P\in(C^4[0,1])^3$ and $F\in(H^2[0,1])^3$. Then the cubic Hermite FEM solutions $\widetilde{W}_h$ using Simpson's rule have the bounds:*

$$|\widetilde{A}(W,V)-A(W,V)| \le Ch^2\|W\|_4\cdot\|V\|_2, \tag{3.29}$$

$$|\widetilde{F}(V)-F(V)| \le Ch^2\|F\|_2\cdot\|V\|_2. \tag{3.30}$$

THEOREM 3.3. *Let the conditions in Lemma 3.2 hold, then the cubic Hermite FEM solutions $\widetilde{W}_h$ using Simpson's rule have the error bound:*

$$\|W - \widetilde{W}_h\|_2 \le Ch^2(|W|_4 + |F|_2). \tag{3.31}$$

From the Nitsche techniques and the Sobolev interpolatory theory (see Strang and Fix [36]), we may obtain the following corollary.

COROLLARY 3.4. *Let all conditions in Lemma 3.2 hold true. Then*

$$\|W - \widetilde{W}_h\|_\ell = O(h^{4-\ell}), \quad \ell = 0,\ 1,\ 2,\ 3. \tag{3.32}$$

COROLLARY 3.5. *Let $P = Const$, $F = \vec{0}$, and all conditions in Theorem 3.3 hold. Then the numerical solutions W_h are exact to the true solution W.*

Proof. When $P = Const$ and $F = \vec{0}$, the true solutions W are entire cubic polynomials on $[0,1]$. Hence $|W|_4 = 0$. It follows from Theorem 3.3:

$$\|W - \widetilde{W}_h\|_2 = 0.$$

This implies that the differences $(W - \tilde{W}_h)$ are constant functions. Moreover, since both W and $\tilde{W}_h$ satisfy the boundary conditions (3.8), then $W = \widetilde{W}_h$ on $[0,1]$. ∎

To close this subsection, we provide numerical examples. For simplicity, choose $p_x(s) = p_y(s) = p_z(s) = 1$ and the isotopic loads: $f_x(s) = f_y(s) = f_z(s) = 1, s$ or s^2, which are called Cases I, II and III respectively. Simpson's rule is used in (3.19). Tables 3.1–3.3 list the numerical and true solutions at $s =$ 0, 1/4, 1/2, 3/4 and 1, and Table 3.4 lists the Sobolev error norms of Case III. It can be seen that the numerical solutions are exact at s_i in Tables 3.1 and 3.2, but not in Table 3.3.

Table 3.1 of Case I for $P = Const$ and $F = \vec{0}$ agrees with Corollary 3.5 perfectly. However, for $P = Const$ and $F = Const$ in Case II, the solutions W are entire polynomial of order 4; but $\widetilde{W}_h$ are the piecewise cubic polynomials, $W_1 \neq \widetilde{W}_h$ on $[0,1]$. Surprisingly, at the element nodes $s = s_i$, there always exist the equalities:

$$W(s_i) = \widetilde{W}_h(s_i), \quad W'(s_i) = \widetilde{W}_h'(s_i).$$

These are exact nodal solutions which will be discussed in the next section.

Besides, Fig. 3.1 indicates the asymptotic behaviour:

$$\|\widetilde{W}_h - W\|_\ell = O(h^{4-\ell}), \quad \ell = 0,\ 1,\ 2,\ 3 \tag{3.33}$$

and

$$Max = \|\widetilde{W}_h - W\|_{\infty,[0,1]} = O(h^{4-\delta}),$$

where $0 < \delta << 1$. Eqs. (3.33) agree with Corollary 3.4 perfectly. It is also straightforward to see from Table 3.3 that at $s = s_i$, the solutions

$$|\widetilde{W}_h(s_i) - W(s_i)| = O(h^4)$$

and

$$|\widetilde{W}_h'(s_i) - W'(s_i)| = O(h^4). \tag{3.34}$$

The superconvergence (3.34) of nodal derivatives can be found in Wahlbin [39].

TABLE 3.1
The numerical and true solutions at $s = 0$, 1/4, 1/2, 3/4 and 1 in Case I: $p_x(s) = p_y(s) = p_z(s) = 1$ and $f_x(s) = f_y(s) = f_z(s) = 0$, by using Simpson's rule.

N	x_0	x'_0	y_0	y'_0	z_0	z'_0
4	1.0000000000	0.9542008844	1.0000000000	2.8626026532	3.5000000000	3.8168035376
8	1.0000000000	0.9542008844	1.0000000000	2.8626026532	3.5000000000	3.8168035376
16	1.0000000000	0.9542008844	1.0000000000	2.8626026532	3.5000000000	3.8168035376
True	1.0000000000	0.9542008844	1.0000000000	2.8626026532	3.5000000000	3.8168035376
N	$x_{1/4}$	$x'_{1/4}$	$y_{1/4}$	$y'_{1/4}$	$z_{1/4}$	$z'_{1/4}$
4	1.2783530085	1.2233693936	1.5104775347	1.3395649084	4.5715315066	4.5726073910
8	1.2783530085	1.2233693936	1.5104775347	1.3395649084	4.5715315066	4.5726073910
16	1.2783530085	1.2233693936	1.5104775347	1.3395649084	4.5715315066	4.5726073910
True	1.2783530085	1.2233693936	1.5104775347	1.3395649084	4.5715315066	4.5726073910
N	$x_{1/2}$	$x'_{1/2}$	$y_{1/2}$	$y'_{1/2}$	$z_{1/2}$	$z'_{1/2}$
4	1.5870578016	1.1970151611	1.7289560960	0.5266108654	5.6948831333	4.2313644978
8	1.5870578016	1.1970151611	1.7289560960	0.5266108654	5.6948831333	4.2313644978
16	1.5870578016	1.1970151611	1.7289560960	0.5266108654	5.6948831333	4.2313644978
True	1.5870578016	1.1970151611	1.7289560960	0.5266108654	5.6948831333	4.2313644978
N	$x_{3/4}$	$x'_{3/4}$	$y_{3/4}$	$y'_{3/4}$	$z_{3/4}$	$z'_{3/4}$
4	1.8522336939	0.8751381870	1.8329566093	0.4237405243	6.5957931933	2.7930748579
8	1.8522336939	0.8751381870	1.8329566093	0.4237405243	6.5957931933	2.7930748579
16	1.8522336939	0.8751381870	1.8329566093	0.4237405243	6.5957931933	2.7930748579
True	1.8522336939	0.8751381870	1.8329566093	0.4237405243	6.5957931933	2.7930748579
N	x_1	x'_1	y_1	y'_1	z_1	z'_1
4	2.0000000000	0.2577384713	2.0000000000	1.0309538850	7.0000000000	0.2577384713
8	2.0000000000	0.2577384713	2.0000000000	1.0309538850	7.0000000000	0.2577384713
16	2.0000000000	0.2577384713	2.0000000000	1.0309538850	7.0000000000	0.2577384713
True	2.0000000000	0.2577384713	2.0000000000	1.0309538850	7.0000000000	0.2577384713

3.3. Accurate Nodal Solutions. Note that in Table 3.2 the nodal solutions W_h of Case II are exact at $s = s_i$. This is an interesting result of exact solutions. In this section, we will apply the analysis in Section 2 to explain the validity of a surprising phenomenon, and to provide new general, exact nodal solutions to the blending curves.

Consider

$$A_h(W, V) = F(V), \quad V \in H_0, \tag{3.35}$$

where the true solution $W \in H$,

$$A_h(W,V) = \int_0^1 P\frac{d^2}{ds^2}W \cdot \frac{d^2}{ds^2}V\, ds + \int_0^1 \left(K^2\frac{dW}{ds}\right)\cdot\left(\frac{dV}{ds}\right) ds, \tag{3.36}$$

$$F(V) = \int_0^1 F \cdot V\, ds,$$

and P are constant:

$$P = \begin{pmatrix} p_x & 0 & 0 \\ 0 & p_y & 0 \\ 0 & 0 & p_z \end{pmatrix}, \quad p_x > 0,\ p_y > 0,\ p_z > 0.$$

In (3.36), the second term of the right side is taken into account if the length of blending

TABLE 3.2

The numerical and true solutions at $s = 0$, $1/4$, $1/2$, $3/4$ and 1 in Case II: $p_x(s) = p_y(s) = p_z(s) = 1$ and $f_x(s) = f_y(s) = f_z(s) = 1$, by using Simpson's rule.

N	x_0	x_0'	y_0	y_0'	z_0	z_0'
4	1.0000000000	0.9644662034	1.0000000000	2.8933986102	3.5000000000	3.8578648136
8	1.0000000000	0.9644662034	1.0000000000	2.8933986102	3.5000000000	3.8578648136
16	1.0000000000	0.9644662034	1.0000000000	2.8933986102	3.5000000000	3.8578648136
True	1.0000000000	0.9644662034	1.0000000000	2.8933986102	3.5000000000	3.8578648136
N	$x_{1/4}$	$x_{1/4}'$	$y_{1/4}$	$y_{1/4}'$	$z_{1/4}$	$z_{1/4}'$
4	1.2818141607	1.2367916272	1.5184840517	1.3678915956	4.5793233403	4.5918038666
8	1.2818141607	1.2367916272	1.5184840517	1.3678915956	4.5793233403	4.5918038666
16	1.2818141607	1.2367916272	1.5184840517	1.3678915956	4.5793233403	4.5918038666
True	1.2818141607	1.2367916272	1.5184840517	1.3678915956	4.5793233403	4.5918038666
N	$x_{1/2}$	$x_{1/2}'$	$y_{1/2}$	$y_{1/2}'$	$z_{1/2}$	$z_{1/2}'$
4	1.5924191277	1.1973968204	1.7413057354	0.5307038324	5.7040939540	4.2240471678
8	1.5924191277	1.1973968204	1.7413057354	0.5307038324	5.7040939540	4.2240471678
16	1.5924191277	1.1973968204	1.7413057354	0.5307038324	5.7040939540	4.2240471678
True	1.5924191277	1.1973968204	1.7413057354	0.5307038324	5.7040939540	4.2240471678
N	$x_{3/4}$	$x_{3/4}'$	$y_{3/4}$	$y_{3/4}'$	$z_{3/4}$	$z_{3/4}'$
4	1.8558379684	0.8619067830	1.8424979889	0.3974603206	6.6008410282	2.7702197173
8	1.8558379684	0.8619067830	1.8424979889	0.3974603206	6.6008410282	2.7702197173
16	1.8558379684	0.8619067830	1.8424979889	0.3974603206	6.6008410282	2.7702197173
True	1.8558379684	0.8619067830	1.8424979889	0.3974603206	6.6008410282	2.7702197173
N	x_1	x_1'	y_1	y_1'	z_1	z_1'
4	2.0000000000	0.2459465151	2.0000000000	0.9837860602	7.0000000000	0.2459465151
8	2.0000000000	0.2459465151	2.0000000000	0.9837860602	7.0000000000	0.2459465151
16	2.0000000000	0.2459465151	2.0000000000	0.9837860602	7.0000000000	0.2459465151
True	2.0000000000	0.2459465151	2.0000000000	0.9837860602	7.0000000000	0.2459465151

curves is also our concern. The constant matrix

$$K = \begin{pmatrix} k_x & 0 & 0 \\ 0 & k_y & 0 \\ 0 & 0 & k_z \end{pmatrix}, k_x \geq 0, k_y \geq 0, k_z \geq 0$$

can be regarded as weights.

The cubic Hermite FEM solution $W_h \in V_h$ is given by

$$(3.37) \qquad A_h(W_h, U_h) = F(U_h), \quad U_h \in V_h^0.$$

When involving approximate numerical integration of $F_h(U_h)$, the final solution $\widetilde{W}_h \in V_h$ of the cubic Hermite FEM can be represented as:

$$(3.38) \qquad A_h(\widetilde{W}_h, U_h) = \widetilde{F}_h(U_h), \forall U_h \in V_h^0,$$

where $A_h(W, V)$ is defined by (3.36), but the integration approximation

$$(3.39) \qquad \widetilde{F}_h(V) = \widetilde{\int_0^1} F \cdot V \, ds,$$

and V_h and V_h^0 are defined in Section 3.2. Note that the lemmas and theorems in Section 2 are valid for $x(s)$, $y(s)$ and $z(s)$ satisfying the simple support boundary conditions, as

TABLE 3.3
The numerical and true solutions at $s = 0$, 1/4, 1/2, 3/4 and 1 in Case III: $p_x(s) = p_y(s) = p_z(s) = 1$ and $f_x(s) = f_y(s) = f_z(s) = s^2$, by using Simpson's rule.

N	x_0	x'_0	y_0	y'_0	z_0	z'_0
4	1.0000000000	0.9567908108	1.0000000000	2.8703724324	3.5000000000	3.8271632432
8	1.0000000000	0.9567908976	1.0000000000	2.8703726928	3.5000000000	3.8271635904
16	1.0000000000	0.9567909030	1.0000000000	2.8703727090	3.5000000000	3.8271636120
True	1.0000000000	0.9567909033	1.0000000000	2.8703727100	3.5000000000	3.8271636134
N	$x_{1/4}$	$x'_{1/4}$	$y_{1/4}$	$y'_{1/4}$	$z_{1/4}$	$z'_{1/4}$
4	1.2792529726	1.2272096956	1.5126679558	1.3481233669	4.5735240960	4.5779045267
8	1.2792526887	1.2272051948	1.5126682385	1.3481225132	4.5735238487	4.5779000746
16	1.2792526710	1.2272049135	1.5126682561	1.3481224598	4.5735238332	4.5778997964
True	1.2792526698	1.2272048947	1.5126682573	1.3481224563	4.5735238322	4.5778997778
N	$x_{1/2}$	$x'_{1/2}$	$y_{1/2}$	$y'_{1/2}$	$z_{1/2}$	$z'_{1/2}$
4	1.5886203372	1.1978958332	1.7326648868	0.5291941216	5.6974168913	4.2303027251
8	1.5886195584	1.1978891461	1.7326655756	0.5291902828	5.6974161450	4.2302959729
16	1.5886195098	1.1978887281	1.7326656186	0.5291900428	5.6974160984	4.2302955509
True	1.5886195065	1.1978887003	1.7326656215	0.5291900268	5.6974160953	4.2302955227
N	$x_{3/4}$	$x'_{3/4}$	$y_{3/4}$	$y'_{3/4}$	$z_{3/4}$	$z'_{3/4}$
4	1.8534130475	0.8712906298	1.8360648887	0.4160261028	6.5973367552	2.7867992447
8	1.8534121632	0.8712841577	1.8360656392	0.4160174076	6.5973358832	2.7867926913
16	1.8534121080	0.8712837532	1.8360656861	0.4160168642	6.5973358287	2.7867922817
True	1.8534121043	0.8712837263	1.8360656893	0.4160168280	6.5973358251	2.7867922544
N	x_1	x'_1	y_1	y'_1	z_1	z'_1
4	2.0000000000	0.2537417417	2.0000000000	1.0149669667	7.0000000000	0.2537417417
8	2.0000000000	0.2537378860	2.0000000000	1.0149515441	7.0000000000	0.2537378860
16	2.0000000000	0.2537376451	2.0000000000	1.0149505802	7.0000000000	0.2537376451
True	2.0000000000	0.2537376290	2.0000000000	1.0149505159	7.0000000000	0.2537376290

TABLE 3.4
Error norms $E_k = \|W - \widetilde{W}_h\|_\ell$, $\ell = 0, 1, 2$ and 3 in Case III.

N	E_0	E_1	E_2	E_3	Max error
4	0.244(-5)	0.389(-4)	0.967(-3)	0.309(-1)	0.714(-5)
8	0.157(-6)	0.498(-5)	0.256(-3)	0.159(-1)	0.525(-6)
16	0.980(-8)	0.638(-6)	0.641(-4)	0.802(-2)	0.360(-7)

well as for $x(s)$, $y(s)$ and $z(s)$ under the additional constraints in (3.8) and (3.9). Since the solution W is a linear combination of $x(s)$, $y(s)$ and $z(s)$, we obtain from Section 2 immediately the following theorem for the blending curves.

THEOREM 3.6. *Let $W \in (H^6[0,1])^3$, there exists the bound:*

$$\|W - \Pi^5_{2h}\widetilde{W}_h\|_2 \le C\left\{h^4|W|_6 + k^2h^4|W|_4 + \sup_{V\in V_h^0}\frac{|F(V) - \widetilde{F}(V)|}{\|W\|_2}\right\},$$

where $k^2 = \max_{1\le i\le 3} k_i^2$ and the operator Π^5_{2h} is defined in (2.19).

COROLLARY 3.7. *Let the integration rule (2.37) be exact for $\forall v \in P_4$ and all conditions in Theorem 3.6 hold. Then there exist the superconvergence rates:*

$$\|W - \Pi^5_{2h}\widetilde{W}_h\|_2 = O(h^4). \tag{3.40}$$

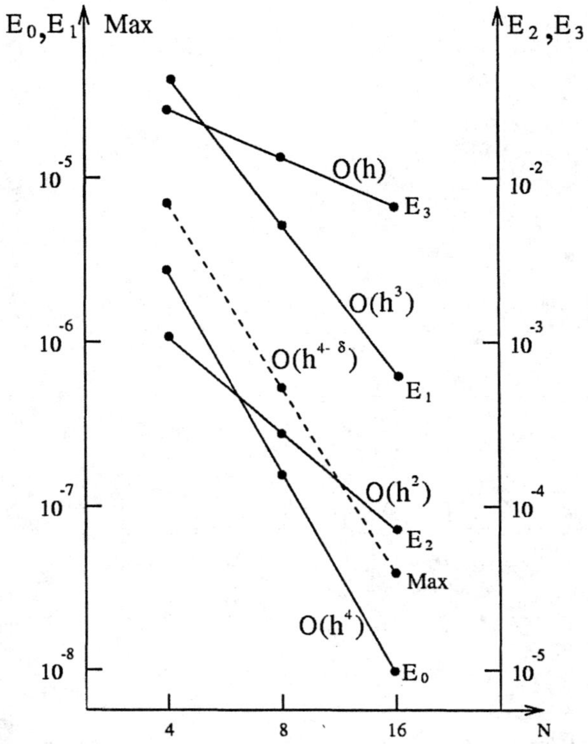

FIG. 3.1. *Curves of error norms E_ℓ and Max, based on Table 3.4.*

COROLLARY 3.8. *Let $k = 0$, all conditions in Theorem 3.6 hold and the integration rule (2.37) be exact $\forall v \in P_\ell$, $\ell > 3$ then*

$$\|W_h - W_I\|_2 = Ch^\ell \|F\|_{\ell+1}.$$

Moreover, if f_x, f_y, $f_z \in P_{\ell-3}$, then the exact nodal solutions of blending curves are obtained:

$$\widetilde{W}_h = W_I. \tag{3.41}$$

If integral $\int_0^1 F \cdot V ds$ can be evaluated exactly based on the tables in [17], we may solicit the FEM, (3.37), directly. Then we give the following corollary.

COROLLARY 3.9. *Let $k = 0$ and all conditions in Theorem 3.6 hold true. Then for the cubic Hermite FEM solutions W_h in (3.37) without integration approximation, we have*

$$W_h = W_I, \quad \text{on } [0, 1].$$

Now let us explain the exactness in Table 3.2 by Corollary 3.8. When $f_x = f_y = f_z = 1$, $F \cdot V = V_x + V_y + V_z$, being cubic polynomials, there exist no errors for Simpson's rule

$$\int_0^1 F \cdot V\, ds = \widetilde{\int_0^1} F \cdot V\, ds.$$

Hence (3.41) holds true from Corollary 3.8.

To further verify Corollary 3.8, we recompute Case III of Table 3.3, where $f_x = f_y = f_z = s^2$. The dot product $F \cdot V$ is a polynomial of order 5. If the integration rules are chosen to be exact for all polynomials P_5, we reach (3.41). Hence we choose the Gaussian rules with three integration nodes, which has accuracy of order 5. The new computed results are listed in Table 3.5 to show that both $\widetilde{W}_h$ and $\widetilde{W}_h'$ at $s = 0,\ 1/4,\ 1/2,\ 3/4$ and 1 are exact! This again confirms Corollary 3.8.

TABLE 3.5

The numerical and true solutions at $s = 0,\ 1/4,\ 1/2,\ 3/4$ and 1 in Case III: $p_x(s) = p_y(s) = p_z(s) = 1$ and $f_x(s) = f_y(s) = f_z(s) = s^2$, by using the Gauss rule of three integration nodes

N	x_0	x_0'	y_0	y_0'	z_0	z_0'
4	1.0000000000	0.9567909033	1.0000000000	2.8703727100	3.5000000000	3.8271636134
8	1.0000000000	0.9567909033	1.0000000000	2.8703727100	3.5000000000	3.8271636134
16	1.0000000000	0.9567909033	1.0000000000	2.8703727100	3.5000000000	3.8271636134
True	1.0000000000	0.9567909033	1.0000000000	2.8703727100	3.5000000000	3.8271636134
N	$x_{1/4}$	$x_{1/4}'$	$y_{1/4}$	$y_{1/4}'$	$z_{1/4}$	$z_{1/4}'$
4	1.279252670	1.2272048947	1.5126682573	1.3481224563	4.5735238322	4.5778997778
8	1.279252670	1.2272048947	1.5126682573	1.3481224563	4.5735238322	4.5778997778
16	1.279252670	1.2272048947	1.5126682573	1.3481224563	4.5735238322	4.5778997778
True	1.279252670	1.2272048947	1.5126682573	1.3481224563	4.5735238322	4.5778997778
N	$x_{1/2}$	$x_{1/2}'$	$y_{1/2}$	$y_{1/2}'$	$z_{1/2}$	$z_{1/2}'$
4	1.5886195065	1.1978887003	1.7326656215	0.5291900268	5.6974160953	4.2302955227
8	1.5886195065	1.1978887003	1.7326656215	0.5291900268	5.6974160953	4.2302955227
16	1.5886195065	1.1978887003	1.7326656215	0.5291900268	5.6974160953	4.2302955227
True	1.5886195065	1.1978887003	1.7326656215	0.5291900268	5.6974160953	4.2302955227
N	$x_{3/4}$	$x_{3/4}'$	$y_{3/4}$	$y_{3/4}'$	$z_{3/4}$	$z_{3/4}'$
4	1.8534121043	0.8712837263	1.8360656893	0.4160168280	6.5973358251	2.7867922544
8	1.8534121043	0.8712837263	1.8360656893	0.4160168280	6.5973358251	2.7867922544
16	1.8534121043	0.8712837263	1.8360656893	0.4160168280	6.5973358251	2.7867922544
True	1.8534121043	0.8712837263	1.8360656893	0.4160168280	6.5973358251	2.7867922544
N	x_1	x_1'	y_1	y_1'	z_1	z_1'
4	2.0000000000	0.2537376290	2.0000000000	1.0149505159	7.0000000000	0.2537376290
8	2.0000000000	0.2537376290	2.0000000000	1.0149505159	7.0000000000	0.2537376290
16	2.0000000000	0.2537376290	2.0000000000	1.0149505159	7.0000000000	0.2537376290
True	2.0000000000	0.2537376290	2.0000000000	1.0149505159	7.0000000000	0.2537376290

To close this paper let us make a few remarks.

REMARK 3.1. *Compared to the numerical methods in ODEs of [1, 6, 20] and in PDEs of [4, 8, 28, 36, 39], the intriguing exact method is proposed in this paper to provide the exact nodal solutions, and also applied to the 3D blending curves. Corollaries 3.8 and 3.9 are significant to the blending curves. Suppose that f_x, f_y and f_z can be represented by piecewise polynomials. Because the uniform analytical expansions of the solutions in this case can not be found, we may choose the intersection as the element nodes. Based on Corollary 3.8, by choosing suitable rules of integration, the solutions x_i, x_i', y_i, y_i', z_i and z_i' can be obtained exactly. Based on these exact nodal solutions, the number n may be chosen substantially small, i.e., $n = 6$; the final solutions $\widetilde{W}_h$ on $[0,1]$ can be obtained by the Lagrange interpolation, to be a polynomial with order 11, accurate enough for engineering applications. Consequently, the new techniques may save the CPU time significantly. Also the numerical stability is not a problem for small n, because the condition number of the* *associate matrix is* $O(h^{-4}) = O(n^4)$.

When f_x, f_y, f_z are piecewise, but not polynomials, we may adopt the FEM (3.37) if the exact integration can be found in [18]. The exact nodal solutions are also obtained.

REMARK 3.2. *It is worthy noting that the optimal convergence rates still occur:*

$$\|\widetilde{W}_h - W\|_2 = O(h^2)$$

even for (3.41) existing. This is because

$$\begin{aligned}\|\widetilde{W}_h - W\|_2 &\le \|\widetilde{W}_h - W_I\|_2 + \|W_I - W\|_2 \\ &= \|W_I - W\|_2 = O(h^2).\end{aligned}$$

In order to raise the global convergence rates, the a posteriori interpolation, as $\Pi^5_{2h}\widetilde{u}_h$, *is necessary.*

REMARK 3.3. *The exact numerical methods described in this paper are confined to specific ODE problems, such as (2.1) as* $k = 0$, *or as* $K = \vec{0}$ *in the blending curve problems (3.15). Moreover, the diagonal constant matrix* P *given in (3.7), may be extended to a positive definite and symmetric constant matrix* P, *or other kinds of boundary conditions involving linear combinations of boundary values and boundary derivatives. The new methods in this paper can be easily simplified to the second order ODE as well.*

Acknowledgements. We are very grateful to Prof. Lin Q. for his lecture [27] in our university, and to Prof. Yan N. N. for her valuables suggestions on the manuscript. We also express our thanks to the referee for his/her careful reading of this manuscript and valuable comments. This work was supported in part by the Natural Science Council of Taiwan, the Natural Sciences and Engineering Research Council of Canada, the Fonds pour la Formation de Chercheurs et l'Aide à la Recherche of Québec, the Ministère de l'Enseignement Superieur et de la Science Action.

REFERENCES

[1] U. M. Ascher, R. M. M. Mattheij and R. D. Russell, *Numerical Solution of Boundary Value Problems for Ordinary Differential Equations*, Prentice Hall, Englewood Cliffs, New Jersey, 1988.

[2] K. E. Atkinson, *An Introduction to Numerical Analysis*, John Wiley and Sons, New York, second edition, 1989.

[3] C. L. Bajaj and I. Ihm, *Algebraic surface design with cubic Hermite interpolation,* ACM Trans. on Graphics, vol. 11, pp. 61-91, 1992.

[4] G. Birkhoff and R. E. Lynch, *Numerical solution of Elliptic Equations*, SIAM Philadelphia, 1984.

[5] M. I. G. Bloor and M. J. Wilson, *Generating blend surface using partial differential equations*, CAD, vol.21, pp.165-171, 1989.

[6] J. C. Butcher, *The Numerical Analysis of Ordinary Differential Equations, Runge-Kutta and General Linear Methods*, John Wiley & Sons,Chichester, 1987.

[7] T. D. Bui, C. S. Chang, D. J. Guan and Z. C. Li, *A separation technique of differential equation systems for blending curves,* Technical Report, Department of Applied Mathematics National Sun Yet Sen University, Kaohsiung, Taiwan, 1998.

[8] P. G. Ciarlet, *The Finite Element Method for Elliptic Problems*, North-Holland, New York, 1978.

[9] G. Celniker and D. Gossard, *Deformable curve and surface finite-elements for free-form shape design,* Computer Graphics, vol. 25(4), pp. 257–265, 1991.

[10] B. K. Choi, *Surface Modelling for CAD/CAM*, Elsevier, Amsterdam, 1991.

[11] R. Courant and D. Hilbert, *Methods of Mathematical Physics*, Interscience Publishers, New York, 1953.

[12] Carmo M. de Do, *Differential Geometry of Curves and Surfaces*, Prentice-Hall Inc., 1976.

[13] G. Farin, *Curves and Surfaces for Computer Aided Geometric Design*, A Practical Guide(Sec. Ed.), Academic Press, 1990.

[14] D. J. Filip, *Blending parametric surfaces,* ACM Transactions on Graphics, vol. 8(3), pp. 164–173, 1989.

[15] R. B. Fisher(Ed.), *Design and Application of Curves and Surfaces*, Oxford University, Press, 1994.

[16] J. D. Foley, A. van Dam, S. K. Feiner and J. F. Hughes, *Computer Graphics, Principle and Practice*(Sec. Ed.), Addison Wesley Publishing Company, Reading, 1991.

[17] I. S. Gradshtey and I. M. Ryzhik, *Table of Integrals, Series and Products*, Academic Press, New York, corrected and enlarged edition, 1980.

[18] J. A. Gregory, (Ed.). *The Mathematics of Surfaces*, Clarendon Press, Oxford, 1986.

[19] H. Hagen, S. Hahmann, T. Schreiber, Y. Nakajima, B. Wördenweber, and P. Hollemann-Grundstedt, *Surface interrogation algorithms,* IEEE Computer Graphics and Applications, vol. 1, pp. 53–59, 1992.

[20] E. Hairer, S.P. Norsett and G. Wanner, *Solving Ordinary Differential Equations, I. Nonstiff Problems*, Springer-Verlag, 1993.

[21] C. Hoffmann and J. Hopcoft, *The potential method for blending surfaces and corners*, In G. Farin, editor, Geometric Modelling, pp. 347–366, SIAM, 1987.

[22] M. Kallay and B. Ravani, *Optimal twist vectors as a tool for interpolating a network of curves with minimum energy surface,* Computer Aided Geometric Design, vol. 7, pp. 465–473, 1990.

[23] J. J. Koendering, *Solid Shape*, MIT Press, Cambridge, 1990.

[24] M. Kosters, *Quadratic blending surfaces for corners*, The Visual Computer, vol.5., pp.134-146, 1989.

[25] M. Krizek and P. Neittaanmaki, *On a global- superconvergence of the gradient of linear triangular elements*, J. Comput. Appl. Math., vol. 18, pp.221-233, 1987.

[26] M. Krizek and P. Neittaanmaki, *Superconvergence phenomenon in the finite element method arising from averaging gradients*, Numer. Math., vol.45, pp.105-116, 1984.

[27] Q. Lin, *High efficient FEM, Seminar of Numerical Analysis* in Department of Applied Mathematics National Sun Yet Sen University, Kaohsiung, Taiwan, April 19, 1997.

[28] Q. Lin and N. N. Yan, *The Construction and Analysis of High Efficient FEM,* Hobel University publishing, 1996.

[29] R. J. MacKinnon and G. F. Carey, *Nodal superconvergence and solution enhancement for a class of finite-element and finite difference methods*, SIAM. J. Sci. Stat. Comput., Vol. 11, pp. 343–353, 1990.

[30] J. T. Marti, *Introduction to Sobolev Spaces and Finite Element Solution of Elliptic Boundary Value Problems*, Academic Press, London, 1986.

[31] H. P. Moreton and C. H. Séquin, *Functional optimization for fair surface design,* Computer Graphics, vol. 26(2), pp. 167–176, 1992.

[32] M. T. Nakao, *Superconvergence of the gradient of Galerkin approximations for elliptic problems*, J. Comput. Appl. Math., vol. 20, pp.341-348, 1987.

[33] A. W. Nutbourne and R. R. Martin, *Differential Geometry Applied to Curves and Surface Design*, volume I: Foundations, John Wiley & Sons, 1988.

[34] K. Ohkura and Y. Kakaxu, *Generalization of the potential method for blending three surfaces*, CAD, vol., pp. 599-610, 1992.

[35] A. I. Pekhlivanov, R. D. Lazarov, G. F. Carey and S. S. Chow, *Superconvergence analysis of approximate boundary-flux calculations,* Numer. Math., vol.63, pp.483-501, 1992.

[36] G. Strang and G. F. Fix, *An Analysis of Finite Element Method*, Prentice-Hall, 1973.

[37] B. Q. Su and D. Y. Liu, *Computational Geometry-Curve and Surface Modelling*, Academic Press, Boston, 1989.

[38] J. Vida, R. R. Martin and T. Varady, *A survey of blending methods that use parametric surfaces*, Computer Aided Design, vol. 26(5), pp. 341–364, 1994.

[39] L. L. Wahlbin, *Local behavior in finite element method, Chap. VII., Superconvergence*, in *Finite Element Methods Part I* (Eds. by P.G. Ciarlet and J.L. Lions), pp. 501-522, North-Holland, 1991.

[40] J. Warren, *Blending algebraic surfaces*, ACM Transaction on Graphics, vol. 8, pp. 263-278, 1989.

[41] M. F. Wheeler and J.R. Whiteman, *Superconvergent recovery of gradients of on subdomains from piecewise linear finite-element approximation*, Num. Methods for PDE., vol. 3, pp.65-82 and pp.357-374, 1987.

TRUNCATED GAUSS-LAGUERRE QUADRATURE RULES*

G. MASTROIANNI† AND G. MONEGATO‡

Abstract. For a certain class of functions, we propose to truncate the classical Gauss-Laguerre quadrature formula, that is, we consider only a fraction of the total number of nodes. For such a new rule we obtain an optimal order error estimate. Numerical examples show that for certain type of functions this new rule performs better than the standard one even for a small or moderate number of points.

1. Introduction. In 1987, during a conversation in Paris with P. Barrucand, this latter described to us the following observation.

Consider the classical Gauss-Laguerre quadrature formula

$$\int_0^\infty e^{-x} f(x)dx = \sum_{i=1}^{n} \lambda_{ni} f(x_{ni}) + R_n(f), \tag{1.1}$$

where we assume $x_{n1} < x_{n2} < \ldots < x_{nn}$. Then, associate with it the following "truncated" rule

$$\int_0^\infty e^{-x} f(x)dx = \sum_{i=1}^{n} \lambda_{mi} f(x_{mi}) + R_{m,n}(f), \qquad m > n; \tag{1.2}$$

that is, the rule one obtains by considering only the first n nodes of a m-point Gauss-Laguerre formulas, with $m > n$.

Numerical experiments show that, at least for certain functions, and taking m large enough, the second rule gives approximations which are better than those given by the first one.

For instance, the following example was considered among several ones which gave similar results:

$$I = \int_0^\infty e^{-x} \frac{e^{\frac{1}{1+x}}}{1+x} dx.$$

This integral was approximated simultaneously by (1.1) and by (1.2) with $m = \lfloor n^{4/3} \rfloor$. Here are the results (relative errors) we obtained at that time[1].

I = 1.197625892150730			
n	(1.1)	m	(1.2)
2	1.2E−01	2	1.2E−01
4	2.5E−02	6	6.9E−03
8	2.2E−03	16	5.4E−05
16	5.3E−05	40	2.4E−08
32	2.1E−07	101	2.1E−13
64	6.5E−11	256	−−

*Work supported by Italian Ministero dell'Università e della Ricerca Scientifica e Tecnologica.

†Dipartimento di Matematica, Università della Basilicata, Potenza, Italy.

‡Dipartimento di Matematica, Politecnico, Torino, Italy.

[1]Here and in Section 3, the sign "−−" means that full accuracy has been achieved, i.e., 15 significant digits.

This behaviour is not surprising, since the coefficients λ_{ni} of the Gauss-Laguerre rule decay exponentially as $x_{ni} \to \infty$ (see, for instance, (2.3)). Therefore, when the function $f(x)$ in (1.1) is bounded or does not increase too fast, in (1.1) the "last" terms of the sum give a neglegible contribution. This is the key observation that led Barrucand to formulate the above mentioned remark.

The question posed by Barrucand was then to understand for which classes of functions $f(x)$ this new formula is better than the classical one, and to find a rule for choosing a proper value of m in (1.2). Of course, one drawback of (1.2) is, at least using standard software, that one has to construct in any case all nodes and weights of the m-point rule. Nevertheless we think that this disadvantage is minor, particularly if we think of using the new rule to solve, for example, integral equations, as done by the authors in [5].

After several attempts to give an answer to the problem, this was put aside till a few months ago when, using recent results of one of us, we were able to give a (partial) answer to the above problem. This will be described in the next section. Several numerical examples are reported in Section 3.

2. Properties of the truncated rule. In the following, the symbol "c" will denote a positive constant taking different values on different occurences. Moreover $A \sim B$ will mean that a positive constant c exists such that $|A/B|^{\pm 1} \leq c$.

Let $\{P_m(w_\alpha; x)\}$ be the sequence of Laguerre orthonormal polynomials associated with the weight function $w_\alpha(x) = x^\alpha e^{-x}, \alpha > -1$ and defined in $(0, \infty)$. Denoting by $x_{m1} < x_{m2} < ... < x_{mm}$ the zeros of $P_m(w_\alpha; x)$, the following estimates are well-known (see for instance [6]):

(2.1) $$x_1 \sim m^{-1},\ x_m < 4m - c(4m)^{1/3},\ c > 0,$$

(2.2) $$\Delta x_k = x_{k+1} - x_k \sim \left(\frac{x_k}{4m - x_k + (4m)^{1/3}}\right)^{1/2}$$

$$\sim \left(\frac{x_k}{4m - x_k}\right)^{1/2},\ k = 0, 1, ..., m,$$

where we have set $x_0 = 0,\ x_{m+1} = 4m$. Moreover, for the Christoffel numbers we have

(2.3) $$\lambda_{mk} = \lambda_{mk}(w_\alpha) \sim w_\alpha(x_k)\Delta x_k.$$

We also need to introduce the space

(2.4) $$W_r(w_\alpha) = W_r = \{f^{(r-1)} \in AC_{loc} : \|f^{(r)}\phi^r w_\alpha\|_1 < \infty\},$$

where $r \geq 1$ and $\phi(x) = \sqrt{x}$. Notice that for functions $g \in W_r$ we have

(2.5) $$\Delta x_k |g(x_k)| \leq c \int_{x_{k-1}}^{x_k} |g(t)| dt + \Delta x_k \int_{x_{k-1}}^{x_k} |g'(t)| dt,\ k = 1, ..., m.$$

This follows from the well-known rectangle formula and its error representation:

$$\int_a^b f(x) dx = f(a)(b - a) + \int_a^b (b - x) f'(x) dx.$$

Further, since from [2, (2.2.14)] we have

$$\|g'\|_{L^1_{(a,b)}} \le M\left[(b-a)^{-1}\|g\|_{L^1_{(a,b)}} + (b-a)^{r-1}\|g^{(r)}\|_{L^1_{(a,b)}}\right],$$

with $M = M(r)$, bound (2.5) can be extended to the form

$$\Delta x_k |g(x_k)| \le c\left\{\int_{x_{k-1}}^{x_k} |g(t)|dt + (\Delta x_k)^r \int_{x_{k-1}}^{x_k} |g^{(r)}(t)|dt\right\}. \tag{2.6}$$

LEMMA 2.1. *Given a real constant $\delta > 0$, for any polynomial $p_m(x)$ of degree $m > 2(2\alpha-1)/\delta$ we have*

$$\int_{2(1+\delta)m}^{\infty} |p_m(x)|w_\alpha(x)dx \le Ae^{-cm}\int_0^\infty |p_m(x)|w_\alpha(x)dx, \tag{2.7}$$

where $A = A(\delta)$ and $c = c(\delta)$ are independent of m and p_m.

Proof. Bound (2.7) follows from an analogous inequality proved in [7] for the Hermite weight function, namely,

$$\int_{(1+\eta)\sqrt{k}}^{\infty} |q_k(x)|e^{-x^2}dx \le Be^{-bk}\int_{-\sqrt{k}}^{\sqrt{k}} |q_k(x)|e^{-x^2}dx, \tag{2.8}$$

where q_k is any polynomial of degree k. Indeed, set $\gamma = 2\alpha+1$, $j = \lceil 2\alpha+1\rceil$, and write

$$I = \int_{2(1+\delta)m}^{\infty} |p_m(x)|x^\alpha e^{-x}dx \le \frac{1}{[2(1+\delta)m]^{\frac{j-\gamma}{2}}}\int_{\sqrt{2(1+\delta)m}}^{\infty} |p_m(x^2)x^j|e^{-x^2}dx.$$

Define

$$q_{2m+j}(x) = p_m(x^2)x^j.$$

Then we have

$$I \le \frac{1}{[2(1+\delta)m]^{\frac{j-\gamma}{2}}}\int_\beta^\infty |q_{2m+j}(x)|e^{-x^2}dx, \tag{2.9}$$

where we have set $\beta = \sqrt{2m+j}\sqrt{\frac{1+\delta}{1+\frac{j}{2m}}}$. Since

$$\sqrt{\frac{1+\delta}{1+\frac{j}{2m}}} \ge \frac{1+\frac{\delta}{2}}{1+\frac{j}{2m}}$$

whenever $m > \frac{2j}{\delta}$, we can apply (2.8) with $\eta = \frac{\delta}{4+\delta}$ and $k = 2m+j$. We obtain:

$$I \le \frac{Ae^{-cm}}{[2(1+\delta)m]^{\frac{j-\gamma}{2}}}\int_{-\sqrt{2m+j}}^{\sqrt{2m+j}} |p_m(x^2)||x|^j e^{-x^2}dx$$

$$= \frac{2A}{[2(1+\delta)m]^{\frac{j-\gamma}{2}}}\int_0^{\sqrt{2m+j}} |p_m(x^2)|x^j e^{-x^2}dx$$

$$\leq \frac{2A(2m+j)^{\frac{j-\gamma}{2}}}{[2(1+\delta)m]^{\frac{j-\gamma}{2}}} \int_0^{\sqrt{2m+j}} |p_m(x^2)| x^\gamma e^{-x^2} dx.$$

Finally, changing back x^2 into x we have (2.7).

□

LEMMA 2.2. *Let $0 < \theta < 1$ be a given constant. For any $f \in W_r$ there exists a sequence of functions $\{f_{m,r}\}_{m \geq m_0}, m_0 > \frac{\alpha}{4\theta}$ such that*

$$\|(f - f_{m,r})w_\alpha\|_1 \leq \frac{c}{(4\theta m)^{\frac{r}{2}}} \|f^{(r)} \phi^r w_\alpha\|_{L^1(4\theta m, \infty)}, \tag{2.10}$$

with c independent of f, θ and m.

Proof. Define

$$f_m(x) = f_{m,r}(x) = \begin{cases} f(x) & \text{if } 0 < x \leq 4\theta m, \\ T_{r-1}(f; 4\theta m; x) & \text{if } x > 4\theta m, \end{cases} \tag{2.11}$$

where $T_{r-1}(f; 4\theta m; x)$ is the Taylor polynomial of degree $r-1$ associated with f and centered at $4\theta m$. Obviously, for $x > 4\theta m > 2\alpha$ we have $x^\alpha e^{-x/2} \leq u^\alpha e^{-u/2}$ whenever $4\theta m \leq u \leq x$; hence

$$|w_\alpha(x)[f(x) - f_m(x)]| \leq w_\alpha(x) \int_{4\theta m}^x \frac{(x-u)^{r-1}}{(r-1)!} |f^{(r)}(u)| du$$

$$\leq e^{-\frac{x}{2}} \int_{4\theta m}^x e^{\frac{u}{2}} \frac{(x-u)^{r-1}}{(r-1)!} |f^{(r)}(u) w_\alpha(u)| du,$$

that is

$$\|(f - f_m)w_\alpha\|_1 \leq \int_{4\theta m}^\infty e^{-\frac{x}{2}} \int_{4\theta m}^x e^{\frac{u}{2}} \frac{(x-u)^{r-1}}{(r-1)!} |f^{(r)}(u)| w_\alpha(u) du dx$$

$$= \int_{4\theta m}^\infty |f^{(r)}(u)| w_\alpha(u) e^{\frac{u}{2}} \int_u^\infty \frac{(x-u)^{r-1}}{(r-1)!} e^{-\frac{x}{2}} dx du$$

$$= \int_{4\theta m}^\infty |f^{(r)}(u)| w_\alpha(u) \left[\int_0^\infty \frac{t^{r-1} e^{-\frac{t}{2}}}{(r-1)!} dt \right] du$$

$$\leq \frac{c}{(4\theta m)^{\frac{r}{2}}} \int_{4\theta m}^\infty |f^{(r-1)}(u) w_\alpha(u) u^{\frac{r}{2}}| du = \frac{c}{(4\theta m)^{\frac{r}{2}}} \|f^{(r)} \phi^r w_\alpha\|_{L^1(4\theta m, \infty)}.$$

□

Now we are ready to prove the main result. By f_n we will always denote the function $f_{n,r}$ defined in (2.11) above.

THEOREM 2.3. *Let $0 < \theta < 1$ be a given constant. For any function $f \in W_r$ we have*

$$t_n(f) := \left| \int_0^\infty f(x) w_\alpha(x) dx - \sum_{i=1}^n \lambda_{ni}(w_\alpha) f_n(x_{ni}) \right| \leq \frac{c}{(\theta n)^{\frac{r}{2}}} \|f^{(r)} \phi^r w_\alpha\|_1,$$

where c is independent of n and f.

Proof. Let $q_{n_s}(x)$ be a polynomial of degree $n_s = \lceil \frac{n}{s} \rceil$, $s \geq \max\{1, \frac{1}{2\theta}\}$ fixed, satisfying the inequality (see [3] and [1, Theorem 3.3]):

$$\|(f_n - q_{n_s})w_\alpha\|_1 \leq \frac{c}{n^{\frac{r}{2}}}\|f_n^{(r)}\phi^r w_\alpha\|_1, \tag{2.12}$$

with c independent of n and f_n. Then,

$$t_n(f) = \left| \int_0^\infty [f(x) - q_{n_s}(x)]w_\alpha(x)dx - \sum_{i=1}^n \lambda_{ni}(w_\alpha)[f_n(x_{ni}) - q_{n_s}(x_{ni})] \right|$$

$$\leq \|(f - q_{n_s})w_\alpha\|_1 + \sum_{i=1}^n \lambda_{ni}|f_n(x_{ni}) - q_{n_s}(x_{ni})|. \tag{2.13}$$

For the first term in (2.13) we have

$$\|(f - q_{n_s})w_\alpha\|_1 \leq \|(f - f_n)w_\alpha\|_1 + \|(f_n - q_{n_s})w_\alpha\|_1,$$

hence, recalling (2.10) and (2.12),

$$\|(f - q_{n_s})w_\alpha\|_1 \leq \frac{c}{(\theta n)^{\frac{r}{2}}}\|f^{(r)}\phi^r w_\alpha\|_{L^1(4\theta n,\infty)} + \frac{c}{(\theta n)^{\frac{r}{2}}}\|f^{(r)}\phi^r w_\alpha\|_{L^1(0,4\theta n)}$$

$$\leq \frac{c}{(\theta n)^{\frac{r}{2}}}\|f^{(r)}\phi^r w_\alpha\|_1.$$

To bound the last term in (2.13) we use (2.3) and (2.6):

$$\lambda_{ni}(w_\alpha)|f_n(x_{ni}) - q_{n_s}(x_{ni})| \sim \Delta x_{ni} w_\alpha(x_{ni})|f_n(x_{ni}) - q_{n_s}(x_{ni})|$$

$$\leq c\left[\int_{x_{n,i-1}}^{x_{ni}} |f_n(x) - q_{n_s}(x)|w_\alpha(x)dx\right.$$

$$\left.+(\Delta x_{ni})^r \int_{x_{n,i-1}}^{x_{ni}} |f_n^{(r)}(x) - q_{n_s}^{(r)}(x)|w_\alpha(x)dx\right].$$

Setting $x_{nk} = \max\{x_{ni} : x_{ni} \leq 4\theta n\}$, we can write

$$I := \sum_{i=1}^n \lambda_{ni}(w_\alpha)|f_n(x_{ni}) - q_{n_s}(x_{ni})| \leq c\int_0^{x_{nn}} |f_n(x) - q_{n_s}(x)|w_\alpha(x)dx$$

$$+\left\{\sum_{i=1}^k + \sum_{i=k+1}^n\right\}(\Delta x_{ni})^r \int_{x_{n,i-1}}^{x_{ni}} |f_n^{(r)}(x) - q_{n_s}^{(r)}(x)|w_\alpha(x)dx$$

$$\leq \|(f_n - q_{n_s})w_\alpha\|_1 + I_1 + I_2, \tag{2.14}$$

where I_1 and I_2 denote the first and the second sum in the last line. Now, for $i \leq k$ we have

$$\Delta x_{ni} \sim \frac{\sqrt{x_{ni}}}{\sqrt{n}} = \frac{\phi(x_{ni})}{\sqrt{n}},$$

and, for $x_{n,i-1} \leq x \leq x_{ni}$,

$$\frac{\phi(x_{ni})}{\sqrt{n}} \sim \frac{\phi(x)}{\sqrt{n}}.$$

Therefore,

$$I_1 \leq \frac{c}{n^{\frac{r}{2}}}\left[\int_0^{4\theta n} |f^{(r)}(x)|\phi^r(x)w_\alpha(x)dx + \int_0^{4\theta n} |q_{n_s}^{(r)}(x)|\phi^r(x)w_\alpha(x)dx\right]. \tag{2.15}$$

For $i \geq k+1$, i.e., $x_{ni} \geq x_{n,k+1} > 4\theta n$,

$$f_n^{(r)}(x_{ni}) - q_{n_s}^{(r)}(x_{ni}) = -q_{n_s}^{(r)}(x_{ni})$$

and, recalling (2.1) and (2.2),

$$\Delta x_{ni} \leq cn^{\frac{1}{3}}\sqrt{\frac{x_{ni}}{n}}.$$

Thus

$$I_2 \leq \frac{cn^{\frac{r}{3}}}{n^{\frac{r}{2}}}\int_{4\theta n}^{4n} |q_{n_s}^{(r)}(x)|w_{\alpha+\frac{r}{2}}(x)dx.$$

Since $q_{n_s}(x)$ has degree $n_s = \lceil\frac{n}{s}\rceil$, and $\theta > 0$ is fixed, we can always choose s such that $4\theta n \geq 2(1+\delta)n_s$ for some $\delta > 0$. Therefore, using Lemma 1 we obtain

$$I_2 \leq \frac{c}{n^{\frac{r}{2}}}\|q_{n_s}^{(r)}\phi^r w_\alpha\|_1. \tag{2.16}$$

Finally, by inserting (2.15) and (2.16) into (2.14) we find

$$I \leq c\left\{\|(f_n - q_{n_s})w_\alpha\|_1 + n^{-\frac{r}{2}}\|q_{n_s}^{(r)}\phi^r w_\alpha\|_1 + n^{-\frac{r}{2}}\|f^{(r)}\phi^r w_\alpha\|_1\right\},$$

hence, recalling (see [1, (3.12)]) that

$$\|(f_n - q_{n_s})w_\alpha\|_1 + \frac{\|q_{n_s}^{(r)}\phi^r w_\alpha\|_1}{n^{\frac{r}{2}}} \leq c\frac{\|f^{(r)}\phi^r w_\alpha\|_1}{n^{\frac{r}{2}}},$$

we obtain

$$I \leq \frac{c}{n^{\frac{r}{2}}}\|f^{(r)}\phi^r w_\alpha\|_1.$$

□

COROLLARY 2.4. *For any function $f \in W_r, r \geq 1$, and any real $0 < \theta < 1$, we have*

$$\left| \int_0^\infty f(x) w_\alpha(x) dx - \sum_{0 < x_{ni} \leq 4\theta n} \lambda_{ni}(w_\alpha) f(x_{ni}) \right|$$

$$\leq c \left(\frac{\|f^{(r)} \phi^r w_\alpha\|_1}{n^{\frac{r}{2}}} + e^{-an} \|f w_\alpha\|_1 \right), \tag{2.17}$$

where the constants c and a are both independent of f and n.

Proof. Denoting by R the left hand side of inequality (2.17) we write

$$R \leq \left| \int_0^\infty f(x) w_\alpha(x) dx - \sum_{i=1}^n \lambda_{ni}(w_\alpha) f_n(x_{ni}) \right| + \sum_{x_{ni} > 4\theta n} \lambda_{ni}(w_\alpha) |f_n(x_{ni})|,$$

where s and n_s are defined as in the proof of Theorem 2.3.

A bound for the first term on the right hand side is given by Theorem 2.3. Moreover, using (2.3) and (2.6) we find that the last sum is bounded by

$$\int_{4\theta n}^{4n} |T_{r-1}(f; 4\theta n; x)| w_\alpha(x) dx,$$

since $f_n(x) \equiv T_{r-1}(f; 4\theta n; x)$ (see (2.11)) for $x > 4\theta n$.

If p_{n_s} is a polynomial of degree n_s, satisfying a property of form (2.12) with respect to $f \in L^1_{w_\alpha}$, then we have

$$\int_{4\theta n}^{4n} |T_{r-1}(f; 4\theta n; x) w_\alpha(x) dx \leq \int_{4\theta n}^{4n} |f - T_{r-1}(f; 4\theta n; x)| w_\alpha(x) dx$$

$$+ \|(f - p_{n_s}) w_\alpha\|_1 + \int_{4\theta n}^\infty |p_{n_s}| w_\alpha dx.$$

The first term on the right hand side is bounded by (2.10), since $f - T_{r-1} \equiv f - f_n$ for $x > 4\theta n$. A bound for the next one is given by Jackson theorem (see [1]), and finally for the last term, using Lemma 1 with $s > \frac{1+\delta}{2\theta}$, we have

$$\int_{4\theta n}^\infty |p_{n_s}(x)| w_\alpha(x) dx \leq A e^{-an} \int_0^\infty |p_{n_s}(x)| w_\alpha(x) dx \leq c e^{-an} \|f w_\alpha\|_1.$$

The proof is then completed.

□

REMARK 1. *Incidentally we notice that bound (2.17) is the best existing in the literature; moreover, it cannot be improved since its order coincides with that of the best polynomial approximation in $W_r, r \geq 1$ (see [1]).*

The surprising aspect of (2.17) is that to obtain this result, in the quadrature sum we need to consider only a fraction of the nodes, namely, those contained in the interval $(0, 4\theta n)$, where $0 < \theta < 1$ is any a priori chosen constant. For instance, if we choose $\theta = \frac{1}{4}$, as we will do in the numerical examples we will present in the next section, then only a little bit more than half of the nodes are included in the above interval.

We also remark that the bound stated in Corollary 2.4 is of the same order of corresponding ones obtained in [4] and [5] for the complete Gauss-Laguerre rule. However (2.17) holds under milder smoothness requirements for the function $f(x)$.

3. Numerical examples. Corollary 2.4 gives a partial answer to the original question arised by Barrucand. It states that for functions $f \in W_r, r \geq 1$, the truncated rule (1.2), including only the (n out of m) nodes belonging to an interval of the form $(0, 4\theta m), 0 < \theta < 1$, has a better order of convergence than the one of the complete formula (1.1) (see [5, Remark 3.2]). This holds no matter how small is chosen the fixed value of θ.

Obviously, in practical computation we are interested in the behaviour of our truncated rule for finite (possibly low or moderate) values of m. Therefore in the following we apply our new rule, with $\alpha = 0$ and taking $\theta = \frac{1}{4}$, to the following test functions,

$$f_1(x) = \cos(\log(1+x)), \qquad f_2(x) = \log(1+x),$$

$$f_3(x) = \frac{x^8}{(x-2)^2+4}, \qquad f_4(x) = e^{\frac{x}{10}},$$

$$f_5(x) = \frac{e^{\frac{x}{10}}}{1+x}, \qquad f_6(x) = \frac{e^{\frac{x}{4}}}{(x-2)^2+1},$$

and we will denote by I_i, $i = 1, \ldots, 6$ the corresponding integrals.

$\mathbf{I_1} = \mathbf{0.7621991262011660}$				
m	n	relative error	n	relative error
2	2	1.9E−02	1	2.9E−03
4	4	2.4E−03	2	1.0E−02
8	8	7.1E−05	5	1.7E−04
16	16	4.2E−07	10	4.5E−07
32	32	2.7E−10	19	2.7E−10
64	64	5.4E−15	39	5.4E−15

$\mathbf{I_2} = \mathbf{0.5963473623231941}$				
m	n	relative error	n	relative error
2	2	2.5E−03	1	3.4E−01
4	4	1.8E−03	2	1.1E−01
8	8	5.1E−05	5	3.3E−04
16	16	3.5E−07	10	2.6E−07
32	32	3.4E−10	19	3.4E−10
64	64	2.1E−14	39	2.1E−14

$\mathbf{I_3} = \mathbf{1223.766216323268}$				
m	n	relative error	n	relative error
2	2	6.3E−01	1	1.0E+00
4	4	8.4E−03	2	1.0E+00
8	8	2.9E−04	5	1.8E−01
16	16	4.5E−05	10	9.4E−04
32	32	6.3E−07	19	6.7E−07
64	64	2.1E−09	39	1.1E−09
128	128	2.1E−13	78	2.1E−13

$\mathbf{I_4} = \mathbf{1.1111111111111}$				
m	n	relative error	n	relative error
2	2	2.0E−05	1	0.2E+00
4	4	2.1E−10	2	5.6E−02
8	8	−−	5	2.4E−04
16	16	−−	10	1.2E−07
32	32	−−	19	7.7E−13
64	64	−−	39	−−

$\mathbf{I_5} = \mathbf{0.7205015065870451}$				
m	n	relative error	n	relative error
2	2	2.7E−02	1	1.4E−01
4	4	3.3E−03	2	3.4E−02
8	8	1.4E−04	5	3.0E−04
16	16	1.5E−06	10	1.7E−06
32	32	2.2E−09	19	2.2E−09
64	64	2.0E−13	39	2.0E−13
128	128	−−	78	−−

$\mathbf{I_6} = \mathbf{0.6241662768900000}$				
m	n	relative error	n	relative error
2	2	2.9E−01	1	4.7E−01
4	4	1.3E−01	2	1.1E−01
8	8	1.5E−02	5	1.5E−02
16	16	6.5E−03	10	6.5E−03
32	32	7.6E−04	19	7.6E−04
64	64	3.5E−05	39	3.5E−05
128	128	2.8E−07	78	2.8E−07
256	256	1.0E−12	156	1.0E−12

By examining the Tables above we notice that the truncated rule performs better when the function $f(x)$ is bounded or mildly increasing, or when it has poles close to the interval of integration. Further investigations are needed.

REFERENCES

[1] M.C.De Bonis, G.Mastroianni, M.Viggiano. K-functionals, moduli of smoothness and weighted best approximation on the semiaxis, *submitted.*

[2] Z.Ditzian, V.Totik. *Moduli of Smoothness*, SCMG Springer-Verlag, New York, 1987.

[3] I.Joò, N.X.Ky. Answer to a problem of Paul Turán, *Ann. Univ. Sci. Budapest Sect. Math.* **31** (1991) 229–241.

[4] G.Mastroianni, G.Monegato. *Error estimates for Gauss–Laguerre and Gauss–Hermite quadrature formulas*, Approximation and Computation (International Series in Numerical Mathematics 119) Basel, Birkhäuser (1994), 421–434, see also the errata corrige in Internal Report n.3/1996, Dipartimento di Matematica, Politecnico di Torino.

[5] G.Mastroianni, G.Monegato. Nyström interpolants based on the zeros of Laguerre polynomials for some Weiner-Hopf equations, *IMA Journal of Numerical Analysis* **17**, no.4 (1997) 621–642.

[6] G.Mastroianni, D.Occorsio. Lagrange interpolation at Laguerre zeros in some weighted uniform spaces, *Acta Math. Hung.* (to appear).

[7] E.B.Saff, V.Totik, *Logarithmic Potentials with External Fields*, Springer-Verlag, New York, 1997.

SPECTRAL FACTORIZATION OF BI-INFINITE BLOCK TOEPLITZ MATRICES WITH APPLICATIONS

CORNELIS V.M. VAN DER MEE[†§], GIUSEPPE RODRIGUEZ[‡§], AND SEBASTIANO SEATZU[†§]

Abstract. In this paper we review some numerical methods for the computation of the spectral factorization of bi-infinite Toeplitz matrices with both scalar and block symbols. We also discuss the spectral factorization of the inverse of bi-infinite block Toeplitz matrices with a finite norm with respect to suitable weight sequences. The results on the spectral factors and their inverses are then applied to the solution of semi-infinite block Toeplitz linear systems and to the identification of the limiting profile in the Gram-Schmidt orthonormalization process, applied to increasing intervals, of a sequence of function vectors generated by integer translates of a given function vector. As applications of the methods proposed, the numerical solution of the Poisson equation on an infinite strip and the identification of the limiting profile in the asymptotic orthonormalization of a vector of B-splines are then illustrated. Finally we discuss the extension of the theory to the spectral factorization of bi-infinite block Toeplitz matrices.

1. Introduction. In this article we give a review of numerical methods for the spectral factorization for bi-infinite block Toeplitz matrices and some of its applications. As a start we list some definitions.

Let $\mathbb{Z}$ be the set of all integers and $\mathbb{Z}_+$ the set of non-negative integers. By a semi-infinite (resp., bi-infinite) block Toeplitz matrix of order k we mean a matrix $A = (A_{i-j})$, indexed by $\mathbb{Z}_+$ (resp., by $\mathbb{Z}$), whose elements are real $k \times k$ matrices where $k = 1, 2, \ldots$ does not depend on (i, j). Block Toeplitz matrices of order 1 are called Toeplitz matrices. Next, let $A = (A_j)_{j\in\mathbb{Z}}$ be a sequence of real $k \times k$ matrices, where $k \in \mathbb{N}$ does not depend on j. Taking a sequence of positive numbers $\boldsymbol{\beta} = (\beta_j)_{j\in\mathbb{Z}}$ satisfying $\beta_{i+j} \leq \beta_i\beta_j$ for all $i, j \in \mathbb{Z}$, by $\ell^k_{1,\beta}(\mathbb{Z})$ we mean the space of all bi-infinite sequences $A = (A_j)_{j\in\mathbb{Z}}$ of real $k \times k$ matrices for which

$$\|A\|_{1,\beta} := \sum_{j\in\mathbb{Z}} \beta_j \|A_j\| < \infty, \tag{1.1}$$

where $\|\cdot\|$ denotes an arbitrary $k \times k$ matrix norm. The most common choices of $\boldsymbol{\beta}$ are $\beta_j = (1 + |j|)^\rho$ for $\rho > 0$ and $\beta_j = g^{|j|}$ for $g > 1$, which correspond to algebraic and exponential weights, respectively. We write $\ell^k_1(\mathbb{Z})$ if $\beta_i = 1$ for all $i \in \mathbb{Z}$.

Spectral factorizations play a crucial role in the solution of semi-infinite Toeplitz systems of the type

$$\sum_{j\in\mathbb{Z}_+} T_{i-j}x_j = b_i, \quad i \in \mathbb{Z}_+. \tag{1.2}$$

Given a weight sequence $\boldsymbol{\beta}$ and a sequence of $k \times k$ matrices $(T_h)_{h\in\mathbb{Z}}$, with finite norm $\|(T_h)_{h\in\mathbb{Z}_+}\|_{1,\beta}$, we seek solutions $(x_h)_{h\in\mathbb{Z}}$ of (1.2) satisfying $\|(x_h)_{h\in\mathbb{Z}_+}\|_{1,\beta} < \infty$ for right-hand sides $(b_h)_{h\in\mathbb{Z}}$ satisfying $\|(b_h)_{h\in\mathbb{Z}_+}\|_{1,\beta} < \infty$. Here b_h and x_h are real column vectors of length k for every $h \in \mathbb{Z}_+$.

[†]Dipartimento di Matematica, Università di Cagliari, viale Merello 92, 09123 Cagliari, Italy. email: `cornelis@krein.unica.it`, `seatzu@unica.it`

[‡]Dipartimento di Matematica, Università di Roma *Tor Vergata*, via della Ricerca Scientifica, 00133 Roma, Italy. email: `rodriguez@mat.uniroma2.it`

[§]Research supported by CNR-GNIM and MURST.

The paper is organized as follows. In Section 2 we define the spectral factorization of bi-infinite block Toeplitz matrices. In Section 3 we discuss various numerical factorization methods for Toeplitz and block Toeplitz matrices. In Section 4 we recall Krein's method for solving the linear system (1.2), emphasizing the crucial role played by the spectral factorization of the inverse of a bi-infinite block Toeplitz matrix. In Section 5 we show that the identification of the limiting profile in the Gram-Schmidt orthonormalization process, applied to the uniform translates of a vector of functions defined on increasing intervals, essentially depends on the spectral factorization of a bi-infinite block Toeplitz matrix. In Section 6 we illustrate the application of the spectral factorization methods to the numerical solution of some semi-infinite block Toeplitz systems. In Section 7 we discuss generalizations to multi-index Toeplitz and block Toeplitz matrices.

2. Preliminaries. In this section we first introduce the Banach algebra $\mathcal{W}^k$ and indicate its invertible elements. We then go on to define spectral factorizations of bi-infinite block Toeplitz matrices and relate them to factorization properties in $\mathcal{W}^k$. Finally, we discuss factorization results for certain classes of bi-infinite and semi-infinite block Toeplitz matrices.

The class of matrix functions

$$\widehat{T}(z) = \sum_{j\in\mathbb{Z}} z^j T_j \tag{2.1}$$

on the unit circle $\mathcal{T}$ for which

$$\|\widehat{T}\|_{1,\beta} := \|(T_h)_{h\in\mathbb{Z}}\|_{1,\beta} = \sum_{j\in\mathbb{Z}} \beta_j \|T_j\| < \infty,$$

is a Banach algebra [10] denoted as $\mathcal{W}^k_\beta$. This means that its norm satisfies the inequality

$$\|\widehat{T}\|_{1,\beta} \le \|\widehat{T}^{(1)}\|_{1,\beta}\|\widehat{T}^{(2)}\|_{1,\beta}; \qquad T_i = \sum_{j\in Z} T^{(1)}_{i-j} T^{(2)}_j, \quad i \in \mathbb{Z}.$$

We write $\mathcal{W}^k$ if $\beta_i = 1$ for all $i \in \mathbb{Z}$; this particular algebra is called the Wiener algebra (of order k). We call $\widehat{T}(z)$ the *symbol* of the block Toeplitz matrix $T = (T_{i-j})_{i,j\in\mathbb{Z}}$.

For $k = 1$ the invertible elements of $\mathcal{W}^k_\beta$ are completely described by the Gelfand theory of commutative Banach algebras [20, 10]. These results extend to matrix functions [10]. We have the following result.

THEOREM 2.1. *Let $\hat{T}(z)$ belong to $\mathcal{W}^k_\beta$ and have the form* (2.1). *Put $\beta_\pm = \lim_{j\to\pm\infty} \beta_j^{1/|j|}$ Then if $\hat{T}(z)$ is a nonsingular $k \times k$ matrix for all z satisfying $(1/\beta_-) \le |z| \le \beta_+$, $\hat{T}(z)^{-1}$ belongs to $\mathcal{W}^k_\beta$.*

In particular, if $\beta_j = g^{|j|}$ for some $g > 1$, then $\hat{T}(z)$ is an invertible element of $\mathcal{W}^k_\beta$ if and only if $\hat{T}(z)$ is a nonsingular matrix for $1/g \le |z| \le g$. Analogously, if $\beta_j = (1 + |j|)^\alpha$ for some $\alpha \ge 0$, then $\hat{T}(z)$ is an invertible element of $\mathcal{W}^k_\beta$ if and only if $\hat{T}(z)$ is a nonsingular matrix for $|z| = 1$.

By a *spectral factorization* of T (with respect to the weight sequence β) we mean a representation of T in any of the two forms

$$T = LDM^T \quad \text{and} \quad T = UDV^T, \tag{2.2}$$

where the superscript T denotes matrix transposition and L, M, D, U and V are bi-infinite Toeplitz matrices of order k having the following properties:

1. $L_0 = M_0 = U_0 = V_0 = I_k$ (the $k \times k$ identity matrix);

2. $D_s = 0$ for $s \neq 0$ and $L_s = M_s = 0$ for $s < 0$ and $U_s = V_s = 0$ for $s > 0$;

3. the inverses L^{-1}, M^{-1}, U^{-1} and V^{-1} of L, M, U and V are bi-infinite block Toeplitz matrices satisfying $[L^{-1}]_s = [M^{-1}]_s = 0$ for $s < 0$ and $[U^{-1}]_s = [V^{-1}]_s = 0$ for $s > 0$;

4. the matrix sequences with entries L_s, M_s, U_s, V_s, $[L^{-1}]_s$, $[M^{-1}]_s$, $[U^{-1}]_s$ and $[V^{-1}]_s$ belong to $\ell^k_{1,\beta}(\mathbb{Z})$.

A given bi-infinite block Toeplitz matrix has at most one spectral factorization of the type $T = LDM^T$ and at most one spectral factorization of the type $T = UDV^T$ as in (2.2).

Given a positive definite bi-infinite block Toeplitz matrix, setting $\mathcal{L}_s = L_s D_0^{1/2}$ or $\mathcal{U}_s = U_s D_0^{1/2}$, we obtain the *block Cholesky factorization* $T = \mathcal{L}\mathcal{L}^T$ or the *block Wiener-Hopf factorization* $T = \mathcal{U}\mathcal{U}^T$, where $\mathcal{L} = (\mathcal{L}_{i-j})_{i,j\in\mathbb{Z}}$ and $\mathcal{U} = (\mathcal{U}_{i-j})_{i,j\in\mathbb{Z}}$ are invertible on $\ell^k_{1,\beta}(\mathbb{Z})$ with inverses $\mathcal{L}^{-1}$ and $\mathcal{U}^{-1}$, the sequences $(\mathcal{L}_s)_{s\in\mathbb{Z}}$, $([\mathcal{L}^{-1}]_s)_{s\in\mathbb{Z}}$, $(\mathcal{U}_s)_{s\in\mathbb{Z}}$ and $([\mathcal{U}^{-1}]_s)_{s\in\mathbb{Z}}$ belong to $\ell^k_{1,\beta}(\mathbb{Z})$, and $\mathcal{L}_0$ and $\mathcal{U}_0$ are positive definite hermitian.

Passing to the symbols $\widehat{L}(z)$, $\widehat{D}(z) \equiv D_0$ and $\widehat{M}(z)$ of L, D and M, [resp., $\widehat{U}(z)$, $\widehat{D}(z) \equiv D_0$ and $\widehat{V}(z)$ of U, D and V], we have

$$\widehat{T}(z) = \widehat{L}(z) D_0 \widehat{M}(z)^T, \quad \left[\widehat{T}(z) = \widehat{U}(z) D_0 \widehat{V}(z)^T\right], \quad |z| = 1, \tag{2.3}$$

where $\widehat{L}(z)$ and $\widehat{M}(z)$ [$\widehat{U}(z)$ and $\widehat{V}(z)$] extend to matrix functions that are continuous on the closed unit disk $\{z \in \mathbb{C} : |z| \leq 1\}$ and analytic on the open unit disk. Further, $\widehat{L}(z)$ and $\widehat{M}(z)$ [resp., $\widehat{U}(z)$ and $\widehat{V}(z)$] are nonsingular matrices for $|z| \leq 1$ and $\widehat{L}(0) = \widehat{M}(0) = I_k$ [resp., $\widehat{U}(0) = \widehat{V}(0) = I_k$].

Positive definite matrix functions in $\mathcal{W}^k$, matrix functions $\hat{T}(z)$ in $\mathcal{W}^k$ such that $\hat{T}(z) + \hat{T}(z)^*$ is positive definite for all $z \in \mathcal{T}$ (where the asterisk denotes the conjugate transpose), and matrix functions $\widehat{T}(z)$ in $\mathcal{W}^k$ such that $\|I - \widehat{T}(z)\| < 1$ for all $z \in \mathcal{T}$ (the norm representing the largest singular value of $I - \widehat{T}(z)$) have factorizations of the type (2.3) (cf. [7]). When the Toeplitz matrix T of order k is banded (i.e., if $T_{i-j} = 0$ whenever $|i-j| \geq m$ for some integer m), necessary and sufficient conditions for the existence of the spectral factorization (2.3) as well as a numerical procedure for its calculation have been given [24], based on matrix polynomial theory (e.g., [29]).

We have the following result.

THEOREM 2.2. *Suppose the bi-infinite block Toeplitz matrix* $T = (T_i)_{i\in\mathbb{Z}}$ *with symbol in* $\mathcal{W}^k$ *has a spectral factorization of the type* (2.1) *and* $\hat{T}(z)$ *is a nonsingular matrix for* $(1/\beta_-) \leq |z| \leq \beta_+$. *Then the factors and their inverses have symbols belonging to* $\mathcal{W}^k_\beta$.

The above theorem is most relevant to banded matrices T where $T_i = 0$ for $|i| > m$. Suppose one can find an annulus $1/g \leq |z| \leq g$ with $g > 1$ on which $\det \hat{T}(z) \neq 0$. Then if T has a spectral factorization, the factors and their inverses have symbols belonging to $\mathcal{W}^k_\beta$ for $\beta_j = g^{|j|}$.

3. Numerical factorization methods. In this section we discuss five well-known numerical factorization methods for bi-infinite banded Toeplitz matrices with positive definite symbol. We then go on to describe the corresponding methods for block Toeplitz matrices. Finally, we illustrate two factorization methods for computing the inverses of block Toeplitz matrices that we will use later to solve a differential boundary value problem on a semi-infinite strip and to find the limiting profile in the Gram-Schmidt orthonormalization process applied to a semi-infinite sequence of function vectors.

3.1. The scalar Toeplitz case. Let us review five methods for the spectral factorization of banded and real symmetric Toeplitz matrices [18, 23].

Let A be a bi-infinite, real symmetric and banded Toeplitz matrix

$$A = \text{Toeplitz}(A_{-m}, \ldots, A_0, \ldots, A_m),$$

where $A_j = \bar{A}_j = A_{-j}$, $j = 0, 1, 2, \ldots, m$. When its symbol $\hat{A}(z) > 0$ for $z \in \mathcal{T}$, the problem is to find real numbers $\Gamma_0, \Gamma_1, \ldots, \Gamma_m$ such that

$$\hat{\Gamma}(z) = \sum_{j=0}^{m} \Gamma_j z^j$$

has all of its roots outside the unit disk and such that

$$\hat{\Gamma}(z)\hat{\Gamma}(z^{-1}) = \hat{A}(z), \quad z \in \mathcal{T}. \tag{3.1}$$

To guarantee the uniqueness of the factorization we impose the normalization $\hat{\Gamma}(0) > 0$ on the polynomial $\hat{\Gamma}(z)$. The existence of such a factorization is well-known and generally attributed to Fejér [cf. [27], Sec. 53].

The first method we wish to describe is due to F. Bauer [2, 3]. As $\hat{A}(z) > 0$ for $z \in \mathcal{T}$, the matrix A is positive definite on $\ell_2(\mathbb{Z})$. Moreover, the factorization (3.1) corresponds to the factorization

$$A = \Gamma\Gamma^T, \qquad \Gamma = (\Gamma_{i-j})_{i,j\in\mathbb{Z}}.$$

Let A_+ be the semi-infinite compression of A given by

$$A_+ = (A_{ij})_{i,j\in\mathbb{Z}_+}$$

and A_n, $n \in \mathbb{Z}_+$, be the sequence of finite compressions of A given by

$$A_n = (A_{ij})_{i,j=0,1,\ldots,n}, \quad n = 0, 1, 2, \ldots.$$

Then each A_n has a unique Cholesky factorization $L_n L_n^T = A_n$, where L_n is a lower triangular matrix with positive diagonal elements. As we replace n by $n+1$, the matrix A_{n+1} agrees with A_n on its first $n+1$ rows and columns. Likewise L_{n+1} has its first $n+1$ rows and columns equal to those of L_n. Thus we may consider L_n as the n-th finite section of a semi-infinite matrix L which is the unique Cholesky factorization of A_+, i.e. $A_+ = LL^T$. Bauer proved that $(L_n)_{ij} \to \Gamma_{i-j}$ as $n \to \infty$, and in [17] it has been proved that these elements approach those of Γ exponentially, i.e., there exist $c > 0$ and $\rho \in (0, 1)$ such that for all $i, j \in \mathbb{Z}_+$

$$|L_{ij} - \Gamma_{i-j}| \leq c\rho^i.$$

Based on this observation, it is straightforward to justify the corresponding algorithm.

The second method we consider is due to Wilson [31] (cf. [18]). His idea is to write equation (3.1) as the equivalent system of quadratic equations

$$\sum_{j=0}^{m-i} \Gamma_j \Gamma_{j+i} = A_i, \quad i = 0, 1, \ldots, m, \tag{3.2}$$

and to use a Newton-Raphson method for solving the system (3.2). Wilson proved that, if $\hat{A}(z)$ has no zeros on the unit circle, for a suitable and easily made choice of the starting values $\Gamma_0^{(0)}, \Gamma_1^{(0)}, \ldots, \Gamma_m^{(0)}$, the iteration is self-correcting and converges quadratically to the required solution. For details on the algorithm we refer to [18].

The third method is based on the computation of the zeros of the Laurent polynomial $\hat{A}(z)$. To improve the effectiveness of the method, in view of the symmetry of its coefficients, we write $\hat{A}(z)$ in the form

$$\hat{A}(z) = A_0 + \sum_{j=1}^{m} A_j \left(z^j + z^{-j}\right)$$

and make the change of variable $w = z + z^{-1}$ to obtain the polynomial

$$\hat{C}(w) := \hat{A}(z(w)) = \sum_{j=0}^{m} C_j w^j$$

whose coefficients $C_0, C_1, \ldots, C_m$ can be expressed recursively in the original coefficients $A_0, A_1, \ldots, A_m$. The zeros w_i, $i = 1, 2, \ldots, p$, of the polynomial $\hat{C}(w)$ can then be evaluated by computing the eigenvalues of its companion matrix by a QR method [16]. Next, we solve the m quadratic equations

$$z + z^{-1} = w_i, \quad i = 1, 2, \ldots, m,$$

and then, taking into account their multiplicity, we obtain the $2m$ roots z_i, $i = 1, 2, \ldots, 2m$, of $\hat{A}(z)$ ordered by decreasing modulus. The polynomial $\hat{\Gamma}(z)$ we are looking for can then be constructed by multiplying the linear factors corresponding to all the roots z_j (including $\overline{z}_j$) outside the unit circle.

The fourth method tested in [18] is the minimum phase factorization. The basic idea of the algorithm is based on the following observation. Let $\hat{A}(z)$ have no zeros on the unit circle. We consider the minimization problem

$$\min \int_{-\pi}^{\pi} \frac{|G(e^{i\theta})|^2}{\hat{A}(e^{i\theta})} \, d\theta,$$

where G runs through all polynomials of degree $\leq m$ and it is normalized by $G(0) = 1$. Then the solution of this minimization problem is given by

$$G_{\text{opt}}(z) = \frac{\hat{\Gamma}(z)}{\hat{\Gamma}(0)},$$

where $\hat{\Gamma}$ is the polynomial in (3.1) we seek.

The final method we review is very popular in signal processing and goes by the name of the *cepstral* algorithm [4, 26]. It is based on some results on the factorization of an absolute convergent Fourier series on the unit circle, discovered independently by M.G. Krein [22] and by A. Calderón et al. [6]. Let us discuss the basic idea underlying this method.

As the Laurent polynomial $\hat{A}(z)$ is strictly positive on the unit circle, by the Wiener-Lévy theorem on trigonometric series, the function $\log \hat{A}(z)$ can be written as an absolutely convergent Laurent series

$$B(z) := \log \hat{A}(z) = \sum_{j \in \mathbb{Z}} B_j z^j.$$

Moreover, as those of $\hat{A}(z)$, the coefficients of $B(z)$ are even in j and decay exponentially, since $\log \hat{A}(z)$ has an extension as an analytic function in some annulus $r^{-1} < |z| < r$, $r > 1$. Splitting the Laurent series in the form

$$B(z) = B_-(z) + B_+(z),$$

where

$$B_\pm(z) := \sum_{j=0}^{\infty} \left(1 - \frac{1}{2}\delta_{j,0}\right) B_{\pm j} z^{\pm j}, \quad |z^\pm| \leq 1,$$

we obtain the factorization (3.1) of the required type, where

$$\hat{\Gamma}(z) := \exp(B_+(z))$$

is analytic and zero free in the disk $\{z : |z| \leq r\}$ and normalized so that $\hat{\Gamma}(0) > 0$. Note that $\hat{\Gamma}$ above is indeed a polynomial of degree at most m. A Matlab implementation of this method has been proposed in [1].

In [18] it was found that there is a huge disparity between the methods, and that all of them except for the Wilson method are significantly affected by the variation of the coefficients of the Laurent polynomial in magnitude, by the vicinity of the zeros of this polynomial to the unit circle and by their spacing.

3.2. The block Toeplitz case. Among the spectral factorization methods for scalar positive definite symbols, only Bauer's method allows for a straightforward generalization to bi-infinite block Toeplitz matrices with positive definite symbols. The generalization of Wilson's method to the block Toeplitz case, as formulated in [32], cannot be implemented in a straightforward way, since the algebraic equations for symbols appearing in [32] cannot be easily converted to linear matrix equations for their coefficients. Further, no extension of the fourth method above to the block Toeplitz case has ever been numerically implemented. Finally, the cepstral method cannot be generalized to the block Toeplitz case either, since it is ultimately based on the additivity property of the logarithm.

Let us now discuss the UDL factorization of the banded bi-infinite block Toeplitz matrix A of the form

$$A = L^T DR, \tag{3.3}$$

where $L = (L_{i-j})_{i,j\in\mathbb{Z}}$, $R = (R_{i-j})_{i,j\in\mathbb{Z}}$ and $D = (D_{i-j})_{i,j\in\mathbb{Z}}$ are banded block Toeplitz matrices with $L_0 = R_0 = I_k$ (the $k \times k$ identity matrix), $L_s = -\mathcal{L}_s$ ($s = 1, \ldots, m$) and $R_s = -\mathcal{R}_s$ ($s = 1, 2, \ldots, n$), $D_0 = \mathcal{D}$, $D_s = 0$ for $s \neq 0$, $L_s = 0$ for $s \neq 0, 1, \ldots, m$, and $R_s = 0$ for $s \neq 0, 1, \ldots, n$. In other words, L and R are lower block-triangular matrices and D is a block-diagonal matrix. Further, let the matrix function

$$\Sigma(z) = \sum_{j=-m}^{n} z^j A_j \tag{3.4}$$

be the *symbol* associated to the matrix A, and let us assume A_n to be nonsingular. Then $\mathcal{P}(z) = z^m A_n^{-1}\Sigma(z)$ is a matrix polynomial of degree $m + n$ whose leading coefficient is the identity matrix I_k.

Let us consider a simple closed positively oriented rectifiable Jordan curve Γ, with $0 \notin \Gamma$, dividing the complex plane into an *interior* bounded domain Ω_+ with $0 \in \Omega_+$ and an *exterior* domain Ω_-, and assume that $\det(z^m\Sigma(z))$ does not vanish for $z \in \Gamma$. Then [29] the matrix factorization (3.3) is equivalent to the factorization

$$\Sigma(z) = \left(I_k - \sum_{i=1}^{m} z^{-i}\mathcal{L}_i\right)^T \mathcal{D}\left(I_k - \sum_{j=1}^{n} z^j\mathcal{R}_j\right), \quad z \in \Gamma, \tag{3.5}$$

of the symbol of A. The nonsingularity of A_n implies that both $\mathcal{R}_n$ and $\mathcal{D}$ are nonsingular.

Before describing the algorithm to factorize A, we recall some basic properties of matrix polynomials [13, 14, 29]. Consider the matrix polynomial

$$\mathcal{P}(z) = z^\ell I_k + z^{\ell-1}A_{\ell-1} + \ldots + zA_1 + A_0,$$

of degree ℓ whose coefficients are $k \times k$ matrices. We call $z_j \in \mathbb{C}$ an *eigenvalue* of $\mathcal{P}$ if $\det\mathcal{P}(z_j) = 0$. The corresponding *eigenvector* $\mathbf{x}_{j1}$ is a nontrivial vector in $\mathbb{C}^k$ satisfying $\mathcal{P}(z_j)\mathbf{x}_{j1} = 0$. Then, obviously, $\mathcal{P}$ has a *spectrum* of exactly $k\ell$ eigenvalues, taking into account their multiplicities as zeros of $\det\mathcal{P}(z)$. The complex vectors $\{\mathbf{x}_{j1}, \mathbf{x}_{j2}, \ldots, \mathbf{x}_{jr}\}$ constitute a *Jordan chain* at z_j of length r, if $\mathbf{x}_{j1} \neq 0$ and the lower triangular linear system of equations

$$\sum_{v=1}^{u} \frac{\mathcal{P}^{(u-v)}(z_j)}{(u-v)!}\mathbf{x}_{jv} = 0, \quad u = 1, 2, \ldots, r,$$

is valid. The lengths of the Jordan chains at the eigenvalue z_j in a system of maximal Jordan chains are called the *partial multiplicities* of $\mathcal{P}$ at z_j. The sum of the partial multiplicities at z_j coincides with the order of z_j as a zero of $\det\mathcal{P}(z)$ and is called its *algebraic multiplicity*; the dimension of the kernel of $\mathcal{P}(z_j)$ is called its *geometric multiplicity* and is equal to the number of linearly independent maximal Jordan chains at z_j.

A pair of matrices (X, T), where X is of size $k \times k\ell$ and T is of size $k\ell \times k\ell$, is called a *right spectral pair* for the polynomial $\mathcal{P}(z)$ if the matrix

$$\text{col}[XT^j]_{j=0}^{\ell-1} := \begin{bmatrix} X \\ XT \\ \vdots \\ XT^{\ell-1} \end{bmatrix}$$

is invertible and the following equality holds

$$XT^\ell + \sum_{j=0}^{\ell-1} A_j XT^j = 0.$$

By a *left spectral pair* for the polynomial $\mathcal{P}(z)$ we mean a right spectral pair for $\mathcal{P}(z)^T$. The *right canonical form* for $\mathcal{P}(z)$ is given by

$$\mathcal{P}(z) = z^\ell I_k - XT^\ell(V_1 + zV_2 + \ldots + z^{\ell-1}V_\ell),$$

where $V_1, \ldots, V_\ell$ are the $k\ell \times k$ matrices defined by

$$\begin{bmatrix} V_1 & V_2 & \ldots & V_\ell \end{bmatrix} = \left(\text{col}[XT^j]_{j=0}^{\ell-1}\right)^{-1}.$$

For a right spectral pair we have the inversion formula

$$\mathcal{P}(z)^{-1} = X(z-T)^{-1}V_\ell. \tag{3.6}$$

A similar definition holds for the *left canonical form.*

Denoting by $z_1, \dots, z_p$ the *distinct* eigenvalues of $\mathcal{P}(z)$, each of algebraic multiplicity m_j, it can be shown [29] that a right spectral pair (X, T) for $\mathcal{P}(z)$ is given by

$$X = \begin{bmatrix} X_1 & X_2 & \dots & X_p \end{bmatrix}, \quad T = T_1 \oplus T_2 \oplus \dots \oplus T_p, \tag{3.7}$$

where the matrices X_j, of size $k \times m_j$, and T_j, of size $m_j \times m_j$, are given by

$$X_j = \begin{bmatrix} \mathbf{x}_{11}^{(j)} & \dots & \mathbf{x}_{1r_{j1}}^{(j)} & \mathbf{x}_{21}^{(j)} & \dots & \mathbf{x}_{2r_{j2}}^{(j)} & \dots & \mathbf{x}_{q_j1}^{(j)} & \dots & \mathbf{x}_{q_jr_{jq_j}}^{(j)} \end{bmatrix},$$
$$T_j = J_{r_{j1}}(z_j) \oplus J_{r_{j2}}(z_j) \oplus \dots \oplus J_{r_{jq_j}}(z_j).$$

Here $\mathbf{x}_{s1}^{(j)}, \mathbf{x}_{s2}^{(j)}, \dots, \mathbf{x}_{sr_{js}}^{(j)}$, $s = 1, \dots, q_j$, are the maximal Jordan chains for $\mathcal{P}(z)$ corresponding to z_j, $J_{r_s}(z_j)$ is the $r_s \times r_s$ upper triangular Jordan block with eigenvalue z_j, and $r_{j1} + r_{j2} + \dots + r_{jq_j} = m_j$ $(j = 1, \dots, p)$.

To obtain the factorization the following theorem is crucial.

THEOREM 3.1. *Let $z_1, \dots, z_s$ be the distinct zeros of* $\det(z^m\Sigma(z))$ *in Ω_- and let $z_{s+1}, \dots, z_{s+t}$ be its distinct zeros in Ω_+. Moreover, let (X, T) be a right spectral pair of the matrix polynomial $\mathcal{P}(z) = z^m A_n^{-1}\Sigma(z)$. Then there exists a factorization of $\Sigma(z)$ of the type* (3.5) *where* $\det(I_k - \sum_{i=1}^m z^{-i}\mathcal{L}_i) \neq 0$ *for $z \in \Omega_-$ and* $\det(I_k - \sum_{j=1}^n z^j\mathcal{R}_j) \neq 0$ *for $z \in \Omega_+$, if and only if*

$$m_1 + \dots + m_s = nk, \quad m_{s+1} + \dots + m_{s+t} = mk,$$

and the restriction of $\mathrm{col}[XT^j]_{j=0}^{n-1}$ *to the linear span of the eigenvectors and generalized eigenvectors of T in Ω_+ is invertible. This factorization is unique and is called the spectral factorization of A.*

Following the proof of Theorem 2.1 of [24], the explicit construction of the factorization (3.5) has been given in [25] and will be presented here in abbreviated form. Indeed, as specified in (3.7), we construct the right spectral pair (X_R, T_R) of the matrix polynomial $\mathcal{P}(z) = z^m A_n^{-1}\Sigma(z)$ by using its Jordan chains, which coincide with the Jordan chains of $z^m\Sigma(z)$, and its left spectral pair (X_L, T_L) in a similar fashion. Partitioning the $k \times (m+n)k$ matrices X_L and X_R into a $k \times nk$ block and a $k \times mk$ block, we have

$$X_L = \begin{bmatrix} ? & V \end{bmatrix}, \quad X_R = \begin{bmatrix} W & ? \end{bmatrix},$$
$$T_L = T_R = \bigoplus_{j=1}^{s+t} \left(J_{r_{j1}}(z_j) \oplus \dots \oplus J_{r_{jq_j}}(z_j)\right),$$

where the matrices at the question marks are irrelevant and $r_{j1} + \dots + r_{jq_j} = m_j$ $(j = 1, \dots, s+t)$. We now set

$$\Lambda_L = \bigoplus_{j=s+1}^{s+t} \left(J_{r_{j1}}(z_j) \oplus \dots \oplus J_{r_{jq_j}}(z_j)\right),$$
$$\Lambda_R = \bigoplus_{j=1}^{s} \left(J_{r_{j1}}(z_j) \oplus \dots \oplus J_{r_{jq_j}}(z_j)\right),$$

and construct the $mk \times k$ matrices $W_1, \dots, W_m$ and the $nk \times k$ matrices $V_1, \dots, V_n$ by putting

$$\begin{bmatrix} W_1 & W_2 & \dots & W_m \end{bmatrix} = \left(\mathrm{col}[V\Lambda_L^j]_{j=0}^{m-1}\right)^{-1},$$
$$\begin{bmatrix} V_1 & V_2 & \dots & V_n \end{bmatrix} = \left(\mathrm{col}[W\Lambda_R^j]_{j=0}^{n-1}\right)^{-1}.$$

The coefficients of the factorization in (3.3) are then given by

$$\begin{aligned}
\mathcal{L}_i &= (A_n^T)^{-1}\, V\Lambda_L^m W_{m-i+1}\, A_n^T, & i &= 1, \dots, m,\\
\mathcal{R}_j &= -\,(W\Lambda_R^n V_1)^{-1}\, W\Lambda_R^n V_{j+1}, & j &= 1, \dots, n-1,\\
\mathcal{R}_n &= (W\Lambda_R^n V_1)^{-1}, & &\\
\mathcal{D} &= -(\mathcal{L}_m^T)^{-1} A_{-m} = -A_n \mathcal{R}_n^{-1}, & &
\end{aligned}$$

which corresponds to taking right spectral pairs (V, Λ_L) and (W, Λ_R) for the matrix polynomials $\mathcal{L}(z)(A_n^T)^{-1}$ and $\mathcal{R}(z)$, respectively, and recovering the spectral factors from their canonical forms.

Assuming $\Gamma = \{z \in \mathbb{C} : |z| = \rho\}$, $\Omega_+ = \{z \in \mathbb{C} : |z| < \rho\}$ and $\Omega_- = \{z \in \mathbb{C} : |z| > \rho\}$, $\Sigma(z)$ has the factorization (3.5) for $z \in \Gamma$, where

$$\begin{aligned}
\left(I_k - \sum_{i=1}^{m} z^{-i}\mathcal{L}_i\right)^{-1} &= z^m (A_n^T)^{-1} V (z - \Lambda_L)^{-1} W_m A_n^T\\
&= \sum_{\mu=0}^{\infty} z^{m-\mu-1} (A_n^T)^{-1} V \Lambda_L^{\mu} W_m A_n^T
\end{aligned}$$

for $z \in \Omega_-$ and

$$\left(I_k - \sum_{j=1}^{n} z^j \mathcal{R}_j\right)^{-1} = W(I_k - z\Lambda_R^{-1})^{-1}\Lambda_R^{-1} V_n \mathcal{R}_n^{-1} = \sum_{\mu=0}^{\infty} z^{\mu} W \Lambda_R^{-(\mu+1)} V_n \mathcal{R}_n^{-1}$$

for $z \in \Omega_+$.

When $\Sigma(z)$ is the symbol corresponding to a real matrix A which is symmetric ($m = n$, $A_{-j} = A_j^T$), positive definite, block Toeplitz and banded, we can assume the curve Γ to be the unit circle, so that Theorem 3.1 assures that there exists only one factorization of A of the form

$$A = UU^T,$$

where $U = L^T D^{1/2}$. This is often referred to as the *Wiener-Hopf factorization* of A. In this case the matrix D in (3.3) is real positive definite, as well as A_0 and $\mathcal{D}$. Moreover, $\mathcal{L}_j = \mathcal{R}_j$ and the symbol

$$\mathcal{U}(z) := \sum_{j=0}^{m} z^{-j}\mathcal{U}_j$$

of the matrix U has coefficients defined by

$$\mathcal{U}_0 = \mathcal{D}^{1/2},$$
$$\mathcal{U}_j = -\mathcal{R}_j^T \mathcal{D}^{1/2} = -\mathcal{L}_j^T \mathcal{D}^{1/2}, \qquad j = 1, \ldots, m.$$

The factorization method just discussed has the disadvantage that it can only be used for banded block Toeplitz matrices. We now present a method based on band extension (cf. Section XXXV.3 of [10]) which applies only to positive definite semi-infinite block Toeplitz matrices.

Given the real $k \times k$ matrices $A_{-p}, A_{-p+1}, \ldots, A_p$, we seek a bi-infinite block Toeplitz matrix $\Phi = (\Phi_{i-j})_{i,j \in \mathbb{Z}}$ which is positive definite, satisfies $\Phi_j = A_j$ for $|j| \le p$, and has the property $\sum_{j \in \mathbb{Z}} \|\Phi_j\| < +\infty$. In other words, we seek a positive definite band extension Φ of the banded bi-infinite block Toeplitz matrix with symbol

$$\sum_{j=-p}^{p} z^j A_j. \tag{3.8}$$

The following result indicates when such a so-called *Carathéodory-Toeplitz extension* exists. Moreover, in the theorem we present an algorithm [10] for constructing one particular such extension as well as all such extensions.

THEOREM 3.2. *Given the real $k \times k$ matrices $A_{-p}, A_{-p+1}, \ldots, A_p$, there exists a Carathéodory-Toeplitz extension of the banded bi-infinite block Toeplitz matrix with symbol* (3.8) *if and only if the $k(p+1) \times k(p+1)$ matrix*

$$\Gamma = \begin{bmatrix} A_0 & A_{-1} & \ldots & A_{-p} \\ A_1 & A_0 & \ldots & A_{-p+1} \\ \vdots & \vdots & & \vdots \\ A_p & A_{p-1} & \ldots & A_0 \end{bmatrix} \tag{3.9}$$

is positive definite. In that case there exists a unique solution Φ_b with the property that the inverse of the corresponding bi-infinite block Toeplitz matrix is banded. Defining the $k(p+1) \times k$ matrices

$$\begin{bmatrix} X_0 \\ X_1 \\ \vdots \\ X_p \end{bmatrix} = \Gamma^{-1} \begin{bmatrix} I_k \\ 0 \\ \vdots \\ 0 \end{bmatrix}, \qquad \begin{bmatrix} Y_{-p} \\ \vdots \\ Y_{-1} \\ Y_0 \end{bmatrix} = \Gamma^{-1} \begin{bmatrix} 0 \\ \vdots \\ 0 \\ I_k \end{bmatrix}, \tag{3.10}$$

the bi-infinite block Toeplitz matrices $U = (X_{i-j})_{i,j \in \mathbb{Z}_+}$ and $V = (Y_{i-j})_{i,j \in \mathbb{Z}_+}$ using $X_j = 0$ for $j < 0$ and $j > p$ and $Y_j = 0$ for $j < -p$ and $j > 0$, and the block-diagonal block Toeplitz matrices $\mathcal{D}_X$ and $\mathcal{D}_Y$ with respective diagonal entries X_0 and Y_0, we have

$$\Phi_b^{-1} = U \mathcal{D}_X U^T = V \mathcal{D}_Y V^T. \tag{3.11}$$

Moreover, each real solution Φ of the Carathéodory-Toeplitz extension problem has the property

$$\Phi^{-1} = (VG + U)(I - G^T G)^{-1}(G^T V^T + U^T), \tag{3.12}$$

where $G = (G_{i-j})_{i,j\in\mathbb{Z}_+}$ *is a bi-infinite real block Toeplitz matrix satisfying* $G_j = 0$ *for* $j \leq p$ *which is a strict contraction on* $\ell_2(\mathbb{Z}_+)$. *There is a* $1,1$-*correspondence between real solutions* Φ *and such matrices* G.

The above theorem leads to the following factorization of positive definite bi-infinite block Toeplitz matrices. Let A be a positive definite bi-infinite block Toeplitz matrix. For $p \in \mathbb{Z}_-$ large enough, the bi-infinite block Toeplitz matrix with symbol (3.8) has a positive definite banded extension Φ such that the bi-infinite inverse Φ^{-1} is banded. The factors of the inverse Φ^{-1}, as given by (3.11), are approximations of the factors of the inverse of the original positive definite block Toeplitz matrix.

4. Spectral factorization and the solution of semi-infinite block Toeplitz systems. Let us now recall Krein's method [22] for solving the semi-infinite block Toeplitz system (1.2). Denoting by $\widehat{T}(z)$ the symbol associated to the Toeplitz matrix T, the system (1.2) has a unique solution with discrete Fourier transform in $\mathcal{W}_\beta^k$ if and only if $\widehat{T}(z) \neq 0$ for all z with $(1/\beta_-) \leq |z| \leq \beta_+$ and $\beta_\pm$ as above. Then $\widehat{T}(z)^{-1}$ has the factorization

$$\widehat{T}(z)^{-1} = \Gamma_+(z)\Gamma_-(z) \tag{4.1}$$

in $\mathcal{W}_\beta^k$, where

$$\Gamma_+(z) = \sum_{j\in\mathbb{Z}_+} z^j\Gamma_j^{(1)}, \qquad \Gamma_-(z) = \sum_{j\in\mathbb{Z}_+} z^{-j}\Gamma_j^{(2)}, \tag{4.2}$$

with $\|\Gamma^{(1)}\|_{1,\beta} < \infty$ and $\|\Gamma^{(2)}\|_{1,\beta} < \infty$. One then has the solution formula

$$x_\ell = \sum_{s\in\mathbb{Z}_+} \Gamma_{\ell s} b_s, \quad \ell \in \mathbb{Z}_+, \tag{4.3}$$

for the system of equations (1.2), where

$$\Gamma_{\ell s} = \sum_{h=0}^{\min(\ell,s)} \Gamma_{\ell-h}^{(1)}\Gamma_{s-h}^{(2)}. \tag{4.4}$$

Note that the mentioned properties of the sequences $\{\Gamma_j^{(1)}\}$ and $\{\Gamma_j^{(2)}\}$ allows us to approximate properly x_ℓ by equation (4.3). If T is block banded, for example, then $\Gamma^{(1)}$, $\Gamma^{(2)}$ and Γ are exponentially decaying and, as a result, we can obtain a good approximation by considering a small number of terms in the series representation.

We now have the following result.

THEOREM 4.1. *Consider the semi-infinite block Toeplitz system* (1.2), *where, for some fixed weight sequence* β, $\|(T_h)_{h\in\mathbb{Z}_+}\|_{1,\beta} < \infty$ *and* $\left\|(b_i)_{i\in\mathbb{Z}_+}\right\|_{1,\beta} < \infty$. *Suppose* $\widehat{T}(z)$ *is invertible for every* z *with* $(1/\beta_-) \leq |z| \leq \beta_+$. *Then each solution* $(x_i)_{i\in\mathbb{Z}_+}$ *of* (1.2) *with* $\left\|(x_i)_{i\in\mathbb{Z}_+}\right\|_1 < \infty$ *satisfies* $\left\|(x_i)_{i\in\mathbb{Z}_+}\right\|_{1,\beta} < \infty$.

Let us now illustrate some asymptotic properties of the factors of a sequence of increasing compressions of an infinite block Toeplitz matrix. Consider the bi-infinite banded block Toeplitz matrix A with symbol given by (3.8), where each A_i is a $k \times k$ matrix. We define the semi-infinite block Toeplitz matrices A_+ and A_- as follows:

$$A_+ = (A_{i-j})_{i,j\in\mathbb{Z}_+}, \qquad A_- = (A_{j-i})_{i,j\in\mathbb{Z}_+}, \tag{4.5}$$

where $A_{i-j} = 0$ if $|i-j| > p$. Then A_+ (resp. A_-) are boundedly invertible on the Hilbert space ℓ_2 of square summable complex sequences $(x_i)_{i\in\mathbb{Z}_+}$ if and only if the symbol $\Sigma(z) = \sum_{j=-p}^{p} z^j A_j$ (resp., $\Sigma(z^{-1}) = \sum_{j=-p}^{p} z^j A_{-j}$) has a spectral factorization. Now we define on ℓ_2 the projections P_n of rank nk as follows: $P_n(x_i)_{i=0}^{\infty} = (x_0, x_1, \ldots, x_{nk}, 0, 0, 0, \ldots)$. Define the $(nk)\times(nk)$ matrices A_{n+} and A_{n-} as the corresponding left upper blocks of A_+ and A_-, respectively. Then, according to Theorem 2.1 of [11] (also [9, 28]), the matrices A_{n+} and A_{n-} are nonsingular for sufficiently large n and for every $\boldsymbol{y} \in \ell_2$ the vector $(A_{n+}^{-1}\boldsymbol{y}, 0, 0, 0, \ldots)$ converges to $A_+^{-1}\boldsymbol{y}$ in the norm of ℓ_2. Similarly, for every $\boldsymbol{y} \in \ell_2$ the vector $(A_{n-}^{-1}\boldsymbol{y}, 0, 0, 0, \ldots)$ converges to $A_-^{-1}\boldsymbol{y}$ in the norm of ℓ_2. In fact, these two convergence properties together are equivalent to the bounded invertibility of both A_+ and A_- on ℓ_2.

The semi-infinite matrices $A_\pm$ given by (4.5) have the LDU-factorizations $A_\pm = L_\pm D_\pm M_\pm^T$ if and only if they are boundedly invertible on ℓ_2. Then, for n large enough, the matrices $A_{n\pm}$ corresponding to their $(nk)\times(nk)$ left upper corners have LDU-factorizations of the form

$$A_{n+} = L_{n+}D_{n+}M_{n+}^T, \quad A_{n-} = L_{n-}D_{n-}M_{n-}^T,$$

where D_{n+} and D_{n-} are block diagonal matrices and L_{n+}, L_{n-}, M_{n+} and M_{n-} are lower block triangular matrices having I_k as their diagonal blocks. Moreover, writing $\mathcal{J}_n[T]$ as the semi-infinite matrix obtained from the $(nk)\times(nk)$ matrix T by adding zero entries, the extended $(nk)\times(nk)$ matrices $\mathcal{J}_n[L_{n\pm}]$, $\mathcal{J}_n[L_{n\pm}^{-1}]$, $\mathcal{J}_n[M_{n\pm}]$, $\mathcal{J}_n[M_{n\pm}^{-1}]$, $\mathcal{J}_n[D_{n\pm}]$ and $\mathcal{J}_n[D_{n\pm}^{-1}]$ converge to the respective semi-infinite matrices $L_\pm$, $L_\pm^{-1}$, $M_\pm$, $M_\pm^{-1}$, $D_\pm$ and $D_\pm^{-1}$ in the strong operator topology as $n \to \infty$. To prove these convergence properties, one applies the theory of multiplicative LU-decompositions [10] with respect to the chain of orthogonal projections $\{0, P_0, P_1, \ldots, I_{\ell_2}\}$ to the semi-infinite matrices $A_\pm$, where I_{ℓ_2} is the identity operator on ℓ_2. Using the same multiplicative LU-decompositions with respect to the same chain of projections in the Hilbert space of complex sequences $(x_i)_{i\in\mathbb{Z}_+}$ endowed with the weighted norm $\left[\sum_{i\in\mathbb{Z}_+} r^{2i}|x_i|^2\right]^{1/2}$ for a suitable $r > 1$, one proves that the above convergence properties also hold with respect to the weighted norm.

Similar results hold for the LDM^T-factorization of a bi-infinite banded block Toeplitz matrix A. Indeed, let $\ell_2(\mathbb{Z})$ stand for the Hilbert space of square summable sequences indexed by the integers and let us define on $\ell_2(\mathbb{Z})$ the projection $P_n^{\leftrightarrow}$ of rank $(2n+1)k$ as follows: $P_n^{\leftrightarrow}(x_i)_{i=-\infty}^{\infty} = (\ldots, 0, 0, x_{-n}, \ldots, x_n, 0, 0, \ldots)$. Next, define the $(2n+1)k\times(2n+1)k$ matrices A_n as the corresponding central blocks of A. Then, according to [11], Theorem 4.1, applied for $R = S$, A_n is nonsingular for sufficiently large n and for every $\boldsymbol{y} \in \ell_2(\mathbb{Z})$ the vector $(\ldots, 0, 0, A_n^{-1}\boldsymbol{y}, 0, 0, \ldots)$ converges to $A^{-1}\boldsymbol{y}$ in the norm of $\ell_2(\mathbb{Z})$. Actually, this convergence property is equivalent to the invertibility of both A_+ and A_- on ℓ_2. Moreover, for sufficiently large n the matrices A_n have an LDU-factorization of the form

$$A = L_n D_n M_n^T,$$

where D_n is a block diagonal matrix and L_n and M_n are lower triangular banded block Toeplitz matrices having I_k as their diagonal blocks. Moreover, writing $\mathcal{J}_n^{\leftrightarrow}[T]$ as the bi-infinite matrix obtained from the $(2n+1)k\times(2n+1)k$ matrix T by adding zero entries, the extended matrices $\mathcal{J}_n^{\leftrightarrow}[L_n]$, $\mathcal{J}_n^{\leftrightarrow}[L_n^{-1}]$, $\mathcal{J}_n^{\leftrightarrow}[M_n]$, $\mathcal{J}_n^{\leftrightarrow}[M_n^{-1}]$, $\mathcal{J}_n^{\leftrightarrow}[D_n]$ and $\mathcal{J}_n^{\leftrightarrow}[D_n^{-1}]$ converge to the respective bi-infinite matrices L, L^{-1}, M, M^{-1}, D and D^{-1} in the strong operator topology as $n \to \infty$. The proof of these convergence properties follows from the theory of multiplicative LU-decompositions [10] with respect to a suitable chain of orthogonal projections.

When A is real positive definite and hence A_0 is real positive definite and $A_1^T = A_{-1}$, A_+ and A_- are both boundedly invertible on ℓ_2, since their LDU-factorizations both exist. In this case the unique solution of the equation $Ax = y$ can be approximated as indicated in the previous two paragraphs.

5. Spectral factorization and limiting profile. The properties of the Cholesky factorization of positive definite bi-infinite block matrices, that we use here, are a generalization of analogous results for scalar matrices obtained in [19], where they were applied to studying the asymptotic behavior of shifts of an exponentially decaying function. In this section we summarize the results from [24] on the asymptotic behavior of shifts of a vector of exponentially decaying functions.

Let $\varphi(x) = (\varphi_1(x), \varphi_2(x), \ldots, \varphi_\ell(x))^T$, $x \in \mathbb{R}$, be a vector of real functions defined on $\mathbb{R}$. We say that φ is an exponentially decaying vector function if there exist constants $\kappa > 0$ and $\lambda \in (0,1)$ such that

$$\|\varphi(x)\|_\infty := \max_{i=1,\ldots,\ell} |\varphi_i(x)| \le \kappa\lambda^{|x|}, \quad x \in \mathbb{R}. \tag{5.1}$$

Now consider the integer translates of φ, i.e., the function vectors

$$\varphi_i := \varphi(\cdot - i), \quad i \in \mathbb{Z},$$

where $\varphi_0 := \varphi$. Furthermore, for $r, s \in \mathbb{Z}$, let the symbol $\langle \varphi_r, \varphi_s \rangle_{[a,b]}$ denote the $\ell \times \ell$ matrix

$$\left(\langle \varphi_{i+r}, \varphi_{j+s} \rangle_{[a,b]} \right)_{i,j=1,\ldots,\ell},$$

where $\langle \cdot, \cdot \rangle_{[a,b]}$ is the usual scalar product in $L^2[a,b]$. According to this definition, φ_r is orthonormal to φ_s on the interval $[a,b]$ if and only if $\langle \varphi_r, \varphi_s \rangle_{[a,b]} = I_\ell \, \delta_{rs}$.

For any such function vector φ, the bi-infinite block matrix sequence

$$T_h := \langle \varphi_0, \varphi_h \rangle_{\mathbb{R}}, \quad h \in \mathbb{Z},$$

is even in the sense that T_h and T_{-h} are each other's transposes, $h \in \mathbb{Z}$, and it is exponentially decaying since

$$\|T_h\|_\infty := \max_{i=1,\ldots,\ell} \sum_{j=1}^{\ell} |[T_h]_{ij}| \le \kappa^2 \ell \, \lambda^{|h|} \left(|h| + \frac{1}{\log(\lambda^{-1})} \right).$$

Hence ([24], Lemma 3.1) the bi-infinite block Toeplitz matrix of order ℓ

$$T = (T_{rs})_{r,s\in\mathbb{Z}} = (T_{r-s})_{r,s\in\mathbb{Z}},$$

is a bounded operator on $\ell_2(\mathbb{Z})$.

Now consider the semi-infinite block Gram matrix

$$G = (G_{rs})_{r,s\in\mathbb{Z}}, \qquad G_{rs} := \langle \varphi_r, \varphi_s \rangle_{\mathbb{R}_+}, \quad r, s \in \mathbb{Z}_+,$$

generated by φ.

LEMMA 5.1. *If φ is exponentially decaying, then there exist $c > 0$ and $\lambda \in (0,1)$ such that*

$$\|G_{ij} - T_{ij}\| \le c\lambda^{i+j}, \quad i, j \in \mathbb{Z}_+.$$

Now consider the finite block Gram matrices

$$[\widehat{G}_n]_{rs} := \langle \varphi_r, \varphi_s \rangle_{[0,n]}, \quad r, s = 0, 1, \dots, n, \quad n \in \mathbb{Z}_+,$$

whose block entries for large enough n, r and s are very close to the corresponding entries of G, as specified by the following lemma.

LEMMA 5.2. *Take $\rho \in (0,1)$ and consider the truncation $G_n = (G_{ij})_{i,j=0}^n$ of G. Then there exist $C > 0$ and $\mu \in (0,1)$ such that*

$$\|[G_n]_{rs} - [\widehat{G}_n]_{rs}\| \le C\mu^{r+s}, \qquad 0 \le r, s \le [\rho n].$$

Moreover, the sequence of inverses to the matrices $\left([\widehat{G}_n]_{rs}\right)_{r,s=0}^{[\rho n]}$ is bounded.

Now suppose that the matrices T and G are positive definite on $\ell_2(\mathbb{Z})$ and $\ell_2(\mathbb{Z}_+)$, respectively. As specified in Theorem 3.11 of [24], under this hypothesis there exist the block Cholesky factorizations

$$T = LL^T \quad \text{and} \quad G = MM^T$$

of T and G where L and L^{-1} are bi-infinite lower triangular block Toeplitz matrices with blocks of order ℓ, M and M^{-1} are semi-infinite lower triangular block matrices, and L, L^{-1}, M and M^{-1} decay exponentially. Consider the block Cholesky factorization

$$\widehat{G}_n = \widehat{M_n}\widehat{M_n^T}, \quad n \in \mathbb{Z}_+.$$

It is straightforward to prove that for all $i = 0, 1, \dots, n$ and $x \in \mathbb{R}$ the function vectors ψ_i^n defined by

$$\psi_i^n(x) = \sum_{j=0}^{i} \left[\widehat{M_n^{-1}}\right]_{ij} \varphi_j(x) \tag{5.2}$$

are the ones generated by the block Gram-Schmidt orthonormalization process applied to $\{\varphi_j\}_{j=0,1,\dots,n}$ on $[0, n]$.

Under the above hypothesis on the Toeplitz matrix T, the function vector

$$\psi(x) := \sum_{j \in \mathbb{Z}_+} [L^{-1}]_j \varphi_{-j}(x), \quad x \in \mathbb{R}, \tag{5.3}$$

decays exponentially. Furthermore, its integer translates are orthonormal on $\mathbb{R}$, i.e.,

$$\langle \psi_r, \psi_s \rangle_{\mathbb{R}} = I_\ell\, \delta_{rs}, \quad r, s \in \mathbb{Z},$$

with $\psi_h := \psi(\cdot - h)$, $h \in \mathbb{Z}$.

The main result of Subsection 5, which is a generalization of Theorem 2.1 of [19], is that, under the conditions claimed in Theorem 5.3, the function vector ψ defined in (5.3) supplies the limiting profile of the block Gram-Schmidt orthonormalization process applied to the interval $[0, n]$, with large enough n.

Using formulae (5.1) and (5.2)-(5.3) we easily obtain that, for $x \in \mathbb{R}$ and all $i = 0, 1, \dots, n$,

$$\|\psi_i^n(x) - \psi_i(x)\| \le \kappa \sum_{j=0}^{i} \|[\widehat{M_n^{-1}}]_{ij} - [L^{-1}]_{ij}\| + \kappa \sum_{j=-\infty}^{-1} \|[L^{-1}]_{ij}\|.$$

Since $[L^{-1}]_{ij} = [L^{-1}]_{i-j}$ and the sequence $[L^{-1}]_h$, $h \in \mathbb{Z}_+$, decays exponentially as $h \to \infty$, there are constants $c > 0$ and $\alpha \in (0,1)$ such that

$$\|[L^{-1}]_{ij}\| \leq c\alpha^{|i-j|}, \quad i,j = 0,1,\ldots,n.$$

This inequality implies that, for $x \in \mathbb{R}$,

$$\|\psi_i^n(x) - \psi_i(x)\| \leq \kappa \sum_{j=0}^{i} \|[\widehat{M}_n^{-1}]_{ij} - [L^{-1}]_{ij}\| + \kappa \frac{\alpha^i}{1-\alpha}, \quad i = 0,1,\ldots,n.$$

THEOREM 5.3. *Suppose φ is a real function vector on $\mathbb{R}$ which satisfies the inequality* (5.1) *and has the property that the corresponding block Toeplitz matrix T and the Gram matrix G are positive definite. Then for every constant $\rho \in (0,1)$ there are constants $C > 0$, $\mu \in (0,1)$ and $N \in \mathbb{Z}_+$ such that for all $i, n \in \mathbb{Z}_+$, with $0 \leq i \leq \rho n$ and $n \geq N$ we have*

$$\|\psi_i^n(x) - \psi_i(x)\| \leq C\mu^i, \quad x \in \mathbb{R},$$

where ψ is defined by (5.3).

Let us now give a practical illustration of the procedure for the identification of the limiting profile proposed above. To this end we consider the following two B-splines, already introduced in [24]:

$$\begin{aligned} B_1(x) &= -\frac{11}{36}(-x)_+^3 - \frac{1}{2}(-x)_+^2 + \frac{1}{2}(1-x)_+^3 \\ &\quad -\frac{1}{4}(2-x)_+^3 + \frac{1}{18}(3-x)_+^3, \\ B_2(x) &= \frac{1}{12}(3-x)_+^4 - 2(2-x)_+^3 + \frac{3}{4}(2-x)_+^4 \\ &\quad -\frac{3}{4}(1-x)_+^4 - 2(1-x)_+^3 - \frac{1}{12}(-x)_+^4. \end{aligned}$$

Now, take $\varphi(x) = (B_1(x), B_2(x))^T$, $x \in \mathbb{R}$, and consider its integer translates $\varphi_i(x) := \varphi(x-i)$, $i \in \mathbb{Z}$.

Further, let T, with entries $T_{ij} = \langle \varphi_i, \varphi_j \rangle$, $i,j \in \mathbb{Z}$, be the Gram matrix associated to them. It is a 5-diagonal block Toeplitz matrix of order 2, whose nonzero blocks $T_{ij} = T_{i-j}$, $|i-j| \leq 2$, are

$$T_0 = \frac{1}{\alpha}\begin{bmatrix} 13176 & 10179 \\ 10179 & 11304 \end{bmatrix}, \quad T_1 = \frac{1}{\alpha}\begin{bmatrix} 4634 & 6573 \\ 1275 & 1688 \end{bmatrix}, \quad T_2 = \frac{1}{\alpha}\begin{bmatrix} 124 & 111 \\ 6 & 4 \end{bmatrix}$$

with $\alpha = 362880$, $T_{-2} = T_2^T$ and $T_{-1} = T_1^T$.

Hence, the symbol corresponding to T is

$$\Sigma(z) = \frac{1}{\alpha}\left(T_2 z^2 + T_1 z + T_0 + T_1^T z^{-1} + T_2^T z^{-2}\right),$$

where the superscript T denotes the matrix transpose. It is positive for $|z| = 1$ and further $\det(z^2\Sigma(z))$ has four zeros inside the unit circle and four outside of it, so that T is positive definite. Since in this case $G = T_+$, the hypotheses of Theorem 5.3 are satisfied.

Now let $\psi_i^n(x)$, $i = 0,1,\ldots,n$, be the function vectors (5.2) generated by the Gram-Schmidt process applied to the translates of the function $\varphi(x)$ on the interval $[0,n]$. Then Figure 5.1 depicts the behavior of the two components of $\psi_i^{20}(x)$ for some values of i and Figure 5.2 shows the graph of the limiting profile $\psi(x)$ arising from the asymptotic orthonormalization process applied to the aforementioned vector functions ϕ_i, $i \in \mathbb{Z}_+$.

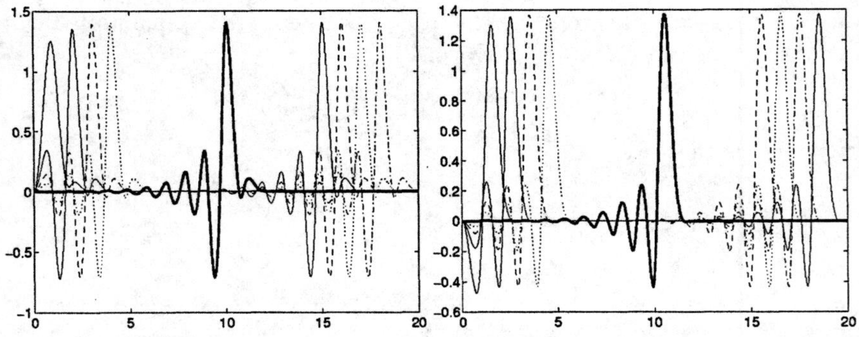

FIG. 5.1. *Components of* $\psi_i^{20}(x)$, $i = 0, 1, 2, 3, \mathbf{9}, 14, 15, 16, 17$

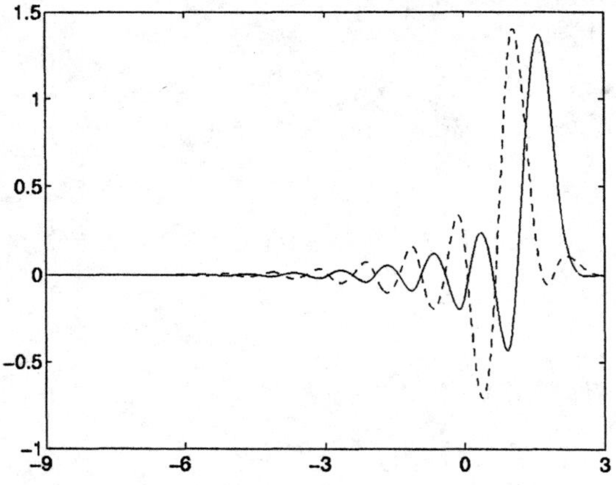

FIG. 5.2. *Limiting profile* $\psi(x)$

6. Applications. As a first application of the previous method, let us consider the numerical solution of the Poisson equation on a semi-infinite strip. More precisely, setting $\Omega = \{(x,y) : 0 < x < \infty,\ 0 < y < 1\}$ and denoting by $\partial\Omega$ the boundary of Ω, we consider the numerical solution, by a finite differences method, of the differential boundary value problem

$$\begin{cases} \Delta u(x,y) = -f(x,y), & (x,y) \in \Omega, \\ u(x,y) = 0, & (x,y) \in \partial\Omega, \end{cases} \tag{6.1}$$

where Δ is the Laplace operator.

Discretizing the differential problem (6.1) by a 5-points scheme on the mesh points

$$(x_i, y_j) = (ih, jh), \quad j = 0, 1, \ldots, n+1, \quad i = 0, 1, \ldots$$

with the stepsize $h = 1/(n+1)$ and choosing the usual order for the unknowns u_{ij}, we obtain a semi-infinite linear system of block Toeplitz type whose matrix is of the form

$$T = \begin{bmatrix} T_0 & -I_n & & \\ -I_n & T_0 & \ddots & \\ & \ddots & \ddots & \ddots \\ & & \ddots & \ddots \end{bmatrix}, \tag{6.2}$$

where I_n is the identity matrix of order n and T_0 is the $n \times n$ tridiagonal Toeplitz matrix

$$\begin{bmatrix} 4 & -1 & & \\ -1 & 4 & \ddots & \\ & \ddots & \ddots & -1 \\ & & -1 & 4 \end{bmatrix}.$$

The symbol associated to the matrix T is the Laurent matrix polynomial

$$\widehat{T}(z) = -z^{-1}I_n + T_0 - zI_n.$$

In this case the eigenvalues $\{\lambda_j\}$ of the corresponding monic matrix polynomial $\mathcal{P}(z) = -z\widehat{T}(z)$ can be obtained analytically. Indeed, it is straightforward to prove that

$$\lambda_j = \begin{cases} \frac{1}{2}\left(\mu_j + \sqrt{\mu_j^2 - 4}\right), & j = 1, \ldots, n \\ 1/\lambda_{2n-j+1}, & j = n+1, \ldots, 2n, \end{cases}$$

where

$$\mu_j = 4 + 2\cos\frac{j\pi}{n+1}, \quad j = 1, \ldots, n.$$

The first n eigenvalues lie outside the unit circle and the last n inside, as shown in Figure 6.1 for $n = 20$. As a consequence, a very accurate spectral factorization of $\mathcal{P}(z)^{-1}$ can be obtained using the first method, even for moderately high values of n.

Denoting by u_{ij}, for $i = 0, 1, \ldots$ and $j = 0, 1, \ldots, n+1$, the solution of the discretized problem, we assess the accuracy of the results, with respect to n, by the following two error estimates:

$$E_n^{(k)} = \max\left\{ |u(x_i, y_j) - u_{ij}^{(k)}|,\ i \in \mathcal{I}_{10n},\ j \in \mathcal{I}_n \right\} \tag{6.3}$$

$$E_{r,n}^{(k)} = \max\left\{ \frac{|u(x_i, y_j) - u_{ij}^{(k)}|}{|u(x_i, y_j)|},\ i \in \mathcal{I}_{10n},\ j \in \mathcal{I}_n,\ |u(x_i, y_j)| > 10^{-16} \right\}, \tag{6.4}$$

where $\mathcal{I}_n = \{1, \ldots, n\}$ and $k = 1, 2$ specifies if the spectral factorization has been carried out by the first or the second method, respectively.

As a first example, take

$$f(x, y) = \left[2\pi\cos\pi x + (2\pi^2 - 1)\sin\pi x\right] \mathrm{e}^{-x}\sin\pi y,$$

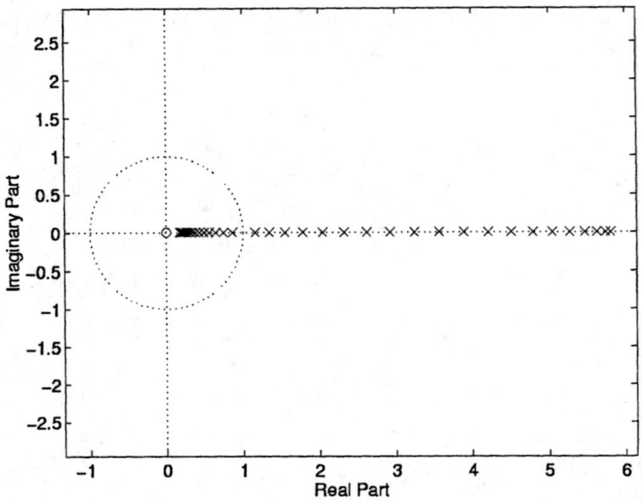

Fig. 6.1

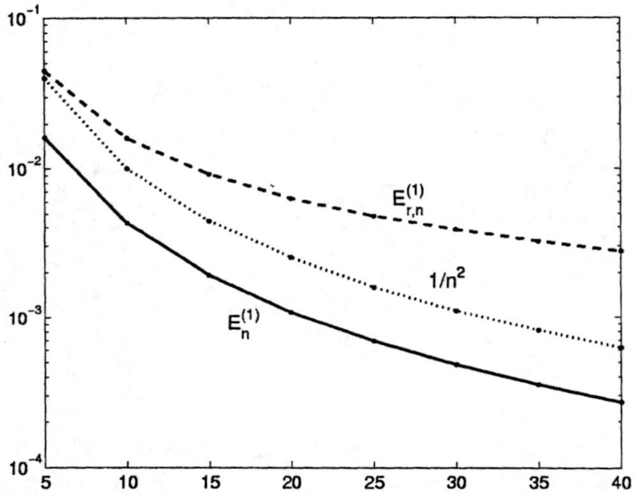

Fig. 6.2

so that the exact solution of (6.1) is

$$u(x, y) = e^{-x} \sin \pi x \ \sin \pi y.$$

Figure 6.2 shows $E_n^{(1)}$, $E_{r,n}^{(1)}$ and $1/n^2$ in the range $5 \leq n \leq 40$. Note that, as we expect solving the Poisson equation by the 5-points discretization method [30], $E_n^{(1)} = O(n^{-2})$.

The high level of accuracy of the results essentially depends on the exact knowledge of the eigenvalues of the matrix polynomial $\mathcal{P}(z) = -z\widehat{T}(z)$. Indeed, our experience suggests that the precision attainable by the first method primarily depends on the accurate evaluation of the eigenvalues, on their separation and on their distance from the curve Γ mentioned in Section 3.

Figure 6.3 reports the error $E_n^{(k)}$, $k = 1, 2$, for several values of n and for two values

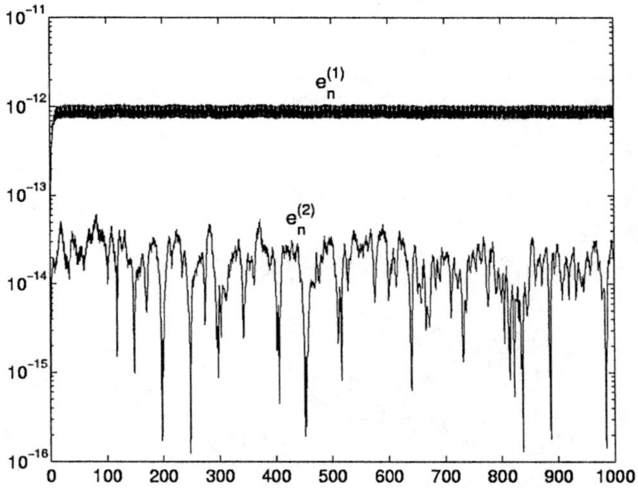

Fig. 6.3

of the extension parameter p. It shows, in particular, that for small values of n the two methods are equally effective. Furthermore, it highlights that the value of the extension parameter depends on the size of the blocks n. In fact, as n gets larger it is necessary to increase the value of p to obtain, by the band extension method, the same accuracy as in the first factorization method. For example, when $n = 25$ and $p = 60$ we need to solve a linear block Toeplitz system of dimension $n(p+1) = 1525$. This fact poses no particular problem from the point of view of computational complexity, the matrix being banded, but implies a larger propagation of roundoff errors with respect to the first method. However, as we will see soon, the first algorithm is not always more accurate than the second one.

As a second example on the solution of the same boundary value problem, let us consider

$$u(x,y) = \mathrm{e}^{-x/2}xy(1-y)$$

so that

$$f(x,y) = \mathrm{e}^{-x/2}\left[(1-y)(1-x/4)y + 2x\right].$$

To show the accuracy of the results, we depict in Figures 6.4 and 6.5 the errors $u(x_i, y_j) - u_{ij}$ corresponding to the first and the second example, respectively. In both cases, the results are obtained via the first spectral factorization method.

Now, as a second application, let T be the 5-diagonal block Toeplitz matrix of order 2 introduced in the previous section, and let T_+ be the semi-infinite block Toeplitz matrix given by $(T_+)_{ij} = T_{ij}$, $i, j \in \mathbb{Z}_+$. Furthermore, set $\mathbf{b} = T_+\mathbf{x}$, with $x_n = \left(\frac{8}{9}\right)^n$, $n \in \mathbb{Z}_+$, and solve the semi-infinite block Toeplitz linear system

$$T_+\mathbf{x} = \mathbf{b}$$

by the two spectral factorization methods illustrated above.

Denoting by $x^{(1)}$ and $x^{(2)}$ the solution vectors obtained by using the two spectral factorization methods, let

$$e_n^{(k)} = \frac{|x_n - x_n^{(k)}|}{|x_n|}, \tag{6.5}$$

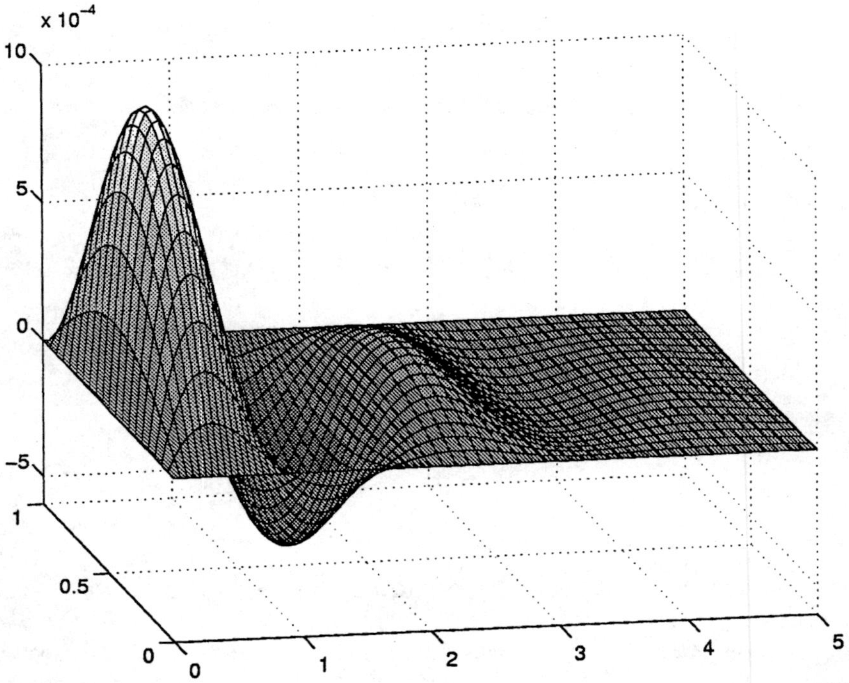

Fig. 6.4

with $k = 1, 2$ and $n \in \mathbb{Z}_+$, be the corresponding componentwise relative error. Figure 6.6 gives the values of $e_n^{(k)}$ in the range $0 \leq n \leq 1000$. Note that the thick black line representing $e_n^{(1)}$ is due to a small oscillation between the even and the odd components of the error.

Figure 6.6 clearly shows that in both cases the results are very accurate and that the second method gives better results than the first one for all the values of n. We note that for this example the application of the band extension method is computationally less expensive than for the previous example. Indeed, for $p = 60$, which is a good extension parameter value, we have to solve a linear system of dimension 240. On the contrary, using the first method we must estimate the 8 eigenvalues of the matrix polynomial $z^2\widehat{T}(z)$, a task which requires a greater computational effort and does not guarantee the same precision of the results.

7. Future research. As we have seen above, the orthonormalization of vectors of B-splines can be reduced to the spectral factorization of a bi-infinite block Toeplitz matrix. An important generalization is the orthonormalization of vectors of m-dimensional splines, with $m \geq 2$, i.e., spline functions defined on $\mathbb{R}^m$ with values in $\mathbb{R}^k$ where k is the order of the blocks. This problem can in principle be reduced to the spectral factorization of block Toeplitz matrices with m indices. In this section we study the definition and existence of the spectral factorization of m-index bi-infinite block Toeplitz matrices.

By a bi-infinite block Toeplitz matrix of order k and m indices, we mean a matrix $A = (A_{i-j})$, indexed by $\mathbb{Z}^m$, whose elements are real $k \times k$ matrices where $k = 1, 2, \ldots$ does not depend on (i, j). As in the one index case ($m = 1$), block Toeplitz matrices of order 1 are called Toeplitz matrices. Next, let $A = (A_j)_{j \in \mathbb{Z}^m}$ be a sequence of real $k \times k$ matrices, where $k \in \mathbb{N}$ does not depend on j. Taking a sequence of positive numbers $\boldsymbol{\beta} = (\beta_j)_{j \in \mathbb{Z}^m}$

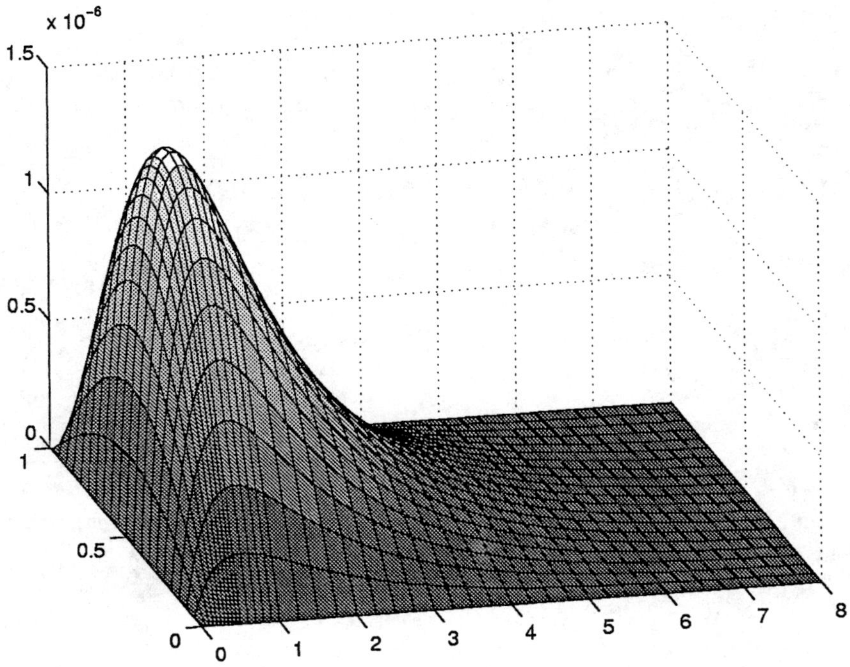

Fig. 6.5

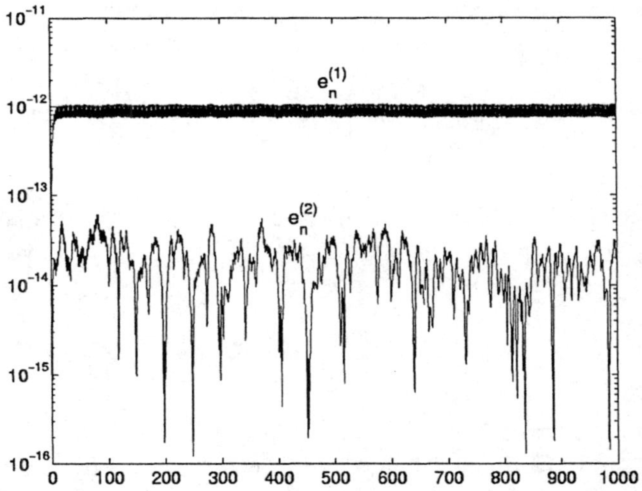

Fig. 6.6

satisfying $\beta_{i+j} \le \beta_i \beta_j$ for all $i, j \in \mathbb{Z}^m$, by $\ell^k_{1,\beta}(\mathbb{Z}^m)$ we mean the space of all bi-infinite sequences $A = (A_j)_{j \in \mathbb{Z}^m}$ of real $k \times k$ matrices for which

$$\|A\|_{1,\beta} := \sum_{j \in \mathbb{Z}^m} \beta_j \|A_j\| < \infty, \tag{7.1}$$

where $\|\cdot\|$ denotes an arbitrary $k \times k$ matrix norm. The most common choices* of β are $\beta_j = (1+|j|)^\rho$ for $\rho > 0$ and $\beta_j = g^{|j|}$ for $g > 1$, which correspond to algebraic and exponential weights, respectively. We write $\ell_1^k(\mathbb{Z}^m)$ if $\beta_i = 1$ for all $i \in \mathbb{Z}^m$. Further, for $k, m \in \mathbb{N}$, let $\ell_2^k(\mathbb{Z}^m)$ be the Hilbert space of all square summable functions $x : \mathbb{Z}^m \to \mathbb{R}^k$ with norm

$$\|x\|^2 = \sum_{j\in\mathbb{Z}^m} \|x_j\|^2, \qquad x = (x_j)_{j\in\mathbb{Z}^m},$$

and scalar product

$$\langle x, y\rangle = \sum_{j\in\mathbb{Z}^m} \langle x_j y_j\rangle, \qquad x = (x_j)_{j\in\mathbb{Z}^m}, \qquad y = (y_j)_{j\in\mathbb{Z}^m},$$

where $\|\cdot\|$ and $\langle\cdot,\cdot\rangle$ denote the Euclidean norm and the usual scalar product in $\mathbb{R}^k$, respectively.

For any m-index block Toeplitz matrix $(A_{i-j})_{i,j\in\mathbb{Z}^m}$ with $k \times k$ matrix entries for which the expression in (7.1) is finite for a suitable weight sequence $(\beta_j)_{j\in\mathbb{Z}^m}$, we define the *symbol* by

$$\hat{A}(z) = \sum_{j\in\mathbb{Z}^m} z^j A_j, \qquad z = (z_1, \cdots, z_m) \in \mathbb{T}^m, \tag{7.2}$$

where $z^j = z_1^{j_1} \dots z_m^{j_m}$ for $j = (j_1, \dots, j_m)$ and $\mathbb{T}^m = \{z = (z_1, \dots, z_m) \in \mathbb{C}^m : |z_1| = \dots = |z_m| = 1\}$ is the m-dimensional torus. Clearly, since necessarily $\beta_j \geq 1$ for any $j \in \mathbb{Z}^m$, the symbol of such a bi-infinite block Toeplitz matrix is a continuous $k \times k$ matrix function defined on $\mathbb{T}^m$. The symbols of the bi-infinite block Toeplitz matrices $(A_{i-j})_{i,j\in\mathbb{Z}^m}$ such that

$$\sum_{j\in\mathbb{Z}^m} \beta_j \|A_j\| < +\infty,$$

form a Banach algebra with respect to the pointwise matrix product, denoted as $\mathcal{W}_\beta^k$; here we note that $\hat{A}(z)\hat{B}(z)$ is the symbol of the bi-infinite Toeplitz matrix given by the convolution product

$$(A * B)_j = \sum_{l\in\mathbb{Z}^m} A_{j-l} B_l.$$

This Banach algebra is commutative if and only if $k = 1$.

In the Banach algebra $\mathcal{W}_\beta^k$ the multiplicative linear projections (in the sense of [15]) [or, for $k = 1$, the multiplicative linear functionals] are exactly the maps $A \mapsto \hat{A}(z)$ for which z belongs to the set

$$\begin{aligned} \mathcal{C}_\beta &= \left\{ z \in \mathbb{Z}^m : \sup_{j\in\mathbb{Z}^m} \frac{|z|^j}{\beta_j} < +\infty \right\} \\ &= \left\{ (z_1, \dots, z_m) \in \mathbb{Z}^m : \sup_{j_1,\dots,j_m\in\mathbb{Z}} \frac{|z_1|^{j_1} \dots |z_m|^{j_m}}{\beta_{(j_1,\dots,j_m)}} < +\infty \right\}. \end{aligned}$$

*The addition in $\mathbb{Z}^m$ is defined componentwise. For $j = (j_1, \cdots, j_m) \in \mathbb{Z}^m$ we put $|j| = |j_1| + \dots + |j_m|$.

Note that C_β contains $\mathbb{T}^m$ (as a result of $\beta_j \geq 1$ for each $j \in \mathbb{Z}^m$), does not contain $(0,\ldots,0) \in \mathbb{C}^m$, is compact, and is closed with respect to the operation $z \mapsto (c_1z_1,\ldots,c_mz_m)$ for any $(c_1,\ldots,c_m) \in \mathbb{T}^m$. When

$$\beta_{(j_1,\ldots,j_m)} = \beta^{(1)}_{j_1}\cdots\beta^{(m)}_{j_m}$$

for m weight sequences $\beta^{(l)}_j$ of positive numbers satisfying $\beta^{(l)}_{i+j} = \beta^{(l)}_i\beta^{(l)}_j$ for $i,j \in \mathbb{Z}$ and $l = 1,\ldots,m$, we have

$$C_\beta = \prod_{l=1}^{m}\left\{z \in \mathbb{C} : (1/g^{(l)}_-) \leq |z| \leq g^{(l)}_+\right\},$$

where $g^{(l)}_\pm = \lim_{j\to+\infty}(\beta^{(l)}_{\pm j})^{1/j}$. In particular, if $\beta_j = (1+|j|)^\rho$ for some $\rho \geq 0$, we have $C_\beta = \mathbb{T}^m$, and if $\beta_j = g^{|j|}$ for some $g \geq 1$, we have $C_\beta = \{z = (z_1,\ldots,z_m) \in \mathbb{C}^m : (1/g) \leq |z_l| \leq g, l = 1,\ldots,m\}$.

For $k = 1$ the following result follows using standard Banach algebra theory [10], Chapter XXX. When $k \geq 2$, it follows instead from the main result of [15].

THEOREM 7.1. *Let $\hat{A}(z)$ belong to $\mathcal{W}^k_\beta$ and have the form* (7.1). *Then if $\hat{T}(z)$ is a nonsingular $k \times k$ matrix for all $z \in C_\beta$, $\hat{T}(z)^{-1}$ belongs to $\mathcal{W}^k_\beta$.*

To define the spectral factorization of an m-index bi-infinite block Toeplitz matrix with $m \geq 2$ one needs a linear order $\preceq$ on $\mathbb{Z}^m$ which allows one to call a Toeplitz matrix A lower (resp. upper) triangular if $A_j = 0$ for all $j \in \mathbb{Z}^m$ with $j \prec (0,\cdots,0)$ (resp. $j \succ (0,\cdots,0)$). This linear order must have the following properties:

(i) $i \preceq j \Longrightarrow i + l \preceq j + l$ for all $l \in \mathbb{Z}^m$, and

(ii) ($i \preceq j$ and $c \geq 0$)$\Longrightarrow ci \preceq cj$.

The main problem is that for $m \geq 2$ such a so-called term ordering is by no means unique. In fact, the lexicographical order on $\mathbb{Z}^m$ with respect to any order of the "letters" $1,\cdots,m$ within the "alphabet" $\{1,\cdots,m\}$ will do. With respect to $\preceq$, we call $A = (a_{i-j})_{i,j\in\mathbb{Z}^m}$ *lower triangular* if $a_j = 0$ for all $j \in \mathbb{Z}^m$ with $j \prec (0,\cdots,0)$, *upper triangular* if $a_j = 0$ for all $j \in \mathbb{Z}^m$ with $j \succ (0,\cdots,0)$, and *diagonal* if $a_j = 0$ for all $j \in \mathbb{Z}^m$ different from $(0,\cdots,0)$.

Consider the linear order $\preceq$ of $\mathbb{Z}^m$ as specified above. Let us study representations of A in the form

$$A = LDU, \tag{7.3}$$

where $L = (L_{i-j})_{i,j\in\mathbb{Z}^m}$ is lower triangular with $\ell_{(0,\cdots,0)} = 1$, $U = (U_{i-j})_{i,j\in\mathbb{Z}^m}$ is block upper triangular with $U_{(0,\cdots,0)} = I_k$, and $D = (D_{i-j})_{i,j\in\mathbb{Z}^m}$ is a block diagonal matrix. This factorization, which can be proven to exist under very general conditions on A, is not unique. To make it unique, we also require D to be invertible (i.e., D_0 to be nonsingular) and L and U to be boundedly invertible on $\ell^k_2(\mathbb{Z}^m)$ with inverses L^{-1} and U^{-1} that are block lower and block upper triangular matrices, respectively. A representation of A in the form (7.3) where L, D and U have the above properties, is called an LDU-factorization of A. In that case, A has to be boundedly invertible on $\ell^k_2(\mathbb{Z}^m)$, but in general this is not sufficient for the existence of an LDU-factorization. It is now easily seen that LDU-factorization of A with the symbols of its factors and inverses in the Banach algebra $\mathcal{W}^k_\beta$ amounts to the factorization

$$\hat{A}(z) = \hat{L}(z)D_0\hat{U}(z), \qquad z \in C_\beta, \tag{7.4}$$

in terms of the corresponding symbols, where D_0 is a nonsingular $k \times k$ matrix, $\hat{L}(z) - 1$ and $\hat{L}(z)^{-1} - 1$ belong to $\mathcal{W}_\beta^k$ and have Fourier series where all terms proportional to z^j with $j \prec (0, \cdots, 0)$ vanish, and $\hat{U}(z) - 1$ and $\hat{U}(z)^{-1} - 1$ belong to $\mathcal{W}_\beta^k$ and have Fourier series where all terms proportional to z^j with $j \succ (0, \cdots, 0)$ vanish.

If A is a bi-infinite Toeplitz matrix with symbol in the so-called Wiener class (i.e., in $\mathcal{W}_\beta^k$ for $\beta_j \equiv 1$) and its symbol $\hat{A}(z)$ is positive definite selfadjoint for all $z \in \mathbb{T}^m$, i.e., if A is positive definite selfadjoint on $\ell_2^k(\mathbb{Z}^m)$, the factorization (7.4) exists. In that case, D_0 is positive definite selfadjoint, $\hat{U}(z) = \hat{L}(\bar{z})^*$, and therefore

$$\hat{A}(z) = \text{Ł}(z)\text{Ł}(\bar{z})^*, \qquad \text{Ł}(z) = \hat{L}(z)D_0^{1/2},$$

leads to a Cholesky factorization of A. Here $D_0^{1/2}$ is the (unique) positive definite selfadjoint square root of D_0. It is easily seen that if $\hat{A}(z)$ is positive definite selfadjoint for all $z \in \mathbb{T}^m$ and belongs to $\mathcal{W}_\beta^k$ for some weight sequence $(\beta_j)_{j \in \mathbb{Z}^m}$, then its LDU factors and their inverses belong to $\mathcal{W}_\beta^k$.

In fact, we have the following result.

THEOREM 7.2. *Suppose the bi-infinite block Toeplitz matrix $A = (A_i)_{i \in \mathbb{Z}^m}$ with symbol in $\mathcal{W}^k$ has a spectral factorization of the type* (7.4) *and $\hat{A}(z)$ is a nonsingular matrix for all $z \in C_\beta$ for some weight sequence $(\beta_j)_{j \in \mathbb{Z}^m}$. Then the factors and their inverses have symbols belonging to $\mathcal{W}_\beta^k$.*

Generalizing the existence theory for LDU-factorizations by using the Wiener-Hopf factorization theory for their symbols is not obvious if $m \geq 2$. First of all, there is no obvious geometrical criterion for the existence of the factorization (7.4) and the construction of the factors as in the case $m = 1$ where the winding number turned out to be the key to the solution of the problem. Secondly, the factorization (7.4) depends in an essential way on the choice of the linear order $\preceq$ on $\mathbb{Z}^m$.

The generalization of the previous results for $m = 1$ can be based on the closely related theories of chains of projections in a Hilbert space [12, 10] and of nest algebras [8], which have the additional advantage of yielding an LDU factorization theory for arbitrary bounded linear operators on $\ell_2^k(\mathbb{Z}^m)$ or on general separable Hilbert spaces. For $k = 1$ these concepts have been introduced in [23], where the main results have been discussed. The definitions and results generalize in a straightforward way to the case $k \geq 2$.

Finally, when the term ordering $\preceq$ is lexicographical, m-index block Toeplitz matrices can be identified in a natural way with one index block Toeplitz matrices, where the entries themselves are $(m-1)$-index block Toeplitz matrices, as observed in [33] for *finite m-index* Toeplitz matrices. Indeed, let us consider $m = 2$ for simplicity and let $\preceq$ be the lexicographical order defined by

$$(i_1, i_2) \preceq (j_1, j_2) \iff \begin{cases} i_1 \leq j_1, & \text{or} \\ \text{if } i_1 = j_1, \text{ then } i_2 \leq j_2. \end{cases}$$

Now consider the two-index bi-infinite block Toeplitz matrix $A = (A_{i-j})_{i,j \in \mathbb{Z}^2}$ where the entries are real $k \times k$ matrices. Then A can be identified in a natural way with the one-index block Toeplitz matrix $B = (B_{i-j})_{i,j \in \mathbb{Z}}$, where B_l is the one-index block Toeplitz matrix defined by $(B_l)_{i-j} = A_{(l,i-j)}$, i.e., the blocks of B have infinite order. An LDU-factorization of A corresponds exactly to an LDU-factorization of B, where the "central" block B_0 itself (which is defined by $(B_0)_{i-j} = A_{(0,i-j)}$) undergoes an LDU-factorization. Hopefully, this observation can be used to obtain numerical spectral factorizations of multi-index block Toeplitz matrices with respect to a lexicographical order in the near future.

REFERENCES

[1] K. Amaratunga. A fast Matlab routine for calculating Daubechies filters. *Wavelet Digest* 4(4), 1995.

[2] F.L. Bauer. Ein direktes Iterationsverfahren zur Hurwitz-zerlegung eines Polynoms. *Arch. Elektr. Uebertragung* 9:285–290, 1955.

[3] F.L. Bauer. Beiträge zur Entwicklung numerischer Verfahren für programmgesteuerte Rechenanlagen. II. Direkte Faktorizierung eines Polynoms. Sitz. Ber. Bayer. Akad. Wiss. 163–203, 1956.

[4] B.P. Bogert, M.J.R. Healy, and J.W. Tukey. The quefrency alanysis of time serier for echoes: cepstrum pseudo-autocovariance, cross-sepstrum and saphe cracking. In: M. Rosenblatt (ed.), *Proc. Symposium Time Series Analysis.* John Wiley and Sons, New York, 1963, pp. 209–243.

[5] A. Böttcher and B. Silbermann. Operator-valued Szegő-Widom theorems. In: E.L. Basor and I. Gohberg (Eds.), *Toeplitz Operators and Related Topics*, volume 71 of *Operator Theory: Advances and Applications.* Birkhäuser, Basel-Boston, 1994, pp. 33–53.

[6] A. Calderón, F. Spitzer, and H. Widom. Inversion of Toeplitz matrices. *Illinois J. Math.*, 3:490–498, 1959.

[7] K. Clancey and I. Gohberg. *Factorization of Matrix Functions and Singular Integral Operators*, volume 3 of *Operator Theory: Advances and Applications.* Birkhäuser, Basel-Boston, 1981.

[8] K.R. Davidson. *Nest Algebras.* Pitman, Research Notes in Mathematics 191, London, 1988.

[9] I.C. Gohberg and I.A. Feldman. *Convolution Equations and Projection Methods for their Solution.* Transl. Math. Monographs 41, Amer. Math. Soc., Providence, R.I., 1974.

[10] I. Gohberg, S. Goldberg, and M.A. Kaashoek. *Classes of Linear Operators, Vol. II*, volume 63 of *Operator Theory: Advances and Applications.* Birkhäuser, Basel-Boston, 1993.

[11] I. Gohberg and M.A. Kaashoek, Projection method for block Toeplitz operators with operator-valued symbols. In: E.L. Basor and I. Gohberg (Eds.), *Toeplitz Operators and Related Topics*, volume 71 of *Operator Theory: Advances and Applications.* Birkhäuser, Basel-Boston, 1994, pp. 79–104.

[12] I.C. Gohberg and M.G. Krein, *Theory and Applications of Volterra Operators in Hilbert Space.* AMS, Providence, R.I., Transl. Math. Monographs 24, 1970.

[13] I. Gohberg, P. Lancaster, and L. Rodman. Spectral analysis of matrix polynomials. I. Canonical forms and divisors. *Linear Algebra and its Applications*, 20:1–44, 1978.

[14] I. Gohberg, P. Lancaster, and L. Rodman. *Matrix Polynomials*, Academic Press, New York, 1982.

[15] I.C. Gohberg and J. Leiterer. Factorization of operator functions with respect to a contour. II. Canonical factorization of operator functions close to the identity. *Math. Nachrichten* 54:41–74, 1972 [Russian] (appendix).

[16] G.H. Golub and C.F. Van Loan. *Matrix Computations.* Third ed., The John Hopkins University Press, Baltimore, 1996.

[17] T.N.T. Goodman, C.A. Micchelli, G. Rodriguez, and S. Seatzu. On the Cholesky factorization of the Gram matrix of locally supported functions. *BIT* 35:233–257, 1995.

[18] T.N.T. Goodman, C.A. Micchelli, G. Rodriguez, and S. Seatzu. Spectral factorization of Laurent polynomials. *Adv. in Comp. Math.*, 7:429–454, 1997.

[19] T.N.T. Goodman, C.A. Micchelli, G. Rodriguez, and S. Seatzu. On the limiting profile arising from orthonormalizing shifts of exponentially decaying functions. *IMA J. Num. Anal.* 18:331–354, 1998.

[20] I. Gelfand, D. Raikov, and G. Shilov. *Commutative Normed Rings*, AMS Chelsea Publ., Providence, R.I., 1964.

[21] S. Jaffard. Propriétés des matrices "bien.localisées" près de leur diagonale et quelques applications. *Ann. Inst. H. Poincaré Anal. Non Linéaire*, 7(5):461–476, 1990.

[22] M.G. Krein. Integral equations on the half-line with kernel depending upon the difference of the arguments. *Uspehi Mat. Nauk.*, 13(5):3–120, 1958. (Russian, translated in *AMS Translations*, 22:163–288, 1962).

[23] C.V.M. van der Mee, G. Rodriguez, and S. Seatzu. *LDU* factorization results for bi-infinite and semi-infinite scalar and block Toeplitz matrices. *Calcolo*, 33:307–335, 1996.

[24] C.V.M. van der Mee, G. Rodriguez, and S. Seatzu. Block Cholesky factorization of infinite matrices, and orthonormalization of vectors of functions. In Z. Chen, Y. Li, C.A. Micchelli, and Y. Xu, editors, *Advances in Computational Mathematics*, volume 202 of *Lecture Notes in Pure and Applied Mathematics*, pages 423–455, New York and Basel, 1998. M. Dekker Inc.

[25] C.V.M. van der Mee, G. Rodriguez, and S. Seatzu. Solution methods for perturbed semi-infinite linear systems of block-Toeplitz type. In: D. Bini, E.E. Tyrtyshnikov and P. Yalamov (eds.), *Structured*

Matrices: Recent Developments in Theory and Computations. Nova Science Publisher Inc., to appear in January 2000.

[26] A.V. Oppenheim and R.W. Schafer. *Discrete-Time Signal Processing.* Prentice Hall, Englewood Cliffs, NJ, 1989.

[27] F. Riesz and B. Sz.-Nagy. *Functional Analysis.* Frederick Ungar, New York, 1955 [also: Dover Publ., New York, 1990].

[28] S. Roch and B. Silbermann. C^*-algebra techniques in numerical analysis. *J. Operator Theory* 35:241–280, 1996.

[29] L. Rodman. *An Introduction to Operator Polynomials*, volume 38 of *Operator Theory: Advances and Applications.* Birkhäuser, Basel-Boston, 1989.

[30] J. Stoer and R. Bulirsch *Introduction to Numerical Analysis, second edition*, volume 12 of *Texts in Applied Mathematics.* Springer-Verlag, New York, 1991.

[31] G. Tunnicliffe Wilson. Factorization of the covariance generating function of a pure moving average process. *SIAM J. Numer. Anal.* 6:1–7, 1969.

[32] G. Tunnicliffe Wilson. A convergence theorem for spectral factorization. *J. Multivariate Anal.* 8:222–232, 1978.

[33] E.E. Tyrtyshnikov and N.L. Zamarashkin. Spectra of multilevel Toeplitz matrices: Advanced theory via simple matrix relationships. *Linear Algebra and its Applications* 270:15–27, 1998.

PARALLEL COMPUTATION AND PROBLEM SOLVING METHODOLOGIES: A VIEW FROM SOME EXPERIENCES

ALMERICO MURLI[†*], PASQUA D'AMBRA[†], AND LUISA D'AMORE[†*]

Abstract. This paper resumes a description of Computational Problem Solving Methodology (PSM), as the methodic process which, given a problem, leads to a software system for its solution. In this context is collocated Scientific Computing and is discussed the influence of computing environments on the development of effective algorithms and software. Parallel computers emphasized the above influence and led to the development of a new framework of principles and methodologies permeating each step of the overall process underlying PSM. The authors discuss the above topics relying on experiences arising from their research activities.

Key words. Parallel Computing, Scientific Computing, Numerical Analysis.

AMS subject classifications. 65Y05, 68U01, 68W10

1. Introduction.

It is widely recognized that there is a feedback process among advances in computer technologies, advances in Scientific Computing and needs to solve new and highly computational demanding problems arising from the real world. In this paper we address this topic and discuss how computing environments influence the overall methodic process which, given a problem, leads to its solution. To explain this point of view we resume a description of *Computational Problem Solving Methodology (PSM)*, including the various steps generally needed to solve a problem using computer machines. In the above process we collocate Numerical Analysis and, more generally, Scientific Computing and its role of ... *providing the scientific community with effective software tools* ... [31] to solve problems and ... *to compute quantities* ... *with lightning speed* ... [52].

The two previuos statements refer to the specific role of Numerical Analysis in the development of theory and methodological approaches for building efficient, accurate and reliable numerical algorithms and software. In this context, the importance and the influence of computing environments on the performance (accuracy and efficiency) of algorithms and the related software is recognized from many years [28, 37]. Furthermore, the advent of parallel machines has introduced new issues and emphasized the above influence on the development and analysis of effective algorithms and software [16, 29, 38]. Today, high complexity of physical and mathematical models of fundamental problems in science and engineering, due to the involved variety of phenomena and physical parameters and the resulting high-computational demanding tasks require an effective exploitation of teraflop-scale massively parallel computers expected to be available in the next years [54, 55]. This means that mathematical models, in which physical processes are adequately described, must be developed; accurate and efficient algorithms have to be designed for solving these complex mathematical models; effective software must be developed to implement the algorithms on advanced computer architectures [41, 56]. Starting from the assumption that attainable final results depend on the way the different components (models, algorithms and software) realize the mapping between the problem and the computing machine [21, 23, 45], our persuasion is that the influence of Parallel Computing concepts and methodologies permeats the overall process underlying PSM. Some examples arising from the research activity in which we have

*University of Naples "Federico II"

†Center for Research on Parallel Computing and Supercomputers (CPS) of the National Research Council (CNR)

been or are currently involved will aid us to explain this point of view.

In Section 2 we introduce PSM and an historical perspective of the influence of computing environments in the development of algorithms and software. In Section 3 we outline main concepts, principles and methodologies of Parallel Computing. Sections from 4 to 6 are devoted to the description of some application problems.

2. Computational Problem Solving Methodology. There is no doubt that what we now call **Numerical Analysis** vastly predates the advent of modern computers. Indeed, as pointed out by Heath in [19], most of the concepts and many of the algorithms that are in use today were first formulated by pre-twentieth century giants - Gauss, Pascal, Babbage and many others. However, the development of digital computers in the middle of the twentieth century was the impetus for the flowering of Numerical Analysis into the ever growing field and has given to *computing* the role now it plays throughout modern science and engineering (**Computational Science and Engineering-CSE**) [26, 51]. It is nowadays widely recognized that computational inquire, mainly in the form of numerical simulation, complements theory and experimentations in scientific research. This approach represents a new paradigm for scientific exploration and visualization of real world phenomena, that allows new approach to problems previously inaccessible. Indeed, it provides both qualitative and quantitative insights into many phenomena that are too complex and too expensive to study by analytical methods (wind tunnels, climate or galactic evolution, ...) or too dangerous to solve via experiments (weapon nuclear safety, drug design, ...).

Since the advent of first electronic computers, every new generation of computers has influenced the design of the algorithm focusing on the most efficient way to compute the solution. More generally, there has always been a feedback process among advances in computer technologies, advances in computing methodologies and, finally, needs to solve new and highly computational demanding problems arising from the real world. The feedback process between advances in computer technologies and advances in computing methodologies was already recognized by Babbage in 1864: ... *As soon as Analytical Engine exists, it will necessarily guide the future course of science. Whenever any result is sought by its aid, the question will then arise - by what course of calculation can these results be arrived at by the machine in the shortest time?* More recently, in 1998, Trefethen reports as fundamental law of Computer Science the following sentence: *As machines become more powerful, the efficiency of algorithms grows more important, not less* [53]. The previous statements highlight an important aspect concerning speeding a computation by improving efficiency of the underlying algorithm: *the speed up factor gets better as the problem gets larger.* The same is not completely true when speeding has been obtained only using a computing machine which runs faster than the previous one. Generally, both increasing computing power and improving efficiency of algorithms enable solution of problems of increasing complexity. On the other hand, in most cases faster algorithms are obtained using more suitable discretization and approximation methods in formulation of numerical schemes exploiting specific features in the structure of the problem to be solved (see for example multilevel algorithms).

As is known, given a real-world problem, its solution can be provided by accurate, efficient, reliable and robust software. The development of mathematical software is a complex process ranging from design of algorithms, to their implementation on a specific computing environment, till to intensive testing and quality assurance of final products. Software, *which can be seen as a map of a real-world problem onto a computer* [45], is the final product of the overall process needed to solve a problem on a computer. We refer to this methodic process as **Computational Problem Solving Methodology (PSM)**. The first step of the process is the formulation of a suitable **mathematical model**, that represents the physical world in terms of rules and mathematical relations. Once mathematical model is given, through approximations and discretizations, we get **numerical model**. The next step is the design of the **algorithm** which, encoded in a specific computing environment, gives rise to the **software**. PSM has already been well described in the context of concentric layers introduced in [26] and reported in FIG. 2.1. Real world that we observe is situated exterior to the layers; the outermost layer contains physical models (constructed to explain observations) while the next inner layer contains mathematical models. Then, going to the innermost layer which include computing systems, we find numerics, algorithms and software, i.e. the focus of **Scientific Computing**.

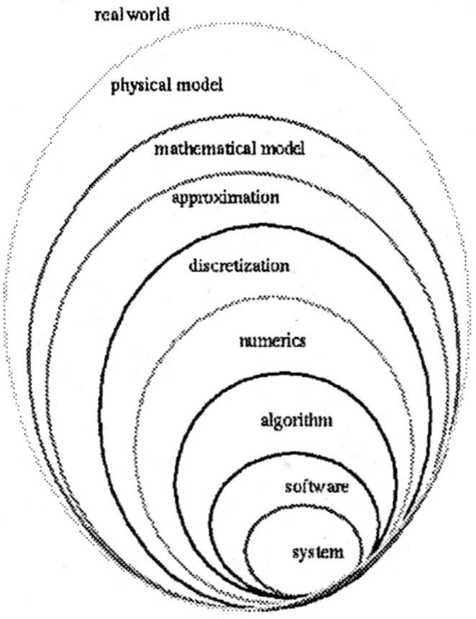

FIG. 2.1. *Computational Problem Solving Methodology*

2.1. Influence of Computing Environment: a historical perspective. In the 1960's was introduced the idea of having a numerical computation adapting itself to the problem in ODE solvers, where variable stepsizes were chosen using local error estimators, and in numerical integration algorithms, where intervals to be subdivided were again chosen using local error estimators. The idea of *adaptivity* has spread through numerical computation and had led to dramatic advances both in theory and in practice [27, 44]. Closely related to adaptivity is the concept of *polyalgorithm*, introduced in attempt to achieve both efficiency and reliability. In this case, the principle is to combine several algorithms along with various rules as when to switch from one to another. The rules are based on information gathered by the polyalgorithm during its computation. Polyalgorithms are predecessors of expert systems and of the more recent *Problem Solving Environments (PSEs)* [46], which are software systems providing all the computing facilities needed to solve a target class of problems efficiently.

Available PSEs provide a concrete example of effective maps between a target class of problems and a target class of computing environments and nowadays we are not surprised at such kind of software tool. Much more surprising is the fact that already in 1925 Whittaker, in its book titled *Calculus of Observation* [58], describes the student's mathematical laboratory needed to perform Fourier analysis as follows: ... *each student should have a seven-place table of logarithms, a stock of computing paper, a computing form for practical Fourier analysis*, and in this environment he proposed the computation of Fourier coefficients in such a way that was the most efficient at that time. The main difficulty in computing Fourier coefficients is the evaluation of trigonometric functions. The basic idea applied by Witthaker was to reduce trigonometric functions to the first quadrant only, i.e. to use known angles by applying symmetry properties of trigonometric functions. Finally, the computation of the global sum in the case of $N = 12$ was performed in 3 steps, as illustrated in FIG. 2.2, requiring a total number of only 44 arithmetic operations. The main drawback of Witthaker algorithm was that the value of N had to be small and fixed (in particular, Whittaker considered $N = 6, 12, 24$).

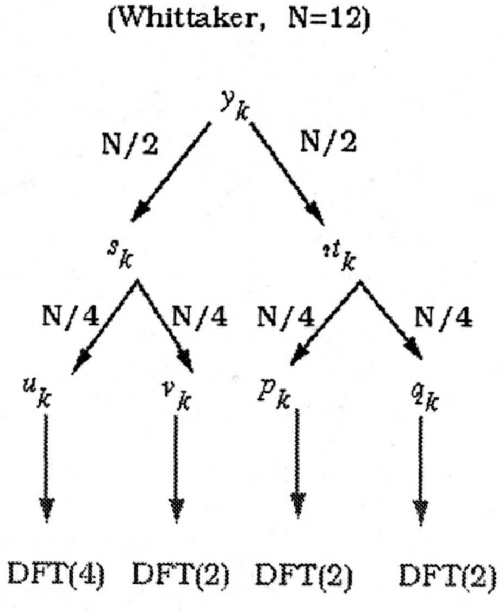

FIG. 2.2. *Witthaker algorithm.*

With the advent of electronic computers, using a computing environment based on an IBM 1041 (0.2 MFLOPS), Goertzel proposed the use of a recurrence formula [30], later stabilized by Reinsch [49], to evaluate trigonometric functions. This formula led to an algorithm which performs more operations than that described by Whittaker but with a minimun number of function evaluations (only 2). Furthermore, Goertzel algorithm does not require any bounds on N. On the other hand, as N grows, Goertzel algorithm appeared to be too time consuming, due to the quadratical behaviour of its computational complexity.

We have to wait until 1965, the year of the publication of Cooley and Tukey FFT algorithm [10] to gain the same efficiency of Witthaker algorithm for any N. In the essence Cooley and Tukey realized that the straightforward approach of Fourier calculations had the computer doing the same multiplications over and over. By reorganizing computation, taking into account algebraic properties of trigonometric functions, they found that most of redundant calculations could be eliminated.

Nowadays, FFT algorithm is widely recognized as *the most valuable algorithm in our lifetime.* Its application is continuing to expand as computing machines become more powerful, and computing scientists still tend to redesign their PSM in order to adapt it to Fourier analysis. Let us mention that, on July 1999 the third Wilkinson prize for the best quality numerical software was awarded to M. Frigo and S. Johnson of MIT for a library of routines (FFTW package) devoted to the computation of DFT of real and complex data by FFT algorithm on both sequential and parallel machines [24]. The winning idea of FFTW software is its innovative way to *adapt itself to the computing environment*, implementing the idea of *machine-oriented adaptivity.* This approach represents the most recent trend in the development of highly efficient software. A similar idea underlies ATLAS (Automatically Tuned Linear Algebra Software) project aimed at the automatic generation and optimization of the basic linear algebra kernels on high-performance computing environments [57].

3. General principles, Basic Issues and Strategies in Parallel Computing. Parallel Computing is sometimes viewed as a subarea of computing, interesting but generally of little relevance in order to solve problems. An analysis of new trends in Information Technology and in the most challenging real world application problems [41, 54, 55, 56] confirms, if it were still needed, that this vision is no longer tenable. Parallelism surfaces in all areas of computing (architectures, operating systems and programming languages, numerical and symbolic algorithms, software, and so on) and is ubiquitous in addressing many scientific problems. Furthermore, many times people do not view that at the back of computer tools used to realize parallel programs there are basic concepts and specific methodologies. As opposed to this vision, we emphasize that *Parallel Computing is a tool-independent framework of general concepts, principles and methodologies, permeating the overall process underlying Computational Problem Solving Methodology.* However, it is worth to note that we do not think concepts and principles can be studied or analyzed independently of the practical experience of using existing parallel computers and software tools to obtain effective solution of problems (... *using real hardware and real software to solve real problems*, Fox [21, 23]), but we point out that, though the details of the computer systems vary widely, there are basic general issues we must deal with. Furthermore, these issues should be taken into account in each phase of the solution process of real world application problems.

The history of trends in hardware and software technology for building faster and faster computers is beyond the scope of this paper and we refer to [35, 36]. The only thing we want to observe here is that the leading character of high-performance computing story is **parallelism**. Computers of all kinds, including single-processor architectures, have harnessed parallelism to gain performance from the raw technology, and the level at which parallelism is exploited continues to raise. From pipelining (different stages of several instruction execute concurrently) and multiple function units (several multipliers, adders, etc, controlled by a single control unit) or multiple processors inside one computer, till to several computers available across a network, *the way to obtain performance is concurrency, i.e. the ability of performing many operations at once* [11].

Another key point in the design of high-performance computers is *data storage and the problem of the proximity of the arithmetic and memory components.* All computer machines, even the simplest PCs, have memory hierarchies, from fast and small memories, like registers and caches, to slower and larger main memory, and finally down to disks.

Multi-processor computers have even more levels, if we look at local memory of a processor as remote memory for other processors. Arithmetic or logical operations can occur only on data stored at the top of the memory hierarchy and in the local memory of a processor, then data must be moved from the lower, slower levels in the hierarchy to the top level and must be communicated among processors to participate in computation. This clock time for communicating data from remote memory to function unit far exceeds the time to perform operations. Therefore, **data locality** in order to minimize the frequency with which data are moved between different levels of the memory hierarchy and function unit and among memories of processors is another fundamental issue to gain performance [11, 13, 14].

In examining current trends, from the technological and application perspectives, we finally see that processor counts to obtain greater computer power will continue to increase and the use of large-scale parallelism in applications is broadening. In this context, the general principle of **scalability** or *the resilience to increasing processor counts as dimensions of application problems increase* becomes central [11, 13, 41].

Issues related to general concepts of concurrency, locality and scalability have driven in the last years the development of a methodological approach in designing and building numerical methods, algorithms and software for high-performance computing. The introduction and the use of block-partitioned algorithms in Numerical Linear Algebra and the evolution of widely used dense Linear Algebra software packages [14, 17] are a well known example of research field in which that approach has been successfully applied.

Decomposition concept plays a central role in the Parallel Computing methodological approach. As claimed by Fox in [23], **decomposition or partitioning is a universal source of scalable parallelism**. The identification of a way to partition a problem into subproblems, a computation that is to be performed and the data operated on by this computation into different tasks, is the main strategy of Parallel Computing.

In this paper, we illustrate, through some experiences, the idea proposed in different ways from several authors, [13, 20, 21, 22, 23, 25], for which parallelism and locality can be searched at many levels of Computational Problem Solving Methodology and final performance can depend on the structure of the real world problem, on the choice of the continuous and discrete models as well as on the algorithm and software design, in the global and interdisciplinary vision that a "good" software system is a "good" mapping between the application problem and the computing machine [21, 23, 45].

4. Parallelism and locality introduced at software development level using mathematical software as building blocks. Many times, especially in industrial applications, there is the need of exploiting high-performance computing by rapid restructuring of existing software. Indeed, the main goal of industrial users is reducing design simulation times in order to have significantly reductions of the time-to-market of new products without the need of complete rewriting of their simulation codes and without dealing with details related to parallel algorithms and their implementations. In this context, the availability of mathematical software providing solutions to main computational kernels and the ability to encapsulate it into existing application codes allow *the exploitation of parallelism inside some modules representing the main computational bottlenecks on the way to gain performance.* This approach realizes a local optimization strategy operating at software development level of PSM process. For a review paper describing the role of high-performance mathematical software we refer to [16].

We summarize the main issues and some design choices involved in the development of a low-cost parallel version of the industrial software KIVA-3 [2]. In this case, the local optimization strategy previously described has been realized by the integration of parallel general-purpose library routines in the original code.

4.1. Application Problem. KIVA-3 is a CFD software package, developed at Los Alamos National Laboratories, for numerical simulation of chemically reactive flows with sprays. This software, widely used in engine applications, is also used at the Engine Technical Innovation Department of Piaggio & C. S.p.A., an industrial partner of PINEAPL (Parallel Industrial NumErical Applications and Portable Library) project [15, 43], for simulations of the scavenging process in two-stroke spark ignition engines, in order to predict performance of new engines in terms of consumption and pollution emission in the pre-design phase.

The mathematical model considered in KIVA-3 is the complete system of *unsteady Navier-Stokes equations* for compressible and turbulent flows, coupled with chemical kinetic and spray droplet dynamic models.

The numerical method for solving the complete system of equations is based on a variable step backward Euler temporal finite difference scheme, where the timesteps are chosen using accuracy criteria. Each time step defines a cycle divided in three phases, corresponding to a physical splitting technique; the last two phases are devoted to the solution of fluid motion equations. The spatial discretization of these equations is based on a *finite volume method*.

In the second phase, the diffusion terms and the terms associated with pressure wave propagation are implicitly solved by a modified version of the SIMPLE (Semi-Implicit Method for Pressure-Linked Equations) algorithm [39]. This algorithm, well known in the CFD community, is an iterative procedure to solve velocity, temperature and pressure equations, where the main computational kernel is the solution of large and sparse linear systems. Sparse linear systems arising from implicit solution of diffusion terms inside SIMPLE algorithm are solved, in the original KIVA-3 code, by a version of *diagonally preconditioned Conjugate Residual method* in a matrix free approach.

4.2. Parallelization approach and results. Conjugate Residual method is a rather old method for symmetric matrices, whereas the sparse linear systems arising from the application of original KIVA-3 solution procedure on non-rectangular and non-uniform grids, such as the block-structured grids used by Piaggio and reported in FIG. 4.1, are unsymmetric; newer, more effective methods for these systems are now available and most of them have been implemented into PINEAPL library. On the other hand, profiling of test cases provided by Piaggio shown that approximately 60% of the total simulation time was spent in the linear solvers (see Table 4.1).

Taking into account the above considerations, parallelism has been essentially introduced in the SIMPLE algorithm, modyfing original code for using PINEAPL sparse solvers, while the computation outside that algorithm have been left sequential. This strategy appeared a good compromise between the quality of obtainable results, end-user requests as well as constraints on the available resources.

Parallel sparse solvers from PINEAPL library are designed to execute on distributed memory multi-processor machines and are based on Krylov subspace projection methods. The implemented parallel algorithms rely on *a data parallelism* approach, in which the same computation is carried out concurrently on different data. There are computation phases, in which each processor computes on it its own portion of the data, and communication phases, in which nearest-neighbor processors exchange local data using a message passing approach in order to continue the computation. Therefore, the integration of that routines in the original KIVA-3 code have involved the main issue of an overall *data partitioning strategy among processors, corresponding to a decomposition of the block-structured computational*

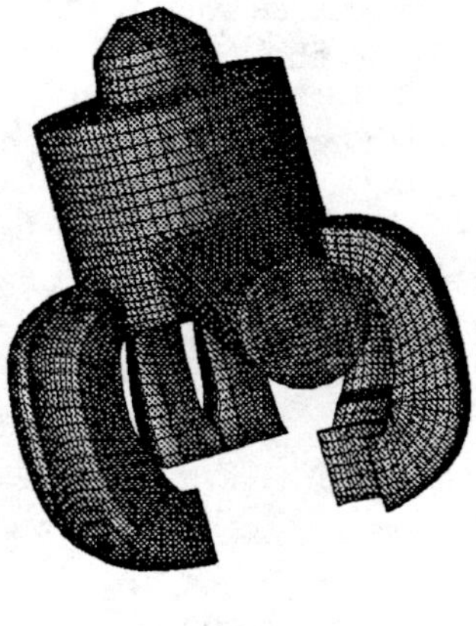

FIG. 4.1. *Test case grid*

grid, in which subgrids of the global grid are assigned to different processors.

The choice of the most suitable decomposition of a computational grid for solving PDE problems involves the two critical and more general Parallel Computing issues strongly influencing final performance of algorithms and software: *balance the computational and the storage loads among the processors and obtain the best ratio between computation and communication*, in order to use available resources at the best. More precisely, a good data/task decomposition strategy among processors have to realize the following state: *each processor have an approximately equal amount of data and work to do on it (load balancing) and the amount of communication among processors is minimized.* General techniques for having data/task decompositions so as to balance the loads and minimize communication among processors derive from algorithms for solving the *graph partitioning problem* [13, 34]. Many times, partitioning the physical or the computational domain requires communication only of data near the boundaries of the partitions, because data dependency is local and it requires interaction among adjacent data. In this case a way to realize load balancing and the best computation to communication ratio is producing the so called *surface to volume effect*, i.e. producing partitions in which the surface area (communication) for a given volume (computation) is reduced [13, 20]. This effect has been realized in parallel KIVA-3 procedure, where a three-dimensional decomposition of the block-structured grid has been adopted, introducing a uniform block partitioning of the grid data which assignes the same amount of data at each processor.

Note that the structure of original KIVA-3 computational grid changes during the computation to simulate the movement of the piston within the engine cylinder. More precisely, in the algorithm there are some points, known in advance, in which horizontal layers of the grid are cut away (or added back) to preserve the cell aspect ratio. This a priori knowledge of the way the grid changes dynamically during the computation guides in the definition of *grid partitioning changes at runtime, in order to have balanced executions in all steps of simulations.*

There are many problems in which the amount of computational load per each partition of possible partitioning varies dynamically during the computation. This can be related both to problem features (see Section 5), and computational approach (see this Section and Section 5 and, more generally, problems for which an effective solution requires the use of adaptive algorithms). In such cases determining and applying new partitioning dynamically during the computation can become crucial to obtain performance. One of the key question in this type of problems, also known *dynamic load-balancing problems*, is when information on changes in computational loads is known. If all information are known at the beginning of the computation or in well-defined points, as in the case of original KIVA-3, problems are referred as *static or semi-static problems.* In these cases definition and the application of new partitioning at runtime can be based on *load balancing algorithms* which many times are simple and no onerous. An example problem in which information about load balancing changes is not known a priori or in well-defined points is in Section 5.

Experiments for testing the parallel KIVA-3 code, named PINEAPL code, have been carried out to reconstruct the behaviour of the flow field into the three-dimensional geometry of a two-stroke engine during the complete gas exchange process, corresponding to a simulated time of 1.5635×10^{-2} *s*. Table 4.1 shows the timings obtained on an IBM SP2[1], by using three computational grids with increasingly finer mesh on the same engine design. The three cases are marked as 15k, 30k and 40k, according to the number of cells in the complete block-structured grid, respectively of about 15000, 30000 and 40000 cells.

TABLE 4.1

Simulation timings in seconds total (linear solvers)			
	Original code		
	15k	30k	40k
	10377	57633	96934
	(5982)	(34578)	(54123)
	PINEAPL code		
# Procs	15k	30k	40k
1	9512	39200	71791
	(4650)	(14620)	(24670)
2	8403	35500	65500
	(3760)	(11100)	(18670)
4	7200	31700	58500
	(2470)	(7400)	(11900)

The PINEAPL code shows a consistent speedup of more than 2 on 4 processors in the linear solver parts, i.e in the parts in which parallelism has been introduced[2]. On the other hand, as Amdahl's law claims, speedup on the overall execution is limited, essentially due to the fact that some parts of the code are still sequential and their computational costs provide a lower bound on the execution time of the parallel code.

Amdahl's law [1] is one of the main performance evaluation models in the framework of Parallel Computing methodology. It quantifies the importance of introducing the most part of parallelism in a computation. Suppose a computation has a fraction $0 < f < 1$ of its

[1] 6 RS/6000 66.7 MHz Wide Nodes, with 256 MB of memory/node, connected by a switch with a peak bidirectional bandwidth of 40 MB/s

[2] The uni-processor speedup observed in a comparison between original code and PINEAPL code is due to the use, in our code, of more appropriate linear solvers with respect to the solver employed in the original code.

operations which have been effectively parallelized, while the remaining fraction $s = 1 - f$ has been kept sequential. Then with p processors, the execution time at most can decrease from $s + f = 1$ to $s + f/p$. Amdahl's law, defining parallel speedup T_1/T_p in terms of s and f, gives

$$T_1/T_p = (s + f)/(s + f/p) = 1/(s + f/p) \leq 1/s. \tag{4.1}$$

Therefore, the sequential fraction s limits the speedup, no matter how many processors can be used.

In FIG. 4.2 speedup lines of PINEAPL code on the three different grids are compared with predicted speedup lines arising from the application of Amdahl's law. It can be seen that in the case of the coarsest grid (15k grid) only a fraction $f = 0.489$ (the sparse solver execution time) of the overall execution time has been parallelized, then the maximum speedup predicted by Amdahl's performance model on 4 processors is 1.58. According with this upper bound, PINEAPL code realized a speedup of 1.32. The further reduction in performance is essentially due to communication and replicated computation costs, also known as *overheads* of parallel computation [20], which are not explicitely considered in the formulation of Amdahl's law used in (4.1). Note that parallel fractions of the computations appear reducing if finer grids are considered. This behaviour is related to specific computations involved in the parts of KIVA-3 procedure which have been left sequential. More precisely, it is related to increasing costs of explicit solution of convection terms in the Navier-Stokes equations when finer grid are considered.

The results of the above analysis clearly indicate that this approach of a partial parallelization strategy of existing software generally can be effective if parallel computers with a limited number of processors are used.

In the past years, Amdahl's law stimulated interesting discussions in the Scientific Computing community on opportunity and validity of using massively parallel machines in the solution of problems, starting from the assumption that for each computation there is an intrinsic sequential part which cannot be parallelized. These discussions converged on the conclusion that *Amdahl's law suggests only large problems can use effectively massively parallel machines.* More precisely, since generally sequential part s of computation shrinks as the problem size grows [40], if number of processors is considered growing also the problem size have to increase for no severe limitation in speedup. In this context have been developed new evaluation parameters and performance models, [32, 33], based on scalability concept, and methodological approaches in the design of scalable algorithms and software using massively parallel computers for effective solution of large scale problems (see Section 6).

5. Parallelism and locality introduced at discrete model and algorithm development level. This section briefly summarizes the main issues and some design choices involved in developing operational software for parallel air quality simulations in the area of Southern Italy centered on Campania Region.

Operational models for automatic control of air quality at various scale are expected to become, in a very near future, routinely used tools for scenario simulations and real time forecasts of atmospheric pollution. The main goal of these systems is providing, in a reasonable response time, a large number of simulations with different scenarios, in order to study and to assess the influence of variations of pollutant emission sources, land use and atmospheric circulation to the related levels of concentrations of pollutants and deposition rates. In the USA and some countries of Europe there are a lot of research programme devoted to the development of these simulation systems. In this context is collocated the

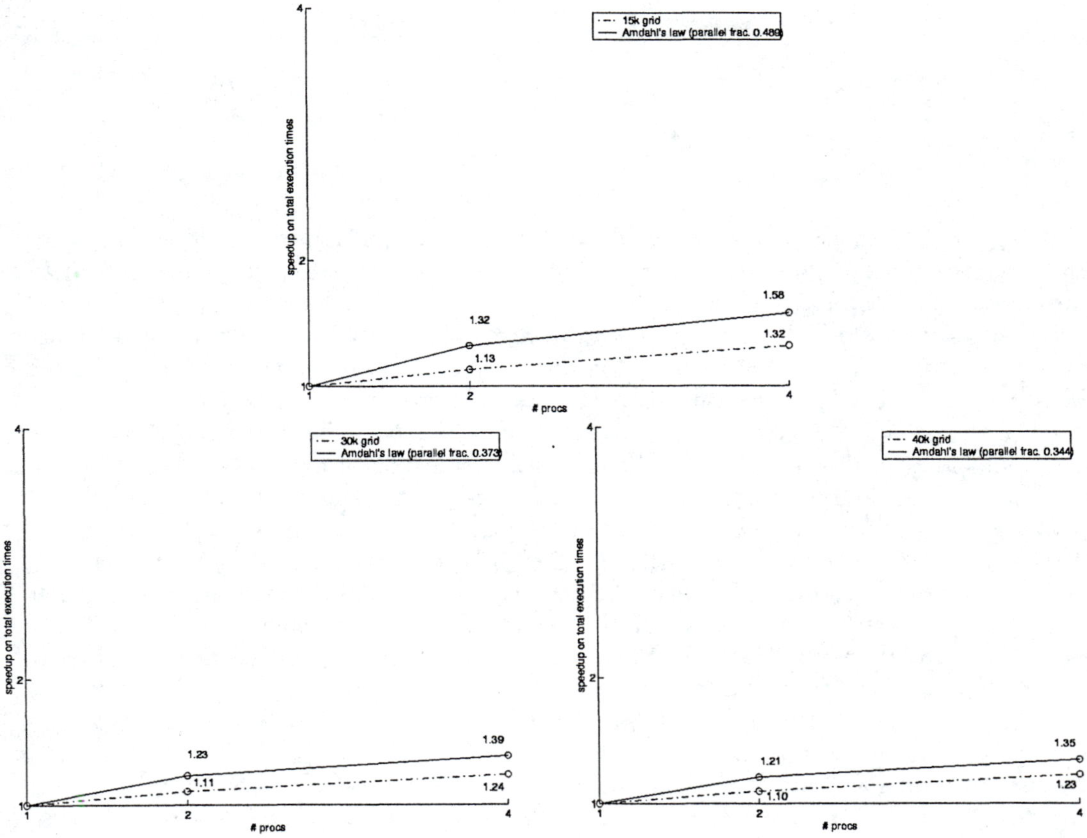

FIG. 4.2. *Amdahl's law. Upper bounds versus measured values for speedup of PINEAPL code on different grids*

project which has the final goal to develop a parallel software for numerical simulation of transport and photochemical transformation of air pollutants in some areas of Southern Italy. The interdisciplinary research activity, involving some environmental chemists and computational mathematicians, started from the individuation of the chemical, physical and meteorological phenomena which are significative for the model definition, till to the development and the selection of efficient and reliable numerical methods, parallel algorithms and software tools [3, 4].

The first result obtained in the context of the project has been a prototypal parallel software, named PNAM (Parallel Naples Airshed Model), for numerical simulation of the dynamics and the chemical transformations of air pollutants in urban-scale domains, using parallel MIMD distributed-memory machines. We focus on the way *parallelism and locality have been searched and introduced in PNAM at the level of numerical model and algorithm definition*, obtaining interesting results in terms of performance.

5.1. The application problem. Modern mathematical models of the transport and chemical transformation phenomena of atmospheric pollutants, namely *Air Quality Models (AQMs)*, are based on mass balances of the chemical species in the form of the following system of time-dependent, three-space dimensional, partial differential equations (*advection–*

diffusion–reaction eqs.) [48]:

$$\frac{\partial c_i}{\partial t} = -\nabla \cdot (c_i \mathbf{u}) + \nabla \cdot (\mathbf{K} \cdot \nabla c_i) + R_i(t, \mathbf{c}) + D_i(t, \mathbf{c}) + E_i(t, \mathbf{c}) \qquad i = 1, \ldots, N, \tag{5.1}$$

where $c_i(t, \mathbf{x})$ is the mean concentration of chemical species i, $\mathbf{c}$ is the vector of all the chemical species $(c_1, c_2, \ldots, c_N)$, $\mathbf{u}(t, \mathbf{x})$ is the wind velocity, $\mathbf{K}(t, \mathbf{x})$ is the turbulent diffusivity tensor, and R_i, D_i and E_i are the reaction, dry deposition and emission terms, respectively. These equations are decoupled from the Navier-Stokes equations modelling the meteorological phenomena, i.e. any dependence of the turbulent diffusivity, the wind velocity and the radiative properties of the atmosphere on the chemical processes is neglected.

The choice of an effective solution method for (5.1) have to be driven by several considerations, taking into account both the physical-chemical features of the problem and the features and size of numerical tasks. Although the system of PDEs is formally parabolic, in most atmospheric flows the transport in the horizontal plane is dominated by the advection, leading to hyperbolic-like characteristic. Furthermore, the terms R_i introduce a very high degree of *stiffness*, reflecting the very different reaction times of the chemical species, that can differ by a factor of more than 10^{10}. Besides taking into account these features, numerical solution methods must guarantee *mass conservation* with respect to the transport process and *positiveness* of concentrations.

The dimensions of the computational tasks for solution of the previous equations are related at the number of chemical species involved in the model and at the dimension of the computational grid imposed on the physical domain. In the case of PNAM, the model involves 42 chemical species and a suitable grid for the physical domain has an horizontal grid spacing of no more than 6 Km for up to 16 vertical levels, for a total number of more than 65,000 grid points. Therefore, the discrete problem have about 3 millions of variables. If it considered that typical period for mesoscale simulations are two days, this lead to a number of time steps, also regulated by numerical stability constraints on solution algorithms, ranging from 3,000 to 30,000. The previous sizes and the need to obtain results in a suitable response time requires the high-performance computational resources of parallel computers.

All the previous considerations regulated the main choices in the definition of the computational approach on which PNAM is based.

To solve system (5.1) a symmetric time-splitting technique [50] that decouples advection and horizontal diffusion from vertical turbulent diffusion and chemistry has been used. Note that the vertical diffusion and the chemistry have been coupled since they have similar time scales and it is generally not advisable to split them, in order to avoid large splitting errors. On the other hand *this coupling largely influenced the way in which parallelism and locality have been introduced in the solution approach.* The computational grid used for the spatial discretization of the equations was a 3D Cartesian rectangular grid and the semi-discretization of the different operators has been based on cell-centered finite-difference methods. An explicit time differencing scheme has been used to solve the system of ODEs arising from the semi-discretization of the advective transport.

The solution of the one-dimensional vertical diffusion and reaction operator is the dominant part of the computational work in air quality simulations, due to the stiffness introduced by reactions. One of the main features of PNAM is the coupled implicit solution of diffusion and chemistry operators, using a modified version of general-purpose VODE package [5]. VODE is a well-known ODE solver for stiff and non-stiff problems; for stiff problems it implements variable-coefficient BDF methods of variable order, with an adaptive selection of order and step size [6].

5.2. Parallelization approach and results. *Parallelism has been introduced in PNAM by a domain decomposition approach at the discrete level*, i.e. as in Section 4.2, a grid partitioning technique which divides the computational grid into subgrids and assignes a subgrid to each of the processors available for computation has been used. However, in this case a bi-dimensional decomposition has been applied: the computational three-dimensional grid has been partitioned only along the x and y directions such that *the computational domain has been decomposed into vertical air columns.*

The two-dimensional decomposition has been mainly driven by the features of algorithmic choices described in the previous section. The coupled implicit solution of vertical diffusion and chemistry introduces *a global coupling among points in the vertical direction*, while requires no interaction between points that lie in different vertical columns. Therefore, *a two-dimensional decomposition of the 3D grid realizes a good tradeoff between parallelism and locality, reducing data communication costs.* Indeed, no data communication among processors must be performed during the vertical diffusion-chemistry steps. Moreover, *this data decomposition allows to reuse available efficient and reliable sequential software*, such as VODE, to solve the stiff ODE systems arising from the semi-discretization of the vertical diffusion and chemistry operators. On the other hand, only a local nearest-neighbour communication is required in the advection steps, to update the boundary points of each subgrid.

Details on setup of the test case and on a comparison analysis between simulation results and available mesured data can be found in [4].

First experiments have been performed using an IBM SP2 machine[3] using a static load balancing algorithm, where grid partitioning assigned the same subgrid at each processor during all the simulation. A uniform partitioning, in which each processor has the same number of grid cells, have been chosen. Table 5.1 shows the execution times, in seconds, the speedup and the efficiency of PNAM on 1, 2, 4, 8 and 12 processors. Note that, in going from 1 to 12 processors, the execution time reduces from 13.17 hours to 1.87 hours and the corresponding speedup is about 7, with a parallel efficiency of about 60%. Higher efficiency values have been obtained on 2, 4 and 8 processors.

TABLE 5.1
Execution times (sec.), speedup, and efficiency of PNAM

Procs	1	2	4	8	12
Time	47415	27385	14694	8232	6737
Speedup	—	1.73	3.23	5.76	7.04
Efficiency	—	0.87	0.81	0.72	0.59

In the analysis of parallel performance results of PNAM it has been detected, as expected, that *a static and uniform data decomposition does not lead to load balanced executions*, limiting performance of the software. The main source of load imbalance was related both to the problem and to the numerical algorithm features. As pointed out in Section 5.1,

[3] 12 Power2 Super chip 160MHz Thin Nodes, with 512 MBytes of memory/node, connected by a switch with a peak bidirection bandwidth of 110 MB/s

one of the main features of system (5.1) is the stiffness introduced by the chemical reaction operator. This stiffness varies in time and space, in a not a priori known manner. Furthermore, an adaptive ODE sequential solver has been applied in separate processors, where different time steps are generally used in the solution of diffusion-reaction operator in different vertical air columns. A complete analysis of load imbalance in the execution of PNAM with static data decomposition is reported in [3], where the following problem-specific *load imbalance measure* has been introduced:

$$LB_{nproc} = \frac{\max_p T_p^{DC} - \frac{1}{nproc}\sum_p T_p^{DC}}{\max_p T_p^{DC}}.$$

T_p^{DC} is the total execution time spent by processor p in the coupled solution of diffusion and chemistry when *nproc* processors are used. A general-purpose measure of load imbalance in parallel environments can be found in [29]. The values of LB_{nproc}, reported in Table 5.2, have shown that different processors spent different times in the diffusion-chemistry stages and that the relative difference between the measured maximum time and the ideal one increased with the number of processors.

TABLE 5.2
Load imbalance in PNAM with a static data decomposition.

Procs	2	4	8	12
LB_p	0.19	0.22	0.33	0.48

To address the above problem, as introduced in Section 4.2, the application of *dynamic load balancing techniques* is required, in which a load balancing algorithm is executed periodically to determine a new and a more balanced data partitioning among processors.

A wide variety of both general-purpose and problem-specific load balancing algorithms have been proposed for use in parallel algorithms based on domain decomposition techniques, see for example [13]. These algorithms usually require further data communication among processors, and may or may not be effective if communication is too expensive. Therefore, also in this case a good choice is a good tradeoff between parallelism and locality.

First results related to the application of a problem-specific dynamic load balancing algorithm in PNAM can be found in [3].

6. Parallelism and locality introduced at continuous and discrete model development level. This section briefly summarizes the main issues and some design choices involved in developing parallel software for solution of *Image Restoration (IR)* problems.

Mathematical formulation of IR problems leads to inverse and ill posed problems. This depends on the use of Fredholm integral equation of first kind as generally adopted degradation model. From the computational point of view, ill position of the mathematical model means that observed images which are only very slightly different from one to another could come from radically different images to be reconstructed. Therefore, the problem is to single out a solution which is, in some sense, the "nearest" to the desidered solution. The usual approach to do it is to *regularize* the mathematical problem. This requires to find a suitable operator (usually named *filter*) which approximates, in some sense to be defined according to the application problem, the inverse of the *blurring* operator. In literature there are many regularization methods which differ essentially in how filtering is introduced on the computed solution. For a review of regularization methods and the state of the art for numerical treatment of such kind of problems we refer to [18].

The activity on this project has started from the analysis and the selection of efficient and reliable filters which appear to be more suitable for efficient mapping on high-performance computing environment. In this context, taking into account data dependencies in the operations required for filtering, a first classification has been made between *local* and *global* filters: in local filters updating of each image pixel requires information only from the nearest-neighbour pixels; in global filters, on the contrary, updating of each pixel requires information on the global image.

We focus on the way *intrinsic data locality features of a local filter for image restoration has been preserved in the choice of a suitable continuous model, for which an effective solution can be obtained by an efficient parallel algorithm.* As local filter we consider the Total Variation-based regularization for denoising problems. Note that in this case locality of the filtering operation is obtained formulating the variational problem related to the regularization scheme in terms of Euler-Lagrange equations. Moreover, as described in the next sections, locality can be preserved both by a suitable choice of the discretization scheme and the decomposition approach. A complete description of the work can be found in [8]. Furthermore, first results on parallel algorithms for a global filter based on transformation on the wavelet domain can be found in [7]. We just note that transformation in frequency domain, both using the Fast Fourier Transform and the Fast Wavelet Transform, generally lead to a global filter due to the intrinsic data coupling of these operations.

6.1. The application problem. Total Variation (TV)-based regularization schemes have been recognized to be very effective for recovering blocky images from noisy observations (*denoising problem*). The Euler-Lagrange equations corresponding to the TV-based regularization scheme are the following:

$$\mathcal{L}_\beta(u) - \lambda(u - u_0) = 0, \tag{6.1}$$

where

$$\mathcal{L}_\beta(u) = \frac{u_{xx}}{\cosh^2(u_x/\beta)} + \frac{u_{yy}}{\cosh^2(u_y/\beta)}$$

derives from a suitable approximation of the usual TV functional, introduced to avoid non differentiability of the last one at the points in which gradient is zero (see [47]), parameter λ is Lagrangian parameter, while β is a non-negative parameter which controls smoothness of $\mathcal{L}_\beta$.

Equation (6.1) is a non-linear diffusion equation where the amount of diffusion in the directions x and y is weighted by the terms $1/\cosh^2(\cdot)$, which depends on the magnitude of the gradient in each direction. The effect is therefore similar to that of a classical anisotropic diffusion [42] with the advantage that the values of the gradients are considered separately. For this reason equation (6.1) has been denoted as *Decoupled Anisotropic Diffusion (DAD)* equation. Ellipticity of DAD equation introduces a global coupling among image data. A reformulation of the above continuous model, based on the *time-marching (tm)* approach, well known in the CFD community, allows to obtain numerical algorithms simply preserving intrisic data locality of the TV-based regularization scheme. Using a *tm* approach, solutions of equation (6.1) are obtained as steady states of the unsteady equation:

$$\frac{\partial u}{\partial t} = \mathcal{L}_\beta(u) - \lambda(u - u_0). \tag{6.2}$$

An effective solution of equation (6.2), reformulated in terms of hyperbolic conservation law, can be based on explicit time differencing schemes which does not require solution of coupled non-linear algebraic systems and are well suited for implementations in parallel environments. For details on applied numerical scheme refer to [8].

6.2. Parallelization approach and results. Parallel algorithm for solution of equation (6.2) has been based on a domain decomposition approach at the discrete level, as in the case of the applications of previous Sections. A *grid partitioning* technique which divides the computational grid[4] into subgrids and assignes a subgrid to each of the processors available for computation has been used. Taking into account that an explicit scheme has been adopted for solution of (6.2), updating of solution on each subgrid requires communication only of data near the boundaries of the subgrids. Therefore, as already observed in Section 4.2, a generally advisable data decomposition is such that realizes the so-called *surface-to-volume effect.* In order to obtain this effect, a bi-dimensional data decomposition, which divides physical images both along x and y directions, has been applied.

Note that applications of high resolution visualization are typically *large-scale* applications. It needs increasing more and more resolution of the involved images (think at the HTV, High Resolution Tele Vision), and managing more and more big and complex images (such an in Astronomy). Therefore, feasibility of the general principle of scalability, underlying Parallel Computing methodology, is central in the development of high-performance software for image processing applications.

To take into account dimension of problems to be solved as well as processor counts, a newer performance evaluation model of parallel algorithms and software than Amdahl's law have to be considered. It was proposed by Gustafson et al. [33] as *scaled-size parallel performance evaluation model.* Their basic idea was to invert the Amdahl's paradigm: rather than ask how fast a given sequential program would run on a parallel machine it is more appropriate to ask how long a given parallel program would have taken to run on a sequential machine. Using s' and $f' = 1 - s'$ to represent the sequential and the parallel fractions of the execution time spent on the parallel machine, speedup T_1/T_p can be written as:

$$T_1/T_p = (s' + f'p)/(s' + f') = p + (1-p)s' = s' + (1-s')p, \tag{6.3}$$

and as p grows the sequential part does not limit the asymptotic behaviour of speed up.

Observe that, in (6.3) the amount of execution time and not the problem size, as in the case of (4.1), is considered fixed, in such a way that when more computing power is available, problem size can be expanded in order to use available computing resource for solving in the same time a bigger problem than that previously solved. Given a parallel software, a measure of scaled-size speedup SS_p can be obtained as:

$$SS_p = \frac{pT_1(n)}{T_p(pn)},$$

while the related scaled efficiency is

$$SE_p = \frac{T_1(n)}{T_p(pn)}.$$

$T_i(j)$ indicates execution time of the parallel program on i processors in the solution of a problem of dimension j. In this measure, dimension of the problem *scales* with the number

[4]Note that a given dimension of the computational grid corresponds to a given resolution of the original image.

of processors, such that, increasing processor counts and problem size, each processor ever solves a fixed-size subproblem[5]. High values of scaled efficiency as number of processors increases reveal that the parallel software has scalability properties and it is referred as *scalable software*, i.e. a software for which performance does not degrade as the processor counts and problem size grows [9].
In FIG. 6.1 scaled efficiency obtained running the software on a massively MIMD-DM machine, an Intel Paragon L38[6] has been shown. A uniform and static decomposition has been applied. On the right of the plot legend shows, for each graphic line, the size $N = n \times n$, of the subimage assigned to each processor. These results confirm that a "good" mapping between the data and the parallel machines has been realized.

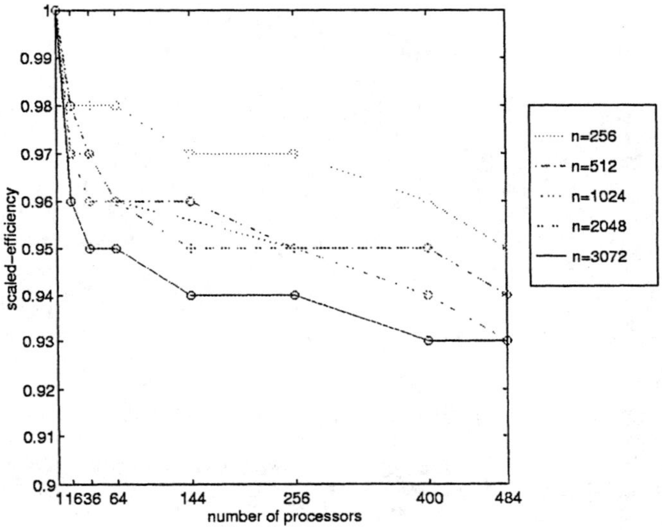

FIG. 6.1. *Scaled efficiency*

As can be observed, the scaled efficiency of the parallel software is not less than 93 %. Furthermore, the efficiency increases as the subimage dimension per processor increases, clearly showing that load balancing and ever growing ratios between computation to communication are obtained as problem dimension increases.

7. Concluding Remarks. Computational Science is often used to identify the discipline relating to the *third mode* of scientific investigation approach. It can be well described in the context of the concentric layers which represent the overall process and information flow of the real world problem solving methodology (PSM) exploiting computing as main tool. High complexity of todays fundamental problems in science and engineering require

[5] In the case of the application problem considered in this section, a measure of scaled-size efficiency on p processors can be obtained as:

$$SE_p = \frac{T_1(n)}{T_p(\sqrt{p} \cdot n)},$$

where $N = n \times n$ is the size of image.

[6] 512 Intel i860XP processors with peak performance of 75 MFlops and 32 MBytes of memory/node arranged in a 2D mesh

effective exploitations of teraflop-scale massively parallel computers expected to be available in the next years. Therefore, Parallel Computing has a predominant role in the PSM process.

Decomposition is the basic concept underlying Parallel Computing methodology. The identification of a way to partition a problem into subproblems, a computation that is to be performed and the data operated on by this computation into different tasks, is the main strategy of parallel computations. Taking into account the description of the PSM, we discuss the way in which decomposition can be introduced on different layers of the PSM, emphasizing that there is no the best methodology for Parallel Computing, but there is a framework that provides strategies and evaluation parameters within which parallelism can be explored.

REFERENCES

[1] G. AMDAHL, *Validity of the Single Processor Approach to Achieving Large Scale Computing Capabilities*, in Proc. of the AFIPS Conference, 30, 1967, pp. 483–485.

[2] L. ARNONE, P. D'AMBRA AND S. FILIPPONE, *A Parallel Version of KIVA-3 based on General Purpose Numerical Software and its Use in Two-stroke Engine Applications*, CPS Tech-Rep. N.99-4, 1999.

[3] G. BARONE, P. D'AMBRA, D. DI SERAFINO, G. GIUNTA, A. MURLI AND A. RICCIO, *Parallel Numerical Simulation of Air Pollution in Southern Italy*, in Large-Scale Computations in Air Pollution Modelling, Z. Zlatev et al. eds., Kluwer Academic Publishers, NATO Science Series, 57, 1999, pp. 39–52.

[4] G. BARONE, P. D'AMBRA, D. DI SERAFINO, G. GIUNTA, A. MURLI AND A. RICCIO, *PNAM: Parallel Software for Air Quality Simulations in the Naples Area*, Environmental Management and Health, 10 (1999), pp. 209–215.

[5] G. BARONE, P. D'AMBRA, D. DI SERAFINO, G. GIUNTA AND A. RICCIO, *A Comparison of Numerical Methods for Solving Diffusion-Reaction Equations in Air Quality Models*, Computing and Visualization in Science, 2 (1999), pp. 1–13.

[6] P. N. BROWN, G. D. BYRNE AND A. C. HYNDMARSH, *VODE: a Variable Coefficient ODE Solver*, SIAM J. Sci. Stat. Comput., 10 (1989), pp. 1038–1051.

[7] L. CARRACCIUOLO, L. D'AMORE AND A. MURLI, *A Fast Wavelet Packet Transform Based Algorithm for Image Restoration Problems in a Parallel Environment*, in Wavelet Application in Signal and Image Processing VI, A. F. Laine, M. Unser and A. Aldroubi eds., SPIE Publication, 3458 (1998), pp. 232–244.

[8] M. CECCARELLI, L. D'AMORE, V. DE SIMONE AND A. MURLI, *Numerical Methods for Total Variation-based Deblurring: a Comparison in a Parallel Environment*, in Notes on Numerical Fluid Mechanics, Vieweg, to appear.

[9] J. CHOI, J. J. DONGARRA AND D. W. WALKER, *The Design of Scalable Software Libraries for Distributed Memory Concurrent Computers*, in Environments and Tools for Parallel Scientific Computing, J. J. Dongarra and B. Tourancheau eds., Elsevier Science Publishers B.V., 1993.

[10] J. W. COOLEY, J. W. TUKEY *An algorithm for the machine calculation of complex Fourier series*, Math. Comput., 19 (1965), pp. 297-301.

[11] D. E. CULLER, J. P. SINGH AND A. GUPTA, *Parallel Computer Architecture: a Hardware/Software Approach*, Morgan Kaufmann Publishers, Inc., San Francisco, California, 1998.

[12] L. D'AMORE, V. DE SIMONE AND A. MURLI, *Design and Implementation Issues for Image Restoration Algorithms on a Parallel and Distributed Memory Environment*, CPS Tech-Rep. N.98-17, 1998.

[13] J. W. DEMMEL, *Lecture Notes on Application of Parallel Computers*, 1999. Available at http://www.cs.berkeley.edu/demmel/cs267_Spr99/

[14] J. W. DEMMEL, M. T. HEATH AND H. A. VAN DER VORST, *Parallel Numerical Linear Algebra*, Acta Numerica, 2 (1993), pp. 111–197.

[15] M. DERAKHSHAN, D. DI SERAFINO, AND A. MURLI, *The PINEAPL Library: A Parallel Numerical Library for Industrial Applications*, in Concepts of Numerical Software, W. Hackbusch, G. Wittum, eds., Notes on Numerical Fluid Mechanics, Vieweg Verlag, to appear.

[16] D. DI SERAFINO, L. MADDALENA, P. MESSINA AND A. MURLI, *Some Perspectives on High-Performance Mathematical Software*, in High-Performance Algorithms and Software in Nonlinear Optimization, R. De Leone, A. Murli, P. M. Pardalos and G. Toraldo, eds., Kluwer Academic Publishers, Boston, 1998, pp. 1–23.

[17] J. J. DONGARRA AND D. W. WALKER, *Software Libraries for Linear Algebra Computations on High Performance Computers*, SIAM Review, 37 (1995), pp. 151–180.

[18] H. W. ENGL, M. HANKE AND A. NEUBAUER, *Regularization of Inverse Problems*, Kluwer, Dordrecht, 1996.

[19] M. HEATH, *Scientific Computing: An Introductory Survey*, McGraw Hill, 1996.

[20] I. FOSTER, *Designing and Building Parallel Programs*, Addison-Wesley Publishing Company, Inc., 1995.

[21] G. C. FOX, *Achievements and Prospects for Parallel Computing*, in Practical Parallel Computing, Status and Prospects, P. C. Messina and A. Murli eds, Special Issue of Concurrency: Practice and Experience, 3 (1991), pp. 725–739.

[22] G. C. FOX, M. A. JOHNSON, G. A. LYZENGA, S. W. OTTO, J. K. SALMON AND D. W. WALKER, *Solving Problems on Concurrent Processors*, Prentice Hall, Englewood Cliffs, NJ, 1998.

[23] G. C. FOX, R. D. WILLIAMS AND P. C. MESSINA, *Parallel Computing Works!*, Morgan Kaufmann Publishers, Inc., San Francisco, California, 1994.

[24] M. FRIGO AND S. G. JOHNSON, *FFTW: An adaptive software architecture for the FFTW*, in Proceedings of ICASSP Conference, 3 (1998), pp. 1381–1384.

[25] I. GALLIGANI *Il calcolo parallelo e l'analisi numerica* in La Matematica per una nuova industria, M. Morelli e A. Murli eds., ETASLIBRI, 1994.

[26] E. GALLOPOULOS AND A. SAMEH, *CSE: Content and Product*, IEEE CSE (1997), pp. 39–43.

[27] C. W. GEAR *Numerical Initial Value Problems in Ordinary Differential Equations*, Prentice-Hall, New Jersey, 1971.

[28] G. GIUNTA, G. LACCETTI AND A. MURLI, *Il Problema della Valutazione del Software Matematico*, Rivista di Informatica, XIII (1983), pp. 45–60.

[29] ———, *Software Matematico e Nuove Architetture*, Rivista di Informatica, XXI (1991), pp. 129–144.

[30] G. GOERTZEL, *Fourier analysis*, in Mathematical Methods for Digital Computers, A. Ralston, H. S. Wilf, Eds., John Wiley and Sons, New Jork, 1960, pp. 258-262.

[31] G. H. GOLUB AND C. F. VAN LOAN, *Matrix Computations*, Second Edition, The Johns Hopkins University Press, Baltimore, 1989.

[32] J. L. GUSTAFSON, *Reevaluating Amdahl's Law*, Comm. of ACM, 31 (1988), pp. 532–533.

[33] J. L. GUSTAFSON, G. R. MONTRY, AND R. E. BENNER, *Development of Parallel Methods for a 1024-Processor Hypercube*, SIAM J. Sci. Stat. Comput., 9 (1988), pp. 609–638.

[34] G. KARYPIS, *METIS: a family of programs for partitioning unstructured graphs and hypergraphs and computing fill-reducing orderings of sparse matrices.*
Available at http://www-users.cs.umn.edu/ karypis/metis/

[35] P. C. MESSINA, *High-Performance Computers: the Next Generation (Part I)*, Computers in Physics, 11 (1997), pp. 454–466.

[36] ———, *High-Performance Computers: the Next Generation (Part II)*, Computers in Physics, 11 (1997), pp. 597–610.

[37] P. C. MESSINA AND A. MURLI(EDS.) *Problems and Methodologies in Mathematical Software Production*, in Lecture Notes in Computer Science, 142, Springer Verlag, 1982.

[38] ———, *Practical Parallel Computing: Status and Prospects*, Special Issue of Concurrency: Practice and Experience, Vol. 3, N. 6, 1991.

[39] S.V. PATANKAR, *Numerical Heat Transfer and Fluid Flow*, Hemisphere Publ. Corp., 1980.

[40] P. C. PATTON, *Performance Limits for Parallel Processors*, in Parallel Supercomputing. Methods, Algorithms and Applications, G. F. Carey ed., John Wiley & Sons, 1989.

[41] *PETAFLOPS II, 2nd Conference on Enabling Technologies for Peta(fl)ops Computing*, February 1999. Presentation materials available at http://www.cacr.caltech.edu/pflops2

[42] P. PERONA AND J. MALIK, *Scale-Space and Edge Detection Using Anisotropic Diffusion*, IEEE Transaction on PAMI, 12 (1990), pp. 629–639.

[43] PINEAPL Library - Routine documents.
Available at http://www.nag.co.uk/projects/PINEAPL.

[44] J. R. RICE, *A Metalgorithm for Adaptive Quadrature*, J. Assoc. Comp. Mach., 22 (1975), pp. 61–82.

[45] J. R. RICE ED., *Mathematical Aspects of Scientific Software*, Springer Verlag, 1988.

[46] J. R. RICE AND R. BOISVERT, *From Scientific Software Libraries to Problem Solving Environments*, IEEE Comp. Sci. Engr., 3 (1996), pp. 44–53.

[47] L. RUDIN, S. OSHER AND E. FADEMI, *Nonlinear Total Variation Noise Removal Algorithms*, Physica D., 60 (1992), pp. 259–268.

[48] J. H. SEINFELD AND S. N. PANDIS, *Atmospheric Chemistry and Physics, from Air Pollution to Climate Change*, John Wiley & Sons, Inc., 1998.

[49] J. STOER *Introduzione all'Analisi Numerica*, Zanichelli, 1974.

[50] G. Strang, *On the Construction and Comparison of Difference Schemes*, SIAM J. Numer. Anal., 5 (1968), pp. 506–517.

[51] ———, *CS and CS&E*, SIAM News, 32 (1999).

[52] L. N. Trefethen, *The Definition of Numerical Analysis*, SIAM News, 25 (1992).

[53] ———, *Maxims about numerical mathematics, science, computers, and life on earth.* SIAM News, 31(1), 1998.

[54] *USA Accelerated Strategic Computing Initiative (ASCI)* http://www.llnl.gov/asci/overview

[55] USA Department of Energy and National Science Foundation, *Final report of the DOE/NSF National Workshop on Advanced Scientific Computing*, July 1998. Available at http://www.er.doe.gov/ssi/

[56] USA National Science and Technology Council, *High Performance Computing and Communications: Information Technology Frontiers for a New Millennium*, National Coordination Office for Computing, Information and Communications ed., April 1999. Available at http://www.ccic.gov

[57] R. C. Whaley and J. J. Dongarra, *Automatically Tuned Linear Algebra Software.* Available at http://www.netlib.org/atlas/

[58] Sir E. Whittaker and G. Robinson, *The Calculus of Observation*, Blackie & Son Limited, London and Glasgow, 1944.

IMPLICIT-EXPLICIT RUNGE-KUTTA SCHEMES FOR STIFF SYSTEMS OF DIFFERENTIAL EQUATIONS

LORENZO PARESCHI* AND GIOVANNI RUSSO†

Abstract. We present new implicit-explicit (IMEX) Runge Kutta methods suitable for time dependent partial differential systems which contain stiff and non stiff terms (i.e. convection-diffusion problems, hyperbolic systems with relaxation). Here we restrict to diagonally implicit schemes and emphasize the relation with splitting schemes and asymptotic preserving schemes. Accuracy and stability properties of these schemes are studied both analytically and numerically.

Key words. Runge-Kutta methods, splitting schemes, stiff systems.

AMS subject classifications. 65C20, 82D25.

1. Introduction. Many stiff systems of differential equations can be written in the form

$$y' = f(y) + \frac{1}{\varepsilon} g(y), \tag{1}$$

where $y = y(t) \in R^N$, $f, g : R^N \to R^N$ and $\varepsilon > 0$ is the stiffness parameter.

System (1) may represent a system of N Ode's or a discretization, obtained by the method of lines, of a system of Pde's, such as, for example, convection-diffusion equations, discrete velocity models in kinetic equations, hyperbolic systems with relaxation.

In this work we consider the latter case. These systems have a special structure, that allows us to rewrite them into two coupled subsystems as

$$\begin{aligned} u' &= f_1(u, v), \\ v' &= f_2(u, v) + \frac{1}{\varepsilon} g_2(u, v), \end{aligned} \tag{2}$$

where $y = (u, v)$, $u = u(t) \in R^m$, $v = v(t) \in R^{N-m}$, $m < N$, $f = (f_1, f_2)$, and $g = (0, g_2)$.

In the stiff limit $\varepsilon \to 0$, system (2) reduces to a differential-algebraic system. We shall assume that the algebraic equation $g_2(u, v) = 0$, can be uniquely solved in v, giving

$$v = e(u), \tag{3}$$

and therefore system (2) reduces to

$$u' = f_1(u, e(u)) \equiv F(u). \tag{4}$$

A very general and commonly used approach to the solution of this problem is based on splitting methods. A simple splitting of system (2) consists in solving separately the non-stiff problem

$$\begin{aligned} u' &= f_1(u, v), \\ v' &= f_2(u, v), \end{aligned} \tag{5}$$

*Department of Mathematics, University of Ferrara, via Machiavelli 35, I-44100 Ferrara, ITALY. pareschi@dm.unife.it

†Department of Mathematics, University of L'Aquila L'Aquila, Italy russo@univaq.it

applying an explicit scheme and, using an implicit scheme, the stiff problem

$$\begin{aligned} u' &= 0, \\ v' &= \frac{1}{\varepsilon} g_2(u, v). \end{aligned} \tag{6}$$

This splitting is restricted to first order accuracy in time, nevertheless its simple structure presents several advantages:

- Some properties of the solution are easily maintained (positivity, TVD property, other physically relevant properties).
- Consistency with the stiff limit (4) as $\varepsilon \to 0$ can be easily checked (asymptotic preservation).
- In many cases the implicit scheme for g can be explicitly solved thanks to the fact that u is unchanged during this step. This is of paramount importance in many applications since it avoids the use of iterative solvers.

Higher order splitting can be constructed using suitable combinations of the two previous steps [23, 8, 9]. Unfortunately all these higher order extensions present a severe loss of accuracy when the g term is stiff [15]. Second order Runge-Kutta splitting which maintain the accuracy in the stiff limit have been constructed recently [15, 5]. There is also a third-order scheme that uses splitting and extrapolation but it is rather inefficient [6].

The aim of this paper is to have a unified approach of splitting schemes, and to provide a framework for the derivation of more general, accurate and efficient schemes. In particular, we show that these schemes are strictly related with the recently developed IMEX Runge-Kutta schemes [1]. An IMEX scheme consists of applying an implicit discretization for g and an explicit one for f. Since systems of the form (1) arise from a Pde in more than one space variable, the simplicity and efficiency of solving the algebraic equations corresponding to the implicit part of the discretization at each step is of fundamental importance. To this aim, it is natural to consider diagonally implicit Runge-Kutta (DIRK) schemes for g.

We show that most of the splitting schemes can be written in the formalism of IMEX Runge-Kutta schemes, where the solver for g is a DIRK scheme. Similarly it is easy to write an IMEX Runge-Kutta scheme in splitting form. In particular, we show that the asymptotic preserving requirement [18], i.e. the consistency of the scheme with (4), is satisfied every time we use an L-stable scheme for g and that the implicit step can be solved, in many cases, every time we use a DIRK scheme for g. Thanks to the IMEX Runge-Kutta formalism we derive new second and third order schemes which maintain the order in the stiff limit.

The rest of the paper is organized as follows. In section 2 we introduce the class of IMEX Runge-Kutta schemes under study, and emphasize the analogy with splitting schemes. Section 3 is devoted to second order schemes which maintain the order of accuracy in the limit. Accuracy and stability analysis are also presented. In section 4 we present some third order extension of the schemes. Finally in section 5 we investigate the performance of the schemes for a simple test equation.

2. IMEX Runge-Kutta schemes. An Implicit-Explicit Runge-Kutta scheme for system (1) is of the form

$$Y_i = y_0 + h \sum_{j=1}^{i-1} \tilde{a}_{ij} f(t_0 + \tilde{c}_j h, Y_j) + h \sum_{j=1}^{\nu} a_{ij} \frac{1}{\varepsilon} g(t_0 + c_j h, Y_j), \tag{7}$$

$$y_1 = y_0 + h \sum_{i=1}^{\nu} \tilde{w}_i f(t_0 + \tilde{c}_i h, Y_i) + h \sum_{i=1}^{\nu} w_i \frac{1}{\varepsilon} g(t_0 + c_i h, Y_i). \tag{8}$$

The matrices $\tilde{A} = (\tilde{a}_{ij})$, $\tilde{a}_{ij} = 0$ for $j \geq i$ and $A = (a_{ij})$ are $\nu \times \nu$ matrices such that the resulting scheme is explicit in f, and implicit in g. An IMEX Runge-Kutta scheme is characterized by these two matrices and the coefficient vectors $\tilde{c} = (\tilde{c}_1, \ldots, \tilde{c}_\nu)^T$, $\tilde{w} = (\tilde{w}_1, \ldots, \tilde{w}_\nu)^T$, $c = (c_1, \ldots, c_\nu)^T$, $w = (w_1, \ldots, w_\nu)^T$. They can be represented by a double *tableau* in the usual Butcher notation,

$$\begin{array}{c|c} \tilde{c} & \tilde{A} \\ \hline & \tilde{w}^T \end{array}, \qquad \begin{array}{c|c} c & A \\ \hline & w^T \end{array}.$$

A sufficient condition to guarantee that f is always evaluated explicitly is that the scheme for g is diagonally implicit i.e. $a_{ij} = 0$, for $j > i$. More general conditions will be considered in [4].

2.1. DIRK schemes and splitting methods. As already observed in the introduction, here we will restrict to diagonally implicit Runge-Kutta (DIRK) schemes. One step of such a scheme can be performed in splitting form as follows:

For $i = 1, \ldots, \nu$,

- Evaluate

$$Y_i^* = y_0 + h \sum_{j=1}^{i-2} \tilde{a}_{i,j} f_j + h\, \tilde{a}_{i,i-1} f(t_0 + \tilde{c}_{i-1}h, Y_{i-1}), \tag{9}$$

where

$$f_j = f(t_0 + \tilde{c}_j h, Y_j).$$

- Solve for Y_i

$$Y_i = Y_i^* + h \sum_{j=1}^{i-1} a_{i,j} \frac{1}{\varepsilon} g_j + h\, a_{i,i} \frac{1}{\varepsilon} g(t_0 + \tilde{c}_i h, Y_i), \tag{10}$$

where

$$g_j = g(t_0 + \tilde{c}_j h, Y_j).$$

Finally evaluate

$$y_1 = y_0 + h \sum_{i=1}^{\nu} \tilde{w}_i f_i + h \sum_{i=1}^{\nu} w_i \frac{1}{\varepsilon} g_i. \tag{11}$$

Note that in many cases the implicit step (10) based on a DIRK solver for g can be solved explicitly for a system of the form (2). This is of paramount importance for the simplicity and efficiency of the resulting schemes.

Similarly, every Runge-Kutta splitting scheme can be written as an IMEX Runge-Kutta scheme based on a DIRK solver for g. For example we report in Tables 1-3 the tableau corresponding to the standard splitting scheme SP(1,1,1), Jin [15] splitting Jin(2,2,2) defined only in the stiff regime $h \gg \varepsilon$ and the L-stable Caflisch-Jin-Russo [5] splitting CJR(2,2,2). We will also report the IMEX Midpoint(1,2,2) scheme [1] in Table 4 which has the nice feature of being second order accurate in the nonstiff regime using only one evaluation of the stiff source term.

As usual, the triplet (s, σ, p) characterizes the number s of stages of the implicit scheme, the number σ of stages of the explicit scheme and the order p of the scheme.

$$\begin{array}{c|c} 0 & 0 \\ \hline & 1 \end{array} \qquad \begin{array}{c|c} 1 & 1 \\ \hline & 1 \end{array}$$

TABLE 1
Tableau for the explicit (left) implicit (right) L-stable SP(1,1,1) scheme

$$\begin{array}{c|cc} 0 & 0 & 0 \\ 1 & 1 & 0 \\ \hline & 1/2 & 1/2 \end{array}, \qquad \begin{array}{c|cc} -1 & -1 & 0 \\ 2 & 1 & 1 \\ \hline & 1/2 & 1/2 \end{array}$$

TABLE 2
Tableau for the explicit (left) implicit (right) Jin(2,2,2) scheme

2.2. Order conditions. These schemes can be viewed as a particular class of partitioned Runge-Kutta schemes [14], and therefore their order conditions can be derived from the general theory of order conditions for partitioned schemes [11], however for low order schemes it is probably preferable to treat these systems differently, since partitioned schemes are usually applied to systems where the stiffness is associated to some components of the unknown vector, while in our case the stiffness is associated to the additive term of the right hand side.

The general technique to derive order conditions for Runge-Kutta schemes is based on the Taylor expansion of the exact and numerical solution. In particular, conditions for schemes of order p are obtained by imposing that the solution of a system

$$y' = f(t, y), \quad y(t_0) = y_0$$

at time $t = t_0 + h$ agrees with the numerical solution obtained by one step of a Runge-Kutta scheme up to order h^{p+1}. For a detailed account of the order conditions for Runge-Kutta schemes see, for example, [3] or [12]. Order conditions for partitioned Runge-Kutta schemes have been derived by Hairer [11].

Here we construct the order conditions for IMEX Runge-Kutta schemes up to order 3. The general treatment of the order condition for IMEX Runge-Kutta will be discussed in [4].

We apply scheme (7)-(8) to system (1), with $\varepsilon = 1$. We assume that the coefficients $\tilde{c}_i$, c_i, $\tilde{a}_{ij}$, a_{ij}, and the weights $\tilde{w}_i$, w_i satisfy the conditions

$$\tilde{c}_i = \sum_j \tilde{a}_{i,j}, \quad c_i = \sum_j a_{i,j}, \quad \sum_i \tilde{w}_i = 1, \quad \sum_i w_i = 1.$$

In this way it is enough to analyze the order conditions for autonomous systems,

$$y' = f(y) + g(y), \quad y(0) = y_0. \tag{12}$$

The Taylor expansion of the exact solution of system (12) can be written in terms of the elementary differentials as

$$y(h) = y_0 + (f+g)h + \frac{1}{2}(f'f + f'g + g'f + g'g)h^2$$

$$
\begin{array}{c|cccc}
0 & 0 & 0 & 0 & 0 \\
\tilde{\alpha} & \tilde{\alpha} & 0 & 0 & 0 \\
\tilde{\alpha} & \tilde{\alpha} & 0 & 0 & 0 \\
\eta(\tilde{\alpha}+\tilde{\beta}) & \eta\tilde{\alpha} & \eta\tilde{\beta} & 0 & 0 \\
\hline
\eta(\tilde{\alpha}+\tilde{\beta}) & \eta\tilde{\alpha} & \eta\tilde{\beta} & 0 & 0
\end{array}
\qquad
\begin{array}{c|cccc}
0 & 0 & 0 & 0 & 0 \\
0 & 0 & 0 & 0 & 0 \\
\gamma & \gamma & 0 & \beta & 0 \\
\eta(\beta+\gamma) & \gamma\eta & 0 & \beta\eta & \mu \\
\hline
\eta(\beta+\gamma) & \gamma\eta & 0 & \beta\eta & \mu
\end{array}
$$

$$\beta = \frac{2\mu-1}{2(\mu-1)}, \gamma = -\frac{2\mu^2-2\mu+1}{2\mu(\mu-1)}, \tilde{\alpha} = \frac{1}{2\mu}, \tilde{\beta} = -\frac{1}{2(\mu-1)}, \eta = -2\mu(\mu-1)$$

TABLE 3
Tableau for the explicit (left) implicit (right) L-stable CJR(3,2,2) scheme

$$
\begin{array}{c|cc}
0 & 0 & 0 \\
1/2 & 1/2 & 0 \\
\hline
 & 0 & 1
\end{array},
\qquad
\begin{array}{c|cc}
0 & 0 & 0 \\
1/2 & 0 & 1/2 \\
\hline
 & 0 & 1
\end{array},
$$

TABLE 4
Tableau for the explicit (left) implicit (right) Midpoint(1,2,2) scheme

$$
\begin{aligned}
(13) \qquad & + \frac{1}{6}(f''ff + 2f''fg + f''gg + g''ff + 2g''fg + g''gg + f'f'f \\
& + f'g'f + g'f'f + g'g'f + f'f'g + f'g'g + g'f'g + g'g'g)h^3 + O(h^4).
\end{aligned}
$$

The Taylor expansion of the Runge-Kutta solution after one step is given by

$$y_1 = y_0 + y_1^1 h + y_1^2 h^2 + y_1^3 h^3 + O(h^4),$$

where the coefficients of the elementary differentials appearing in the expansion depend on the matrices $\tilde{A}$, A and on the vectors $\tilde{c}$, c, $\tilde{w}$, w. By equating the coefficients of the elementary differentials one obtains the following order conditions:

First order:.

$$(14) \qquad \sum_i \tilde{w}_i = 1, \quad \sum_i w_i = 1,$$

Second order:.

$$(15) \qquad \sum_i \tilde{w}_i \tilde{c}_i = 1/2, \quad \sum_i w_i c_i = 1/2,$$

$$(16) \qquad \sum_i \tilde{w}_i c_i = 1/2, \quad \sum_i w_i \tilde{c}_i = 1/2,$$

Third order:.

$$
(17) \qquad
\begin{aligned}
& \sum_{ij} \tilde{w}_i \tilde{a}_{ij} \tilde{c}_j = 1/6, \quad \sum_i \tilde{w}_i \tilde{c}_i \tilde{c}_i = 1/3, \\
& \sum_{ij} w_i a_{ij} c_j = 1/6, \quad \sum_i w_i c_i c_i = 1/3,
\end{aligned}
$$

$$\begin{array}{c|ccc} 0 & 0 & 0 & 0 \\ \gamma & \gamma & 0 & 0 \\ 1 & \delta & 1-\delta & 0 \\ \hline & \delta & 1-\delta & 0 \end{array}, \quad \begin{array}{c|ccc} 0 & 0 & 0 & 0 \\ \gamma & 0 & \gamma & 0 \\ 1 & 0 & 1-\gamma & \gamma \\ \hline & 0 & 1-\gamma & \gamma \end{array}, \quad \gamma = 1 - \frac{\sqrt{2}}{2}, \delta = 1 - \frac{1}{2\gamma}$$

TABLE 5
Tableau for the explicit (left) implicit (right) L-stable ARS(2,2,2) scheme

$$\begin{array}{c|ccc} 0 & 0 & 0 & 0 \\ \gamma & \gamma & 0 & 0 \\ 1 & \delta & 1-\delta & 0 \\ \hline & 0 & 1-\gamma & \gamma \end{array}, \quad \begin{array}{c|ccc} 0 & 0 & 0 & 0 \\ \gamma & 0 & \gamma & 0 \\ 1 & 0 & 1-\gamma & \gamma \\ \hline & 0 & 1-\gamma & \gamma \end{array}, \quad \gamma = 1 - \frac{\sqrt{2}}{2}, \delta = -2\frac{\sqrt{2}}{3}$$

TABLE 6
Tableau for the explicit (left) implicit (right) L-stable ARS(2,3,2) scheme

$$\begin{gathered} \sum_{ij} \tilde{w}_i \tilde{a}_{ij} c_j = 1/6, \quad \sum_{ij} \tilde{w}_i a_{ij} \tilde{c}_j = 1/6, \quad \sum_{ij} \tilde{w}_i a_{ij} c_j = 1/6, \\ \sum_{ij} w_i \tilde{a}_{ij} c_j = 1/6, \quad \sum_{ij} w_i a_{ij} \tilde{c}_j = 1/6, \quad \sum_{ij} w_i \tilde{a}_{ij} \tilde{c}_j = 1/6, \\ \sum_i \tilde{w}_i c_i c_i = 1/3, \quad \sum_i \tilde{w}_i \tilde{c}_i c_i = 1/3, \\ \sum_i w_i \tilde{c}_i \tilde{c}_i = 1/3, \quad \sum_i w_i \tilde{c}_i c_i = 1/3. \end{gathered} \tag{18}$$

Conditions (14), (15), (17) are the standard order conditions for the two *tableau*, each of them taken separately. Conditions (16) and (18) are new conditions that arise because of the coupling of the two schemes.

The order conditions will simplify a lot if $\tilde{c} = c$. For this reason only such schemes are considered in [1] and [4]. In particular, we observe that, if the two tableau differ only for the value of the matrices A, i.e. if $\tilde{c}_i = c_i$ and $\tilde{w}_i = w_i$, then the standard order conditions for the two schemes are enough to ensure that the combined scheme is third order. Note, however, that this is true only for schemes up to third order.

3. L-stable second order DIRK schemes. In this section we deal with the accuracy of IMEX schemes. Since the schemes are used to solve a system that depends on the stiffness parameter ε, L-stability of the implicit Runge-Kutta integrator is a highly desirable property. In particular we expect that the accuracy of the resulting IMEX schemes depends on this stiffness parameter. Actually, even the regularity properties of the solution of the system (2) may depend on ε. In many cases of physical relevance, the solution of the initial value problem for system (2) is smooth, uniformly in ε, provided the initial condition is a *local equilibrium*, i.e. satisfies

$$g_2(u_0, v_0) = 0. \tag{19}$$

In such cases one might require that the numerical solution after one time step remains "close" to the exact solution uniformly in the stiffness parameter ε.

$$
\begin{array}{c|cccc}
0 & 0 & 0 & 0 & 0 \\
1/2 & 1/2 & 0 & 0 & 0 \\
1/3 & 1/3 & 0 & 0 & 0 \\
1 & 0 & 1 & 0 & 0 \\
\hline
 & 0 & 1 & 0 & 0
\end{array},
\qquad
\begin{array}{c|cccc}
0 & 0 & 0 & 0 & 0 \\
1/2 & 0 & 1/2 & 0 & 0 \\
1/3 & 0 & 0 & 1/3 & 0 \\
1 & 0 & 0 & 3/4 & 1/4 \\
\hline
1 & 0 & 0 & 3/4 & 1/4
\end{array}
$$

TABLE 7

Tableau for the explicit (left) implicit (right) L-stable LRR(3,2,2) scheme

$$
\begin{array}{c|cc}
0 & 0 & 0 \\
1 & 1 & 0 \\
\hline
 & 1/2 & 1/2
\end{array},
\qquad
\begin{array}{c|cc}
1-C & 1-C & 0 \\
C & C-\delta & \delta \\
\hline
 & 1/2 & 1/2
\end{array},
\qquad \delta = 1 - \frac{1}{2C}
$$

TABLE 8

Tableau for the explicit (left) implicit (right) L-stable PR(2,2,2) scheme

Here we shall limit to analyze the accuracy properties of the schemes for $\varepsilon = 1$ and for small values of ε. Uniform accuracy of the schemes will be investigated numerically.

3.1. Accuracy analysis. In this section we perform the accuracy analysis of the different L-stable schemes presented in Table 3 and Tables 5-8. Schemes 5-6 were presented in [1], scheme 7 was derived in [20], whereas scheme 8 is new.

All these schemes have second order accuracy in the nonstiff limit, i.e. for $\varepsilon = 1$. Here we are interested in the stiff accuracy, i.e. in the accuracy of the numerical solution in the limit as $\varepsilon \to 0$. In addition, we consider the asymptotic behavior of the solution for small values of ε.

Let us consider the system (2), in the particular case where $g_2(u, v) = s(u) - v$. In this case, the asymptotic behavior of the solution for small values of the parameter ε can be obtained by the so called Chapman-Enskog expansion. According to this expansion, the function u satisfies the equation

$$u' = f_1 + \varepsilon \frac{\partial f_1}{\partial v}(f_2 - s'(u) f_1) + O(\varepsilon^2),$$

where f_1, f_2, and $\partial f_1/\partial v$ are evaluated at $(u, s(u))$. The function v is given by

$$v = s(u) + \varepsilon(f_2 - s'(u) f_1) + O(\varepsilon^2).$$

From this expansion it is possible to derive a Taylor expansion of the solution after one time step, in terms of ε and h. For simplicity we shall apply the different IMEX to system (2), with f_1, f_2, and g_2 linear, i.e.

$$
\begin{aligned}
u' &= a_1 u + b_1 v, \\
v' &= a_2 u + b_2 v + \frac{1}{\varepsilon}(cu - v).
\end{aligned}
\tag{20}
$$

We shall assume that the initial state is in "equilibrium", i.e. it satisfies the condition

$$v_0 = s(u_0) = c u_0.$$

After one time step, the exact solution of the system satisfies

$$(21)\qquad \begin{aligned} u &= u_0(1+\tilde{a}h+\frac{1}{2}\tilde{a}^2h^2+\varepsilon b_1\tilde{b}h)+O(h^3,\varepsilon^2 h),\\ v &= u_0(c(1+\tilde{a}h+\frac{1}{2}\tilde{a}^2h^2)+\varepsilon\tilde{b}(1+(\tilde{a}+b_1c)h))+O(h^3,\varepsilon^2 h), \end{aligned}$$

where $\tilde{a}=a_1+b_1c$, and $\tilde{b}=a_2+(b_2-a_1)c-b_1c^2$.

Now we apply the different schemes for one time step. The results are summarized in the sequel.

CJR(3,2,2). The Taylor expansion of the first component is

$$u_1 = u_0\left(1+\tilde{a}h+\frac{1}{2}\tilde{a}^2h^2+\frac{1-\mu}{1-2\mu}\varepsilon b_1\tilde{b}h\right)+O(h^3,\varepsilon^2 h).$$

Therefore the scheme is second order accurate in the limit $\varepsilon=0$, but it does not provide the correct limit for small values of ε, unless one takes $\mu=1$. The value $\mu=1$, however, is excluded, because for this value of the parameter the scheme is no longer L-stable (this can be seen by the expression of the stability function, and its limit as $z\to\infty$).

This can be expressed by the formula

$$u_1(\varepsilon,h)-u(\varepsilon,h)=O(h^3,\varepsilon h).$$

Similarly, for the second component one has

$$v_1(\varepsilon,h)-v(\varepsilon,h)=O(h^3,\varepsilon h).$$

ARS(2,2,2). The Taylor expansion of the first component is

$$u_1 = u_0\left(1+\tilde{a}h+\frac{1}{2}\tilde{a}^2h^2+\frac{1}{2-\sqrt{2}}\varepsilon b_1\tilde{b}h\right)+O(h^3,\varepsilon^2 h).$$

Therefore the scheme is second order accurate in the limit $\varepsilon=0$, but it is not accurate for small values of ϵ. The result of the analysis can be summarized by the formulas

$$\begin{aligned} u_1(\varepsilon,h)-u(\varepsilon,h)&=O(h^3,\varepsilon h),\\ v_1(\varepsilon,h)-v(\varepsilon,h)&=O(h^3,\varepsilon h). \end{aligned}$$

ARS(2,3,2). In this case the Taylor expansion of the numerical solution for u agrees with the expansion of the exact solution up to terms $O(h^2,\varepsilon h)$. However, the scheme suffers a degradation of accuracy in the second variable. The result may be summarized by the formulas

$$\begin{aligned} u_1(\varepsilon,h)-u(\varepsilon,h)&=O(h^3,\varepsilon^2 h),\\ v_1(\varepsilon,h)-v(\varepsilon,h)&=O(h^2,\varepsilon h). \end{aligned}$$

Note, however, that the loss of accuracy in the variable v does not affect the accuracy in the variable u. This is easily understood in the limit $\varepsilon\to 0$. In this case the scheme becomes a second order scheme for the limit equation in u. In the exact solution the second variable is just a function of the first. The numerical scheme does not provide the correct projection, but computes the second variable by solving an equation for it. Although the local truncation error in v is $O(h^2)$, the scheme is essentially second order even in v in the limit $\varepsilon\to 0$, because the variable u is second order accurate, and so is v. This analysis is in agreement with the numerical results shown in Section 5.

PR(2,2,2). The Taylor expansion of the numerical solution for u is given by

$$u_1 = u_0(1 + \tilde{a}h + \frac{1}{2}\tilde{a}^2h^2 + \frac{C}{2C-1}\varepsilon b_1\tilde{b}h) + O(h^3, \varepsilon h).$$

From this expression it follows that the scheme, similarly to CJR(3,2,2), is second order in u, in the limit $\varepsilon = 0$, but it is not accurate for small values of ϵ

$$u_1(\varepsilon, h) - u(\varepsilon, h) = O(h^3, \varepsilon h).$$

The expansion of v_1 gives

$$v_1(\varepsilon, h) - v(\varepsilon, h) = O(h^2, \varepsilon h),$$

therefore the local truncation error is second order in h. Note that, because u is accurate, we expect the global truncation error in v to be of the same order of the local one.

LRR(3,2,2). The scheme provides second order accuracy, both in u and v, in the limit $\varepsilon = 0$, and the first order term in ε is first order in u. The results are summarized by the expressions

$$\begin{aligned} u_1 - u(\varepsilon, h) &= O(h^3, \varepsilon h^2), \\ v_1 - v(\varepsilon, h) &= O(h^3, \varepsilon h). \end{aligned}$$

We remark that the above analysis is performed by assuming that the stiffness parameter is either $O(1)$ or quite small. Probably a more satisfactory approach to the same problem, in the linear case, is obtained by using the concept of B-consistency and B-convergence [7]. In this case, in fact, uniform bounds in ε for the local truncation error could be derived.

3.2. L-stability and asymptotic preservation. In this section we will show that the so called asymptotic preservation property for hyperbolic systems with relaxation, i.e. consistency of the scheme in the stiff limit $\varepsilon \to 0$, is guaranteed if the implicit step is solved by a L-stable scheme.

To this aim let us consider the general structure of an IMEX Runge-Kutta scheme written in splitting form (9)-(10) and (11) with $Y = (U, V)$, $f = (f_1, f_2)$ and $g = (0, g_2)$. We will also assume that the algebraic equation $g_2(u, v) = 0$ can be uniquely solved in v giving (3).

The L-stability property together with our assumptions implies that the implicit step (10) as $\varepsilon \to 0$ yields

$$g_2(t_0 + \tilde{c}_j h, U_j, V_j) = 0 \quad \Longleftrightarrow \quad V_j = e(U_j). \tag{22}$$

Hence, denoting by $F(u) = f_1(U, e(U))$, the explicit step (9) gives the explicit scheme for the limiting equilibrium equation (4)

$$U_i^* = u_0 + h\sum_{j=1}^{i-2} \tilde{a}_{ij} F_j + h\,\tilde{a}_{ii-1} F(t_0 + \tilde{c}_{i-1}h, U_{i-1}), \tag{23}$$

where

$$F_j = F(t_0 + \tilde{c}_j h, U_j).$$

Finally we need to evaluate

$$u_1 = u_0 + h \sum_{i=1}^{\nu} \tilde{w}_i F_i. \tag{24}$$

It is remarkable that, under our assumptions, the tableau of the limiting scheme for (4) is always the tableau of the explicit part of the IMEX Runge-Kutta scheme. Note that in the limit the value of the v component can be recovered directly from u using (3).

3.3. Stability analysis. In this section we perform the stability analysis of the schemes. In particular, we use a generalization of the concept of A-stability to systems of equations which are the sum of a stiff and a non-stiff part.

Let us consider a system of the form (1) with $f(y) = \lambda_1 y$, $g(y) = \varepsilon \lambda_2 y$, and $y_0 = 1$. When applying scheme (7)-(8) to the above system, one obtains

$$Y_i = 1 + h \sum_{j=1}^{\nu} \tilde{a}_{ij} \lambda_1 Y_j + h \sum_{j=1}^{\nu} a_{ij} \lambda_2 Y_j, \tag{25}$$

$$y_1 = 1 + h \sum_{i=1}^{\nu} \tilde{w}_i \lambda_1 Y_i + h \sum_{i=1}^{\nu} w_i \lambda_2 Y_i. \tag{26}$$

Using a vector notation with $Y = (Y_1, \ldots, Y_\nu)^T$, Eq. (25) becomes,

$$Y = e + \lambda_1 h \tilde{A} Y + \lambda_2 h A Y$$

where $e = (1, \ldots, 1)^T$.

Solving for Y and substituting in Eq. (26), one has

$$y_1 = \mathcal{R}(z_1, z_2) \equiv 1 + (z_1 \tilde{w}^T + z_2 w^T)(I - z_1 \tilde{A} - z_2 A)^{-1} e, \tag{27}$$

where $z_1 = \lambda_1 h$, $z_2 = \lambda_2 h$. The function $\mathcal{R}(z_1, z_2)$ is the *function of absolute stability.*

As an example, we report the function of absolute stability of some of the schemes we discussed before.

The function of absolute stability of scheme PR(2,2,2) is given by

$$\mathcal{R}(z_1, z_2) = 1 - \frac{(z_1 + z_2)\,(2\,C - 3\,z_2\,C + z_2 + z_1\,C + 2\,z_2\,C^2)}{(1 - z_2 + z_2\,C)\,(-2\,C + 2\,z_2\,C - z_2)}.$$

For $C = 1$ and $z_1 = 0$ this function becomes

$$\mathcal{R}(0, z_2) = -\frac{2 + z_2}{-2 + z_2},$$

which is not the stability function of an L-stable scheme, since $R(0, \infty) = 1$.

The function of absolute stability of scheme ARS(2,2,2) is given by

$$\mathcal{R}(z_1, z_2) = 1 - \frac{{z_2}^2\gamma^2 - z_1\,\gamma\,z_2 + z_1\,\gamma^2 z_2 + {z_1}^2\gamma\,\delta - z_1 - z_2 + z_1\,\delta\,z_2\,\gamma - {z_1}^2\gamma}{(-1 + z_2\,\gamma)^2},$$

and the one of scheme ARS(2,3,2) is given by

$$\mathcal{R}(z_1, z_2) = 1 - \frac{(z_1 + z_2)\,(-2\,z_1\,\gamma + z_1\,\gamma^2 - {z_1}^2\gamma^2 + {z_1}^2\gamma^2\delta + z_1\,\gamma^2\delta\,z_2 - 1 + \gamma^2 z_2)}{(-1 + z_2\,\gamma)^2}.$$

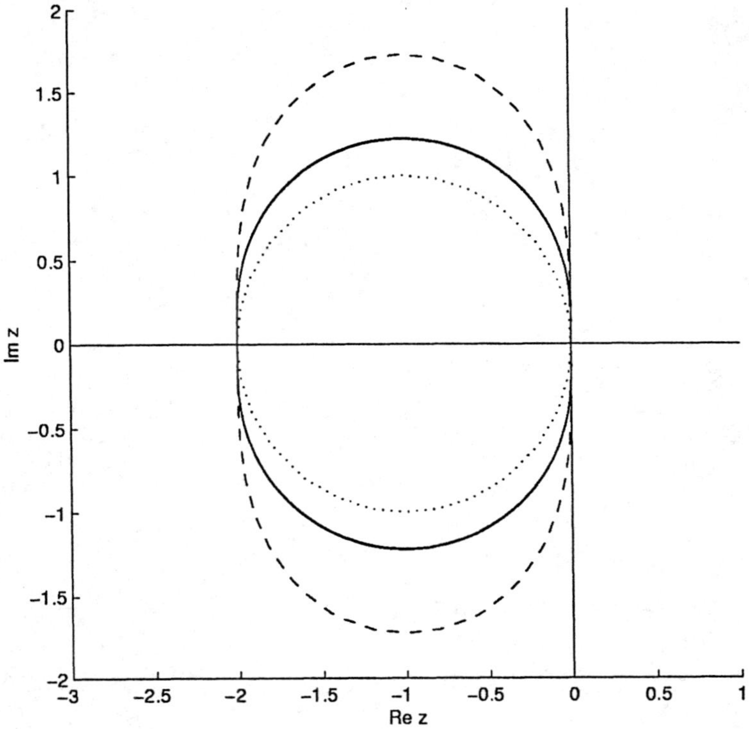

FIG. 1. Stability region S_1 for scheme CJR(3,2,2) with $\mu = 1/3$, ARS(2,3,2), PR(2,2,2) with $C = 1/\sqrt{2}$ (continuous line). The dotted and the dashed lines are the stability region of explicit Euler and explicit second order Runge-Kutta respectively.

The *region of absolute stability* S_A associated to scheme (7)-(8) is defined as

$$S_A = \{(z_1, z_2) \in \mathbf{C}^2 : |\mathcal{R}(z_1, z_2)| \leq 1\}.$$

It is evident that the region does not contain the set $\mathbf{C}^- \times \mathbf{C}^-$. Our goal is to show that there exist two regions of the complex plane, $S_1 \subset \mathbf{C}$, $S_2 \subset \mathbf{C}$, with the following properties:

$$S_A \supset S_1 \times S_2, \tag{28}$$

$$S_2 \supset \mathbf{C}^- \equiv \{z \in \mathbf{C} : \Re(z) \leq 0\}, \tag{29}$$

and to compute them. It is clear that such two sets, if they exist, are not unique. We shall compute explicitly the largest set S_1 for which $S_2 \supset \mathbf{C}^-$. Such region is defined by

$$S_1 = \{z_1 \in \mathbf{C} : \sup_{z_2 \in \mathbf{C}^-} |\mathcal{R}(z_1, z_2)| \leq 1\}.$$

In order to proceed we make use of the following

LEMMA 3.1. *For any fixed $z_1 \in \mathbf{C}$, the function $|\mathcal{R}(z_1, z_2)|$ assumes its maximum value in $\mathbf{C}^-$ for some z_2 belonging to the imaginary axis.*

Proof.

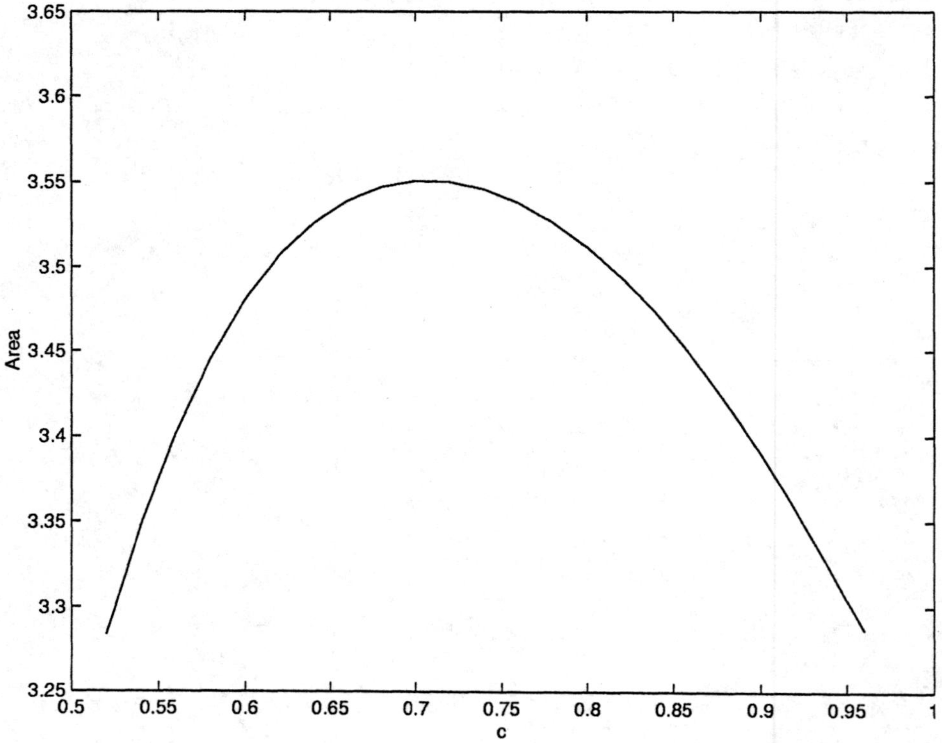

FIG. 2. Area of the stability region S_1 for scheme PR, as a function of the parameter c.

For any fixed value z_1, the function $\mathcal{R}(z_1, z_2)$ is analytic in z_2, olomorphic in $\mathbf{C}^-$. Therefore the maximum value of the function in $\mathbf{C}^-$ is obtained for some value of $z_2 \in \partial\mathbf{C}^-$. Furthermore, we observe that

$$\lim_{z_2 \to \infty} \mathcal{R}(z_1, z_2) = 0,$$

and hence the maximum of $|\mathcal{R}(z_1, z_2)|$ occurs on the imaginary axis.

From the above lemma, we can write

$$S_1 = \{z_1 \in \mathbf{C} : \max_{y \in \mathbf{R}} |\mathcal{R}(z_1, iy)| \leq 1\}.$$

The boundary of this region (boundary locus) will be obtained as

$$\partial S_1 = \{z_1 \in \mathbf{C} : \max_{y \in \mathbf{R}} |\mathcal{R}(z_1, iy)| = 1\}.$$

In order to compute the region S_1 we act as follows. Let us define the function

$$F(\theta, \rho, y) = |\mathcal{R}(-1 + \rho e^{i\theta}, iy)|^2, \quad \theta \in [-\pi, \pi), \quad y \in \mathbf{R}.$$

This function can be written in the form

$$F(\theta, \rho, y) = \frac{N(\theta, \rho, y)}{D(\theta, \rho, y)},$$

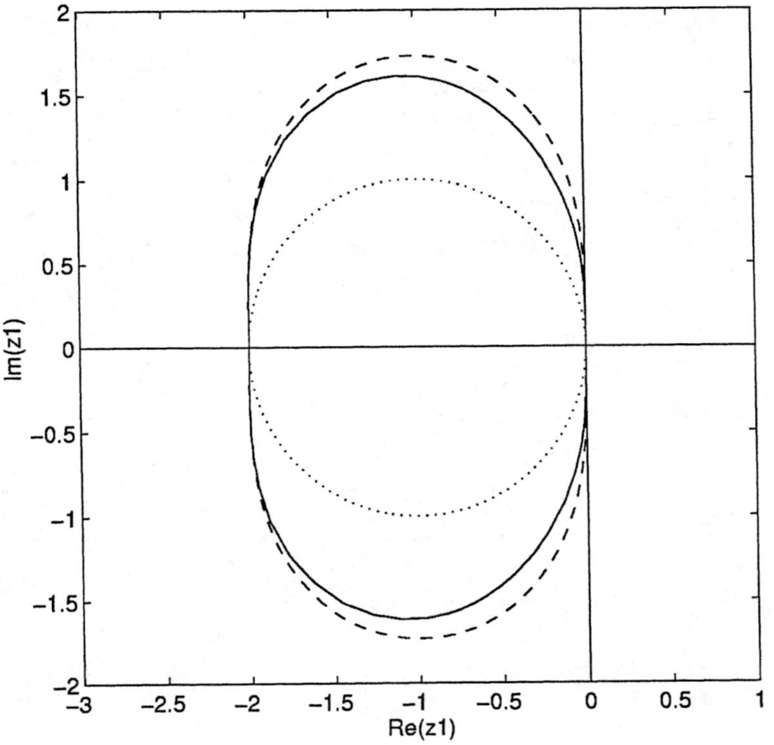

FIG. 3. Stability region S_1 for scheme LRR(3,2,2).

where N and D are polynomials in y and ρ, whose coefficients are trigonometric functions in θ. Now, let $F_s(\theta, \rho, y) = N(\theta, \rho, y) - D(\theta, \rho, y)$ and

$$\mathcal{H}(\theta, \rho) = \max_{y \in \mathbf{R}} F_s(\theta, \rho, y).$$

The boundary of S_1 can therefore be expressed in terms of $\mathcal{H}(\theta, \rho)$,

$$\partial S_1 = \{(\theta, \rho) : \mathcal{H}(\theta, \rho) = 0\}.$$

The boundary locus can be efficiently computed by a technique similar to the one used to compute the stability region of Runge-Kutta schemes. First observe that $\mathcal{H}(\theta, 0) < 0$, therefore $-1 \in S_1$. Now let us denote by $\hat{y}$ the value of y for which $\mathcal{H}(\theta, \rho) = F_s(\theta, \rho, y)$, i.e.

$$\hat{y}(\theta, \rho) = \arg \max_{y \in \mathbf{R}} F_s(\theta, \rho, y),$$

and let $\hat{\rho}$ denote the value of ρ for which $F_s(\theta, \rho, \hat{y}(\theta, \rho)) = 0$. Then the boundary locus satisfies the equations

$$F_s(\theta, \rho, y) = 0, \tag{30}$$

$$G(\theta, \rho, y) \equiv \frac{\partial F_s}{\partial y}(\theta, \rho, y) = 0. \tag{31}$$

The second equation is a consequence of the fact that $\hat{y}$ is an extremal, and the function $F_s(\theta,\rho,y)$ is smooth in y.

By differentiating equations (30) and (31) one obtains a set of differential equations that defines the boundary locus. By assuming that the quantities θ, ρ, and y are function of a parameter τ, one can derive the equations for the trajectories

$$\frac{d\theta}{d\tau} = -\alpha(\theta,\rho,y)F_1(\theta,\rho,y), \tag{32}$$

$$\frac{d\rho}{d\tau} = \alpha(\theta,\rho,y)F_2(\theta,\rho,y), \tag{33}$$

$$\frac{dy}{d\tau} = \alpha(\theta,\rho,y)F_3(\theta,\rho,y), \tag{34}$$

where the function α is arbitrary, and has the effect of changing the parametrization, and

$$F_1(\theta,\rho,y) = \frac{\partial F_s}{\partial \rho}\frac{\partial G}{\partial y}, \tag{35}$$

$$F_2(\theta,\rho,y) = \frac{\partial F_s}{\partial \theta}\frac{\partial G}{\partial y}, \tag{36}$$

$$F_3(\theta,\rho,y) = \frac{\partial G}{\partial \theta}\frac{\partial F_s}{\partial \rho} - \frac{\partial F_s}{\partial \theta}\frac{\partial G}{\partial \rho}. \tag{37}$$

The initial condition for the system is

$$\rho(0) = 1, \quad \theta(0) = 0, \quad y(0) = 0.$$

We decided to use the arc-length as parameter. In this case

$$\alpha = \frac{1}{\sqrt{F_1^2 + F_2^2 + F_3^2}}.$$

The system is solved forward in τ to obtain the upper branch of the boundary locus, up to a point so that $\theta = \pi$. Then the boundary locus is extended symmetrically below the real axis.

By using the above mentioned technique, we computed the stability region for the schemes shown in the previous section.

First we show the stability region S_1 for schemes CJR(3,2,2) with $\mu = 1/3$, ARS(2,3,2), PR(2,2,2) with $C = 1/\sqrt{2}$. The result is shown in Fig. 1. Note that the stability regions of the three schemes are almost identical and are smaller than the one of standard explicit RK2, but they include the stability region of explicit Euler.

The size of the stability region of the family of schemes PR(2,2,2) depends slightly on the parameter C, while its shape is almost unchanged. In Fig. 2 we plot the area of the stability region S_1 as a function of the parameter C. We observe that the maximum of the area is obtained near the value $C = 1/\sqrt{2}$.

We report also the stability region S_1 for scheme LRR(3,2,2) in Fig. 3. Note that this scheme has a slightly wider stability region.

4. Third order schemes. Third order IMEX Runge-Kutta schemes can be obtained directly using the order conditions we have seen in Section 2. We will report here, as examples, the ARS(2,3,3) and the L-stable ARS(3,4,3) and ARS(4,4,3) IMEX Runge-Kutta

$$
\begin{array}{c|ccc}
0 & 0 & 0 & 0 \\
\gamma & \gamma & 0 & 0 \\
1-\gamma & \gamma-1 & 2-2\gamma & 0 \\
\hline
 & 0 & 1/2 & 1/2
\end{array}, \qquad
\begin{array}{c|ccc}
0 & 0 & 0 & 0 \\
\gamma & 0 & \gamma & 0 \\
1-\gamma & 0 & 1-2\gamma & \gamma \\
\hline
 & 0 & 1/2 & 1/2
\end{array}, \qquad \gamma = \frac{3+\sqrt{3}}{6}
$$

TABLE 9

Tableau for the explicit (left) implicit (right) third order ARS(2,3,3) scheme

$$
\begin{array}{c|cccc}
0 & 0 & 0 & 0 & 0 \\
\gamma & \gamma & 0 & 0 & 0 \\
(1+\gamma)/2 & (1+\gamma)/2-\eta & \eta & 0 & 0 \\
1 & 1-2\mu & \mu & \mu & 0 \\
\hline
 & 0 & 1-\delta-\gamma & \delta & \gamma
\end{array}, \qquad
\begin{array}{c|cccc}
0 & 0 & 0 & 0 & 0 \\
\gamma & 0 & \gamma & 0 & 0 \\
(1+\gamma)/2 & 0 & (1-\gamma)/2 & \gamma & 0 \\
1 & 0 & 1-\delta-\gamma & \delta & \gamma \\
\hline
 & 0 & 1-\delta-\gamma & \delta & \gamma
\end{array}
$$

$$\gamma = 0.4358665215, \delta = -0.644373171, \eta = 0.3966543747, \mu = 0.5529291479.$$

TABLE 10

Tableau for the explicit (left) implicit (right) third order L-stable ARS(3,4,3) scheme

schemes (see [1]). In all these schemes the implicit scheme is a DIRK scheme and hence they can be written in splitting form accordingly to (9)-(10) and (11). In particular scheme ARS(3,4,3) is defined using the root of a third order polynomial. We report here its numerical equivalent.

5. Test equation. A simple prototype of stiff system of the form (2) is obtained taking

$$f_1(u,v) = \alpha u, \quad f_2(u,v) = \beta v, \quad g_2(u,v) = e(u) - v,$$

which corresponds to the system of Ode's

$$u' = \alpha v, \tag{38}$$

$$v' = \beta u + \frac{1}{\varepsilon}(e(u) - v). \tag{39}$$

The choice $\alpha = -1$, $\beta = 1$ guarantees that the eigenvalues of the explicit part are $\pm i$.

The convergence plots have been obtained first computing the relative error E_h by comparing the numerical solution with step h to the numerical solution with step $h/2$ with different values of ε ranging from 10^{-5} to 1. The value $h = 0.05$ has been used, and the system has been integrated for $t \in [0, 5]$. Then, the convergence rates have been computed as usual through the error curves E_h and $E_{h/2}$. We present the results by plotting the L_2 norm of the relative error versus ϵ using a log-scale on the x-axis.

To examine the behavior of the different IMEX Runge-Kutta schemes we have considered $e(u) = \sin(u)$ and two different set of initial data:

Equilibrium data.

$$u(0) = \pi/2, \quad v(0) = \sin(u(0)) = 1.$$

0	0	0	0	0	0	0	0	0	0	0	0
1/2	1/2	0	0	0	0	1/2	0	1/2	0	0	0
2/3	11/18	1/18	0	0	0	2/3	0	1/6	1/2	0	0
1/2	5/6	−5/6	1/2	0	0 ,	1/2	0	−1/2	1/2	1/2	0
1	1/4	7/4	3/4	−7/4	0	1	0	3/2	−3/2	1/2	1/2
	1/4	7/4	3/4	−7/4	0		0	3/2	−3/2	1/2	1/2

TABLE 11

Tableau for the explicit (left) implicit (right) third order L-stable ARS(4,4,3) scheme

In Fig. 4 we report the convergence profiles of the different L-stable second order schemes together with scheme Midpoint(2,2,2) in the case of equilibrium initial data. All the L-stable schemes gives similar results with second order accuracy both for large and small values of ε, and with an intermediate region where $h = O(\varepsilon)$ in which we have a deterioration of the accuracy, in particular for the v component of the solution. Note that, in the stiff limit the accuracy on the v component in the PR(2,2,2) deteriorates to first order. This is not a strong drawback of the scheme since an accurate value of v can be always computed from u in the limit $\varepsilon \to 0$. As expected the Midpoint(2,2,2) scheme is unstable for very small values of ε and the v component has a slightly oscillating behavior in the stiff limit. It is remarkable, however, that for this scheme the accuracy on u is uniformly second order with respect to ε.

The corresponding results for the third order schemes are given in the left column of Fig. 6. Again the behavior of the schemes is rather similar although it can be observed that the accuracy of ARS(4,4,3) deteriorates in a wider region.

Non-equilibrium data.

$$u(0) = \pi/2, \quad v(0) = 1/2.$$

In this case $g_2(u(0), v(0)) \neq 0$ and hence the problem presents an initial layer on the v component when $\varepsilon \to 0$.

In Fig. 5 we present the same convergence profiles as before of the different second order IMEX Runge-Kutta schemes in the case of non equilibrium initial data. The accuracy of the schemes deteriorates dramatically in the intermediate region $h = O(\varepsilon)$ and again this is more evident in the v component. In the stiff limit, in particular, both CJR(3,2,2) and ARS(2,2,2) schemes deteriorate to first order. Note that scheme PR(2,2,2) seems to outperform the other schemes since the results are similar to the case of equilibrium initial data. For this initial data, the Midpoint(2,2,2) scheme has a strongly oscillating behavior on the v component for ε small and the scheme is clearly unstable in the stiff regime.

Finally in the right column of Fig. 6 we give the results of the third order schemes for non-equilibrium initial data. Scheme ARS(3,4,3) clearly outperforms the other two schemes. Note that scheme ARS(2,3,3) is not L-stable and hence, although no instability is observed, v does not have the correct asymptotic limit. Furthermore scheme ARS(4,4,3) reduces to first order accuracy in the stiff limit.

REFERENCES

[1] U. Asher, S. Ruuth, and R. J. Spiteri, *Implicit-explicit Runge-Kutta methods for time dependent Partial Differential Equations*, Appl. Numer. Math. 25, (1997), pp. 151-167.

[2] U. Asher, S. Ruuth, and B. Wetton, *Implicit-explicit methods for time dependent PDE's*, SIAM J. Numer. Anal., 32, (1995) pp. 797-823.

[3] J. Butcher, *The numerical analysis of Ordinary differential equations. Runge-Kutta and general linear methods.*. John Wiley & Sons, Chichester and New York (1987).

[4] J. Butcher, N. Guglielmi, and G. Russo, in preparation.

[5] R. E. Caflisch, S. Jin and G. Russo, *Uniformly accurate schemes for hyperbolic systems with relaxation*, SIAM J. Numer. Anal., 34, (1997) pp. 246-281.

[6] M. Cecchi Morandi, M. Redivo-Zaglia, G.Russo, *Extrapolation methods for hyperbolic systems with relaxation*, Journal of Computational and Applied Mathematics 66 (1996) pp. 359-375.

[7] K. Dekker, and J.G. Verwer, *Stability of Runge-Kutta Methods for Stiff Nonlinear Differential Equations*, North-Holland, Amsterdam (1984).

[8] Boun Oumar Dia and Michelle Schatzman, *Estimation sur la formule de Strang*, C. R. Acad. Sci. Paris, t.320, Série I (1995) pp. 775-779.

[9] Boun Oumar Dia and Michelle Schatzman, *Commutateur de certains semi-groupes holomorphes et applications aux directions alternées*, Mathematical Modelling and Numerical Analysis 30 (1996) pp. 343-383.

[10] S.Gottlieb, C.-W. Shu, *Total Variation Diminishing Runge-Kutta schemes*, Math. Comp. 67 (1998) pp. 73-85.

[11] E. Hairer, *Order conditions for numerical methods for Partitioned ordinary differential equations*, Numerische Mathematik 36 (1981) pp. 431-445.

[12] E. Hairer, S.P.Nørset and G. Wanner, *Solving ordinary differential equations, Vol.1 Nonstiff problems*, Springer-Verlag, New York (1987).

[13] E. Hairer and G. Wanner, *Solving ordinary differential equations, Vol.2 Stiff and differential-algebraic problems*, Springer-Verlag, New York (1987).

[14] E. Hofer *A partially implicit method for large stiff systems of Ode's with only few equations introducing small time-constants*, SIAM J. Numer. Anal. 13, (1976) pp. 645-663.

[15] S. Jin, *Runge-Kutta methods for hyperbolic systems with stiff relaxation terms* J. Comput. Phys., 122 (1995) pp. 51-67.

[16] S. Jin and C. D. Levermore, *Numerical Schemes for hyperbolic conservation laws with stiff relaxation terms*, J. Comp. Physics, 126 (1996) pp. 449-467.

[17] S. Jin, L. Pareschi and G. Toscani *Diffusive relaxation schemes for multiscale discrete-velocity kinetic equations*, SIAM J. Num. Anal., 35 (1999) pp. 2405-2439.

[18] S.Jin, A.Klar and L.Pareschi *Asymptotic preserving schemes for multiscale kinetic equations: a unified approach*, work in preparation.

[19] R.J. LeVeque, *Numerical Methods for Conservation Laws*, Lectures in Mathematics, Birkhauser Verlag, Basel (1992).

[20] S.F.Liotta, V.Romano, G.Russo, *Central schemes for balance laws of relaxation type*, submitted to SIAM J. Numer. Anal.

[21] G.Naldi and L.Pareschi, *Numerical schemes for hyperbolic systems with stiff diffusive relaxation*, SIAM J. Numer. Anal. to appear.

[22] L.Pareschi and G.Russo, *Implict-Explicit Runge-Kutta schemes for hyperbolic systems with stiff relaxation*, work in progress

[23] G.Strang, *On the construction and comparison of difference schemes*, SIAM J. Numer. Anal. **5**, (1968) pp. 506.

[24] Tadmor E., *Approximate Solutions of Nonlinear Conservation Laws*, Cockburn, Johnson, Shu and Tadmor Eds., Lecture Notes in Mathematics, N. 1697 (1998).

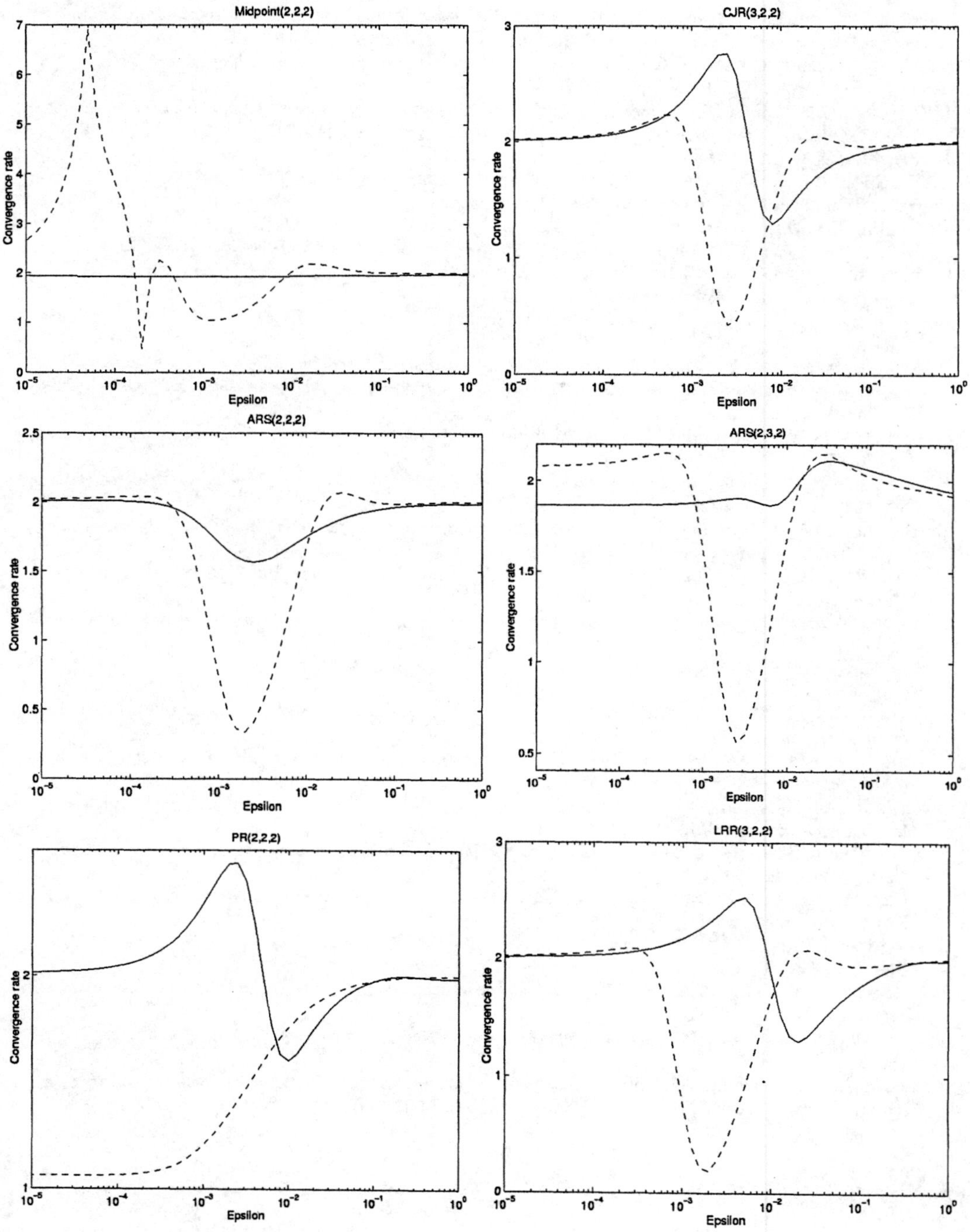

FIG. 4. Convergence rates of second order IMEX schemes for equilibrium initial data (dashed line: v component, continuous line: u component).

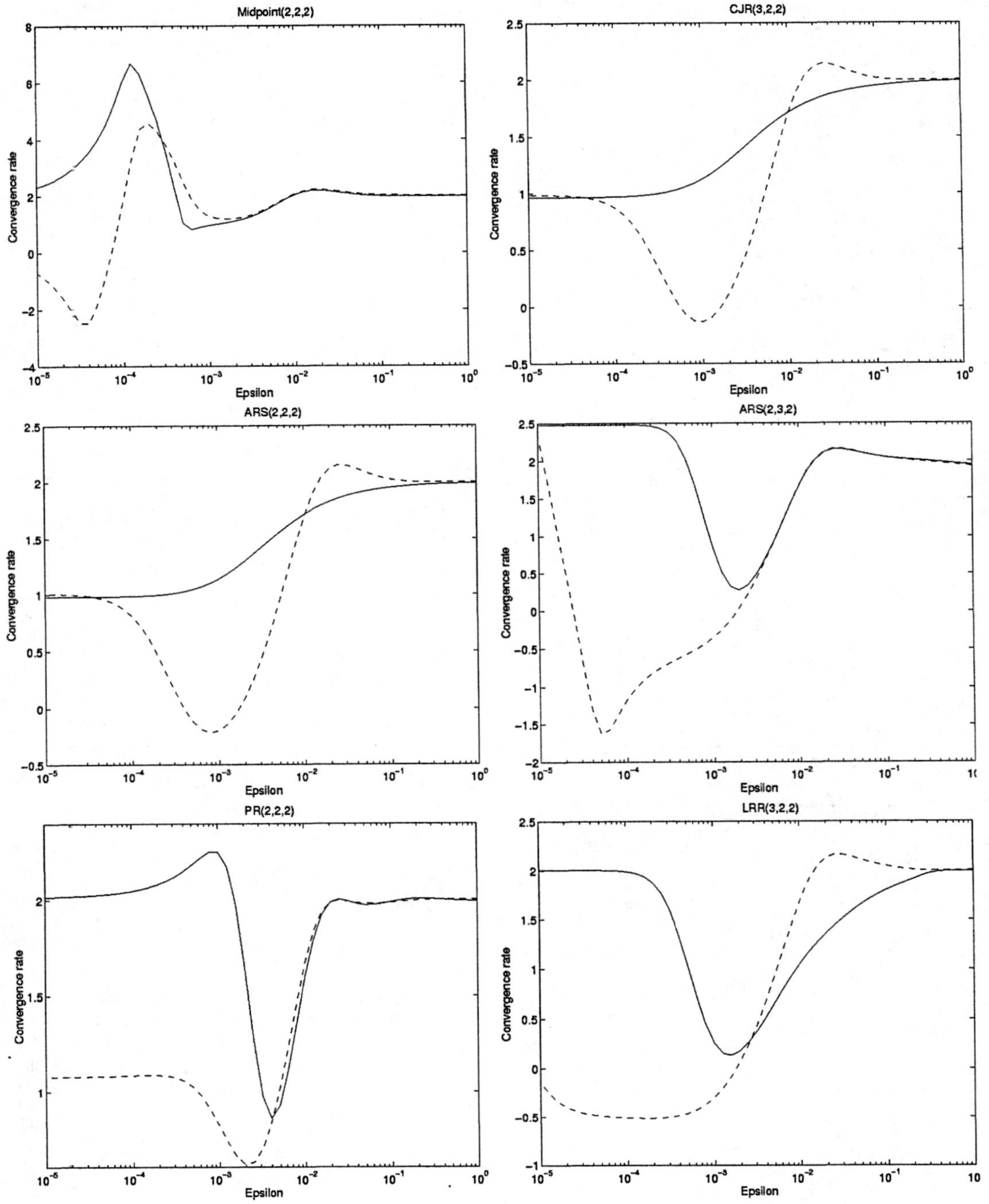

FIG. 5. Convergence rates of second order IMEX schemes for non equilibrium initial data (dashed line: v component, continuous line: u component).

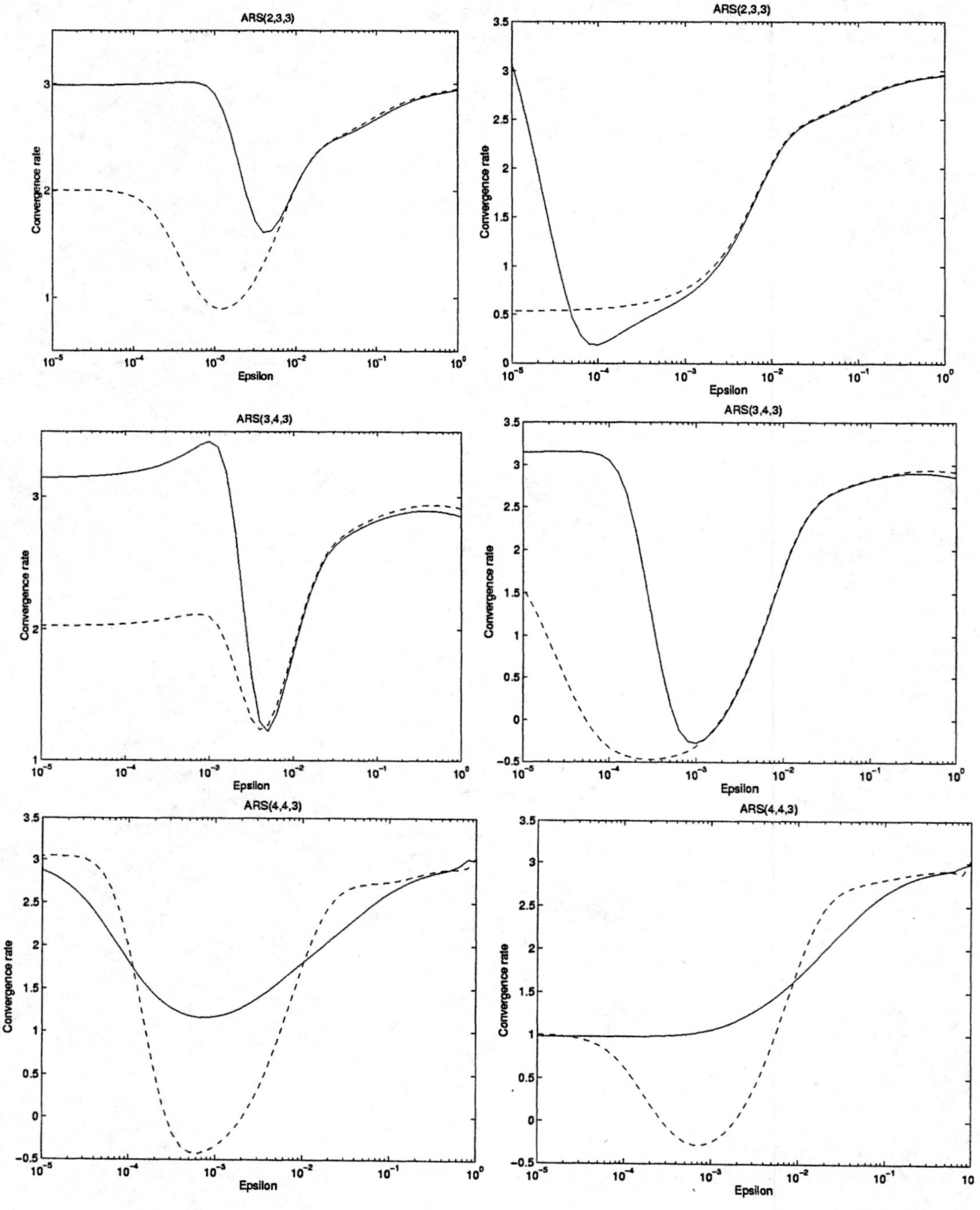

FIG. 6. Convergence rates of third order IMEX schemes for equilibrium initial data (left column) and non-equilibrium initial data (right column). Dashed line: v component, continuous line: u component.

NEWTON METHOD FOR SYMMETRIC INVERSE EIGENVALUE PROBLEMS

R. PELUSO* AND I. SGURA†

Abstract. A *Symmetric Inverse Eigenvalue Problem* arises when a real symmetric matrix has to be constructed from a given set of eigenvalues. In this paper, we restrict our attention to the case of distinct eigenvalues.

An algebraic approach is followed to solve the problem by means of the Newton method and two new algorithms using traces of matrices are introduced. Moreover, it is shown that they are less expensive than other methods for these problems based on a similar approach (see for example [8]). The additive inverse eigenvalue problem (AIEP) is analysed in detail.

Numerical experiments for additive, multiplicative and Toeplitz inverse eigenvalues problems are performed, showing rather satisfactory results.

1. Introduction. The inverse eigenvalue problem concerns the reconstruction of a matrix from prescribed spectral data. Inverse eigenvalue problems arise in different applications such as geophysics, particle physics, circuit theory and control design [1]. In fact, the entries of the required matrix usually represent physical parameters to be determined.

We restrict our attention to the *symmetric inverse eigenvalue problem (IEP, hereafter)*, where a real symmetric matrix has to be constructed from a given set of distinct eigenvalues.

In the literature, sufficient conditions for the existence of solutions are given when the structure of the matrix to be constructed is fixed (see for example [3, 7, 5]). Recently, the computational problem of numerically determining such matrices has been accomplished by using isospectral flows [1, 2].

A different approach to the computability question is investigated in [8] by solving a nonlinear system of equations $F(c) = 0$, with $F : R^n \to R^n$, suitably defined. The Newton method is then applied for solving the above equation so that the locally quadratic convergence is guaranteed. Moreover, quasi-Newton methods preserving quadratic convergence are used to reduce computational costs. In the same paper the inverse problem for multiple eigenvalues is also well investigated.

In this paper, we consider the use of the classical Newton method for two new choices of the function F involving traces of matrices. For the general symmetric case, each Newton step in our procedures is less expensive than those of the quasi-Newton methods in [8].

In Section 2, after setting a general mathematical formulation of the problem, we show that specific applications can be obtained by fixing together with the eigenvalues the form of the required matrix, too. The methods using the traces of matrices, along with their complexity analysis, are presented in Section 3. In Section 4, the application of one of the methods to the *additive inverse eigenvalue problem (AIEP)* is considered in more detail. Finally, some numerical tests are presented in Section 5 to show the actual performances of the first method here introduced for *additive, multiplicative* and *Toeplitz inverse eigenvalue problems*.

2. Formulation of the problem. Let $A_0, A_1, A_2, \ldots, A_n \in \mathbb{R}^{n\times n}$ be real and symmetric matrices and $c = (c_1, \ldots, c_n)^T \in \mathbb{R}^n$. Let us define

*Dipartimento Interuniversitario di Matematica, Politecnico di Bari, Via E. Orabona, 4, 70125 Bari irmarp07@area.ba.cnr.it.

†Istituto per Ricerche in Matematica Applicata - I.R.M.A. - CNR, Via Amendola, 122-I, 70126 Bari, irmais16@area.ba.cnr.it.

$$A(c) = A_0 + \sum_{i=1}^{n} c_i A_i. \tag{1}$$

The symmetric inverse eigenvalue problem (IEP) can be formulated as follows.

PROBLEM 1. *Let $\lambda_1^* < \lambda_2^* < \ldots < \lambda_n^*$ be real and distinct eigenvalues. Find $c^* \in I\!R^n$ such that the $\{\lambda_i^*\}$ are the eigenvalues of $A(c^*)$.*

Two questions about such problem arise:

- its *solvability*, concerning the existence of solution(s), and
- its *computability*, concerning the numerical determination.

In a recent paper of Chu [1] many results about both questions are summarized for a collection of IEP. However, the previous formulation (1) is too general to induce an existence theorem without further assumptions. In fact, several results are present in the literature when some features of the IEP are specified. Let us give some examples.

Structured IEP (SIEP). The matrix A(c) has a given structure. For example, A(c) is a real and symmetric Toeplitz matrix (ITEP). ITEP can be formulated in the form (1) by considering $A_0 = 0$ and, for all $j = 1, .., n$, the basis $A_j = H^{j-1} + (H^T)^{j-1}$, where

$$H = \begin{pmatrix} 0 & \cdots & \cdots & \cdots & 0 \\ 1 & \ddots & & & \vdots \\ 0 & \ddots & \ddots & & \vdots \\ \vdots & \ddots & \ddots & \ddots & \vdots \\ 0 & \cdots & 0 & 1 & 0 \end{pmatrix}.$$

Additive IEP (AIEP). Let be $D(c) = \mathrm{diag}(c_1, ..., c_n)$ and $M \in \mathrm{I\!R}^{n \times n}$ a given matrix. In such a case, we consider

$$A(c) = M + D(c), \tag{2}$$

which can be expressed in the general formulation (1) by taking $A_0 = M$ and $A_j = e_j e_j^T$, $j = 1, \ldots, n$, where as usual e_j is the jth unit vector in $\mathrm{I\!R}^n$.

Multiplicative IEP (MIEP). By using the same notation introduced for AIEP, we now have

$$A(c) = D(c)\, M, \tag{3}$$

with $c > 0$, and M is positive definite. In order to cast such problem in the general formulation (1), it must be symmetrized through the following tranformation:

$$\hat{A}(c) = L^T D(c)\, L \equiv L^T A(c)\, L^{-T}, \tag{4}$$

where $M = LL^T$ is the Cholesky factorization of M. In such a way, MIEP is obtained from (1) by setting $\lambda_j^* > 0$, $j = 1, \ldots, n$, $A_0 = 0$, and $A_j = L^T \mathrm{diag}(e_j) L$.

For example, the last two problems respectively arise in the discretization of the Inverse Sturm-Liouville problem and the inverse vibrating string problem [8]. Moreover, particular applications of AIEP and MIEP can correspond to the research of a suitable splitting and preconditioning of a matrix, respectively [1].

Concerning ITEP, in [5] Landau gives a non constructive existence theorem for real distinct eigenvalues. Besides, he conjectures that in the subset of regular Toeplitz matrices, ITEP solution is unique.

The computational question has been faced by Chu [1] by using isospectral flows whose equilibria are symmetric Toeplitz matrices. In the case of distinct eigenvalues, and when the starting matrix is tridiagonal symmetric and centro-symmetric, the global convergence of the flow to a regular Toeplitz matrix has been proved in [2] for $n = 3$ and numerically shown for $n > 3$.

Hence, a class of SIEP has been solved using ODE matrix as a continuation method globally convergent to a matrix solution. Moreover, when isospectral methods are used for the numerical approximation of the ODE solution, the given eigenvalues are *exactly* preserved (see [2]).

Instead for the AIEP (2) and MIEP (4), Friedland [3] showed that both problems are always solvable on the complex field with a finite number of solutions. But, in the real field the existence has yet to be settled and only some sufficient conditions for the solvability are known for Hermitian matrices (see for example [1, 7]).
From the computational point of view, starting by the paper [8], IEP is transformed in a nonlinear system of algebraic equations:

$$F(c) = 0, \qquad F : \mathbb{R}^n \to \mathbb{R}^n. \tag{5}$$

Hence, the Newton method is applied to the above system, solving IEP in *approximate* way with locally quadratic convergence.

For example, all methods in [8] solve the system of components

$$\lambda_i(c) - \lambda_i^* = 0, \qquad i = 1, \ldots, n,$$

where $\lambda_i(c)$ are the eigenvalues of $A(c)$ to be calculated at each iteration. Moreover, quasi-Newton methods preserving the quadratic convergence are used in the same paper to reduce computational costs.

Note that in [8] an algorithm using determinants was rejected by the authors because its cost was higher than that of the other algorithms proposed.

Starting from this idea, in this paper we consider the classical Newton method for two new choices of the function $F(c)$ involving traces of matrices instead of determinants.

Let us remark that, in practical applications, this algebraic approach can be preferred to the ODE approach if the data λ_i^* are affected by error measurements and there is information on the region where the vector solution c^* belongs. In the following we will refer to the case of simple eigenvalues, because for IEP with multiple eigenvalues a further analysis is needed.

3. The methods of traces. In this section we introduce two new methods using traces of matrices and based on the same approach as in [8]. Our aim is to show that both the new methods are less expensive than those proposed in [8]. As we said, the idea is to find $c^* \in \mathbb{R}^n$ as a solution of a nonlinear system of equations in the form (5).

We recall that the Newton method applied to (5) can be formulated as

$$c^{k+1} = c^k - J_F^{-1}(c^k)F(c^k),$$

starting from a given vector c^0, where J_F is the Jacobian of F. Equivalently, by setting $\delta_k = c^{k+1} - c^k$ as the Newton step, we have to solve at each iteration the linear system

$$J_F(c^k)\delta_k = -F(c^k). \tag{6}$$

Moreover, we also recall some properties of the trace of a matrix.

PROPOSITION 3.1. *For $A, B \in I\!R^{n\times n}$, the following properties hold:*

1) $tr(A^T) = tr(A)$;
2) $tr(AB) = tr(BA)$;
3) Let $A_i \in I\!R^{n\times n}$ be nonsingular for all $i = 1, \ldots, m$, then $tr(A_1A_2\ldots A_m) = tr(A_iA_{i+1}\ldots A_mA_1\ldots A_{i-1})$ for all i.

Moreover, if for all $t \in I\!R$, $A(t) \in I\!R^{n\times n}$ is a differentiable function, then:

4) $\frac{d}{dt}tr(A^m(t)) = tr(\frac{d}{dt}A^m(t)) = m \cdot tr((\frac{d}{dt}A(t))A^{m-1}(t))$;
5) $\frac{d}{dt}tr(A^{-1}(t)) = tr(\frac{d}{dt}A^{-1}(t)) = -tr(A^{-1}(t)(\frac{d}{dt}A(t))A^{-1}(t)) = -tr(\frac{d}{dt}A(t)A^{-2}(t))$.

Let us now introduce the following:

METHOD I. Let us consider

$$b_i \equiv \sum_{k=1}^{n}(\lambda_k^*)^i = tr(A^i(c^*)), \qquad i = 1, \ldots, n.$$

Hence, we are looking for $c^* \in \mathrm{I\!R}^n$ such that $F(c^*) = 0$, where

$$F_i(c) = tr(A^i(c)) - b_i, \qquad i = 1, \ldots, n. \tag{7}$$

For our choice the elements $J_F(c)_{ij}$ of the Jacobian matrix are given by

$$\frac{\partial F_i}{\partial c_j} = tr(\frac{\partial A^i(c)}{\partial c_j}) = i \cdot tr(A_jA^{i-1}(c)), \qquad i, j = 1, \ldots, n. \tag{8}$$

Complexity. In terms of Matlab *flops*, we recall the operation count for some basic computations of Linear Algebra (see, for example, [4]):

(i) x^Ty, $x, y \in \mathrm{I\!R}^n$: $2n$ flops;
(ii) $z = Ax$, $A \in \mathrm{I\!R}^{m\times n}$: $2mn$ flops;
(iii) $C = A + x^Ty$, $A \in \mathrm{I\!R}^{n\times n}$: $2n^2$ flops;
(iv) $C = A\cdot B$, with A, B symmetric and commuting matrices of dimension n : $n^2(n+1)$ flops.

Moreover, the following property holds also true:

(v) $tr(A \cdot B)$, where A and B are symmetric matrices of dimension n : $(n+1)^2$ flops.

The proof of the last statement follows by considering the splittings $A = D + L + L^T$ and $B = \Gamma + M + M^T$, where D and Γ are diagonal matrices, and L and M are strictly lower triangular matrices. We then obtain that $tr(AB) = 2tr(LM^T) + tr(D\Gamma)$. Hence, $n(n-1)$ flops are required for computing the diagonal entries of LM^T, with additional n flops for computing the trace. Moreover $2n$ flops are needed for computing $tr(D\Gamma)$, and 1 flop for the sum of the partial results, thus resulting in a total of $(n+1)^2$ flops.

The above results allows us to derive the cost of the whole algorithm at each Newton iteration, which is due to the computation of:

1. the matrix powers $A^i = AA^{i-1}$, $i = 2, \ldots, n-1$: $(n-2)n^2(n+1)$ flops;
2. the Jacobian matrix (8): n^2 flops are needed for the first row, and $(n-1)n(n+1)^2$ for the subsequent $n-1$ rows;
3. the right-hand-side of (6), that is $F(c^k)$ in (7): $tr(A^i(c)) - b_i$, for $i = 1, \ldots, n-1$, requires $(n-1)n$ flops. Moreover, we need to compute $tr(A^n(c)) = tr(A(c)A^{n-1}(c))$, which requires additional $(n+1)^2$ flops;
4. the symmetric matrix $A(c^k)$: $n^3 + n^2(n+1) = 2n^3 + n^2$ flops.

In conclusion, the total cost for each Newton iteration amounts to

$$c_{tot}^I = 2n^4 + n^2 + n + 1 \qquad \text{flops.}$$

Here we are supposing $A(c)$ to be a dense symmetric matrix, but for specific problems, as we will show for AIEP in the next section, the $O(n^4)$ cost can be dramatically reduced.

REMARK 1. *We observe that in [8] only* multiplications *are considered for the complexity analysis. Moreover, the methods presented in the above reference, though based on quasi-Newton techniques, require n^4 multiplications to construct the Jacobian matrix and at least $3n^3$ multiplications for the rest of the algorithms, for a total of at least $2n^4 + 20n^3/3$ flops per iteration.*

On the other hand, we are here using the classical Newton method and no cubic term is present in the total cost of each iteration of Method I.

METHOD II. Let $\mu_1, \mu_2, \ldots, \mu_{n-1} \in \mathbb{R}$ be distinct values and

$$M_i(c) = A(c) + I\mu_{i-1}, \qquad i = 2, \ldots, n.$$

Let us consider

$$\begin{cases} b_1 = \sum_{k=1}^{n} \lambda_k^*, \\ b_i = \sum_{k=1}^{n} \frac{1}{\lambda_k^* + \mu_{i-1}}, \quad \mu_{i-1} \neq -\lambda_k^*, \qquad i = 2, \ldots, n, \end{cases}$$

and the equation $F(c) = 0$ with components

$$F_1(c) = tr(A(c)) - b_1, \qquad F_i(c) = tr(M_i(c)^{-1}) - b_i, \quad i = 2, \ldots, n. \tag{9}$$

The elements $J_F(c)_{ij}$ of the Jacobian matrix are given by

$$\begin{cases} \dfrac{\partial F_1}{\partial c_j} &= tr(A_j), \qquad j = 1, \dots, n, \\ \dfrac{\partial F_i}{\partial c_j} &= tr(\frac{\partial M_i^{-1}(c)}{\partial c_j}) = -tr(M_i^{-1}(c)(\frac{\partial M_i(c)}{\partial c_j})M_i^{-1}(c)) \\ &= -tr(A_j M_i^{-2}(c)), \qquad j = 1, \dots, n,\ i = 2, \dots, n. \end{cases}$$

Complexity. For our purpose, it is convenient to use Householder matrices to transform $A(c)$ to tridiagonal form, thus obtaining the factorization

$$A(c) = Q(c)T(c)Q^T(c), \tag{10}$$

with $Q(c)$ an orthogonal matrix. It is well known that the cost for obtaining the factorization (10) amounts to $\frac{8}{3}n^3$ flops. By posing, for simplicity, $Q = Q(c)$, $T = T(c)$ and $M_i = M_i(c)$, we then obtain for all $i = 2, \dots, n$,

$$M_i = QT_iQ^T, \qquad T_i = T + I\mu_{i-1}.$$

For each iteration of Method II, we need the computation of:

1. $C_i = M_i^{-2} = QT_i^{-2}Q^T = S_i^T S_i$, where for all $i = 2, \dots, n$, S_i is the solution of the tridiagonal systems $T_iS_i = Q^T$: $(n-1)(n^2(n+1) + 4n^2 + 4n)$ flops;
2. $tr(A_jC_i)$, $i = 2, \dots, n$, $j = 1, \dots, n$, and $tr(A_j)$ $j = 1, \dots, n$, for computing the Jacobian matrix: $n(n-1)(n+1)^2 + n^2$ flops;
3. the right-hand-side (9), namely:
   $tr(A(c))$ and the b_i : $2n$ flops;
   $tr(M_i^{-1}) = tr(T_i^{-1})$, hence for all $i = 2, \dots, n$, the inverse of T_i is needed: $(4n^2 + n)(n-1)$ flops.

In conclusion, by adding the cost for the evaluation of $A(c^k)$, for each Newton iteration we have a total cost of

$$c_{tot}^{II} = 2n^4 + \frac{41}{3}n^3 - 3n^2 - 4n \qquad \text{flops.}$$

From the previous considerations (see Remark 1), also Method II turns out to be competitive with the algorithms in [8], because the slight higher flop count per iteration is well balanced by the more favorable convergence properties of the Newton method. Nevertheless, Method II is more expensive than Method I. In a future research we would like to exploit the presence of the parameters μ_i in order to enlarge the convergence domain of the Newton method, when sufficient conditions for the existence of solutions will be known, so that the global cost could result to be competitive.

In the following we continue the analysis of the first method introduced for a specific application.

4. Additive case: AIEP. In such a case, the problem is to find a diagonal matrix $D(c) = \mathrm{diag}(c_1, \ldots, c_n)$, such that the matrix

$$A(c) = A_0 + D(c)$$

has the given spectrum $\{\lambda_1^*, \ldots, \lambda_n^*\}$.

In this case, we can simplify the procedure for computing the Jacobian matrix. In fact, as shown in Section 2, AIEP is obtained when in (1) we have $A_j = e_j e_j^T$, $j = 1, \ldots, n$, where e_j is the jth unit vector. Hence, by considering the application of Method I, the entries of the Jacobian matrix (8) become:

$$\frac{\partial F_i}{\partial c_j}(c) = i \cdot tr(A_j A^{i-1}(c)) = i \cdot tr(A^{i-1}(c) e_j e_j^T) = i \left(A^{i-1}(c)\right)_{jj}, \qquad i, j = 1, \ldots, n.$$

Therefore, at the kth iteration of the Newton method, we need only to compute

$$d^{(i)} = (d_1^{(i)}, d_2^{(i)}, \ldots, d_n^{(i)})^T, \qquad i = 1, \ldots, n,$$

where

$$d_j^{(i)} = \left(A^{i-1}(c^k)\right)_{jj}, \qquad j = 1, \ldots, n. \tag{11}$$

As before, it is useful to consider Householder tridiagonalization. Hence, by (10) it follows that

$$A^{i-1}(c) = (A_0 + D)^{i-1} = QT^{i-1}Q^T, \qquad i = 2, \ldots, n.$$

In such a way, we can compute cheaply the powers $A^{i-1}(c)$, by considering that

$$d_j^{(i)} = q_j^T T^{i-1} q_j, \qquad j = 1, \ldots, n, \ i = 2, \ldots, n,$$

where q_j is the jth column of Q.

Complexity. From (11), we obtain that

$$d_j^{(1)} = 1, \quad d_j^{(2)} = A(c^k)_{jj}, \qquad j = 1, \ldots, n.$$

For this reason, the effective cost only concerns the calculation of

$$d_j^{(l+1)} = q_j^T T^l q_j, \qquad l = 2, \ldots, n-1, \ j = 1, \ldots, n.$$

Let us now assume, for sake of brevity, n odd and $l = 2p$. Therefore, for $j = 1, \ldots, n$, we can calculate recursively

$$u_j^{(0)} = q_j, \qquad u_j^{(p)} = T u_j^{(p-1)}, \quad p = 1, \ldots, \frac{n-1}{2}. \tag{12}$$

In this way, it is easy to realize that

a) $d_j^{(2p)} = (u_j^{(p)})^T u_j^{(p)}$,
b) $d_j^{(2p+1)} = (u_j^{(p)})^T u_j^{(p+1)}$.

A similar procedure can be devised for the case n even.

The cost of (12) is then given by $n(n-1)(3n-2)$ flops, whereas the cost for a) and b), $j = 2, \dots, n$, is $n^2(n-1)$ flops for each of them.

We observe that the evaluation of the right-hand-side (7) also requires to compute $\text{diag}(A^n(c^k))$, that is b) for $p = \frac{n-1}{2}$ is needed. Then, additional $2n^2$ flops are needed. Moreover, the computational cost for the tridiagonalization of $A(c)$ has to be considered. By summing all the above costs, we then obtain a partial cost of

$$\frac{8}{3}n^3 + (3n-2)(n-1)n + 2n^2(n-1) + 2n^2 \qquad \text{flops,}$$

so that the global cost for the method becomes

$$c_I^A = \frac{23}{3}n^3 - 5n^2 + 2n \qquad \text{flops.}$$

Therefore, the $O(n^4)$ cost of the general case can be indeed reduced to a value slightly smaller than $8n^3$ flops.

5. Numerical results. We here report some examples taken from the literature (see [7, 8, 2]). For all of them, the existence of solutions is guaranteed. All examples are solved by means of Method I, which has been implemented in Matlab for the AIEP, MIEP and ITEP cases.

In all tests the starting points are taken in a neighbourhood of the solution, so that the convergence of the Newton method is guaranteed. In practical applications a suitable strategy would be needed to better locate the domain of convergence. By fixing a tolerance *tol*, the stop criterion for Newton iterations is $\|F(c^k)\|_2 \le tol$. The quadratic convergence of the method is numerically confirmed for all tests.

5.1. Additive case: AIEP. Let be $A(c) = A_0 + \text{diag}(c)$ and $tol = 1e{-8}$. The first and second examples are taken from [7] and the third from [8]. The residuals of the Newton method are shown for all additive examples in Table 5.1.

EXAMPLE 1. For this problem, we have

$$A_0 = \begin{pmatrix} 0 & 1 & 1 \\ 1 & 0 & 1 \\ 1 & 1 & 0 \end{pmatrix}, \qquad \lambda^* = (-2, 0, 3)^T.$$

the starting point is $c_0 = (-3, -1, 4)^T$. Method I converges in $k = 5$ iterations, giving the solution $c^* = (-1.48119, 0.31111, 2.17008)^T$.

EXAMPLE 2. For this problem, we have

$$A_0 = \begin{pmatrix} 0 & 4 & 4 \\ 4 & 0 & 4 \\ 4 & 4 & 0 \end{pmatrix}, \qquad \lambda^* = (-10, 0, 6)^T.$$

The starting point is $c_0 = (-1, 5, 3)^T$, and in $k = 8$ iterations we obtain $c^* = (-1.51735, -7.01262, 4.52998)^T$.

TABLE 1
Method I - Additive case: Newton residuals.

Iteration	0	1	2	3	4	5	6	7	8
Ex. 1	2.61e0	8.04e−1	9.97e−2	2.25e−3	2e−6	3e−12			
Ex. 2	14.5e0	19.5e0	8.9e0	3.4e0	1.1e0	5.7e−2	2.9e−4	1.8e−8	1e−16
Ex. 3	4.85e0	1.86e−1	8.04e−3	6e−6	4e−12				

EXAMPLE 3. We now have

$$A_0 = \begin{pmatrix} 0 & 4 & -1 & 1 & 1 & 5 & -1 & 1 \\ 4 & 0 & -1 & 2 & 1 & 4 & -1 & 2 \\ -1 & -1 & 0 & 3 & 1 & 3 & -1 & 3 \\ 1 & 2 & 3 & 0 & 1 & 2 & -1 & 4 \\ 1 & 1 & 1 & 1 & 0 & 1 & -1 & 5 \\ 5 & 4 & 3 & 2 & 1 & 0 & -1 & 6 \\ -1 & -1 & -1 & -1 & -1 & -1 & 0 & 7 \\ 1 & 2 & 3 & 4 & 5 & 6 & 7 & 0 \end{pmatrix}, \qquad \lambda^* = \begin{pmatrix} 10 \\ 20 \\ 30 \\ 40 \\ 50 \\ 60 \\ 70 \\ 80 \end{pmatrix}.$$

Starting from $c_0 = (11, 19, 30, 40, 51, 63, 70, 72)^T$, Method I in $k = 4$ iterations gives

$$c^* = (11.90787, 19.70555, 30.54549, 40.06265, 51.58715, 64.70213, 70.17067, 71.31849)^T.$$

5.2. Multiplicative case: MIEP. We are looking for $D(c) = \mathrm{diag}(c_1, \dots, c_n)$ such that $A(c) = D(c)M$. The problem is solved after symmetrization (see (3)-(4)), with tolerance $tol = 1e{-}12$. We have

$$M = \begin{pmatrix} 1 & 0.15 & 0.15 \\ 0.15 & 1 & 0.15 \\ 0.15 & 0.15 & 1 \end{pmatrix}, \qquad \lambda^* = (4, 10, 19)^T.$$

Starting from $c_0 = (1, 10, 20)^T$, in $k = 5$ iterations the method converges to $c^* = (4.22727, 10.44248, 18.33025)^T$. The residuals are

$$res = (2.49e0, 3.53e{-}1, 2.92e{-}2, 3.33e{-}4, 3e{-}8, 1e{-}15).$$

5.3. Toeplitz case: ITEP. This example was solved in [2] by means of isospectral ODE matrix approach. We use the tolerance $tol = 1e{-}12$. Given

$$\lambda^* = (2.2361, 1.5858, 4.4142, -2.2361)^T,$$

we are now looking for $c^* \in \mathbb{R}^4$ such that $A(c^*)$ is a real symmetric Toeplitz matrix. For the chosen basis $A_0 = 0$ and c^* corresponds to the first row (or column) of $A(c^*)$.

Starting from $c_0 = (1, 2, -0.2, 1)^T$, in $k = 6$ iterations Method I finds $c^* = (1.5, 1.79911, -0.41689, 1.20089)^T$. The residuals are

$$res = (5.6e{-}1, 7.95e{-}2, 1.54e{-}2, 7.94e{-}4, 4e{-}6, 1e{-}10, 1e{-}15).$$

6. Conclusions. An algebraic approach using traces of matrices has been proposed to solve the Inverse Eigenvalue Problem in the symmetric case by means of the classical Newton method. The computational cost of the two algorithms introduced is competitive with respect to the other methods for this problem known in the current literature.

Moreover, the algorithms proposed can be applied to different kind of IEP and in particular the additive problem has been investigated. Numerical experiments, performed on AIEP, MIEP and ITEP, seem to be rather satisfactory.

Future developments will concern the research of alternative sufficient conditions for the existence of solutions of the problem, as well as the ways for assuring larger convergence domains for the new methods here introduced. Finally, the case of multiple eigenvalues has not been considered, even though this is a very important case arising in many practical applications. For this reason, a further trend of investigation will concern a suitable analysis of our methods also for IEP with multiple eigenvalues.

Acknowledgements. The authors are very grateful to the referee for the many useful comments and the helpful suggestions for improving the original version of the paper.

REFERENCES

[1] M.Chu. Inverse eigenvalue problem, *SIAM Review* **40** (1998) 1–39.

[2] F.Diele, I.Sgura. Isospectral flows and the inverse eigenvalue problem for Toeplitz matrices, *Jour. Comp. Applied Math.* **110** (1999) 25–43.

[3] S.Friedland. Inverse eigenvalue problems, *Linear Algebra Appl.* **17** (1977) 15–51.

[4] G.H.Golub, C.F. Van Loan. *Matrix Computations, 3rd ed.*, The John Hopkins University Press, 1996.

[5] H.J.Landau. The inverse eigenvalue problem for real symmetric Toeplitz matrices, *J. Amer. Math. Soc.* **7** n.3 (1994) 749–767.

[6] D.P.Laurie. Solving the inverse eigenvalue problem via the eigenvector matrix, *J. Comp. Appl. Math.* **35** (1991) 277–289.

[7] L.L.Li. Some sufficient contions for the solvability of inverse eigenvalue problems, *Linear Algebra Appl.* **148** (1991) 225–236

[8] S.Friedland, J.Nocedal, M.L.Overton. The formulation and analysis of numerical methods for inverse eigenvalue problems, *SIAM J. Numer. Anal.* **24** (1987) 634–667.

VARIABLE PROJECTION METHODS FOR LARGE CONVEX QUADRATIC PROGRAMS*

VALERIA RUGGIERO† AND LUCA ZANNI‡

Abstract. We propose two projection-type methods for solving large convex quadratic programs. The main feature of these iterative schemes consists in using, at each iteration, a variable projection parameter instead of a fixed one as in the classical projection methods. The convergence may be obtained without restrictive conditions on the projection parameters by using appropriate correction rules that imply, at each iteration, a sufficient decrease in the objective function. The first method uses a correction rule on the descent direction produced by the projection step, while in the second method, the correction formula works adaptively on the value of the variable projection parameter. We give convergence results for the general case of inexact solution of the inner subproblems. The numerical behaviour of the methods is strictly dependent on the sequence of the projection parameters. We introduce a practical nonexpensive updating rule for these parameters and evaluate its effectiveness on large scale-test problems.

Key words. convex quadratic programs, large-scale problems, projection-type methods.

AMS subject classifications. 65K05, 90C06, 90C20, 90C25.

1. Introduction. We consider the numerical solution of the following linearly constrained convex quadratic programming (QP) problem:

$$\begin{array}{ll} \textit{minimize} & f(x) = \frac{1}{2}x^TGx + q^Tx \\ \textit{subject to} & Cx = d, \qquad Ax \geq b, \end{array} \tag{1}$$

where G is a symmetric positive semidefinite matrix of order n, C is an $m_e \times n$ matrix of full row rank ($m_e \leq n$) and A is an $m_i \times n$ matrix. We assume that the feasible region $K = \{x \mid Cx = d,\ Ax \geq b\}$ is nonempty and that $f(x)$ is bounded from below on K; then, the set of optimal solutions of (1), denoted by K^*, is a nonempty polyhedral set, given by $K^* = \{x \in K \mid E(x - x^*) = 0,\ q^T(x - x^*) = 0\}$, where x^* is a solution of (1) and E is a matrix such that $G = E^TE$. For any $x \in K^*$, $f(x) = f^*$, where f^* denotes the optimal value of (1).

When the matrices G, C and A are large, sparse and without a particular structure, an approach for solving the problem (1) is represented by the projection methods, that are largely developed also for linear complementarity and variational inequality problems. In the framework of the QP problems, they are known as scaled gradient projection methods "with constant stepsize" [1, p. 207]. The iterative scheme of the projection methods consists in to generate, from an arbitrary vector $x^{(0)} \in R^n$, a sequence of vectors $\{x^{(k)}\}$, $k = 1, 2, \ldots$, by solving the QP subproblems

$$\begin{array}{ll} \textit{minimize} & \frac{1}{2}x^T\frac{D}{\rho}x + (q + (G - \frac{D}{\rho})x^{(k-1)})^Tx \\ \textit{subject to} & Cx = d, \qquad Ax \geq b, \end{array} \tag{2}$$

where ρ is a positive parameter, called projection parameter, and D is a prefixed symmetric positive definite matrix of order n, that, from the practical point of view, must be an easily solvable matrix (diagonal or block diagonal matrix).

*This work was supported by MURST Project "Numerical Analysis: Methods and Mathematical Software" and by CNR Research Contribution N. 98.01029.CT01, Italy.

†Dipartimento di Matematica, Università di Ferrara, Via Machiavelli 35, Ferrara, Italy (rgv@dns.unife.it).

‡Dipartimento di Matematica, Università di Modena e Reggio Emilia, Via Campi 213/b, 41100 Modena, Italy (zanniluca@unimo.it).

The iterative scheme (2) appears convenient for the solution of large-scale sparse QP problems, because it allows any sparsity of the objective and constraint matrices to be readily exploited, uses little storage and is well suited to implementation on parallel computers. Indeed, each iteration consists essentially in computing the matrix-vector product $(G - \frac{D}{\rho})x^{(k-1)}$ and solving the QP subproblem (2). For this subproblem, according to the structure of the constraints of the application in question, we can use specialized sequential or parallel solvers for separable or nearly separable QP problems (see [17] and [19] as examples of solvers for separable quadratic programs with special constraints). When the constraints do not present a particular structure, we can formulate each inner subproblem as a mixed linear complementarity problem (LCP) that can be solved by sequential or parallel splitting methods (see, for example, the classical projected SOR method of Cryer [3] or the parallel solvers in [9, 14, 15]).

The convergence of the projection method can be immediately derived from the projection schemes for the variational inequality problems. In the case of strictly convex QP problems, by proceeding as in [4], the convergence of $\{x^{(k)}\}$ can be obtained by a contraction argument for $\rho < \frac{2\lambda_{min}(G)}{\lambda_{max}(GD^{-1}G)}$; in this case, the method has a geometrical convergence with rate

$$(1 - \frac{\rho}{\lambda_{max}(D)}(2\lambda_{min}(G) - \rho\lambda_{max}(GD^{-1}G)))^{\frac{1}{2}}.$$

For convex QP problems, following the proof of Theorem 2.1 of [16] and using the Theorem 2.1 of [11], the R-linear convergence ([18]) of the projection method may be proved under the sufficient condition $\rho < \frac{2}{\lambda_{max}(D^{-\frac{1}{2}}GD^{-\frac{1}{2}})}$.

When $\rho = 1$, the iterative scheme (2) is known as splitting method; the R-linear convergence of the splitting method is proved in [11] under the hypothesis that $(D, G - D)$ is a P–regular splitting of G. For strictly convex QP problems, the splitting method has a geometrical convergence with rate of convergence equal to the spectral radius of the matrix $D^{-\frac{1}{2}}(I - G)D^{-\frac{1}{2}}$ [8].

From the viewpoint of numerical effectiveness, in the splitting methods the practical requirement to select an easily solvable matrix D is very often in conflict with the choice of a splitting of G that implies fast convergence (unless the splitting is suggested by the underlying application; see, for example, the quadratic program arising in the constrained bivariate interpolation problem [6, 7, 22]). On the other hand, the projection schemes allow any positive definite matrix D to be selected, but the values of ρ that satisfy the sufficient convergence conditions are often so small as to determine very slow convergence. Furthermore, in some cases, it is possible to find values of ρ that do not satisfy the sufficient convergence condition and imply a fast convergence (see the numerical experiments in [20]).

In [1], the introduction of a limited minimization rule in the iterative scheme is a way to avoid the difficulty in finding a convenient value for the projection parameter (scaled gradient projection method with a "limited minimization rule"); indeed, the iterate $x^{(k)}$ is obtained by the correction formula:

$$x^{(k)} = x^{(k-1)} + \theta_k(y^{(k)} - x^{(k-1)}), \tag{3}$$

where $y^{(k)}$ denotes the solution of (2) and θ_k is chosen by solving the following problem:

$$\min\{f(x^{(k-1)} + \theta(y^{(k)} - x^{(k-1)})); \quad \theta \in (0, 1]\}. \tag{4}$$

A convenient value for the projection parameter can also be obtained by making a search along the projection arc (scaled gradient projection method with an "Armijo rule along a projection arc" [1]). In particular, given $\overline{\rho} > 0$ and denoted by $x^{(k)}(\rho)$ the solution of the subproblem (2), the iterate $x^{(k)}$ is equal to $x^{(k)}(\beta^m\overline{\rho})$ where m is the first nonnegative integer for which the following Armijo-like inequality holds:

$$f(x^{(k-1)}) - f\left(x^{(k)}(\beta^m\overline{\rho})\right) \geq \sigma \nabla f(x^{(k-1)}) \left(x^{(k-1)} - x^{(k)}(\beta^m\overline{\rho})\right),$$

where $\beta \in (0,1)$ and $\sigma \in (0,1)$ are prefixed scalars.

Another way to improve the projection methods consists in using, at each iteration, a variable projection parameter; on this idea is based the scheme, named Variable Projection Method (VPM), introduced in [22] for strictly convex QP problems. The projection parameter is updated according to a nonexpensive rule, deduced by heuristic considerations, and the convergence of the method is obtained under weak hypotheses by combining the projection step with a correction rule of the form (3), as in the scaled gradient projection methods with a "limited minimization rule".

In this paper, we generalize the VPM to the case of convex QP problems and we propose a variant of this scheme, named Adaptive Variable Projection Method (AVPM). In this last method, the value of the projection parameter given by an updating rule is adaptively reduced until the projection step produces a sufficient decrease in the objective function. In practice, the VPM and the AVPM generalize the above scaled gradient projection schemes with a "limited minimization rule" and with an "Armijo rule along a projection arc", by introducing a rule that determines a convenient value for ρ at each step.

In section 1 and 2, the convergence of the VPM and the AVPM is proved for convex QP problems under the additional hyphothesis that the inner QP subproblems (2) are inexactly solved. Indeed, when an iterative scheme is used for solving the inner QP subproblems (2), it is convenient to consider an inner progressive termination rule, in the sense that the accuracy in the solution of each subproblem depends on the quality of the previous iterate $x^{(k-1)}$: the closer $x^{(k-1)}$ is to satisfying the external stopping criterion for the outer iteration, the more accurately the corresponding subproblem is solved. In this way, unnecessary inner iterations are avoided when $x^{(k-1)}$ is far from the solution. Following [10, 11, 12, 13], the inexact solution of the inner subproblem can be viewed as exact solution of the perturbed subproblem:

$$(5) \qquad \begin{array}{ll} \textit{minimize} & \frac{1}{2}x^T \frac{D}{\rho} x + (q + h^{(k-1)} + (G - \frac{D}{\rho})x^{(k-1)})^T x \\ \textit{subject to} & Cx = d, \qquad Ax \geq b, \end{array}$$

where $h^{(k-1)}$ is an error vector convergent to a null vector as $k \to \infty$.

In section 3, we suggest an updating rule for varying the projection parameter and we report the results of a numerical experimentation concerning the methods described in the previous sections. From these results, we can conclude that, when a convenient updating rule for the choice of the projection parameter is used, the VPM and the AVPM show an efficient numerical behaviour and the corrective step along a descent direction or a projection arc arises in few iterations.

In the sequel of the work, we denote by $\lambda_{min}(G)$ and $\lambda_{max}(G)$ the minimum and the maximum eigenvalues, respectively, of a symmetric matrix G. We indicate by $\|x\|$ the usual Euclidean norm of x and by $\|x\|_G$ the elliptic norm of x defined as $\sqrt{x^T G x}$ where G is a symmetric positive definite matrix. Furthermore, we denote by $\|B\|$ the norm of a matrix B induced by the vector Euclidean norm. Finally, we indicate by $[x]^+$ the orthogonal projection of a vector x onto a closed convex set K, i.e., $[x]^+ = \arg\min_{y \in K} \|y - x\|$, and by $\phi(x)$ the distance from x to K^*, i.e.,

$$\phi(x) = \min_{\overline{x} \in K^*} \|x - \overline{x}\|.$$

2. The Variable Projection Method. Given a symmetric positive definite matrix D, the Variable Projection Method introduced in [22] can be generalized for solving (1) as follows:

1. Let $x^{(0)}$ be an arbitrary vector and ρ_1 be an arbitrary positive constant; $k \leftarrow 1$;
2. Compute the unique solution $y^{(k)}$ of the subproblem

(6)
$$\begin{array}{ll} \textit{minimize} & \frac{1}{2} x^T \frac{D}{\rho_k} x + (q + h^{(k-1)} + (G - \frac{D}{\rho_k}) x^{(k-1)})^T x \\ \textit{subject to} & Cx = d, \qquad Ax \geq b; \end{array}$$

3. Set $d^{(k)} = y^{(k)} - x^{(k-1)}$;
4. If $(Gx^{(k-1)} + q)^T d^{(k)} < 0$ and $k \neq 1$, compute the solution θ_k of the problem

(7)
$$\min\{f(x^{(k-1)} + \theta d^{(k)}); \quad \theta \in (0, 1]\};$$

else

$$\theta_k = 1;$$

5. Compute

(8)
$$x^{(k)} = x^{(k-1)} + \theta_k d^{(k)};$$

6. Terminate if $x^{(k)}$ satisfies a stopping rule, otherwise update ρ_{k+1} by an appropriate rule; then $k \leftarrow k + 1$ and go to step 2.

The aim of the initial step of the scheme is to generate, from an arbitrary point $x^{(0)}$, a vector $x^{(1)} \in K$. When $\lambda_{min}(D)$ is easily computable, a convenient value for ρ_1 is

$$\rho_1 = \lambda_{min}(D) / \left(\sum_{i=1}^{n} \sum_{j=1}^{n} g_{ij}^2 \right)^{1/2},$$

where g_{ij} is the ij-th element of G; this value satisfies the sufficient convergence condition for the step size of the projection methods and guarantees that $\|x^{(1)} - x^*\|_D \leq \|x^{(0)} - x^*\|_D$, with $x^* \in K^*$.

When the computation of $Dx^{(k-1)}$ is nonexpensive, the computational complexity of one iteration of the VPM is essentially equal to that of the projection methods. Indeed, if we keep stored from the previous iteration the vector $t = Gx^{(k-1)}$, at the k-th iteration the computation of θ_k requires the matrix-vector product $\overline{t} = Gy^{(k)}$ and some less expensive vector operations. Now, the vector t is updated by the following rule:

$$t \leftarrow t + \theta_k(\overline{t} - t).$$

We will prove the convergence of the VPM under the following assumptions:

AS1. G is a symmetric positive semidefinite matrix, K is a nonempty set and $f(x)$ is bounded from below on K:

AS2. D is a symmetric positive definite matrix and the sequence of scalars $\{\rho_k\}$ is bounded from below and above by positive constants: $0 < \rho_m \leq \rho_k \leq \rho_M$;
AS3. the sequence of the error vectors $\{h^{(k)}\}$ satisfies the following condition:

$$\|h^{(k-1)}\| \leq \left(\frac{\lambda_{min}(D)}{\rho_k} - \epsilon_k\right) \|y^{(k)} - x^{(k-1)}\| \qquad k > 1 \tag{9}$$

with $\epsilon_k \in [\frac{\lambda_{min}(D)}{\rho_M}, \frac{\lambda_{min}(D)}{\rho_k}]$.

The assumption AS2 on $\{\rho_k\}$ is very general; the choice of a particular bounded sequence $\{\rho_k\}$ is based only on its numerical effectiveness. The case of a constant projection parameter is a special case of this.

The following Lemmas state some basic assertions for the convergence of the VPM.

LEMMA 2.1. *If, for a given $k > 1$, $x^{(k-1)}$ is feasible to (1) and $x^{(k-1)} \neq y^{(k)}$, then the following assertions hold:*

a. *$y^{(k)}$ is well-defined and is a feasible point of (1); furthermore, $d^{(k)}$ is a descent direction for the objective function $f(x)$ in $x^{(k-1)}$;*

b. *the sequence $\{\theta_k\}$ satisfies the following bounds:*

$$0 < \theta_{min} = \min\left\{1, \frac{\lambda_{min}(D)}{\rho_M \lambda_{max}(G)}\right\} \leq \theta_k \leq 1, \qquad \forall\, k > 1. \tag{10}$$

Proof. **a.** For the uniqueness of the solution of the linearly constrained strictly convex subproblem (6), $y^{(k)}$ is well-defined and it is a feasible point of (1). Furthermore, we consider the first order optimality condition for the solution $y^{(k)}$ of the subproblem (6):

$$\left(\frac{D}{\rho_k}(y^{(k)} - x^{(k-1)}) + Gx^{(k-1)} + q + h^{(k-1)}\right)^T (x - y^{(k)}) \geq 0, \quad \forall\, x \in K. \tag{11}$$

Taking $x = x^{(k-1)}$ in (11), since $\frac{D}{\rho_k}$ is a symmetric positive definite matrix and $x^{(k-1)} \neq y^{(k)}$, we obtain from (9) that $d^{(k)}$ is a descent direction for $f(x)$ in $x^{(k-1)}$:

$$\begin{aligned}(Gx^{(k-1)} + q)^T d^{(k)} &\leq -d^{(k)T} \frac{D}{\rho_k} d^{(k)} - h^{(k-1)T} d^{(k)} \\ &\leq \left(-\frac{\lambda_{min}(D)}{\rho_k} + \frac{\lambda_{min}(D)}{\rho_k} - \epsilon_k\right) \|d^{(k)}\|^2 \\ &\leq -\frac{\lambda_{min}(D)}{\rho_M} \|d^{(k)}\|^2 < 0.\end{aligned} \tag{12}$$

When $x^{(k-1)} = y^{(k)}$, then $x^{(k-1)} \in K^*$.

b. For $k > 1$ the exact solution of the problem (7) is given by the following rule:

$$\theta_k = \begin{cases} \min\left\{-\frac{(Gx^{(k-1)}+q)^T d^{(k)}}{d^{(k)T} G d^{(k)}}, 1\right\} & \text{if } d^{(k)T} G d^{(k)} \neq 0 \\ 1 & \text{otherwise.} \end{cases} \tag{13}$$

From (12) and (13) the assertion **b** immediately follows. □

LEMMA 2.2. *The sequence $\{x^{(k)}\}$ generated by the VPM is well-defined, $x^{(k)} \in K$ for all $k \geq 1$ and $\{f(x^{(k)})\}$ is a sequence monotonically nonincreasing and convergent as $k \to \infty$. In particular, there exists a positive scalar α_1 such that*

$$f(x^{(k-1)}) - f(x^{(k)}) \geq \alpha_1 \|x^{(k)} - x^{(k-1)}\|^2 \tag{14}$$

and then the sequences $\{x^{(k)} - x^{(k-1)}\}$ and $\{d^{(k)}\}$ are convergent to the null vector as $k \to \infty$. Furthermore, $\phi(x^{(k)}) \to 0$ as $k \to \infty$.

Proof. Since $x^{(1)} \in K$, from the assertions of Lemma 2.1 and the convexity of K, it follows that $\{x^{(k)}\}$ is a well-defined sequence of feasible points of (1) for $k \geq 1$.

From (7), (8) and the following equality:

$$f(x^{(k-1)} + \theta d^{(k)}) - f(x^{(k-1)}) = \theta(Gx^{(k-1)} + q)^T d^{(k)} + \frac{1}{2}\theta^2 d^{(k)T} G d^{(k)}, \tag{15}$$

we have

$$f(x^{(k)}) \leq f(x^{(k-1)}) + \theta(Gx^{(k-1)} + q)^T d^{(k)} + \frac{1}{2}\theta^2 d^{(k)T} G d^{(k)}, \tag{16}$$

for any $\theta \in (0, 1]$. If $d^{(k)T} G d^{(k)} = 0$ or $-\frac{(Gx^{(k-1)}+q)^T d^{(k)}}{d^{(k)T} G d^{(k)}} \geq 1$, the solution of the problem (7) is obtained for $\theta_k = 1$; from (10), (12) and (16) we have

$$\begin{aligned} f(x^k) &\leq f(x^{(k-1)}) + \frac{1}{2}(Gx^{(k-1)} + q)^T d^{(k)} \\ &\leq f(x^{(k-1)}) - \frac{\lambda_{min}(D)}{2\rho_M}\|d^{(k)}\|^2 \\ &\leq f(x^{(k-1)}) - \frac{\lambda_{min}(D)\theta_{min}}{2\rho_M}\|d^{(k)}\|^2. \end{aligned} \tag{17}$$

If $d^{(k)T} G d^{(k)} \neq 0$ and $-\frac{(Gx^{(k-1)}+q)^T d^{(k)}}{d^{(k)T} G d^{(k)}} < 1$, the solution of the problem (7) is obtained for $\theta_k = -\frac{(Gx^{(k-1)}+q)^T d^{(k)}}{d^{(k)T} G d^{(k)}}$; from (10), (12) and (16) we have

$$\begin{aligned} f(x^k) &\leq f(x^{(k-1)}) - \frac{1}{2}\frac{((Gx^{(k-1)} + q)^T d^{(k)})^2}{d^{(k)T} G d^{(k)}} \\ &= f(x^{(k-1)}) + \frac{\theta_k}{2}(Gx^{(k-1)} + q)^T d^{(k)} \\ &\leq f(x^{(k-1)}) - \frac{\lambda_{min}(D)\theta_{min}}{2\rho_M}\|d^{(k)}\|^2. \end{aligned} \tag{18}$$

Thus, if $\alpha_1 = \frac{\lambda_{min}(D)\theta_{min}}{2\rho_M}$, from (17), (18) and (8) we conclude that

$$\begin{aligned} f(x^{(k-1)}) - f(x^{(k)}) &\geq \alpha_1 \|d^{(k)}\|^2 = \frac{\alpha_1}{\theta_k^2}\|x^{(k)} - x^{(k-1)}\|^2 \\ &\geq \alpha_1 \|x^{(k)} - x^{(k-1)}\|^2 \geq 0, \end{aligned} \tag{19}$$

where the equality to zero holds when $x^{(k-1)} = y^{(k)} = x^{(k)} \in K^*$.

Since $f(x)$ is bounded from below on K, the sequence $\{f(x^{(k)})\}$ is convergent and $\lim_{k\to\infty} f(x^{(k)}) \geq f^*$. Thus, from (19) we have that the sequences $\{d^{(k)}\}$ and $\{x^{(k)} - x^{(k-1)}\}$ converge to the null vector as $k \to \infty$.

Now, we observe that the solution $y^{(k)}$ of the subproblem (6) can be considered as the orthogonal projection of an appropriate vector onto K, i.e.

$$y^{(k)} = \left[y^{(k)} - \frac{D}{\rho_k}(y^{(k)} - x^{(k-1)}) - (Gx^{(k-1)} + q + h^{(k-1)}) \right]^+ .$$

From the nonexpansive property of the projection operator, we have

$$\begin{aligned} \left\| x^{(k-1)} - [x^{(k-1)} - (Gx^{(k-1)} + q)]^+ \right\| &= \Big\| x^{(k-1)} - [x^{(k-1)} - (Gx^{(k-1)} + q)]^+ + \\ &\quad - y^{(k)} + [y^{(k)} - \frac{D}{\rho_k}(y^{(k)} - x^{(k-1)}) - (Gx^{(k-1)} + q + h^{(k-1)})]^+ \Big\| \\ &\leq \left(2 + \frac{\lambda_{max}(D) + \lambda_{min}(D)}{\rho_m} \right) \|d^{(k)}\|. \end{aligned} \tag{20}$$

Since the sequence $\{\|d^{(k)}\|\}$ converges to 0 as $k \to \infty$, the sequence $\{\|x^{(k-1)} - [x^{(k-1)} - (Gx^{(k-1)} + q)]^+\|\}$ also converges to 0 as $k \to \infty$. Then, given a positive scalar δ, there exists an integer $\bar{k} > 0$ such that, for any $k > \bar{k}$, we have

$$\|x^{(k-1)} - [x^{(k-1)} - (Gx^{(k-1)} + q)]^+\| \leq \delta.$$

Hence, by the Theorem 2.1 in [11], there exists a scalar $\tau > 0$ such that:

$$\phi(x^{(k-1)}) \leq \tau \|x^{(k-1)} - [x^{(k-1)} - (Gx^{(k-1)} + q)]^+\| \qquad \text{for any } k > \bar{k}$$

From (20), for any $k > \bar{k}$, we have

$$\phi(x^{(k-1)}) \leq \alpha_2 \|d^{(k)}\| \tag{21}$$

where $\alpha_2 = \tau \left(2 + \frac{\lambda_{max}(D) + \lambda_{min}(D)}{\rho_m}\right)$; since $\{\|d^{(k)}\|\}$ converges to 0, the sequence $\{\phi(x^{(k-1)})\}$ also converges to 0 as $k \to \infty$. □

THEOREM 2.1. *The sequence $\{x^{(k)}\}$ generated by the VPM converges at least R–linearly to a solution of (1).*

Proof. Let x^* be the element of K^* closest to $x^{(k-1)}$, i.e. $\phi(x^{(k-1)}) = \|x^{(k-1)} - x^*\|$. From the first order optimality conditions for the problems (6) and the Mean Value Theorem, it is possible to write:

$$\begin{aligned} \left(\frac{D}{\rho_k}(y^{(k)} - x^{(k-1)}) + Gx^{(k-1)} + q + h^{(k-1)} \right)^T (x^* - y^{(k)}) &\geq 0, \\ f(y^{(k)}) - f(x^*) = (G\zeta^{(k)} + q)^T (y^{(k)} - x^*), \end{aligned}$$

where $\zeta^{(k)}$ lies on the line segment joining $y^{(k)}$ with x^*. Thus, combining the above relations, we have

$$\begin{aligned} f(y^{(k)}) - f^* &= f(y^{(k)}) - f(x^*) \\ &\leq \left(\lambda_{max}(G)\|\zeta^{(k)} - x^{(k-1)}\| + \|h^{(k-1)}\| + \frac{\lambda_{max}(D)}{\rho_m}\|d^{(k)}\| \right) \|y^{(k)} - x^*\|. \end{aligned}$$

Since $\|\zeta^{(k)} - x^{(k-1)}\| \leq \|x^{(k-1)} - x^*\| + \|d^{(k)}\|$ and $\|y^{(k)} - x^*\| \leq \|x^{(k-1)} - x^*\| + \|d^{(k)}\|$, from (9) we obtain

$$f(y^{(k)}) - f^* \leq \Big(\lambda_{max}(G)\left(\phi(x^{(k-1)}) + \|d^{(k)}\|\right) + \frac{\lambda_{min}(D) + \lambda_{max}(D)}{\rho_m}\|d^{(k)}\|\Big)\left(\phi(x^{(k-1)}) + \|d^{(k)}\|\right).$$

From (21), for any $k > \overline{k}$, we can write

$$f(y^{(k)}) - f^* \leq \alpha_3\|d^{(k)}\|^2$$

where $\alpha_3 = \left(\lambda_{max}(G)(\alpha_2+1) + \frac{\lambda_{min}(D)+\lambda_{max}(D)}{\rho_m}\right)(\alpha_2+1)$. Then, from (7), (8) and (19), we have

$$f(x^{(k)}) - f^* \leq \frac{\alpha_3}{\alpha_1}(f(x^{(k-1)}) - f(x^{(k)})), \qquad \forall\, k > \overline{k}. \tag{22}$$

By rearranging terms in (22), we obtain

$$f(x^{(k)}) - f^* \leq \frac{\alpha_3}{\alpha_3 + \alpha_1}(f(x^{(k-1)}) - f^*), \qquad \forall\, k > \overline{k},$$

and then $\{f(x^{(k)})\}$ converges at least linearly to f^*. Since (14) holds, then it follows from Lemma 3.1 in [11] that $\{x^{(k)}\}$ converges at least linearly. Since $\{\phi(x^{(k)})\}$ converges to zero as $k \to \infty$, the limit point of $\{x^{(k)}\}$ is an element of K^*. □

3. The Adaptive Variable Projection Method. Given a symmetric positive definite matrix D and two positive scalars $\beta \in (0,1)$ and $\eta > \frac{1}{2}$, the Adaptive Variable Projection Method for the solution of (1) is stated as follows:

1. Let $x^{(0)}$ be an arbitrary vector and $\overline{\rho}_1$ be an arbitrary positive constant, $k \leftarrow 1$, $\rho_k = \overline{\rho}_1$, $m_k = 0$;
2. Compute the unique solution $x(\rho_k)$ of the subproblem

$$\begin{array}{ll} \textit{minimize} & \frac{1}{2}x^T\frac{D}{\rho_k}x + (q + h(\rho_k) + (G - \frac{D}{\rho_k})x^{(k-1)})^Tx \\ \textit{subject to} & Cx = d, \qquad Ax \geq b; \end{array} \tag{23}$$

3. Set $d^{(k)} = x(\rho_k) - x^{(k-1)}$;
4. If $-(Gx^{(k-1)} + q)^Td^{(k)} < \eta d^{(k)T}Gd^{(k)}$ and $k \neq 1$, then $m_k = m_k + 1, \rho_k = \beta^{m_k}\overline{\rho}_k$ and go to step 2, otherwise $x^{(k)} = x(\rho_k)$;
5. Terminate if $x^{(k)}$ satisfies a stopping rule, otherwise update ρ_{k+1} by an appropriate rule; then $\overline{\rho}_{k+1} = \rho_{k+1}$, $m_{k+1} = 0$, $k \leftarrow k+1$ and go to step 2.

The computational complexity of each iteration of the AVPM depends on the number m_k+1 of inner subproblems that have to be solved to find a convenient value for ρ_k. Since the test at the step 4 requires the matrix-vector product $Gx(\rho_k)$, each iteration of the AVPM is essentially equivalent to $m_k + 1$ iterations of the projection methods.

As in the iterative scheme of the section 2, the aim of the first step is to find an initial feasible point $x^{(1)}$.

We will prove the convergence of the AVPM under the assumptions AS1 and AS2 of section 2 and the following condition:

AS4. the sequence of the error vectors $\{h(\rho_k)\}$ is such that

$$\|h(\rho_k)\| \le \min\left(\eta\lambda_{max}(G), \left(\frac{\lambda_{min}(D)}{\rho_k} - \epsilon_k\right)\right) \|x(\rho_k) - x^{(k-1)}\|, \quad k > 1, \tag{24}$$

with $\epsilon_k \in [\frac{\lambda_{min}(D)}{\rho_M}, \frac{\lambda_{min}(D)}{\rho_k}]$.

First, we observe that the vector $x(\rho_k)$ satisfying the condition

$$-(Gx^{(k-1)} + q)^T(x(\rho_k) - x^{(k-1)}) \ge \eta(x(\rho_k) - x^{(k-1)})^T G(x(\rho_k) - x^{(k-1)}) \tag{25}$$

can be found after a finite number of trials. Indeed, if we consider the first order optimality conditions of the subproblem (23) computed in $x = x^{(k-1)}$, we have

$$\begin{aligned}(Gx^{(k-1)} + q)^T(x^{(k-1)} - x(\rho_k)) \ \ge\ & \tfrac{1}{\rho_k}(x^{(k-1)} - x(\rho_k))^T D(x^{(k-1)} - x(\rho_k)) \\ & -h(\rho_k)^T(x^{(k-1)} - x(\rho_k)).\end{aligned} \tag{26}$$

If m_k is the first nonnegative integer such that

$$\beta^{m_k}\overline{\rho}_k \le \frac{\lambda_{min}(D)}{2\eta\lambda_{max}(G)}, \tag{27}$$

then the condition (25) is satisfied for $\rho_k = \beta^{m_k}\overline{\rho}_k$.

The inequality (27) states a very strong sufficient condition; from the viewpoint of the numerical effectiveness, it is advisable to define the projection parameter by a nonexpensive rule that gives values of ρ_k greater than that obtained by (27) and such that $m_k = 0$ for almost all k.

Now we prove the convergence of the AVPM for convex QP problems.

THEOREM 3.1. *The sequence $\{x^{(k)}\}$ generated by the AVPM is well-defined and converges at least R-linearly to a solution of (1).*

Proof. Since the solution of the subproblem (23) is unique and, at each iteration, the steps 2, 3 and 4 of the AVPM are repeated a finite number of times, $x^{(k)}$ is a well-defined element of K.

The R-linear convergence of the sequence $\{x^{(k)}\}$ follows from Theorem 3.1 of [11], if the following relations hold:

$$f(x^{(k-1)}) - f(x^{(k)}) \ge \alpha_1\|x^{(k-1)} - x^{(k)}\|^2 \tag{28}$$

$$\|x^{(k-1)} - [x^{(k-1)} - (Gx^{(k-1)} + q)]^+\|^2 \le \alpha_2(f(x^{(k-1)}) - f(x^{(k)})) \tag{29}$$

$$x^{(k)} = [x^{(k-1)} - (Gx^{(k-1)} + q) + e^{(k)}]^+ \tag{30}$$

where α_1 and α_2 are positive scalars and the sequence $\{e^{(k)}\}$ converges to zero as $k \to \infty$.

Now, we prove that (28), (29) and (30) hold.

Since $x^{(k)}$ is such that the condition $-(Gx^{(k-1)}+q)^T(x^{(k)}-x^{(k-1)}) \ge \eta(x^{(k)}-x^{(k-1)})^T G(x^{(k)}$ $x^{(k-1)})$ holds, we can write

$$\begin{aligned} f(x^{(k)}) - f(x^{(k-1)}) &= (Gx^{(k-1)} + q)^T(x^{(k)} - x^{(k-1)}) \\ &\quad + \frac{1}{2}(x^{(k)} - x^{(k-1)})^T G(x^{(k)} - x^{(k-1)}) \\ &\le (Gx^{(k-1)} + q)^T(x^{(k)} - x^{(k-1)}) \\ &\quad - \frac{1}{2\eta}(Gx^{(k-1)} + q)^T(x^{(k)} - x^{(k-1)}) \\ &\le \left(1 - \frac{1}{2\eta}\right)(Gx^{(k-1)} + q)^T(x^{(k)} - x^{(k-1)}). \end{aligned} \tag{31}$$

From (26) with $x(\rho_k) = x^{(k)}$ and (24), it follows that

$$f(x^{(k)}) - f(x^{(k-1)}) \leq -\alpha_1 \|x^{(k)} - x^{(k-1)}\|^2 \tag{32}$$

where $\alpha_1 = \left(1 - \frac{1}{2\eta}\right) \frac{\lambda_{min}(D)}{\rho_M}$ and (28) holds.

Now, we can prove (by proceeding as in the proof of the Lemma 2.2) that

$$\begin{aligned} \left\| x^{(k-1)} - [x^{(k-1)} - (Gx^{(k-1)} + q)]^+ \right\| &= \Big\| x^{(k-1)} - [x^{(k-1)} - (Gx^{(k-1)} + q)]^+ + \\ & -x^{(k)} + [x^{(k)} - \frac{D}{\rho_k}(x^{(k)} - x^{(k-1)}) - (Gx^{(k-1)} + q + h(\rho_k))]^+ \Big\| \\ &\leq \left(2 + \frac{\lambda_{max}(D) + \lambda_{min}(D)}{\rho_m}\right) \|x^{(k)} - x^{(k-1)}\|. \end{aligned} \tag{33}$$

Then, the inequality (29) follows from (33) with $\alpha_2 = \left(2 + \frac{\lambda_{max}(D) + \lambda_{min}(D)}{\rho_m}\right)^2 \frac{1}{\alpha_1}$.

Finally, we observe that

$$x^{(k)} = [x^{(k-1)} - (Gx^{(k-1)} + q) + e^{(k)}]^+$$

where $e^{(k)} = \left(I - \frac{D}{\rho_k}\right)(x^{(k)} - x^{(k-1)}) - h(\rho_k)$ and

$$\begin{aligned} \|e^{(k)}\| \leq \; & \left(\max\left(\left|1 - \frac{\lambda_{min}(D)}{\rho_M}\right|, \left|1 - \frac{\lambda_{max}(D)}{\rho_m}\right|\right) \right. \\ & \left. + \frac{\lambda_{min}(D)}{\rho_m}\right) \|x^{(k)} - x^{(k-1)}\|. \end{aligned} \tag{34}$$

Since the sequence $\{f(x^{(k)})\}$ is monotonically nonincreasing and $f(x)$ is bounded from below on K, $\{f(x^{(k)})\}$ is convergent, as $k \to \infty$; from (28), it follows that $\{x^{(k)} - x^{(k-1)}\}$ and $\{h(\rho_k)\}$ converge to the null vector as $k \to \infty$. Thus, from (34), also $\{e^{(k)}\} \to 0$ as $k \to \infty$ and then (30) holds. □

4. Numerical Remarks and Results. The numerical behaviour of the VPM and the AVPM depends on the choice of the sequence of projection parameters $\{\rho_k\}$. These projection parameters can be chosen according to an updating rule involving the last approximation of the solution.

An updating rule for ρ_k can be considered numerically efficient when, combined with the VPM and the AVPM, produces an approximate solution of (1) by a number of iterations significantly less than that of the projection methods. At the same time, the introduction of this rule must not increase the computational complexity of one iteration and must imply that the corrective step along a descent direction in the VPM and along a projection arc in the AVPM arises only in few iterations, i.e. $\theta_k = 1$ and, above all, $m_k = 0$ for almost all k. An updating rule satisfying these features is given by

$$\rho_k = \begin{cases} \rho_{k-1} & \text{for } \|Gd^{(k-1)}\|^2 \leq \psi \|d^{(k-1)}\|^2 \\ \frac{d^{(k-1)T} G d^{(k-1)}}{d^{(k-1)T} G D^{-1} G d^{(k-1)}} & \text{otherwise} \end{cases} \qquad k = 2, 3, \ldots \tag{35}$$

where $d^{(k-1)} = y^{(k-1)} - x^{(k-2)}$ in the VPM and $d^{(k-1)} = x^{(k-1)} - x^{(k-2)}$ in the AVPM. The positive scalar ψ is a prefixed small tolerance. The rule is a generalization of that proposed in [22] for strictly convex QP problems and it is deduced from the following considerations.

We assume that the subproblem (6) is exactly solved, $h^{(k-1)} = 0$, for any $k \geq 1$ and $\{x^{(k)}\}$ converges to $x^* \in K^*$ as $k \to \infty$. By rearranging the first order conditions for problems (1) and (6) we obtain

$$\|x^* - y^{(k)}\|_D^2 \leq (x^* - x^{(k-1)})^T (D - \rho_k G)(x^* - y^{(k)}).$$

This last inequality holds in the VPM; for the AVPM we can substitute $y^{(k)}$ with $x^{(k)}$. Now, by using the Cauchy–Schwarz inequality, we obtain

$$\|x^* - y^{(k)}\|_D^2 \leq \| (D - \rho_k G)(x^* - x^{(k-1)})\|_{D^{-1}} \|x^* - y^{(k)}\|_D$$

and, after dividing through by $\|x^* - y^{(k)}\|_D$, squaring and expanding, we have

$$\|x^* - y^{(k)}\|_D^2 \leq \tau_k \|x^* - x^{(k-1)}\|_D^2,$$

where

$$\tau_k = 1 - \frac{\rho_k}{\|x^* - x^{(k-1)}\|_D^2}(2(x^* - x^{(k-1)})^T G(x^* - x^{(k-1)}) + \\ -\rho_k (x^* - x^{(k-1)})^T G D^{-1} G (x^* - x^{(k-1)})).$$

If $(x^* - x^{(k-1)})^T G D^{-1} G (x^* - x^{(k-1)}) \neq 0$, the quantity τ_k is minimized for

$$\rho_k = \frac{(x^* - x^{(k-1)})^T G (x^* - x^{(k-1)})}{(x^* - x^{(k-1)})^T G D^{-1} G (x^* - x^{(k-1)})};$$

then, we can try to obtain a "good" value of ρ_k by using $x^{(k-2)}$ in place of the solution x^*. For the VPM, we observe that $x^{(k-1)} - x^{(k-2)} = \theta_{k-1} d^{(k-1)}$. Since the vector $Gd^{(k-1)}$ is already computed, the updating rule does not increase significantly the computational complexity of each iteration of the VPM and the AVPM.

In the context of the projection methods for variational inequalities, a similar value for the step size ρ_k has been deduced, again by heuristic arguments, in [5].

In the VPM, the sequence $\{\rho_k\}$ is bounded from below and above as follows:

$$\min\left(\rho_1, \frac{\lambda_{min}(D)}{\lambda_{max}(G)}\right) \leq \rho_k \leq \max\left(\rho_1, \frac{\lambda_{max}(G)\lambda_{max}(D)}{\psi}\right),$$

where the bounds are obtained by using the inequality $x^T G x \geq \frac{1}{\|G\|}\|Gx\|^2$.

In the AVPM, because of the adaptive procedure for determining ρ_k, the resulting sequence $\{\rho_k\}$ is bounded from below and above in the following way:

$$\min\left(\rho_1, \frac{\lambda_{min}(D)}{\lambda_{max}(G)}, \frac{\beta\lambda_{min}(D)}{2\eta\lambda_{max}(G)}\right) \leq \rho_k \leq \max\left(\rho_1, \frac{\lambda_{min}(D)}{2\eta\lambda_{max}(G)}, \frac{\lambda_{max}(G)\lambda_{max}(D)}{\psi}\right).$$

We show the effectiveness of the updating rule (35) by reporting the results of a set of numerical experiments carried out on a Digital Personal Workstation 500au, using the double precision (macheps= $2.22 \cdot 10^{-16}$) and a set of programs written in Fortran 90.

In Table 1 we describe the test problems used in the numerical experimentation.

These test problems are randomly generated with assigned features by following a technique similar to that introduced in [24], but using Givens rotations instead of Householder elementary trasformations for obtaining the required level of sparsity [21]. In particular,

Table 1 - Features of the test problems of Tables 2 – 4.

$n = 5000$ spars$(G) = 99.8\%$ $\|G\| = 1$ $K(G) = 100$ spars$((A^T\ C^T)^T) = 99.9\%$ $\|(A^T\ C^T)^T\| = 1$ $K((A^T\ C^T)^T) = 10$				
name	m_e	m_i	nac	rank(G)
TP1	1000	1000	500	4900
TP2	2000	2000	1000	4900
TP3	3500	1000	500	4900
TP4	2000	2000	1000	4500
TP5	2000	2000	1000	3500

we prefix the sizes n, m_e, m_i, a solution x^* and the corresponding Lagrange multipliers, the number nac of inequality constraints that are active in the solution x^*, the level of sparsity (denoted by spars$(\cdot)$), the spectral condition number (denoted by $K(\cdot)$), the rank, the euclidean norm and the distribution of singular values of the matrices G and $\begin{pmatrix} A \\ C \end{pmatrix}$.

In the first set of experiments we compare the numerical behaviour of the VPM and the AVPM with respect to the class of the gradient projection methods. In particular, we consider the following schemes:

- the projection method with a limited minimization rule ([1, p. 205]). This scheme is equal to the VPM with $D = I$ and $\rho_k = \rho$ for any k; it is denoted by FPM in Table 2.
- the projection method with an Armijo rule along a projection arc ([1, p. 206]); for the choice of the projection parameter we can use the following strategy: given a positive integer γ and a constant value $\rho > 0$, at the k–th iteration with k such that $mod(k-1,\gamma) = 0$, we start the search procedure with the value ρ and, for the next $(\gamma - 1)$ iterations, we use as projection parameter the value obtained by the search procedure in the previous iteration. In practice, the scheme is equal to the AVPM with $D = I$ and $\overline{\rho}_k = \rho$ when $mod(k-1,\gamma) = 0$ and $\overline{\rho}_{k+i+1} = \rho_{k+i}$ for $i = 0, ..., \gamma - 2$. In the following tables, we denote this scheme by AFPM and use $\gamma = 10$. This value of γ is empirically chosen so that, for large values of ρ, the number of the correction steps on the projection parameter is reduced and an efficient behaviour in terms of the number of outer iterations is obtained.

For all methods, the inner QP subproblems are solved by the projected SOR method of Cryer. We use in the SOR method an empirical optimal value of the relaxation parameter ($\omega = 1.5$) and an inner progressive termination rule.

In the experiments we use the following stopping rule:

$$\frac{\|x^{(k+1)} - x^{(k)}\|_2}{\|x^{(k+1)}\|_2} \le 10^{-6}, \tag{36}$$

and, for all methods, we obtain that $|f(x^{(it)}) - f(x^*)|/|f(x^*)| \le 10^{-11}$.

In Tables 2 and 3 we report for the test problems TP1, TP2, TP3 the results of the FPM, the AFPM, the VPM and the AVPM combined with the rule (35). In the VPM and in the AVPM we put $D = I$.

In the tables we denote by *it* and *time* the number of iterations and the time in seconds to obtain the solution. The total number of inner iterations for the solution of the inner QP subproblems is reported in brackets. Furthermore, we denote by n_c the number of correction steps performed after the line search in the FPM and the VPM. In the AFPM and in the AVPM, n_c is the number $\sum_{k=1}^{it} m_k$ of the additional projection steps performed in the search procedure of the projection parameter. In these last methods, $\beta = 0.5$ and $\eta = 0.9$.

From the Tables 2 and 3, we can make the following remarks:

- in the gradient projection methods (Table 2), as ρ increases, the number it of outer iterations goes down to a minimum value and, then, it begins again to increase. The reduction of it is more consistent in the AFPM. Nevertheless, while the line search procedure and the corrective step of the FPM are nonexpensive, the additional projection steps in the AFPM affect considerably the computational complexity. Generally, since for sufficient large values of ρ the number $it + n_c$ in the AFPM is lower than the number of iterations in the FPM, the AFPM appears more efficient.
- The VPM and the AVPM combined with the updating rule (35) have a similar numerical behaviour (Table 3). They are more efficient than the gradient projection methods: the number of outer iterations is lower and the corrective steps of the VPM or the additional projection steps of the AVPM arise in few iterations. Furthermore, the VPM does not require any prefixed scalar parameter.

The above considerations about the numerical behaviour of the AFPM,[1] the VPM and the AVPM are confirmed in the Table 4, where we consider some test problems with a large number of null eigenvalues in the Hessian matrix G. Here, Z denotes the $n \times (n-(m_e+nac))$ matrix having as columns an orthonormal basis of the null space of the matrix of the equality constraints and of the inequality constraints active in the prespecified solution. Then, $Z^T G Z$ is the reduced Hessian matrix; when this matrix has a full column rank, the problem (1) has a unique solution, otherwise, K^* is not a singleton. We may observe that, on these test problems, the effectiveness of the three methods appears independent on the rank of G.

Finally, Table 5 shows the results obtained by the VPM and the AVPM with the rule (35) on some large-scale and very large-scale test problems of the CUTE library [2]. These results confirm that when, in the VPM and the AVPM, a convenient updating rule for the choice of the projection parameter is used, the corrective step along a descent direction or a projection arc arises in few iterations and the two schemes show an efficient numerical behaviour.

REFERENCES

[1] D. P. Bertsekas, *Nonlinear Programming*, Athena Scientific, Belmont, MA, 1995.

[2] I. Bongartz, A. R. Conn, N. Gould, and P. L. Toint, *CUTE: Constrained and unconstrained testing environment*, ACM Trans. Math. Software, 21 (1995), pp. 123–160.

[3] C. W. Cryer, *The solution of a quadratic programming problem using systematic overrelaxation*, SIAM J. Control, 9 (1971), pp. 385–392.

[4] S. Dafermos, *Traffic equilibrium and variational inequalities*, Transportation Sci., 14 (1980), pp. 42–54.

[5] C. S. Fisk and S. Nguyen, *Solution algorithms for network equilibrium models with asymmetric user costs*, Transportation Sci., 16 (1980), pp. 361–381.

[6] E. Galligani, *C^1 surface interpolation with constraints*, Numer. Algorithms, 5 (1993), pp. 549–555.

[1] In Table 4, for the AFPM we report the results related to an optimal emprirical value of ρ.

Table 2

	FPM			AFPM		
ρ	*it*	n_c	*time*	*it*	n_c	*time*
TP1						
1.4	557 (1747)	1	35.9	557 (1747)	0	34.7
2	403 (1277)	290	26.0	401 (1290)	1	25.5
20	393 (1448)	393	25.3	163 (944)	75	15.9
100	344 (2692)	340	24.7	92 (882)	65	11.4
200	622 (5123)	620	44.6	122 (1214)	96	15.1
1000	1866 (20475)	1864	138.5	146 (2071)	142	20.5
TP2						
1.4	517 (1617)	1	47.4	517 (1616)	0	47.5
2	374 (1185)	72	36.3	373 (1204)	1	36.4
7	333 (1154)	324	33.4	242 (1651)	64	35.1
100	909 (5487)	885	96.4	108 (1575)	72	26.9
200	1951 (11850)	1928	201.8	250 (3754)	193	57.2
TP3						
1.4	481 (1513)	3	54.6	482 (1523)	1	53.7
2	356 (1126)	340	42.1	346 (1120)	1	41.0
7	197 (700)	184	27.1	206 (940)	59	34.2
100	467 (2927)	456	65.7	103 (1136)	71	30.2
200	1606 (9008)	1606	196.9	123 (1477)	97	36.2

Table 3

AVPM – rule (35)			VPM – rule (35)		
it	n_c	*time*	*it*	n_c	*time*
TP1					
71 (391)	18	6.9	82 (403)	21	6.6
TP2					
90 (1184)	17	19.9	71 (850)	14	16.5
TP3					
80 (528)	12	18.9	86 (560)	13	19.3

[7] E. GALLIGANI, V. RUGGIERO, AND L. ZANNI, *Splitting methods for quadratic optimization in data analysis*, Intern. J. Computer Math., 63 (1997), pp. 289–307.

[8] E. GALLIGANI, V. RUGGIERO, AND L. ZANNI, *Splitting methods for constrained quadratic programs in data analysis*, Computers Math. Applic., 32 (1996), pp. 1–9.

[9] E. GALLIGANI, V. RUGGIERO, AND L. ZANNI, *Parallel solution of large scale quadratic programs*, In High Performance Algoritms and Software in Nonlinear Optimization, (R. De Leone, A. Murli, P. Pardalos and G. Toraldo, eds., Kluwer Academic Publishers, Dordrecht, 1998, pp. 189–205.

[10] W. LI, *Remarks on convergence of the matrix splitting algorithm for the symmetric linear complementarity problem*, SIAM J. Optim., 3 (1993), pp. 155–163.

[11] Z. Q. LUO AND P. TSENG, *On the linear convergence of descent methods for convex essentially smooth minimization*, SIAM J. Control Optim., 30 (1992), pp. 408–425.

[12] Z. Q. LUO AND P. TSENG, *Error bounds and convergence analysis of feasible descent methods: a general approach*, Ann. Oper. Res., 46 (1993), pp. 157–178.

[13] O. L. MANGASARIAN, *Convergence of iterates of an inexact matrix splitting algorithm for the sym-*

Table 4

	VPM			AVPM			AFPM		
rank($Z^T GZ$)	*it*	n_c	*time*	*it*	n_c	*time*	*it*	n_c	*time*
TP4									
2000	80 (494)	11	13.8	97 (601)	22	16.5	102 (1145)	71	21.7
1750	83 (521)	12	14.2	74 (523)	18	14.7	94 (856)	62	20.3
1500	89 (399)	16	14.4	79 (421)	13	14.7	109 (937)	69	23.7
TP5									
2000	79 (327)	15	15.9	82 (409)	25	18.5	99 (727)	64	25.0
1250	87 (358)	20	13.7	67 (376)	25	14.2	92 (727)	65	21.6
500	76 (544)	9	13.8	81 (665)	15	15.7	91 (793)	60	19.5

Table 5 - Test problem CVXQP from CUTE

	AVPM – rule (35)			VPM – rule (35)		
n	*it*	n_c	*time*	*it*	n_c	*time*
$m_e = n/4$						
5000	257 (460)	2	5.5	249 (425)	1	5.2
7000	247 (392)	1	7.8	264 (433)	1	8.4
9000	315 (882)	3	20.3	304 (784)	2	18.8
$m_e = n/2$						
11000	135 (245)	0	107.2	135 (245)	0	107.0
13000	148 (232)	3	117.8	142 (229)	2	112.3
17000	129 (206)	6	159.6	128 (218)	3	172.2

metric monotone linear complementarity problem, SIAM J. Optim., 1 (1991), pp. 114–122.

[14] O. L. Mangasarian and R. De Leone, *Parallel successive overrelaxation methods for symmetric linear complementarity problems and linear programs*, J. Optim. Theory Appl., 54 (1987), pp. 437–446.

[15] O. L. Mangasarian and R. De Leone, *Parallel gradient projection successive overrelaxation for symmetric linear complementarity problems and linear programs*, Ann. Oper. Res., 14 (1988), pp. 41–59.

[16] P. Marcotte and J. H. Wu, *On the convergence of projection methods: application to the decomposition of affine variational inequalities*, J. Optim. Theory Appl., 85 (1995), pp. 347–362.

[17] S. N. Nielsen and S. A. Zenios, *Massively parallel algorithms for single constrained convex programs*, ORSA J. on Computing, 4 (1992), pp. 166–181.

[18] J. M. Ortega and W. C. Rheinboldt, *Iterative Solution of Nonlinear Equations in Several Variables*, Academic Press, New York, 1970.

[19] P. M. Pardalos and N. Kovoor, *An algorithm for a single constrained class of quadratic programs subject to upper and lower bounds*, Math. Programming, 46 (1990), pp. 321–328.

[20] V. Ruggiero and L. Zanni, *On the efficiency of splitting and projection methods for large strictly convex quadratic programs*, in Nonlinear Optimization and Related Topics, G. Di Pillo and F. Giannessi, eds, Kluwer Academic Publ. B. V., 1999, pp. 401–413.

[21] V. Ruggiero and L. Zanni, *On a class of iterative methods for large-scale convex quadratic programs*, in Numerical Methods in Optimization, A. Maugeri and E. Galligani, eds, Rend. Circ. Matem. Palermo, Ser. II, Suppl. 58 1999, pp. 213–227.

[22] V. Ruggiero and L. Zanni, *A Modified Projection Algorithm for Large Strictly Convex Quadratic Programs*, J. Optim. Theory Appl., 102 (2000), to appear.

[23] M. V. Solodov and P. Tseng, *Modified projection-type methods for monotone variational inequalities*, SIAM J. Control Optim., 34 (1996), pp. 1814–1830.

[24] G. W. Stewart, *The efficient generation of random orthogonal matrices with an application to condition estimators*, SIAM J. Numer. Anal., 17 (1980), pp. 403–409.

COMPUTER GENERATION OF NUMERICAL METHODS FOR ORDINARY DIFFERENTIAL EQUATIONS

MARK SOFRONIOU * AND GIULIA SPALETTA †

Abstract. In this article we will describe some issues related to the derivation of numerical methods for ordinary differential equations. We will discuss how to establish algebraic constraints on the free parameters of the methods which control the quality of the numerical solution. Furthermore we will describe a *Mathematica* package which essentially automates the derivation process for Runge-Kutta methods. The package incorporates many new facilities to aid in analysis and construction and we will give examples to explain the design and motivation for the functionality. We will also discuss some techniques that are useful for obtaining solutions to the algebraic constraints.

Key words. Ordinary differential equations; initial value problems; Runge-Kutta methods; symbolic computation; computer algebra systems; computer generation of numerical methods.

AMS subject classifications. 65L05; 65L06; 68W30.

1. Historical motivation. There are many wide-ranging facilities in *Mathematica*. The system provides a higher level of functionality than traditional computing environments and is distinguished by its ability to manipulate algebraic expressions. This enables the automation of many mathematical processes. Here we will focus on the construction of a certain class of numerical methods.

Runge-Kutta methods find wide-scale application in the numerical solution of certain types of ordinary differential equations. The derivation of these methods is a very challenging mathematical problem with a long historical background. The origins date back to Euler (1768) and subsequently Runge (1895) and Kutta (1904) who generalized the construction. The more recent formalism of Butcher contributes significantly to the construction of these methods [3]. Other authors such as Albrecht (see §5.13 in [13]) and Munthe-Kaas [16] have investigated alternative formalisms for the construction of Runge-Kutta methods, although Butcher's work is the most recognized.

This article is organized as follows. In section 2 we introduce the class of problems that will be studied. In sections 3, 4, 5 and 6 we discuss the class of methods which will be of interest and the conventional techniques for their construction. Aside from some inherent limitations of the classical approach, the derivation and solution processes can become tedious and error prone. In section 7 we will introduce the elegant formalism of Butcher which considerably simplifies the construction of numerical methods. Then we describe some of the objectives and scope of the package that has been developed to automate the formalism of Butcher. Some specific details of the methods that are distinguished in the package are given in section 8. By drawing on algorithms from the field of algebraic geometry we will demonstrate how to use *Mathematica* to alleviate the process of constructing methods in sections 9 and 10.

All computations in this article were carried out using version 4.0.2 of *Mathematica* under RedHat Linux 5.2 on a Pentium II machine running at 450 Mhz with 256MB of RAM.

2. Discrete vs. continuous. Ordinary differential equations (*ODE*s) describe physical models. They are abundant in applied science, governing such problems as Newtonian

*Wolfram Research Inc., Champaign, Illinois, USA
†Mathematics Department, Bologna University, Italy

motion. When the equations are non-linear there are no guaranteed algorithmic methods of solution. *Mathematica*'s **DSolve** implements many of the known techniques. In general, however, it is necessary to resort to numerical methods of solution.

The most common application of numerical methods for ordinary differential equations is to *initial value problems*. These are systems of ordinary differential equations of dimension $n \in \mathbb{N}$:

$$y'(t) = f(t,\ y(t)), \qquad f : \mathbb{R} \times \mathbb{R}^n \longrightarrow \mathbb{R}^n \tag{2.1}$$

$$y(t_0) = y_0. \tag{2.2}$$

For simplicity the dependence of f on t will often be ignored, so that we are dealing with an *autonomous* system. Equation (2.1) defines a family of solutions. A specific solution is fixed by specifying the initial state, or initial conditions (2.2). Cauchy (1824) proved existence and uniqueness of solutions to (2.1-2.2).

The initial value problem (2.1-2.2) varies continuously with time. In order to solve the problem, an approximate solution is sought at *fixed* output points, within a finite time range. Take the time interval to be $[t_0,\ t_N]$. For simplicity, consider the case when the output points occur at equally spaced intervals: $t_n = t_0 + i\,h,\ i \in \mathbb{N}$. The constant t is referred to as the *time step*, while h is called the *step-length*. A one-step numerical method advances the solution from one time step t_n to a subsequent time step t_{n+1}. Thus information from computations at previous time steps is not required.

3. Taylor series. One of the most established techniques for function approximation in numerical analysis is the *Taylor series*. A Taylor series approximates a function in the neighborhood of a point of interest. It does so by making use of successively higher derivatives. Practically the series is truncated after some finite number of terms.

As an example, consider the truncated Taylor series expansion of order p which determines an approximate solution $\tilde{y}(t_n, p)$ to the true solution $y(t_n) = y_n$. We need to express derivatives of y in terms of partial derivatives of f. First we specify the differential equation and then how the general series is determined. The function **D** computes partial derivatives and a *transformation rule* is used to evaluate the derivatives at a specific value.

In[1] := **y'[t_] := f[y[t]];**

In[2] := $\tilde{\mathbf{y}}$**[t_, p_Integer?Positive] :=** $\sum_{i=0}^{p} \frac{h^i}{i!}$ **D[y[s], {s, i}] /. s → t;**

The series through order three at t_n is then given by:

In[3] := $\tilde{\mathbf{y}}$**[t$_n$, 3]**

$$\text{Out[3]} = h\, f[y[t_n]] + y[t_n] + \tfrac{1}{2} h^2 f[y[t_n]]\, f'[y[t_n]] + \frac{1}{6} h^3 \left(f[y[t_n]]\, f'[y[t_n]]^2 + f[y[t_n]]^2 f''[y[t_n]] \right)$$

The output is a little complicated. Let us simplify it, using a transformation rule to remove the functional dependence of f and its derivatives on $y(t_n)$.

In[4] := **RemoveDependency = {_f :> f, (df : Derivative[_][f])[_] :> df};**

In[5] := $\tilde{\mathbf{y}}$**[t$_n$, 3] /. RemoveDependency**

$$\text{Out[5]} = f\, h + y[t_n] + \frac{1}{2} f\, h^2 f' + \frac{1}{6} h^3 \left(f\,(f')^2 + f^2 f'' \right)$$

Taking high order terms only makes sense if the series converges and the derivatives are well-defined in the neighborhood of the expansion.

4. Generalizing Euler's method. One of the first and simplest methods for solving initial value problems was proposed by Euler:

$$y_{n+1} = y_n + h\, f(y_n).$$

The method is not very accurate. Local accuracy is measured by how high terms are matched in the Taylor expansion of the solution. Comparing with the result of the previous section, we see that Euler's method is first order accurate; errors occur one order higher starting at powers of h^2.

The difficulty in computing increasingly higher derivatives limits the applicability of Taylor series. The idea behind Runge-Kutta methods is to take successive (weighted) Euler steps to approximate the Taylor series. In this way function evaluations (and not derivatives) are used.

For example, consider the *midpoint rule*:

$$\begin{aligned} k_1 &= f(y_n) \\ k_2 &= f(y_n + \frac{1}{2}\, h\, k_1) \\ y_{n+1} &= y_n + h\, k_2\,. \end{aligned} \tag{4.1}$$

Since the intermediary terms k_i are scaled by h, higher order terms in the Taylor expansion can be matched and the error in approximation is improved. To see this, look at the series expansion:

In[6] := $\mathbf{k_1 = f[y[t_n]];}$

In[7] := $\mathbf{k_2 = f[y[t_n] + \frac{1}{2} h\, k_1];}$

In[8] := $\mathbf{y[t_n + h] = y[t_n] + h\, k_2;}$

In[9] := **Series[y[tn + h], {h, 0, 3}] /. RemoveDependency**

Out[9] = $y[t_n] + f\, h + \frac{1}{2} f\, f'\, h^2 + \frac{1}{8} f^2\, f''\, h^3 + O[h]^4$

Comparing with the series expansion of the true solution, the midpoint rule can be seen to be second order accurate. Extending the above approach, repeated function evaluation can be used to obtain higher order methods.

5. Runge-Kutta methods. So far we have only discussed *explicit methods*, that is methods whose solution is given by a simple recursion. In fact the class of Runge-Kutta methods is much wider and a more general framework is needed.

Denote the *s-stage* Runge-Kutta method for the approximate solution to the (non-autonomous) initial value problem (2.1-2.2) at $t_{n+1} = t_n + h$, by:

$$k_i = f\Big(t_n + c_i\, h, \quad y_n + \sum_{j=1}^{s} a_{i,j}\, k_j\Big), \qquad i = 1, 2, \ldots, s,$$

$$y_{n+1} = y_n + h \sum_{i=1}^{s} b_i\, k_i$$

where we have adopted the general convention that the *row-sum conditions* hold:

$$c_i = \sum_{i=1}^{s} a_{i,j} \tag{5.1}$$

Conditions (5.1) are a particularly useful device, appearing naturally in the derivation of high order Runge-Kutta methods.

The *coefficients of the method* are *free parameters* that are usually chosen to satisfy a Taylor series expansion through some order in the time step h. In practice other conditions such as stability can also constrain the coefficients, although this aspect is not dealt with here.

It has become customary to denote the method coefficients $c = [c_i]^T$, $b = [b_i]^T$ and $A = [a_{i,j}]$ using a *Butcher table*:

$$\begin{array}{c|cccc} c_1 & a_{11} & a_{12} & \cdots & a_{1s} \\ c_2 & a_{21} & a_{22} & \cdots & a_{2s} \\ \vdots & \vdots & \vdots & \cdots & \vdots \\ c_s & a_{s1} & a_{s2} & \cdots & a_{ss} \\ \hline & b_1 & b_2 & \cdots & b_s \end{array} \tag{5.2}$$

The row-sum conditions (5.1) can be visualized as summing across the rows of table (5.2). The explicit Runge-Kutta methods are a special case where the matrix A is strictly lower triangular. Notice that a consequence of explicitness is the condition $c_1 = 0$.

6. Classical derivation. The classical way of deriving Runge-Kutta methods was to construct constraints on the coefficients of the method by matching terms in a Taylor expansion of the difference between the approximate and true solutions. The resulting coefficients of the derivatives are then equated to zero to the highest possible power of the time step h. This is the approach used in [6].

Let us illustrate the process for the two-stage explicit method which generalizes the midpoint rule (4.1).

In[10] := $\mathbf{k_1 = f[y[t_n]];}$

In[11] := $\mathbf{k_2 = f[y[t_n] + a_{2,1}\, h\, k_1];}$

In[12] := $\mathbf{y[t_n + h] = y[t_n] + h(b_1\, k_1 + b_2\, k_2);}$

In[13] := $\mathbf{ser = Normal[\ Series[y[t_n + h],\ \{h, 0, 3\}]];}$

In[14] := $\mathbf{ser\ /.\ RemoveDependency}$

Out[14] = $h\ (f\, b_1 + f\, b_2) + y(t_n) + f\, h^2\, a_{2,1}\, b_2\, f' + \dfrac{f^2\, h^3\, {a_{2,1}}^2\, b_2\, f''}{2}$

In[15] := $\mathbf{diff = (ser - \tilde{y}[t_n,\ 3])\ /.\ RemoveDependency}$

Out[15] = $-f\, h + h(f\, b_1 + f\, b_2) - \frac{1}{2} f\, h^2\, f' + f\, h^2\, b_2\, a_{2,1}\, f' +$
$\frac{1}{2} f^2\, h^3\, b_2\, a_{2,1}^2\, f'' - \frac{1}{6} h^3\, (f\, (f')^2 + f^2\, f'')$

We can get a better view of the constraints by examining specific coefficients and factoring out derivatives:

In[16] := **grouping = Alternatives[f, Derivative[_][f]];**

In[17] := **Collect[CoefficientList[diff, h], grouping]**

Out[17] = $\{0,\ f\,(-1+b_1+b_2),\ f\,\bigl(-\frac{1}{2}+b_2\,a_{2,1}\bigr)\,f',$
$-\frac{1}{6}\,f\,(f')^2+f^2\,\bigl(-\frac{1}{6}+\frac{1}{2}\,b_2\,a_{2,1}^2\bigr)\,f''\}$

Taking terms through h^2, we obtain the algebraic constraints:

$$b_1+b_2=1, \quad b_2\,a_{2,1}=\frac{1}{2}.$$

Solutions to these two equations guarantee that the method is of second order accuracy. The midpoint rule (4.1) has

$$b_1=0, \quad b_2=1, \quad a_{2,1}=\frac{1}{2}$$

and it can now be seen as a special case of a second order method.

Notice also that terms involving h^3 cannot be satisfied. To satisfy the condition involving $f^2\,f''$ we obtain the requirement: $b_2\,a_{2,1}^2=1/3$. There is a single solution to the system, namely:

$$b_1=\frac{1}{4}, \quad b_2=\frac{3}{4}, \quad a_{2,1}=\frac{2}{3}.$$

However, there are no parameter values that allow the term $f\,(f')^2$ to be similarly extinguished, since this derivative term does not appear in the series expansion of the method. Hence we arrive at the well known result that any two-stage explicit Runge-Kutta scheme has order at most two.

To obtain a desired order we have seen that we can choose special values for the coefficients of the method. There is a second possibility for certain types of systems. Given a specific structural form, it may be possible that some of the derivatives ensure that certain constraints vanish. This second possibility has traditionally been ruled out, since the focus has been towards numerical integration methods which can be used for *any given problem*. Using an additional package which extends the work outlined in this article, numerical methods for solving quadratic ordinary differential equations are derived in [12].

The first difficulty in deriving Runge-Kutta methods is in establishing the *order conditions* which are the hierarchy of algebraic constraints on the coefficients. These help to ensure that the numerical solution accurately models the underlying physical system. The second difficulty is in obtaining solutions: the equations are nonlinear in the coefficients and the number of equations grows combinatorially.

Aside from being inefficient, the construction of order conditions using Taylor series of a scalar problem is fundamentally flawed. It was often assumed in the early construction of Runge-Kutta methods that nothing significant changes if order conditions are derived from a single equation or a system of equations. This is not the case. Reputedly, some of the Runge-Kutta methods used in the early space shots suffered from this limitation [13].

In order to illustrate why the series approach for a scalar equation is limited, let us examine the terms at fourth order.

In[18] := **Coefficient[$\tilde{\mathbf{y}}[\mathbf{t_n},\ 4]$, $\mathbf{h^4}$] /. RemoveDependency**

Out[18] = $\frac{1}{24}\left(f\,(f')^3 + 4\,f^2\,f'\,f'' + f^3\,f^{(3)}\right)$

Three derivatives appear. In fact there is a fourth derivative which turns up when a series expansion of a system of equations is performed. This additional derivative reduces to $f^2\,f'\,f''$ in the scalar case. Hence two distinct order conditions, needed to solve a system of equations, coincide. Satisfying the sum of several algebraic conditions does not ensure that the conditions hold independently. Had we used a series expansion of a scalar non-autonomous equation, then we would have encountered similar problems at order five instead [13].

A series expansion of a system of equations is therefore necessary to derive the full set of order constraints, but constructing such a multivariate series is extremely inefficient. This is compounded by the combinatorial nature of the problem. For example, there are 1205 order conditions that must be satisfied in order to obtain a tenth order Runge-Kutta method.

The package **Butcher.m** addresses the difficulty of constructing order conditions. The essential ideas will be outlined in the next section. Analytic solution of the order equations is also possible using techniques that will be outlined in sections 9 and 10.

This loads the aforementioned package.

In[19] := **Needs["NumericalMath`Butcher`"];**

7. Butcher trees, elementary differentials and order conditions. As we have seen, each differential in a series expansion has an associated constraint on the coefficients of a Runge-Kutta method. The derivatives are fairly easy to calculate, in the scalar case at least. The constraints do not appear to be so easy to compute. It would be nice to be able to determine a constraint in some way from the derivative with which it is associated. A much more elegant approach to the derivation of Runge-Kutta order conditions is due to Butcher [3]. Butcher has gone on to establish many significant results in the field (see [2] for a recent appreciation).

Butcher's formalism makes use of many elements of graph theory and (rooted) trees. Each tree uniquely corresponds to a derivative and an associated order condition. In *Mathematica, Butcher trees* are represented in functional form. Here are the first few Butcher trees.

In[20] := **ButcherTrees[4]**

Out[20] = $\{\{f\},\ \{f[f]\},\ \{f[f[f]],\ f[f^2]\},$
$\{f[f[f[f]]],\ f[f[f^2]],\ f[f\,f[f]],\ f[f^3]\}\}$

How exactly are trees related to derivatives? Trees correspond to derivatives of a system of equations. For this reason, there is a need to keep track of the variable with respect to which we are differentiating.

Take as an example the fourth order tree **f[f f[f]]**. It is often useful to be able to work with a pictorial representation of a tree. To facilitate this, the command **ButcherPlot** constructs a plot of a tree or set of trees out of graphics primitives.

In[21] := **treeplot = ButcherPlot[f[f f[f]]];**

Out[21] =

Now label each node of the tree with distinct letters *i, j, k, l.* The precise label chosen does not actually matter but is used to distinguish the vertices of the tree.

In[22] := **labels = Graphics[{Text[i, {0.1, 0}], Text[j, {−0.125, 0.5}]
Text[k, {0.35, 0.5}], Text[l, {0.35, 1}]];**

In[23] := **Show[First[treeplot], labels];**

Out[23] =

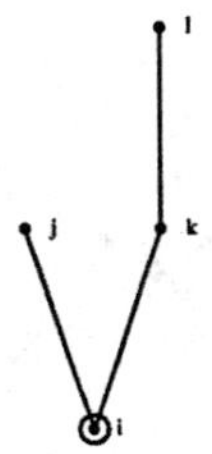

Start with the symbol *f*, which corresponds to the outer part (or head) of the functional tree. The corresponding vertex is encircled on the figure above and is termed the *root.* Attach the letter i to the root as a superscript. Now move up through the tree and form a product of terms in f with corresponding letters as superscripts. We then have:

$$f^j\, f^i\, f^k\, f^l\,.$$

The reason for the ordering is so that we can keep track of which symbols were *neighbors* in the tree (neighboring vertices are attached together by an edge).

Next, go back through the tree and attach subscripts corresponding to neighbors one level further up the tree. These are called *descendents.* If there are no more descendents then we are done. The result of this process is the component of the *elementary differential*

$$f^j\, f^i_{j\,k}\, f^k_l\, f^l\,.$$

Forming the derivative, or elementary differential, is a matter of summing the above component over the indices from one to the dimension of the differential system.

For the scalar case, there is only one component so superscripts can effectively be ignored. Subscripts represent derivatives and are additive. Therefore, as noted in section 6, the scalar derivatives for the two trees $f[f\,f[f]]$ and $f[f[f^2]]$ coincide yielding:

$$f^2\, f'\, f''\,.$$

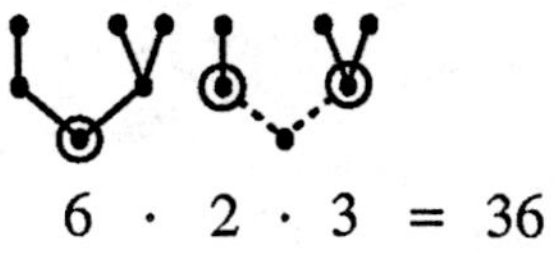

FIG. 7.1. *Computation of the density of the tree* $f[f[f]\,f[f^2]]$

Now that we have seen how to form a derivative from a tree, we turn to the problem of forming the associated order condition. The process of forming the left hand side of an order equation is similar to that of forming the elementary differential. Instead of labeling the symbol f we label the coefficients of the method a, b and c. For the root of the labeled tree we associate the coefficient b with the appropriate label as a subscript. For the terminal vertices we associate the coefficient c with the label of the vertex that is the predecessor as an index. For the remaining vertices we associate the coefficients a and use a double subscript consisting of the labels of the predecessor and the current vertex. The function **ButcherPhi** performs the task of forming a sum of coefficients for an s-stage Runge-Kutta method with a tree or set of trees.

In[24] := **tree = f[f f[f]];**

In[25] := **ButcherPhi[tree, s]**

Out[25] = $\sum_{i=1}^{s} b_i\, c_i \sum_{j=1}^{s} a_{i,j}\, c_j$

The right hand side of an order condition is (the reciprocal of) an integral number associated with a tree, the *density*. An example of the computation of the density is given in figure 7.1. Starting at the root we recursively delete edges and form the product of the order, or number of vertices, in the subtrees that remain. The function **ButcherGamma** calculates this quantity.

In[26] := **ButcherGamma[tree]**

Out[26] = 8

Combining the weight and density for a tree, we obtain an algebraic constraint on the coefficients of the method:

In[27] := $\mathbf{ButcherPhi[tree, s] == \dfrac{1}{ButcherGamma[tree]}}$

Out[27] = $\sum_{i=1}^{s} b_i\, c_i \sum_{j=1}^{s} a_{i,j}\, c_j == \frac{1}{8}$.

8. Classes of method distinguished. Explicit Runge-Kutta methods have received the most attention in the literature, primarily because they are relatively simple to implement. However, in recent years there has been much interest in solving systems of differential equations with components varying on very different time scales, *stiff systems.* Explicit methods are not well suited to the solution of stiff problems since stability considerations place severe restrictions on the size of time steps which can be taken. For this reason, there has been much interest in developing efficient *implicit* Runge-Kutta methods.

One drawback of implicit methods is that they require the solution of *coupled* systems of equations. Generally some form of iterative process such as *Newton* or *modified Newton iteration* must be used. *Diagonally implicit* Runge-Kutta methods have been proposed to address the issue of efficiency. These extend the explicit form of the Butcher table to include the leading diagonal elements of the matrix A in (5.2).

The current version of **Butcher.m** distinguishes the following classes of methods: **Explicit**, **Implicit** and **DiagonallyImplicit**. This distinction facilitates efficient generation of order conditions and related properties for these specific cases. Other more specialized variants can be derived from these using transformation rules [18]. To distinguish between different methods, the option **RungeKuttaMethod** can be specified to any order condition-related function. For example, the order conditions for a third order two-stage diagonally implicit Runge-Kutta method can be derived as:

In[28] := **RungeKuttaOrderConditions[3, 2, RungeKuttaMethod → DiagonallyImplicit]**

Out[28] = $\{\{b_1 + b_2 == 1\},\ \{b_1\, c_1 + b_2\, c_2 == \frac{1}{2}\},$
$\{a_{1,1}\, b_1\, c_1 + b_2\,(a_{2,1}\, c_1 + a_{2,2}\, c_2) == \frac{1}{6},\ b_1\, c_1^2 + b_2\, c_2^2 == \frac{1}{3}\}\}$

In the remainder of this article we will focus on the construction of explicit Runge-Kutta methods. Therefore we set a global environment variable to refer to this class of methods.

In[29] := **$RungeKuttaMethod = Explicit;**

9. Obtaining solutions. The package **Butcher.m** addresses the issue of forming the order equations for Runge-Kutta methods. It does not attempt to solve them. However, the equations can be passed to **Solve** and related functions. In order to get the most out of this process, it is useful to understand a little more about how **Solve** works.

Solve makes use of the method of *Gröbner bases*, which is a technique for solving non-linear algebraic equations [5]. Gröbner bases are formed by reducing a set of equations into a new basis. This new basis is simpler and contains all the solutions of the original system. Finding the best order of variable elimination (which results in the simplest basis) is an ongoing research problem.

Mathematica's **Solve** uses a *heuristic* to determine a default variable ordering. This ordering is then used for subsequent reduction of the system of equations. An example of the solution of a third order three-stage explicit Runge-Kutta scheme is now given. Several issues relating to the order of elimination are highlighted.

In[30] := **rkoc = Flatten[RungeKuttaOrderConditions[3, 3]]**

Out[30] = $\{b_1 + b_2 + b_3 == 1,\ b_2\, c_2 + b_3\, c_3 == \frac{1}{2},\ a_{3,2}\, b_3\, c_2 == \frac{1}{6},$
$b_2\, c_2^2 + b_3\, c_3^2 == \frac{1}{3}\}$

First we solve the equations making use of *Mathematica*'s internal heuristics for variable ordering.

In[31] := **Timing[soln1 = Solve[rkoc];]**

Solve::svars: Equations may not give solutions for all " solve " variables.

Out[31] = {0.32 Second, Null }

There is a linear solution to the order equations, but the default heuristics of **Solve** are not yet able to detect this. As a result the routine ends up finding multiple nonlinear solutions which is considerably slower than necessary.

In[32] := **Length[soln1]**

Out[32] = 4

It is much more efficient if we explicitly specify which variables to solve for and hence which to eliminate. By inspecting the order equations, and in particular noting the Vandermonde type system for b_1, b_2, b_3, the following seems like a reasonable choice.

In[33] := **vars = {b₁, b₂, b₃, a₃,₂};**

In[34] := **Timing[soln2 = Solve[rkoc, vars];]**

Out[34] = {0.12 Second, Null }

It is still more efficient if we specify both the variables and the *order* of elimination. To accomplish this, we make sure that the variables are not sorted canonically.

In[35] := **Timing[soln3 = Solve[rkoc, vars, Sort → False];]**

Out[35] = {0.07 Second, Null }

Solve gives generic solutions only. It discards solutions that are valid only when the parameters satisfy special conditions. However, we can generate all possible solutions to the order equations using **Reduce**.

In[36] := **Timing[Reduce[rkoc, vars, Sort → False];]**

Out[36] = {0.18 Second, Null }

There are three solutions (c.f §5.9 in [13]).

In[37] := **Map[Factor, soln4, {3}]**

Out[37] =
$$b_1 == \frac{1}{4} \;\&\&\; b_2 == \frac{-1+3\,a_{3,2}}{4\,a_{3,2}} \;\&\&\; b_3 == \frac{1}{4\,a_{3,2}} \;\&\&$$
$$c_2 == \frac{2}{3} \;\&\&\; c_3 == \frac{2}{3} \;||$$
$$b_1 == \frac{-1+a_{3,2}}{4\,a_{3,2}} \;\&\&\; b_2 == \frac{3}{4} \;\&\&\; b_3 == \frac{1}{4\,a_{3,2}} \;\&\&$$
$$c_2 == \frac{2}{3} \;\&\&\; c_3 == 0 \;||$$
$$a_{3,2} == \frac{(c_2-c_3)\,c_3}{c_2\,(-2+3\,c_2)} \;\&\&\; b_1 == \frac{2-3\,c_2-3\,c_3+6\,c_2\,c_3}{6\,c_2\,c_3} \;\&\&$$
$$b_2 == -\frac{-2+3\,c_3}{6\,c_2\,(c_2-c_3)} \;\&\&\; b_3 == \frac{-2+3\,c_2}{6\,(c_2-c_3)\,c_3} \;\&\&$$
$$c_2 \neq 0 \;\&\&\; -2+3\,c_2 \neq 0 \;\&\&\; c_2-c_3 \neq 0 \;\&\&\; c_3 \neq 0$$

Some of the drawbacks of Gröbner bases should perhaps be summarized. The reduction process is sensitive to the order of elimination. The time required to obtain solutions grows exponentially with the number of variables and with the number of solutions of the system according to Bézout's theorem [7]. Separate cases arise for each nonlinear branch encountered. For modest sized problems however Gröbner bases are a very powerful technique.

Some recent enhancements to the Gröbner basis implementation used by **Solve** and related functions are described in [14].

10. Optimal methods. The solutions to the order conditions often contain free parameters. In order to obtain specific methods we need a way of imposing additional constraints. One way of eliminating the freedom of choice is to minimize the *principal local truncation error* [3, 10]. This quantity can be derived using the function **ButcherPrincipalError**.

The traditional approach for finding optimal methods uses numerical minimization routines to search for local minima on a predefined grid (e.g. [8]). The problem here is that the objective function can be highly nonlinear and there is no guarantee that a locally found minimum on a finite grid will actually be the best method. In this section we show how to derive methods that minimize the truncation error using a technique based upon *cylindrical algebraic composition* [4]. There are numerous advantages to this approach. In particular it furnishes a *global minimum* for a given search region and furthermore the minimum and parameter values are given in *exact* form.

As illustration let us consider the class of explicit Runge-Kutta methods of order three with three stages. A widely used method of this class has been derived by Ralston [17] (see also [8, 9, 10]). The method is used as a high order formula of a pair with an embedded error estimate of second order derived by Bogacki and Shampine [1]. The pair has been adopted in the Texas Instruments TI85 calculator, in Matlab and is part of the software library RKSUITE. The coefficients of Ralston's method are given in table 10.1. Ralston actually used a variant of the principal local truncation error in the derivation of his methods. Since Ralston's weighting yields a different objective function it is interesting to see how close the minimum in his derivation is to the minimum of the quantity that is used in deriving modern formulae.

$$\begin{array}{c|ccc} 0 & 0 & 0 & 0 \\ \frac{1}{2} & \frac{1}{2} & 0 & 0 \\ \frac{3}{4} & 0 & \frac{3}{4} & 0 \\ \hline & \frac{2}{9} & \frac{1}{3} & \frac{4}{9} \end{array}$$

TABLE 10.1
A third order three-stage explicit Runge-Kutta method of Ralston

Let us use the main branch of the third order solutions of section 9 as a starting point.

In[38] := **soln2**

$$\text{Out[38]} = \left\{\left\{b_1 \to \frac{2-3\,c_2-3\,c_3+6\,c_2\,c_3}{6\,c_2\,c_3},\ a_{3,2} \to \frac{(c_2-c_3)\,c_3}{c_2\,(-2+3\,c_2)},\ b_2 \to \frac{2-3\,c_3}{6\,c_2\,(c_2-c_3)},\ b_3 \to \frac{-2+3\,c_2}{6\,(c_2-c_3)\,c_3}\right\}\right\}$$

Here is the general form for the principal error terms for a third order explicit Runge-Kutta method.

In[39] := **pet = ButcherPrincipalError[3, s]**

$$\text{Out[39]} = \quad \{-\frac{1}{24} + \sum_{i=2}^{s} b_i \sum_{j=2}^{-1+i} a_{i,j} \sum_{k=2}^{-1+j} a_{j,k}\, c_k,\ \frac{1}{2}\Big(-\frac{1}{12} + \sum_{i=2}^{s} b_i \sum_{j=2}^{-1+i} a_{i,j}\, c_j^2\Big),\ -\frac{1}{8} + \sum_{i=2}^{s} b_i\, c_i \sum_{j=2}^{-1+i} a_{i,j}\, c_j,\ \frac{1}{6}\Big(-\frac{1}{4} + \sum_{i=2}^{s} b_i\, c_i^3\Big)\}$$

Next we derive the weighted principal error coefficients for a three stage method, substitute the third order solution of the order equations and simplify the result.

In[40] := **pet = Factor[Flatten[pet /. soln2]]**

$$\text{Out[40]} = \quad \{-\frac{1}{24}, \frac{1}{24}(-1+2c_2), \frac{1}{24}(-3+4c_3), \frac{1}{72}(-3+4c_2+4c_3-6c_2c_3)\}$$

Notice that two of the four principal error coefficients vanish for the method of Ralston in (10.1), which means that the method may be more accurate than expected when applied to certain problems.

To construct an algebraic objective function we take the inner product, which corresponds to forming the square of the Euclidean norm.

In[41] := **pltesq = Dot[pet, pet];**

Now in order to rule out solutions that violate the assumptions used in solving for the order conditions, we form the following set of inequalities.

In[42] := $\mathbf{constraints} = 0 < c_2 \leq 1 \,\&\&\, 0 < c_3 \leq 1 \,\&\&\, c_2 \neq c_3 \,\&\&\, c_3 \neq \frac{2}{3}$;

From a computational standpoint several of these assumptions are desirable. Choosing $c_i, c_j \in \{c_1, c_2, c_3\}$ as distinct ensures that the method samples at distinct intermediary values. This makes it more likely that unexpected changes in the solution, such as discontinuities, will be revealed.

The function **Minimize** is still under development and is not yet part of the official documented set of *Mathematica* commands. Therefore function resides in the **Experimental** context and the precise semantics are subject to change in future versions. **Minimize** returns a list of the minimum value and a list of parameters and values that are obtained at the minimum.

In[43] := **vars = {c2, c3};**

In[44] := **Timing[Experimental`Minimize[pltesq, constraints, vars]]**

$$\text{Out[44]} = \quad \{0.86\ \text{Second}, \{\ \text{Root}[-13301 + 6638640\,\#1 + 534988800\,\#1^2 + 11609505792\,\#1^3\&,\ 1], \{c_2 \rightarrow \text{Root}\,[-20 + 47\,\#1 - 18\,\#1^2 + 9\,\#1^3\&,\ 1], c_3 \rightarrow \text{Root}\,[-37 + 83\,\#1 - 72\,\#1^2 + 36\,\#1^3\&,\ 1]\}\}\}$$

Taking numerical approximations to the algebraic solutions shows that the optimal solution is indeed quite close to the choice of Ralston.

In[45] := **N[Part[%, 2, 2]]**

Out[45] = $\{c_2 \to 0.496505,\ c_3 \to 0.751747\}$

A plot of the Euclidean norm of the local truncation error in the vicinity of the minimum shows that the surface is quite flat. In fact Ralston's method is within about 0.005% of the optimum value.

In[46] := **Plot3D[Evaluate[$\sqrt{\textbf{pltesq}}$], {c2, $\frac{2}{5}$, $\frac{3}{5}$}, {c3, $\frac{7}{10}$, $\frac{4}{5}$}];**

Out[46] =

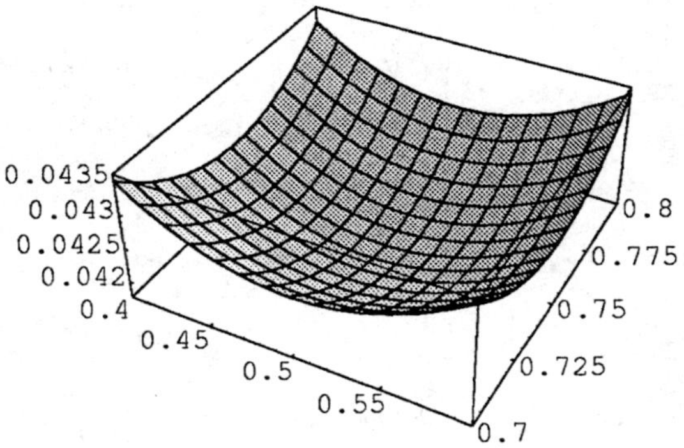

It is not difficult to verify that no other solution branches, obtained using **Reduce** in section 9, yield a method with smaller truncation error.

11. Conclusions. The construction of Runge-Kutta methods is a fairly involved and mathematically challenging topic. Much of this process can be automated and *Mathematica* provides a very powerful environment for doing this. **Butcher.m** provides a fairly comprehensive coverage of tools required in the construction of Runge-Kutta methods and is the first such package to implement Butcher's elegant formalism. Techniques for constructing sub-classes of methods can readily be built upon the functionality [12, 20]. For brevity we have not discussed issues related to the derivation of high order methods and in particular the use of Butcher's simplifying assumptions which reduce the nonlinearity of the algebraic equations [3, 10, 11]. Details relating to the implementation and use of simplifying assumptions in **Butcher.m** are given in [15, 18]). Here we have discussed the local order of approximation of Runge-Kutta methods, but tools for the analysis of stability of numerical methods for differential equations also exist [19].

REFERENCES

[1] P. Bogacki and L. F. Shampine, *A 3(2) pair of Runge-Kutta Formulas*, Appl. Math. Letters, 2, pp. 1–9, 1989.

[2] K. Burrage, *The work of John Butcher: an appreciation*, Annals of Numerical Mathematics, 1, pp. 1–24, 1994.

[3] J. C. Butcher, *The Numerical Analysis of Ordinary Differential Equations: Runge-Kutta and General Linear Methods*, John Wiley and Sons, New York, 1987.

[4] G. E. Collins, *Quantifier elimination for real closed fields by cylindrical algebraic decomposition*, Proc. 2nd GI Conf. Automata Theory and Formal Languages, Springer Lecture Notes in Computer Science 33, MR 55, pp. 134–183, 1977.

[5] D. A. Cox, J. B. Little and D. O'Shea, *Ideals, Varieties and Algorithms*, 2nd ed., Springer, New York, 1997.

[6] W. Gander and D. Gruntz, *Derivation of numerical methods using computer algebra*, Siam Review, Vol. 41, N. 3, pp. 577–593, 1999.

[7] J. von zur Gathen and J. Gerhard, *Modern Computer Algebra*, Cambridge University Press, Cambridge, 1999.

[8] T. E. Hull and R. L. Johnston, *Optimum Runge-Kutta methods*, Math. Comp., 18, pp. 306–310, 1964.

[9] T. E. Hull, *A search for optimum methods for the numerical integration of ordinary differential equations*, SIAM Rev., 9, pp. 647–654, 1967.

[10] E. Hairer, S. P. Nørsett and G. Wanner, *Solving Ordinary Differential Equations I: Nonstiff Problems*, second edition, 2nd ed., Springer Verlag, New York, 1993.

[11] E. Hairer and G. Wanner, *Solving Ordinary Differential Equations II: Stiff and Differential Algebraic Problems*, 2nd ed., Springer Verlag, New York, 1996.

[12] A. Iserles, G. Ramaswami and M. Sofroniou, *Runge-Kutta methods for quadratic ordinary differential equations*, BIT, Vol. 38, N. 2, pp. 315–346, 1998.

[13] J. D. Lambert, *Numerical Methods for Ordinary Differential Systems. The Initial Value Problem*, John Wiley and Sons, New York, 1991.

[14] D. Lichtblau, *Gröbner bases in Mathematica*, The Mathematica Journal, Vol 6, No. 4, pp. 81–88, 1996.

[15] *Mathematica 4.0 Standard Add-On Packages*, Wolfram Media Inc., Champaign, Illinois, 1999.

[16] H. Munthe-Kaas, *Lie-Butcher theory for Runge-Kutta methods*, BIT, Vol. 35, pp. 572–587, 1995.

[17] A. Ralston, *Runge-Kutta methods with minimum error bounds*, Math. Comp., 16, pp. 431–437, 1962.

[18] M. Sofroniou, *Symbolic derivation of Runge-Kutta methods*, J. Symb. Comp., Vol. 18, N. 3, pp. 265–296, 1994.

[19] M. Sofroniou, *Order stars and linear stability theory*, J. Symb. Comp., Vol. 21, N. 1, pp. 101–131, 1996.

[20] M. Sofroniou and W. Oevel, *Symplectic Runge-Kutta schemes I: order conditions*, SIAM J. Numer. Anal., Vol. 34, N. 5, pp 2063–2086, 1997.

ON SOME CLASSES OF ITERATIVE ABS-LIKE METHODS FOR LARGE-SCALE LINEAR SYSTEMS*

EMILIO SPEDICATO† AND ZHENGFENG LI‡

Abstract. The original version of the ABS methods for solving linear systems, proposed by Abaffy, Broyden and Spedicato (1984), is too expensive in storage for large-scale problems. In order to limit memory requirements and computational complexity per iteration step, two storage efficient implementations are considered in this paper. The first one defines the *limited memory ABS methods*. In these methods, the expensive explicit computation of the Abaffian matrices, which are generally dense, is avoided. After the storage available for the iteration is exhausted, the oldest of the stored vectors is replaced by the most recent one. The second approach is based upon *restarting*, i.e. clearing the storage and starting again. Their computational complexity in comparison with other well-known methods is discussed. When the coefficient matrix is nonsingular, a q-linear rate of convergence for a large subclass of limited memory ABS methods is proved under a mild hypothesis. We show that several classical conjugate direction algorithms can be obtained via the ABS approach.

Key words. ABS methods, sparse linear systems, limited memory approach, computational complexity, q-linear convergence.

1. Introduction. Consider solving the linear system of equations

$$Ax = b, \quad x \in R^n, \tag{1.1}$$

where $A \in R^{n,n}$ is nonsingular and $b \in R^n$. Denote A by

$$A = [a_1, \dots, a_n]^T,$$

where $a_j \in R^n$, $j = 1, \dots, n$. The ABS class of methods for solving (1.1), proposed by Abaffy, Broyden and Spedicato (1984), is a quite general finite iteration procedure that solves the system (1.1) in at most n steps. In the standard formulation of ABS methods, matrices H_k's, called Abaffians, are required to be stored. For large n, such methods require large storage. In order to reduce the storage, two new implementations of ABS methods have been considered. One is the *permutated ABS method* due to Spedicato and Tuma (1993), another is the *full rank ABS method* proposed by Spedicato and Zhu (1997). Although these methods are more convenient in term of storage, still the required storage is large and there is a high number of operations per iteration step.

More storage efficient algorithms can be obtained from the ABS methods by relinquishing the property of finite termination, generating large classes of iterative methods. A class of such iterative methods is based upon the so called *limited memory* formulation of ABS methods. In this approach, the expensive explicit computation of the Abaffian matrices, which are generally dense, is avoided and only a number of stored previous direction vectors are used. After the storage available for the iteration is exhausted, the oldest of the stored vectors is replaced by the most recent one. A similar approach can also be found in Deuflhard, Freund and Walter (1990) and Engleman, Strang and Bathe (1981) for Broyden's method and in Eirola and Nevanlinna (1989) for the EN method (see also Yang (1995)). Another approach, *restarting*, clears the storage and starts again.

*Work partially sponsored by MURST 97 program Analisi Numerica e Matematica Computazionale. Dr. Li thanks CNR for support via the Young Mathematicians Program

†Department of Mathematics, University of Bergamo, 24129 Bergamo, Italy

‡ Department of Mathematics, National University, Canberra, Australia

In this paper we consider mainly limited memory ABS methods, though similar results can be obtained for a restarting approach. A q-linear rate of convergence for a subclass of limited memory ABS methods is proved under a mild hypothesis. Our solvers do not require necessarily that A be explicitly stored, a feature which is useful for problems where A is only implicitly given as an operator. Therefore they could be viewed as "matrix-free" solvers.

The paper is organized as follows. First we recall from Abaffy and Spedicato (1989) three formulations of search vectors without the explicit computation of the Abaffian matrices. Based on these formulations, we present three restarted and limited memory formulations of ABS methods. Their computational complexity in comparison with other well-known methods is discussed in Section 3. In Section 4 we discuss some properties of a subclass of limited memory ABS methods and their relation with some existing methods. Under a mild condition its q-linear rate of convergence is proved. Finally, we outline some future work.

For reference convenience, the original scaled ABS methods for the determined nonsingular linear system (1.1) is given below. Notice that ABS methods can also be applied to underdetermined or rank deficient systems.

Algorithm 1 (Scaled ABS methods)

Let $x_1 \in R^n$ be arbitrary. Let $H_1 \in R^{n,n}$ and $V = (v_1, \ldots, v_n) \in R^{n,n}$ be arbitrary nonsingular.

For $k = 1$ to n

$\tau_k = v_k^T A x_k - v_k^T b$

$p_k = H_k^T z_k, \quad (z_k \in R^n, \; z_k^T H_k A^T v_k \neq 0)$

$\alpha_k = \tau_k / (v_k^T A p_k)$

$x_{k+1} = x_k - \alpha_k p_k$

$H_{k+1} = H_k - H_k A^T v_k w_k^T H_k, \quad (w_k \in R^n, \; w_k^T H_k A^T v_k = 1)$

End

One of the disadvantages of Algorithm 1 when n is large is its computational complexity. Even if A is a sparse matrix, H_k is in general dense. Therefore multiplication of H_k with a vector requires $O(n^2)$ number of operations, compared to $O(n)$ for multiplication of a sparse A with a vector. In the following we will assume that n is large and that A is sparse. We will develop methods which do not require the actual computation of H_k and which will have an order(n) storage. To describe such algorithms, we need the following definition:

Definition 1.1 A set of n linearly independent vectors $z_1, \ldots, z_n$ are called feasible if, given H_1 nonsingular and $A^T v_1, \ldots, A^T v_n$ linearly independent vectors, $z_k^T H_k A^T v_k$ is nonzero for $k = 1, \ldots, n$ when H_k is updated by

$$H_{k+1} = H_k - H_k A^T v_k w_k^T H_k, \tag{1.2}$$

where $w_k \in R^n$ is arbitrary save that $w_k^T H_k A^T v_k = 1$.

In the following we will assume, without loss of generality and as is the case for all algorithms considered insofar, that $w_k = z_k / z_k^T H_k A^T v_k$ for all k. Then a necessary and sufficient condition for the vectors $z_1, \ldots, z_n$ to be feasible follows from the following theorem, see Abaffy and Spedicato (1989) for necessity and Galántai (1994) for sufficiency.

Theorem 1.1 Let $H_1 = I$ and $Z = (z_1, \ldots, z_n)$. Then the vectors $z_1, \ldots, z_n$ are feasible and Algorithm 1 is well-defined if and only if the matrix $V^T A Z$ is strongly nonsingular (i.e. all its principle submatrices are nonsingular).

Now, we consider three formulations of the search vector p_k without the actual computation of H_k, taken from Abaffy and Spedicato (1989). We assume that the vectors z_k are feasible.

- The search vectors p_k can be determined in terms of $2k$ vectors p_j, s_j by the relations

$$p_1 = H_1^T z_1, \qquad p_k = H_1^T z_k - \sum_{j=1}^{k-1} p_j s_j^T z_k / v_j^T A p_j, \quad \text{for } k > 1, \tag{1.3}$$

where the s_j are vectors in R^n given by

$$s_1 = H_1 A^T v_1, \qquad s_k = H_1 A^T v_k - \sum_{j=1}^{k-1} s_j p_j^T A^T v_k / v_j^T A p_j, \quad \text{for } k > 1,$$

- The search vector p_k can be determined in terms of $n - k$ vectors by the relation

$$p_k = u_k^k, \tag{1.4}$$

where $u_j^1 = H_1^T z_j$ for $j = 1, \ldots, n$ and for $k = 1, \cdots, n-1$

$$u_j^{k+1} = u_j^k - \frac{v_k^T A u_j^k}{v_k^T A u_k^k} u_k^k, \quad j = k+1, \ldots, n.$$

- By a fundamental ABS property, the search vectors p_k satisfy the following homogeneous system

$$v_j^T A p_k = 0, \quad j = 1, \cdots, k-1,$$

which can be efficiently solved by the ABS method itself. If we take $H_1 = I$, z_k as the starting vector p_k^1, with the z_j feasible vectors, and we use for the first $k-1$ equations of the k-th system the search vectors used in the $(k-1)$th system, then

$$p_k = p_k^k, \tag{1.5}$$

where

$$p_k^{j+1} = p_k^j - \frac{v_j^T A p_k^j}{v_j^T A p_j} p_j, \qquad j = 1, \ldots, k-1.$$

One can exploit the above three properties for a reformulation of ABS methods. The following Algorithms 2, 3, 4 can be found in Abaffy and Spedicato (1989) and are due respectively to Spedicato, Abaffy and Bodon; see also Gu (1988) for a different derivation of Algorithm 3.

Algorithm 2

Let $x_1 \in R^n$ be arbitrary. Let $H_1 \in R^{n,n}$ and $V = (v_1, \ldots, v_n) \in R^{n,n}$ be arbitrary nonsingular.
For $k = 1$ to n
$\tau_k = v_k^T A x_k - v_k^T b$
$p_k = H_1^T z_k - \sum_{j=1}^{k-1} p_j s_j^T z_k / v_j^T A p_j$

$\alpha_k = \tau_k/(v_k^T A p_k)$
$x_{k+1} = x_k - \alpha_k p_k$
$s_k = H_1 A^T v_k - \sum_{j=1}^{k-1} s_j p_j^T A^T v_k / v_j^T A p_j$
End

Algorithm 3

Let $x_1 \in R^n$ be arbitrary. Let $H_1 \in R^{n,n}$ and $V = (v_1, \ldots, v_n) \in R^{n,n}$ be arbitrary nonsingular.
For $j = 1$ to n
$u_j^1 = H_1^T z_j$
End
For $k = 1$ to n
$\tau_k = v_k^T A x_k - v_k^T b$
$p_k = u_k^k$
$\alpha_k = \tau_k/(v_k^T A p_k)$
$x_{k+1} = x_k - \alpha_k p_k$
For $j = k+1$ to n
$u_j^{k+1} = u_j^k - (v_k^T A u_j^k / v_k^T A u_k^k) u_k^k$
End
End

Algorithm 4

Let $x_1 \in R^n$ be arbitrary. Let $H_1 \in R^{n,n}$ and $V = (v_1, \ldots, v_n) \in R^{n,n}$ be arbitrary nonsingular.
For $k = 1$ to n
$\tau_k = v_k^T A x_k - v_k^T b$
If $k > 1$ then
$p_k^1 = H_1^T z_k$
For $j = 1$ to $k-1$
$p_k^{j+1} = p_k^j - \left(v_j^T A p_k^j / v_j^T A p_j\right) p_j$
End
$p_k = p_k^k$
Else
$p_k = H_1^T z_1, \ (z_1 \in R^n)$
Endif
$\alpha_k = \tau_k/(v_k^T A p_k)$
$x_{k+1} = x_k - \alpha_k p_k$
End

2. Restarted and Limited Memory ABS Methods. Table 2.1 lists the total multiplication counts and the amount of memory required in addition to the matrix A and the right hand side of some versions in the ABS class, where Algorithm SZ denotes the method proposed by Spedicato and Zhu. It is assumed that the ABS parameters are given and general. For special choices of the parameters the above counts may be reduced or increased (e.g. if $H_1 = V = I$ and z_k is the $k-$th unit vector, choices giving the implicit LU algorithm, well defined if A is strongly nonsingular) then the number of multiplications reduces to $n^3/3 + O(n)$.

For large n, the operation and memory requirement are too high. Hence, for large-scale sparse linear systems, we consider restarted and truncated versions. An overview

Table 2.1 Operation counts and memory

Method	Operation Counts	Memory Required
Algorithm 1	$6n^3 + O(n^2)$	$n^2 + O(n)$
Algorithm 2	$4n^3 + O(n^2)$	$2n^2 + O(n)$
Algorithm 3	$4n^3 + O(n^2)$	$n^2 + O(n)$
Algorithm 4	$4n^3 + O(n^2)$	$n^2 + O(n)$
Algorithm SZ	$3n^3 + O(n^2)$	$n^2 + O(n)$

of classical restarted and limited memory algorithms can be found in Saad and Schultz (1985). Restarted and truncated ABS methods are a natural modification of the above considered ABS processes. In general such modifications lead to loss of the finite termination, while defining an infinite iterative process whose convergence can be established under mild conditions.

The restarted version of any of the above methods can be formulated as follows:

Algorithm 5

Do, until convergence

(1) perform m iterations of the method, with $m << n$ a given integer

(2) set $x_0 = x_{m+1}$

(3) go back to (1)

End do

Algorithms 2 - 4 are also suitable to truncation, yielding *limited memory iterative ABS methods*. If m is the number of available storage vectors, then a limited memory iterative ABS method is obtained by including only information from the last m iterations, i.e. replacing e.g. all occurrences of $\sum_{j=1}^{k-1}$ by $\sum_{j=\max(1,k-m)}^{k-1}$. Consequently, after the first m iterations one keeps the most recent m vectors in a moving window instead of discarding them as is done in the restarted version. One would expect such approach to lead to better convergence than via restarting, since more recent information is being kept. Of course, one might consider strategies where the kept vectors are not necessarily the last m vectors.

For the restarted and truncated versions the matrix H_1 should be chosen to require a low storage (e.g. as a diagonal matrix or a Householder matrix....). In the following we state the truncated versions of Algorithms 2–4, having chosen $H_1 = I$. Parameters v_k are assumed to be linearly independent and parameters z_k to be feasible.

Algorithm T-ABS2(m) (Limited memory version of Algorithm 2)

Let $x_1 \in R^n$ be arbitrary. Give the integer $m, 1 \leq m \leq n$.

Do $k = 1, 2, \cdots$, until convergence

$t = \max(1, k - m)$

$\tau_k = v_k^T A x_k - v_k^T b$

$p_k = z_k - \sum_{j=t}^{k-1} p_j s_j^T z_k / v_j^T A p_j$

$\alpha_k = \tau_k / (v_k^T A p_k)$

$x_{k+1} = x_k - \alpha_k p_k$

$s_k = A^T v_k - \sum_{j=t}^{k-1} s_j p_j^T A^T v_k / v_j^T A p_j$

End do

Algorithm T-ABS3(m) (Limited memory version of Algorithm 3)

Let $x_1 \in R^n$ be arbitrary and give the integer $m, 1 \leq m < n$.

Do $i = 1, 2, \ldots$, until convergence

For $j = 1$ to m

$u_j^1 = z_j$
End
$x_1 = x_i$
Do $k = 1, 2, \cdots$, until convergence
$\tau_k = v_k^T A x_k - v_k^T b$
$p_k = u_k^k$
$\alpha_k = \tau_k / (v_k^T A p_k)$
$x_{k+1} = x_k - \alpha_k p_k$
For $j = k+1$ to $k+m$
$u_j^{k+1} = u_j^k - (v_k^T A u_j^k)/(v_k^T A u_k^k) u_k^k$
End
End do
$x_i = x_{m+1}$
End do

Algorithm T-ABS4(m) (Limited memory version of Algorithm 4)

Let $x_1 \in R^n$ be arbitrary.
Give the integer $m, 1 \leq m \leq n$.
Do $k = 1, 2, \cdots$, until convergence
$\tau_k = v_k^T A x_k - v_k^T b$
If $k > 1$ then
$t = \max(1, k - m)$
$p_k^t = z_k$
For $j = t$ to $k - 1$
$p_k^{j+1} = p_k^j - \left(v_j^T A p_k^j / v_j^T A p_j \right) p_j$
End
$p_k = p_k^k$
Else
$p_k = z_1, \; (z_1 \in R^n)$
Endif
$\alpha_k = \tau_k / (v_k^T A p_k)$
$x_{k+1} = x_k - \alpha_k p_k$
End do

One can show that the truncated versions with $m = 1$ of the implicit LU algorithm ($H_1 = I,\ v_i = e_i,\ z_i = e_i,\ w_i = e_i / e_i^T H_i a_i$), of the Huang algorithm ($H_1 = I,\ v_i = e_i,\ z_i = a_i,\ w_i = a_i / a_i^T H_i a_i$) and of the implicit QR algorithm ($H_1 = I,\ v_i = A p_i,\ z_i = e_i,\ w_i = e_i / e_i^T H_i a_i$) are respectively the Gauss-Seidel method, the Kaczmarz method, see Kaczmarz (1937) and the De la Garza method, see De la Garza (1951). Notice that in recent years the Kaczmarz method, after a long period of neglect, has again attracted considerable interest.

3. Complexity of Limited Memory ABS Methods. In order to compare the algorithms with each other as well as with existing methods we should consider various issues:

- the total operation count
- the type of operations
- the memory requirements
- the convergence conditions.

Table 3.1 Operation counts per iteration step for the limited memory versions

Method	Operations	$m = 2$
CGS	$15n + 2\ smv + 2\ prec$	$15n + 2\ smv + 2\ prec$
Bi-CGSTAB	$20n + 2\ smv + 2\ prec$	$20n + 2\ smv + 1\ prec$
T-Broyden(m)	$(4m + 11)n + 1\ smv + 1\ prec$	$19n + 1\ smv + 1\ prec$
T-EN(m)	$(12m + 10)n + 2\ smv + 2\ prec$	$34n + 2\ smv + 2\ prec$
T-SEN(m)	$(12m + 16)n + 2\ smv + 2\ prec$	$40n + 2\ smv + 2\ prec$
T-ORTHOMIN(m)	$(6m + 2)n + 1\ smv + 1\ prec$	$14n + 1\ smv + 1\ prec$
T-ABS2(m)	$(4m + 3)n + 2\ smv$	$11n + 2\ smv$
T-ABS3(m)	$(2m + 3)n + 2\ smv$	$7n + 2\ smv$
T-ABS4(m)	$(2m + 3)n + 2\ smv$	$7n + 2\ smv$

Table 3.2 Additional work vectors required

Method	Memory required	$m = 2$	$m = 4$
CGS	8	8	8
Bi-CGSTAB	8	8	8
T-Broyden(m)	$2m + 5$	9	13
T-EN(m)	$2m + 4$	8	12
T-SEN(m)	$2m + 4$	8	12
T-ABS2(m)	$2m + 2$	6	10
T-ABS3(m)	$2m + 2$	6	10
T-ABS4(m)	$2m + 2$	6	10

Since we are dealing with iterative methods, the total operation count consists of the number of operations per iteration step, which we consider in this section, and the number of iterations required to achieve a certain accuracy, which will be the object of a future work. It is important to consider the type of operations involved, since the performance of computational primitives can differ significantly from one to another and depends strongly on the type of machine used. This is particularly true for parallel computers.

In Table 3.1 the types and number of operations per iteration step for the truncated versions are given for a variety of methods. We have included CGS ([18]), Bi-CGSTAB ([21]) and the truncated forms of ORTHOMIN ([22]), Broyden's "good method" ([7,9]), EN and SEN ([24]). We use (smv) to denote multiplications of a sparse matrix with a vector (e.g. Ax_k and $A^T v_k$). The flop count depends strongly on the linear system we are considering. It can be small for a very sparse matrix and may require little time if the matrix is also well structured, such as a banded matrix. It can be very large if the matrix is fairly dense or given in implicit form. Another important factor is the preconditioning step or the construction of the parameters in the ABS class ($prec$). We use the notation T-Method(m) for the limited memory versions via truncation.

Table 3.2 lists the amount of memory in terms of vectors of length n required in addition to the matrix A and the right hand side.

4. Some Convergence Properties of Algorithm T-ABS4(m). In this section, without loss of generality, we consider some convergence properties of Algorithm T-ABS4(m). First we discuss its well-definiteness.

Definition 4.1 Algorithm T-ABS4(m) is well-defined iff the following condition is satisfied for all k:

$$v_k^T A p_k \neq 0.$$

Theorem 4.1 Let $V_k = (v_t, \cdots, v_k)$ and $Z_k = (z_t, \cdots, z_k)$, where $t = \max(1, k-m)$. If $V_k^T A Z_k$ is strongly nonsingular for all k, then Algorithm T-ABS4(m) is well-defined.

Proof The proof is similar to that of Theorem 5 by Galántai (1994).

Let $t = \max(1, k-m)$. By the way the recursion has been obtained and by ABS properties it follows that the search direction p_k satisfies

$$v_i^T A p_k = 0, \quad t \leq i < k. \tag{4.1}$$

For use in the following analysis, we restate the basic recursions of Algorithm T-ABS4(m):

$$p_k = p_k^k, \tag{4.2}$$

where

$$p_k^{j+1} = p_k^j - \frac{v_j^T A p_k^j}{v_j^T A p_j} p_j, \quad j = t, \cdots, k-1, \tag{4.3}$$

and

$$p_k^1 = z_k. \tag{4.4}$$

By setting

$$\beta_k^j = \frac{v_j^T A p_k^j}{v_j^T A p_j}, \tag{4.5}$$

then we have

$$p_k = z_k - \sum_{j=t}^{k-1} \beta_k^j p_j. \tag{4.6}$$

The update of x_k is

$$x_{k+1} = x_k - \alpha_k p_k, \quad \alpha_k = \frac{v_k^T (A x_k - b)}{v_k^T A p_k}. \tag{4.7}$$

From now on, and essentially with no loss of generality, we restrict our discussion to the case where the scaling vectors are given by

$$v_k = A^{-T} Y p_k, \tag{4.8}$$

where Y is a symmetric, positive definite matrix. The following choices for Y define, in the original scaled ABS class, three important subclasses and require a low storage for large sparse systems:

1. $Y = I$, giving the optimally scaled subclass of the ABS algorithms, which includes the implicit Gram-Schmidt algorithm and the Craig algorithm [5]
2. $Y = A^T A$, giving the orthogonally scaled subclass of the ABS algorithms, which includes the implicit QR algorithm and the GMRES algorithm
3. $Y = A^T > 0$, for $A = A^T > 0$, giving the conjugate direction subclass, which includes the Choleski algorithm and the Hestenes-Stiefel algorithm.

In the following, we will show that Algorithm T-ABS4(m) with the scaling parameter choice (4.8) has an important variational property and produces a sequence p_k of mutually Y-conjugate directions ($p_i^TYp_j = 0,\ i \neq j, | i - j | \leq m$).

Denote

$$q(x) = \frac{1}{2}(x - x^*)^T Y (x - x^*), \tag{4.9}$$

where Y is defined as in (4.8) and $x^* = A^{-1}b$ is the solution of problem (1.1). We use the notation

$$\bar{r}_k = -\nabla q(x_k) = h - Yx_k, \tag{4.10}$$

and

$$r_k = b - Ax_k, \tag{4.11}$$

where $h = Yx^*$.

We take $x_1 = 0$ as the initial point. This is not restrictive. If a better approximation to the solution, say $\hat{x}$, is known, the problem $Ax = b$ with initial point $\hat{x}$ can be solved by setting $y = x - \hat{x}$ and solving $Ay = b - A\hat{x}$ with initial point zero.

Theorem 4.2 Let $s_k = x_{k+1} - x_k$ and $t = \max(1, k - m)$. If Y is a symmetric positive definite matrix, then the following statements are true for Algorithm T-ABS4(m) with the parameter choice (4.8):

(a) $p_i^TYp_j = 0,\ i \neq j,\ | i - j | \leq m,\ p_i^TYp_i \neq 0$.

(b) dim $Span\{p_t, \cdots, p_k\} = k - t + 1$.

(c) x_k minimizes the quadratic function $q(x)$ over $\{x_t + Span\{p_t, \cdots, p_k\}\}$, i.e., assuming that $y \in R^n$ is an arbitrary vector of the form

$$y = x_t + \sum_{i=t}^{k} \mu_i p_i, \tag{4.12}$$

where the μ_i are arbitrary scalars, then the quadratic function $\hat{q}(\mu)$ given by

$$\hat{q}(\mu) \equiv q(y) = \frac{1}{2}(y - x^*)^T Y (y - x^*), \tag{4.13}$$

where $\mu = (\mu_t, \cdots, \mu_k)^T$, is minimized with respect to the μ_i by the values $\mu_i = \alpha_i$ for $i = t, \cdots, k$, α_i being the step sizes of Algorithm T-ABS4(m).

(d) $\bar{r}_{k+1}^T p_j = 0,\ t \leq j \leq k$.

(e) $s_j^T Y s_j = s_j^T \bar{r}_t,\ t \leq j \leq k$ for all j.

Proof From (4.8), we have $Yp_i = A^T v_i$. For $t \leq i < k$, using (4.1), we get

$$p_k^T Y p_i = p_k^T A^T v_i = 0.$$

For $i > k$, the same follows from the symmetry of Y. For $i = k$, $p_i^T Y p_i > 0$ since $p_i \neq 0$ and Y is positive definite. Hence, property (a) is true.

Property (b) is a direct corollary of property (a).

To prove property (c), we proceed by induction. The derivative of $\hat{q}(\mu)$ with respect to μ_i has the form

$$\frac{\partial \hat{q}}{\partial \mu_i} = p_i^T Y (y - x^*). \tag{4.14}$$

For $i = t$, we can write

$$\frac{\partial \hat{q}}{\partial \mu_t} = p_t^T Y(x_t - x^* + \mu_t p_t + \sum_{j=t+1}^{k} \mu_j p_j). \tag{4.15}$$

Setting the derivative to zero, we get for the minimizing μ_t^*

$$\mu_t^* = \frac{p_t^T Y(x_t - x^*)}{p_t^T Y p_t} + \sum_{j=t+1}^{k} \frac{\mu_j p_t^T Y p_j}{p_t^T Y p_t}. \tag{4.16}$$

The second term in (4.16) is zero because of property (a), while the first term, from (4.8) and the identity $x_t - x^* = A^{-1} r_t$, takes the form

$$\mu_t^* = \frac{v_t^T r_t}{v_t^T A p_t} = \alpha_t. \tag{4.17}$$

Assuming now that $\mu_j = \alpha_j$ for $j = 1, \cdots, i-1$, we can write (4.14) in the form

$$\frac{\partial \hat{q}}{\partial \mu_i} = p_i^T Y(x_i - x^* + \mu_i p_i + \sum_{j=i+1}^{k} \mu_j p_j), \tag{4.18}$$

where x_i is the estimate of the solution computed by Algorithm T-ABS4(m) at the $(i-1)$-th step, and the summation should be omitted if $i = k$. By zeroing the derivative and proceeding as before, we again find that $\mu_i^* = \alpha_i$. So property (c) is true.

Since x_{k+1} minimizes $q(x)$ on $\{x_t + Span\{p_t, \cdots, p_k\}\}$, we have

$$0 = -\nabla q(x_{k+1})^T p_j = -\bar{r}_{k+1}^T p_j, \quad t \le j \le k, \tag{4.19}$$

which yields property (d).

Note that $x_{k+1} = x_t + \sum_{j=t}^{k} s_j$, so that

$$\bar{r}_{k+1} = \bar{r}_t - \sum_{j=t}^{k} Y s_j.$$

Thus, from (4.19) and property (a), we have

$$0 = \bar{r}_{k+1}^T s_k = s_k^T \bar{r}_t - \sum_{j=t}^{k} s_k^T Y s_j = s_k^T \bar{r}_t - s_k^T Y s_k, \tag{4.20}$$

which yields property (e).

Thus, the proof of Theorem 4.2 is completed.

Remark 4.1 The results in Theorem 4.2 are the reformulation for Algorithm T-ABS4(m) of variational results known for the ABS class, see Abaffy and Spedicato (1989).

The following result is useful for a parallel implementation of Algorithm T-ABS4(m).

Using property (a) of Theorem 4.2, we can simplify the formula for the search direction p_k. From (4.3) and the definition of β_k^j, we have

$$\begin{aligned} p_k^j &= p_k^{j-1} - \beta_k^{j-1} p_{j-1} \\ &= p_k^t - \textstyle\sum_{i=t}^{j-1} \beta_k^i p_i \end{aligned} \tag{4.21}$$

and by (4.8),

$$\begin{aligned}\beta_k^j &= (v_j^T A p_k^j)/(v_j^T A p_j) \\ &= [p_j^T Y(p_k^t - \textstyle\sum_{i=t}^{j-1} \beta_k^i p_i)]/(v_j^T A p_j) \\ &= (p_j^T Y z_k)/(v_j^T A p_j).\end{aligned} \tag{4.22}$$

Hence from (4.6) we get

$$p_k = z_k - \sum_{j=t}^{k-1} \frac{p_j^T Y z_k}{p_j^T Y p_j} p_j. \tag{4.23}$$

Notice that using formula (4.23) the computation of the search direction p_k can be easily implemented on parallel processors, while formula (4.2) has to be implemented in sequential form, due to its recurrence structure.

Remark 4.2 Choice (4.8), Theorems 4.1 and 4.2 imply that Algorithm T-ABS4(m) is well defined if, letting $P_k = (p_t, \cdots, p_k)$, then $P_k^T Y P_k$ is strongly nonsingular, which is true if Z_k has full rank.

According to formula (4.23), Algorithm T-ABS4(m) can be restated as follows:

Algorithm T-ABS4(m)

Set the initial point $x_1 = 0$. Give an integer $m, 1 \le m \le n$.
Let Y be arbitrary symmetric positive definite and the sequence
$z_k \in R^n, \quad k = 1, 2, \cdots$ be arbitrary save that $Z_k = (z_t, \cdots, z_k)$ has full
rank for all k, where $t = \max(1, k - m)$.
Do $k = 1, 2, \cdots$, until convergence
  If $k > 1$ then
    $t = \max(1, k - m)$
    $p_k = z_k$
    For $j = t$ to $k - 1$
      $p_k = p_k - (p_j^T Y z_k / p_j^T Y p_j)\, p_j$
    End
  Else
    $p_k = z_1 \ (z_1 \in R^n)$
  Endif
  $\tau_k = p_k^T (h - Y x_k)$
  $\alpha_k = \tau_k / (p_k^T Y p_k)$
  $x_{k+1} = x_k + \alpha_k p_k$
End do

We close this section by showing that the truncated generalized conjugate direction algorithm (T-GCD(m)), due to Dennis and Turner (1987), can be obtained by specially choosing the parameters in Algorithm T-ABS4(m). Algorithm T-GCD(m) can be phrased as follows:

Algorithm T-GCD(m)

$\overline{x}_1 = 0, \ \overline{r}_1 = h.$
Do $k = 1, 2, \cdots$, until convergence
  Get d_k such that $d_k^T \overline{r}_k \ne 0$, where $\overline{r}_k = h - Y x_k$.
  $t = \max(1, k - m)$

$\overline{x}_{k+1} = arg\min\{q(x) \mid x \in sp\{x_t + \overline{s}_t, \ldots, \overline{s}_k, d_k\}\}$
$\overline{s}_k = \overline{x}_{k+1} - \overline{x}_k$
$\overline{r}_{k+1} = \overline{r}_k - Y\overline{s}_k$

End do

Algorithm T-GCD(m) has the following properties (see [6]):

$$\overline{p}_k = d_k - \sum_{j=t}^{k-1} \overline{\beta}_k^j \overline{p}_j \tag{4.24}$$

and

$$\overline{s}_k = \overline{\alpha}_k \overline{p}_k, \tag{4.25}$$

where

$$\overline{\beta}_k^j = \frac{d_k^T Y \overline{p}_j}{\overline{p}_j^T Y \overline{p}_j} \tag{4.26}$$

and

$$\overline{\alpha}_k = \frac{d_k^T Y \overline{r}_k}{d_k^T Y \overline{p}_k}. \tag{4.27}$$

Theorem 4.3 If the parameter z_k in Algorithm T-ABS4(m) is equal to the parameter d_k in Algorithm T-GCD(m) for all k, then the sequences x_k, α_k, p_k generated by Algorithm T-ABS4(m) are identical with the sequences $\overline{x}_k, \overline{\alpha}_k$, $\overline{p}_i$ generated by Algorithm T-GCD(m).

Proof For Algorithm T-ABS4(m) and Algorithm T-GCD(m), it is easy to see that x_{k+1} minimizes $q(x_k + \alpha p_k)$ and $\overline{x}_{k+1}$ minimizes $q(x_k + \alpha \overline{p}_k)$ with respect to the variable α respectively. Noticing that $x_1 = \overline{x}_1 = 0$, we just need to show that, for all k

$$p_k = \overline{p}_k. \tag{4.28}$$

By relation $z_k = d_k$, (4.28) follows from (4.23) and (4.24) by induction.

Remark 4.3 Theorem 4.3 and results of Dennis and Turner imply the validity of the following statements:

- If $Y = A^T A$ and $z_k = r_k$ for all k, then Algorithm T-ABS4(m) is equivalent to the truncated version of Algorithm ORTHOMIN (i.e. ORTHOMIN(m)) ([22]).
- If $Y = ZA$, $z_1 = r_1$ and $z_k = As_{k-1}$ for $k \geq 2$, then Algorithm T-ABS4(m) is equivalent to the truncated version of Algorithm ORTHODIR (i.e. ORTHODIR(m)) ([25]).
- If $Y = A^T A$ and $d_k = v_k$ for all k, where $\{v_k\}$ are determined by Arnoldi's method (see [17]), then Algorithm T-ABS4(m) is equivalent to the restarted GMRES(m) algorithm.

We now discuss the convergence of the proposed ABS methods. To ensure convergence for limited memory ABS methods, following Dennis and Turner (1987), we slightly strengthen the requirement that p_k be non-orthogonal to the gradient of $q(x)$ at x_k, asking that the parameters z_k are chosen such that the angle between the search direction p_k and $\overline{r}_k$ be uniformly bounded away from 90^0. We use the notation $\kappa(Y) = \lambda_{\max}(Y)/\lambda_{\min}(Y)$ for the l_2 condition number of a symmetric positive definite matrix Y.

The following theorem shows a q-linear rate of convergence for algorithms in the class T-ABS4(m).

Theorem 4.4 Suppose that Y is symmetric positive definite, $p_k \neq 0$ and there exists $\gamma > 0$ such that

$$| p_k^T \bar{r}_k | \geq \gamma \| p_k \|_2 \| \bar{r}_k \|_2 \tag{4.29}$$

for all k. Then the sequence of iterates generated by any member of Algorithm T-ABS4(m) converges to x^* and satisfies

$$\| x_{k+1} - x^* \|_Y \leq (1 - \frac{\gamma^2}{\kappa(Y)})^{1/2} \| x_k - x^* \|_Y, \tag{4.30}$$

where $\| \cdot \|$ denotes the Y-norm of a vector.

Proof If $r_k = 0$ for some k, then $x_k = x^*$ from the assumption of A being nonsingular. Now assume $r_k \neq 0$ for all k. Then $\bar{r}_k \neq 0$ and, with $s_k = x_{k+1} - x_k$ and $h = Yx^*$

$$\begin{aligned} q(x_{k+1}) - q(x_k) &= \tfrac{1}{2}(x_k + s_k - x^*)^T Y (x_k + s_k - x^*) \\ &\quad -\tfrac{1}{2}(x_k - x^*)^T Y (x_k - x^*) \\ &= s_k^T Y x_k + \tfrac{1}{2} s_k^T Y s_k - s_k^T h \\ &= -s_k^T \bar{r}_k + \tfrac{1}{2} s_k^T \bar{r}_k \\ &= -\tfrac{1}{2} s_k^T \bar{r}_k \end{aligned} \tag{4.31}$$

and

$$\begin{aligned} q(x_{k+1}) - q(x_1) &= \textstyle\sum_{j=1}^{k} [q(x_{j+1}) - q(x_j)] \\ &= -\tfrac{1}{2} \textstyle\sum_{j=1}^{k} (s_j^T \bar{r}_j) < 0. \end{aligned} \tag{4.32}$$

Since Y is positive definite, $q(x)$ is bounded below, so we conclude that $\lim_{k\to\infty} \sum_{j=1}^{k} s_j^T \bar{r}_j$ exists. Hence

$$\lim_{k\to\infty} s_k^T \bar{r}_k = 0. \tag{4.33}$$

Notice that

$$\bar{r}_k^T s_k = \bar{r}_k^T (\alpha_k p_k) = \alpha_k \bar{r}_k^T p_k, \tag{4.34}$$

then we have from (4.7) and (4.8)

$$\alpha_k = \frac{p_k^T Y A^{-1} r_k}{p_k^T Y p_k} = \frac{p_k^T \bar{r}_k}{p_k^T Y p_k}. \tag{4.35}$$

Hence

$$\bar{r}_k^T s_k = \frac{(\bar{r}_k^T p_k)^2}{p_k^T Y p_k} \geq \frac{\gamma^2 \| \bar{r}_k \|_2^2 \| p_k \|_2^2}{\lambda_{\max}(Y) \| p_k \|_2^2} = \frac{\gamma^2}{\lambda_{\max}(Y)} \| \bar{r}_k \|_2^2 > 0 \tag{4.36}$$

and

$$0 = \lim_{k\to\infty} \bar{r}_k^T s_k \geq \frac{\gamma^2}{\lambda_{\max}(Y)} \lim_{k\to\infty} \| \bar{r}_k \|_2^2 \geq 0, \tag{4.37}$$

which shows that $\{\bar{r}_k\}$ converges to 0. Since Y is positive definite, it follows that

$$\lim_{k\to\infty} x_k = x^*. \tag{4.38}$$

Moreover, we have

$$q(x_k) - q(x_{k+1}) = \frac{1}{2}s_k^T\bar{r}_k = \frac{1}{2}\alpha_k p_k^T\bar{r}_k \geq \frac{\gamma^2}{\lambda_{\max}(Y)} \| \bar{r}_k \|_2^2 \tag{4.39}$$

and

$$\begin{aligned} \| x_{k+1} - x^* \|_Y^2 &= 2[q(x_{k+1}) - q(x^*)] \\ &= 2[q(x_{k+1}) - q(x_k) + q(x_k) - q(x^*)] \\ &\leq \| x_k - x^* \|_Y^2 - \tfrac{\gamma^2}{\lambda_{\max}(Y)} \| \bar{r}_k \|_2^2 . \end{aligned} \tag{4.40}$$

Noticing that

$$\bar{r}(x) = h - Yx = Y(x^* - x),$$

then we have

$$\| x - x^* \|_Y^2 = \bar{r}(x)^T Y^{-1} \bar{r}(x) \tag{4.41}$$

and

$$\| \bar{r}_k \|_2^2 \geq \lambda_{\min}(Y) \| x_k - x^* \|_Y^2 . \tag{4.42}$$

Thus

$$\begin{aligned} \| x_{k+1} - x^* \|_Y^2 &\leq \| x_k - x^* \|_Y^2 - \tfrac{\gamma^2\lambda_{\min}(Y)}{\lambda_{\max}(Y)} \| x_k - x^* \|_Y^2 \\ &= (1 - \tfrac{\gamma^2}{\kappa(Y)}) \| x_k - x^* \|_Y^2, \end{aligned} \tag{4.43}$$

which yields (4.30).

Remark 4.4 Convergence of ORTHOMIN(m), ORTHODIR(m) and GMRES(m) is easily verified. The analysis is similar to that of Dennis and Turner (1987) word by word.

Remark 4.5 Theorem 4.4 shows the monotone reduction of the Y weighted error norm of the considered methods. Therefore in evaluating the performance of such methods one should refer to the specific Y norm associated to them. It is not appropriate and it is biased to compare, as it is often done, the performance of methods in an arbitrary norm (often taken as the residual norm, i.e. the A^TA weighted norm).

Remark 4.6 Relation (4.30) shows that the error reduction factor depends on $Cond(Y)$. Therefore methods in the optimally scaled subclass are expected to have a better reduction factor than methods in the conjugate direction subclass, while methods in the orthogonally scaled subclass have the worse expected reduction factor. Since (4.30) holds for the entire subclass, it is however possible for specific algorithms in that subclass to have a much better actual reduction factor, possibly even corresponding to a superlinear convergence rate! Notice that for methods in the optimally scaled subclass the reduction factor does not depend on the condition number of A, if γ is independent of it.

5. Conclusions and Further Remarks. In this paper, we have discussed implementations of the ABS methods for large-scale sparse linear systems. In order to limit memory requirement and computational complexity per iteration step, some efficient implementations are considered in this paper. One approach is called limited memory ABS methods. In these methods, the expensive explicit computation of the Abaffian matrices, which are generally dense, is avoided. After the storage available for the iteration is exhausted, the oldest of the stored vectors is replaced by the most recent one. Another approach is *restarting*, i.e. clearing the storage and starting again. Their computational complexity in comparison with other well-known methods is discussed. We use a limited memory ABS methods, though a restarting approach could be used as well. A q-linear rate of convergence for a subclass of limited memory ABS methods is proved under a mild hypothesis.

Comparison of the numerical performance of the standard version of the ABS class and the restarted/limited memory versions with other methods will be the object of a future paper.

REFERENCES

[1] J. Abaffy, C.G. Broyden and E. Spedicato, A class of direct methods for linear equations, *Numer. Math.*, 45, 361-376, 1984.

[2] J. Abaffy and E. Spedicato, A generalization of the ABS algorithm for linear systems, Quaderno DMSIA 4/85, University of Bergamo, 1985.

[3] J. Abaffy and E. Spedicato, ABS Projection Algorithms: Mathematical Techniques for Linear and Nonlinear Equations, Ellis Horwood, Chichester, 1989.

[4] E. Bodon, Biconjugate algorithms in the ABS class I: alternative formulations and theoretical properties, Quaderno DMSIA 3/89, University of Bergamo, 1989.

[5] E. J. Craig, The n-step iterations procedures, *Journal of Mathematical Physics*, 34, 64-73, 1955.

[6] J. E. Dennis, and K. Turner, Generalized conjugate directions, *Linear Algebra and its Applications*, 88/89 (1987), 187-209.

[7] P. Deuflhard, R. Freund and A. Walter, Fast secant methods for the iterative solution of large non-symmetric linear systems, *Impact of Computing in Science and Engineering*, 2 (1990), 244-276.

[8] T. Eirola and O. Nevanlinna, Accelerating with rank-one updates, *Linear Algebra and its Applications*, 121 (1989), 511-520.

[9] M. Engleman, G. Strang and K.J. Bathe, The application of quasi-Newton methods in fluid mechanics, *International Journal of Numerical Methods in Engineering*, 17 (1981), 707-718.

[10] A. Galántai, ABS methods for the solution of linear and nonlinear systems of algebraic equations, Habilitation Dissertation, Miskolc, 1994.

[11] De la Garza, An iterative method for solving systems of linear equations, Oak Ridge Gaseous Diffusion Plant, Report K-731, Oak Ridge, TN, 1951.

[12] S. Gu, The simplified ABS algorithm, Report RABSCA1, Department of Applied Mathematics, Dalian University of Technology, 1988.

[13] S. Kaczmarz, Angenaeherte Auflösung von Systemen Linearen Gleichungen, *Bull. Acad. Polon. Sci. Lett.* A 35(1937), 355-357.

[14] Z. F. Li, Restarted and truncated versions of ABS methods for large linear systems: a basic framework, Report DMSIA 96/8, University of Bergamo, 1996.

[15] Z. F. Li, On the convergence of truncated ABS methods for large linear systems, Report DMSIA 96/29, University of Bergamo, 1996.

[16] Y. Saad and M. Schultz, Conjugate gradient-like algorithms for solving nonsymmetric linear systems, *Math. Comput.*, 44 (1985), 417-424.

[17] Y. Saad and M. Schultz, GMRES: a generalized minimal residual algorithm for solving nonsymmetric linear systems, *SIAM J. Sci. Stat. Comput.*, 7 (1986), 856-864.

[18] P. Sonneveld, CGS, a fast Lanczos-type solver for nonsymmetric linear systems, *SIAM J. Sci. Stat. Comput.*, 10 (1989), 36-52.

[19] E. Spedicato and M. Tuma, Solving sparse unsymmetric linear systems by implicit Gauss algorithm: stability, Quaderno DMSIA 4/93, University of Bergamo, 1993.

[20] E. Spedicato and M. Zhu, Reformulation of the ABS algorithms via full rank Abaffians, Proceedings EAMA Conference, Seville, September 1997

[21] H. Van der Vorst, Bi-CGSTAB: a fast and smoothly converging variant of Bi-CG for the solution of nonsymmetric linear systems, *SIAM J. Sci. Stat. Comput.*, 13 (1992), 631-644.

[22] P.K.W. Vinsome, ORTHOMIN, an iterative method for solving sparse sets of simultaneous linear equations, in: *Proceedings of the Fourth Symposium on Reservoir Simulation*, Soc. Petroleum Engineers of AIME, 1976.

[23] C. Vuik and H. Van der Vorst, A comparison of some GMRES-like methods, *Linear Algebra and its Applications*, 160 (1992), 131-162.

[24] U.M. Yang, A family of preconditioned iterative solvers for sparse linear systems. PhD. Thesis, University of Illinois at Urbana-Champaign, 1995.

INDEX

V

X